Geometrical and Structural Crystallography

Geometrical and Structural Crystallography

Joseph V. Smith

Louis Block Professor of Physical Sciences

University of Chicago

1807 1982

175 YEARS OF PUBLISHING

John Wiley & Sons

New York   Chichester   Brisbane   Toronto   Singapore

Library of Congress Cataloging in Publication Data

Smith, Joseph V.
  Geometrical and structural crystallography.

  (Smith-Wyllie intermediate geology series, ISSN
0196-427X)
  Bibliography: p. 419
  Includes index.
  1. Crystallography. I. Title. II. Series.
QD905.2.S64   549   82-2058
ISBN 0-471-86168-5   AACR2

Printed in the United States of America

10 9 8 7 6 5 4 3 2 1

Preface

Crystallography is a delightful branch of science. The exactness of the mathematical theory of symmetry provides a stable framework, and the unlimited range of crystal structures provides a continual challenge. A crystallographer must be in tune with the abstract world of a mathematician, and yet have an artistic and magical sympathy for the infinitely complex nature of crystalline properties. A crystallographer can explore the disciplines of physics, chemistry, mineralogy, and metallurgy, and can contribute to the understanding of biological processes. Particularly satisfying is the application of crystallography to the practical needs of human life—essentially all industrial, agricultural, and medical processes involve some aspects of crystallography. Finally, a crystallographer can develop a deep understanding of the ways in which things are connected together, and this leads to an integration of the scientific and artistic aspects of life.

The diversity of crystallography and its applications is too great for any crystallographer to encompass all aspects, and the subject is particularly difficult to teach. Most students are given only a few weeks of crystallography, and the illogical hodgepodge of topics is indigestible. This is a great pity because there are many excellent textbooks and monographs on crystallography. So why am I writing yet another book, especially if most students are not being offered a thorough course in crystallography? This book (i) uses the simplest possible mathematics consistent with proper use of the concepts of symmetry theory, packing, and topology; (ii) concentrates on the way in which atoms are connected together in crystal structures; and (iii) offers a systematic set of exercises designed to allow a student to self-discover the nature of crystalline properties. Furthermore, each chapter becomes progressively more difficult, and is designed so that the advanced topics can be omitted without destruction of the logical continuity. A class can be split up into faster and slower sections, and the faster sections can keep step with the slower ones by reading the advanced sections and solving the exercises of greater difficulty. Finally, I hope that the book can be studied without the guidance of an instructor, and that some readers will use their leisure to read this book instead of solving mathematical games or crossword puzzles. Complete answers or references are given for all exercises. Obviously, it is best if a reader has access to many models of crystal structures, but a satisfactory understanding of crystals can be obtained by carrying out the exercises. Most geometrical exercises require only paper, cardboard, drawing instruments, and some balls.

I hope the reader will not feel that the following advice is gratuitous.

Crystallography is a hierarchical subject that collapses if logical or numerical errors are made. It is essential to master each aspect before proceeding. I recommend a 4-point plan: read a chapter, do the exercises, reread the chapter, and then look back at preceding chapters to see how the hierarchy of topics is developing. Be systematic and thorough. Double check your work before testing an answer. Seek help immediately if you obtain incorrect answers. Cooperate with your fellow students. Some drawings require an accuracy near 0.2 mm, and good instruments are needed. Take heart from the knowledge that most students do not know how to think in three dimensions, and that several weeks of frustration are needed before the brain develops the requisite skills; the exercises are designed to encourage this development by proceeding from perspective drawings to abstract projections. Each subject is deliberately treated in several ways to give a reader the maximum opportunity of understanding it. I find that most students profit from this repetitive approach, and I think that the cost of printing a longer book is not unreasonable.

Every aspect of this book is covered in more detail in some other book or article, and I wish to acknowledge my debt to their authors. Dr. F.C. Phillips excited my interest in crystallography at Cambridge University in 1945, an otherwise gray year at the end of a miserable war, and I have continued to benefit from interaction with successive faculty members and students of that august institution. In particular, Drs. P. Gay, N.F.M. Henry, C. McKie, H.D. Megaw, J.F. Nye, and W.A. Wooster (in alphabetical order) have written valuable books. The writings of Dr. A.F. Wells have fascinated me for years, and my colleague Dr. P.B. Moore has profited even more from them than I. Crystallography is truly an international science, and my English heritage has been augmented by study of the contributions by M.J. Buerger and L. Pauling (American), F. Laves (German), and A.V. Shubnikov and N.V. Belov (Russian), to list just a few of the many creative crystallographers who have advanced the subject since the pioneering work of M. von Laue and the father–son team of W.H. and W.L. Bragg. The books by L. Azároff, F.D. Bloss, and M.J. Buerger are highly recommended as supplementary reading. I am especially grateful to my students and research associates, of which G.V. Gibbs, L.S. Dent Glasser, J.M. Bennett, P.B. Moore, J.J. Pluth, and T.J. McLarnan (in order of age) are making particularly notable contributions to crystallography.

My secretary, Irene Baltuska, and younger daughter, Susan, have respectively produced excellent typescript and drawings, and my wife kindly helped with the proofreading. The following members of the Wiley staff have earned my gratitude: Don Deneck, my editor, Deborah Herbert and Catherine Caffrey, copy editing, Lilly Kaufman, production, and Kevin Murphy, design.

I am also indebted to H. T. Evans, Jr., W. M. Meier, D. H. Olson and J. J. Pluth for provision of some figures, and to N. Weber for typing of indexes.

Join me now in a voyage of discovery. I wrote this book to relieve the tensions of the decade of exploration of the planets following the *Apollo 11* landing on the Moon. Enjoy yourself too.

Chicago, 1982   Joseph V. Smith

Contents

CHAPTER 9 ATOMIC PACKING AND SYMMETRY II. OCTAHEDRAL STRUCTURES

Geometrical
and Structural
Crystallography

1
Introduction: Packing and Crystal Morphology in Two Dimensions

Preview This chapter considers two-dimensional crystals because 3D ones are difficult for a beginner. Single atoms represented by circles can pack together in different ways depending on their sizes and numbers. Clusters of atoms such as doublets and triplets fit together in complex ways of lower symmetry than for single atoms. Errors and holes can occur in the packing. Each crystal boundary corresponds to a layer of atoms, and each angle between boundaries depends on the internal packing.

Atoms pack together in a regular pattern to form a *crystal*. Isolated atoms are spherical, and some atoms remain essentially spherical when packed together in a crystal (e.g., argon atoms in solid argon). Other atoms condense into clusters of various shapes that pack together in complex patterns. Atoms are also packed together in a liquid, but the arrangement is irregular and changes continually with time. Atoms move freely in a gas except when they hit each other.

This book assumes that the reader has studied the basic concepts of chemistry. It is important to recall that there are four main types of chemical bonding. First, remember that an atom has a tiny nucleus with positive electric charge equal to the atomic number multiplied by the charge of an electron. The nucleus is surrounded by clouds of electrons, whose position and shape are controlled by quantum numbers.

For *ionic* (or *heteropolar*) *bonding*, electrons are transferred. Gain of one or more electrons turns an atom into a negatively charged ion (*anion*), while loss of one or more electrons gives a positively charged ion (*cation*). In an ionic crystal, anions and cations repeat regularly in the correct ratio to give local balance of electric charge: thus, singly charged Na^+ and Cl^- ions alternate in the *crystal structure* of NaCl, which occurs as table salt or the mineral halite (Fig. 1.1a). Twice as many F^- ions are needed as doubly charged Ca^{++} ions in the crystal structure of CaF_2, which occurs as the mineral

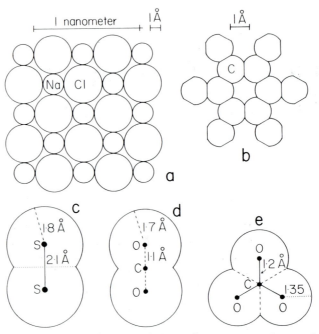

FIGURE 1.1 Shapes and sizes of some atoms, ions and molecules. (*a*) 2D slice of halite crystal showing packing of Na^+ and Cl^- ions. (*b*) Slice through middle of an infinite layer of C atoms in graphite. Note how strong covalent bonding has flattened the spheres (circles in cross section). (*c*) Double ion of S_2 in pyrite FeS_2. (*d*) CO_2 molecule. The strong bonding between C and O results in the C atom becoming surrounded by the two O atoms. (*e*) $(CO_3)^{2-}$ complex ion using C–O distance 1.23 Å and oxygen anion radius 1.35 Å. The O^{2-} ions are strongly distorted by the strong bonding to the C^{4+} ion, and, indeed, the formal ionic radius for C^{4+} is given as –0.08 Å in Table 1.1 to approximately express the distortion.

fluorite. Each ion can be represented fairly accurately by a sphere, whose *ionic radius* (Table 1.1) is estimated from measurement of interatomic distances. A cation becomes progressively smaller as its parent atom loses one or more electrons, while an anion is larger. A magnesium atom always loses two electrons when it becomes ionized, whereas an iron atom can lose either two or three electrons. Ions tend to become larger as the atomic number increases, but most anions are larger than most cations. All ionic crystals can be interpreted in terms of packing of spheres of different radii, but larger ions may be distorted up to about 20 percent of the ionic radius.

In ***metallic bonding,*** some electrons move through the entire crystal as a kind of gas, and the atomic residues pack together like regular or slightly distorted spheres with an ***atomic radius*** given in Table 1.1.

For ***covalent*** (or ***homopolar***) ***bonding,*** electron clouds are shared by adjacent atoms. This imposes strong geometrical controls: thus, each carbon atom in graphite has three equidistant neighbors (Fig. 1.1*b*). Covalent bonding distorts the original spherical shape of isolated atoms, especially when the atoms combine into molecules such as the S_2 and CO_2 molecules (Fig. 1.1*c,* 1.1*d*). Nevertheless, it is still possible to think of the atoms as spheres slightly flattened at the contacts.

Covalently bonded molecules are attracted only by weak residual forces to give ***van der Waals bonding.*** Molecules with complex shapes pack together into complicated patterns, often with a zigzag or herringbone feature.

Many crystals contain ionic complexes linked by spherical ions. Thus a C^{4+} ion surrounds itself by three O^{2-} ions to give the triangular carbonate ion $(CO_3)^{2-}$. Such triangular ions (Fig. 1.1*e*) combine with Ca^{2+} near-spherical

Table 1.1 Atomic and ionic properties of selected elements

Z	Symbol	Element	Atomic[a] Weight	Atomic[b] Radius (Å)	Ionic Radius (Å)[c] and Charge
1	H	Hydrogen	1.0080	0.37	[OH⁻]–1: 1.32(II)
3	Li	Lithium	6.939	1.57^m	+1: 0.59(IV) 0.76(VI)
4	Be	Beryllium	9.012	1.12^m	+2: 0.27(IV)
5	B	Boron	10.811	~0.9	+3: 0.01(III) 0.11(IV)
6	C	Carbon	12.011	0.77	+4: –0.08(III)
7	N	Nitrogen	14.007	0.74	+5: –0.10(III)
8	O	Oxygen	15.999	0.74	–2: 1.35(II) 1.36(III)
9	F	Fluorine	18.998	0.72	–1: 1.28(II) 1.30(III)
11	Na	Sodium	22.990	1.91^m	+1: 1.02(VI)
12	Mg	Magnesium	24.312	1.60^m	+2: 0.57(IV) 0.72(VI) 0.92(VIII)
13	Al	Aluminum	26.981	1.43^m	+3: 0.39(IV) 0.48(V) 0.53(VI)
14	Si	Silicon	28.086	1.17	+4: 0.26(IV) 0.40(VI)
15	P	Phosphorus	30.974	1.10	+5: 0.17(IV)
16	S	Sulfur	32.064	1.04	–2: 1.84(II) / +6: 0.12(IV)
17	Cl	Chlorine	35.453	0.99	–1: 1.81(VI) / +7: 0.08(IV)
19	K	Potassium	39.102	2.35^m	+1: 1.38(VI) 1.51(VIII)
20	Ca	Calcium	40.08	1.97^m	+2: 1.00(VI) 1.12(VIII)
21	Sc	Scandium	44.956	1.64^m	+3: 0.74(VI) 0.87(VIII)
22	Ti	Titanium	47.90	1.47^m	+4: 0.60(VI)
23	V	Vanadium	50.942	1.35^m	+3: 0.64(VI) / +5: 0.46(V)
24	Cr	Chromium	51.996	1.29^m	+3: 0.61(VI) / +2: 0.80(VI)
25	Mn	Manganese	54.938	1.37^m	+2: 0.83(VI) / +3: 0.64(VI)
26	Fe	Iron	55.847	1.26^m	+2: 0.78(VI) / +3: 0.49(IV)
27	Co	Cobalt	58.933	1.25^m	+2: 0.74(VI)
28	Ni	Nickel	58.71	1.25^m	+2: 0.69(VI)

Table 1.1 Atomic and ionic properties of selected elements (continued)

Z	Symbol	Element	Atomic[a] Weight	Atomic[b] Radius (Å)	Ionic Radius (Å)[c] and Charge
29	Cu	Copper	63.54	1.28^m	+1: 0.46(II) / +2: 0.57(IV)
30	Zn	Zinc	65.37	1.37^m	+2: 0.60(IV) 0.74(VI)
31	Ga	Gallium	69.72	1.35^m	+3: 0.47(IV)
32	Ge	Germanium	72.59	1.39^m	+4: 0.39(IV)
33	As	Arsenic	74.92	1.21	+5: 0.33(IV)
34	Se	Selenium	78.96	1.17	+6: 0.28(IV)
35	Br	Bromine	79.91	1.14	−1: 1.96(VI) / +7: 0.39(VI)
37	Rb	Rubidium	85.47	2.50^m	+1: 1.61(VIII)
38	Sr	Strontium	87.62	2.15^m	+2: 1.18(VI) 1.26(VIII)
39	Y	Yttrium	88.90	1.82^m	+3: 0.90(VI) 1.02(VIII)
40	Zr	Zirconium	91.22	1.60^m	+4: 0.72(VI) 0.84(VIII)
41	Nb	Niobium	92.90	1.47^m	+5: 0.64(VI)
42	Mo	Molybdenum	95.94	1.40^m	+6: 0.41(IV) 0.59(VI)
46	Pd	Palladium	106.4	1.37^m	complex 0.6–0.9
47	Ag	Silver	107.87	1.44^m	+1: 1.02(IV) / +2: 0.94(VI)
50	Sn	Tin	118.69	1.58^m	+4: 0.69(VI)
51	Sb	Antimony	121.75	1.61^m	+3: 0.76(IV) / +5: 0.60(VI)
52	Te	Tellurium	127.60	1.37	−2: 2.2(VI) / +4: 0.52(III)
53	I	Iodine	126.90	1.33	−1: 2.20(VI) / +7: 0.53(VI)
55	Cs	Cesium	132.90	2.72^m	+1: 1.74(VIII)
56	Ba	Barium	137.34	2.24^m	+2: 1.35(VI) 1.42(VIII)
57	La	Lanthanum	138.91	1.88^m	+3: 1.02(VI)
58	Ce	Cerium	140.12	1.82^m	+3: 1.01(VI) / +4: 0.97(VIII)
63	Eu	Europium	151.96	2.06^m	+2: 1.17(VI) / +3: 0.95(VI)
74	W	Tungsten	183.85	1.41^m	+6: 0.42(IV)
77	Ir	Iridium	192.2	1.36^m	+4: 0.63(VI)
78	Pt	Platinum	195.09	1.39^m	+4: 0.63(VI)
79	Au	Gold	196.97	1.44^m	+3: 0.68(IV)
80	Hg	Mercury	200.59	1.55^m	+1: 0.97(III) / +2: 0.96(IV)
82	Pb	Lead	204.37	1.75^m	+2: 0.98(IV) 1.19(VI) 1.29(VIII)
92	U	Uranium	238.03	1.56^m	+4: 1.00(VIII) and others

[a]*American Institute of Physics Handbook,* 3rd ed., McGraw-Hill, New York, 1972.

[b]Values labeled with the superscript m are for metallic atoms in 12-fold coordination, tabulated by Wells (1975, Table 29.5). Other values are radii for covalent bonding, tabulated by Wells (1975, Table 7.4). Readers should be cautious for the semimetals: for example, the following covalent radii listed by Wells: Ge 1.22 Å, Sn 1.40 Å, and Sb 1.41 Å, differ considerably from the metallic radii. Boron forms complex structures with variable atomic radius.

[c]Shannon and Prewitt (1969) with revision by Shannon (1976). Only the more useful information is quoted here. The charge (e.g., −1 for Cl) is followed by the distance (e.g., 1.81 Å) for a particular coordination number (e.g., six in Roman numerals). A slash (/) precedes data for a second charge (e.g., +7) and distance (0.20 Å). The negative values for C^{+4} and N^{+5} result from the inapplicability of a hard-sphere ionic model to tightly bonded atoms. High-spin state is used for Cr, Mn, Fe, and Co. Values for the low spin state are: Cr+2: 0.73 (VI); Mn+2: 0.67(VI) / +3: 0.58(VI); Fe+2 0.61(VI) / +3: 0.55(VI); Co+2: 0.65(V1).

ions to give the crystal structure of $CaCO_3$ (calcite). Each Ca^{2+} ion is sur-rounded by six O^{2-} ions at the corners of an octahedron. Furthermore, the calcite structure can be interpreted as the packing of small Ca^{2+} and very small C^{4+} ions into the holes between large O^{2-} ions that touch each other.

This introduction provides justification for studying close packing of spheres and less-regular bodies. Rather than beginning with three-dimensional (3D) crystallography, this chapter is devoted to the simpler packing in two dimensions (2D), in which atoms and ions are represented by circles.

Crystallographers usually express distances in the traditional *angstrom unit (*Å*)*, where 1 Å = 10^{-10} meter = 0.1 nanometer. There is strong psycho-logical resistance against conversion to the nanometer of the new Système Internationale. Actually, atomic and ionic radii are conveniently expressed in angstroms with a range from ~0.2 to ~1.5 Å (i. e., 0.02–0.15 nm). For scale, note that a crystal of length 1 cm will contain about 5×10^7 atoms or ions along each edge, while a cube 1 cm across will contain about 10^{23} atoms or ions.

1.1   Packing of Equal Circles in 2D

Perhaps the most obvious way to pack equal circles is in a square *array* (Fig. 1.2*a*), in which each circle touches four other circles such that the centers of four adjacent circles lie at the corners of a square. This simple diagram al-ready expresses important features of a crystal. The *packing* is *regular* be-cause each circle has the same arrangement of neighbors as any other circle. The regularity of the packing produces *symmetry*. The circles are in contact, as though pulled together by attractive forces.

It is important to emphasize the regularity of the packing because atoms are also in contact in a liquid. In a crystal, the pattern stays the same all the way across, except for the disturbance caused just at the surface. In a liquid, the arrangement of neighbors changes from place to place and, indeed, from time to time. Actually, local mistakes can occur in a real crystal, but we shall deal mainly with an ideal crystal that is perfect mathematically. Strictly speaking, an ideal crystal must be infinite because a surface cannot be allowed.

It is not necessary, however, to show an infinite number of circles (or atoms), because the essential information is available in just the smallest part of the pattern that can yield the entire pattern by regular *translation*. Thus the single circle inside the dotted lines is sufficient to yield the infinite pat-tern when translated (i.e., moved) repeatedly by the distance between adja-cent circles. The translated unit is called a *unit cell*. Choice of its origin is arbitrary, as shown by comparing the dashed lines with the dotted lines. A

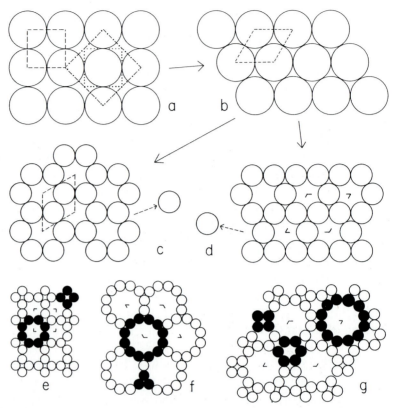

FIGURE 1.2 Packing of equal circles in 2D. (*a*) Square packing with three possible unit cells. (*b*) Closest packing with hexagonal symmetry. (*c* and *d*): Two types of open packing obtained by removing circles from pattern *b*. (*e*, *f*, *g*). Three types of open packing with regular but complex symmetry. Corner symbols mark possible unit cells. In *e*, *f*, and *g* the patterns can be regarded as the condensation of rings of circles as depicted by the filled circles: of course, all circles are really equivalent in each pattern, and there is no distinction between open and filled circles.

larger unit cell can be chosen (dot–dash lines), but a crystallographer usually chooses the smallest unit cell that is convenient for expressing the symmetry of the crystal.

Square packing of equal circles is highly symmetrical. The array can be rotated 90°, 180°, 270°, or 360° about the center of any circle, and the new array is identical to the starting array. Hence, there is a *4-fold rotor* at the center of each circle (and also at each midpoint between four adjacent circles). Symmetry will be developed systematically in later chapters, but readers should look out for **rotation symmetry** in the remaining diagrams of this chapter. There is also **mirror symmetry** (see Chapter 2).

Square packing of circles uses up 78.5 percent of the area (Let r be the radius. The unit cell has edge $2r$ and area $4r^2$. There is one circle of area πr^2 in each unit cell.) Sliding the circles can yield **hexagonal closest packing** of circles (Fig. 1.2*b*), in which each circle has six touching neighbors. The smallest unit cell is a rhombus (i. e., a parallelogram with sides of equal length) with alternate 60° and 120° angles. Its area is $4r^2 \sin 60° = 4r^2 \cdot \sqrt{3}/2$, and it contains one circle of area πr^2 to give a **packing efficiency** of 91 percent. Although this array has 6-fold rotation symmetry, the unit cell is selected to be a 120° rhombus and not a hexagon, which would be three times larger. Equal circles cannot be packed closer than this array.

To understand crystallography it is essential to make drawings and models. Close packing of circles can be readily demonstrated by using coins or metal washers or circles cut out of cardboard.

In graphite, the carbon atoms form a giant molecule consisting of an infinite 2D sheet in which each C atom is covalently bonded to three neighbors (Fig. 1.1*b* with distorted circles, or Fig. 1.2*c* with idealized undistorted circles). This variety of **open packing** is enforced by bonding to three neighbors rather than four or six neighbors. Every third circle of hexagonal closest packing is removed to produce the graphite sheet, and the unit cell is again a 120° rhombus but with three times the area. Although symmetry is not the main concern of this chapter, note that there are both 6-fold and 3-fold rotors. Furthermore, show that the packing efficiency is only 61 percent. The reader will find that removal of circles leads to a less stable pattern of coins or substitutes. Why does graphite not collapse into hexagonal closest packing of carbon atoms? The reason is that the shared electrons are strongly locked into 120° positions. Of course, carbon atoms are spheres, not circles, and Figure 1.2*c* is an idealized section through the center of each distorted sphere.

A second type of hexagonal open packing (Fig. 1.2*d*) is obtained by removing one quarter of the circles. Show that the smallest unit cell has four times the area of that in Figure 1.2*b*, and that the packing efficiency is 68 percent. Another way of removing one quarter of the holes is left for discovery by the reader: see Figure 1.3*i* for the answer.

Three other types of open packing (Figs. 1.2*e*–1.2*g*) have high symmetry and low packing efficiency. Two can be found as components of the 3D structures of zeolite minerals.

These regular patterns of circles have only 3-, 4-, or 6-fold symmetry. Can a regular pattern be made with 5-fold symmetry? The answer will appear in Chapter 2.

1.2 Packing of Two Types of Circles in 2D

Figure 1.1*a* showed how small Na$^+$ ions pack alternately with large Cl$^-$ ions in a slice taken through a crystal of NaCl (halite). Two factors determine the packing of spherical ions or atoms of more than one type: the relative size of the spheres (or circles in 2D) and the relative number. The size factor is expressed by the ***radius ratio,*** which for Na$^+$ and Cl$^-$ is 1.02 Å/1.81 Å = 0.56. The relative number is fixed by the ionic charges in ionic crystals, but may vary in alloys (i. e., mixed metals).

For two types of spheres of essentially the same size (e.g., silver and gold atoms both of radius 1.44Å), there is no significant geometrical control over the occupancy of sites in Figure 1.2*b*. This is why it is possible to grow crystals of mixed silver and gold atoms (denoted Ag$_x$Au$_{1-x}$), where *x* varies all the way from 0 to 1. Mathematically, it is possible to invent many regular arrays for two types of spheres with the same radius. For square packing of equal spheres (Fig. 1.2*a*), there are two simple arrays for a 1 : 1 ratio (Figs. 1.3*a*, 1.3*b*) and two for a 3 : 1 ratio (Figs. 1.3*c*, 1.3*d*). For hexagonal closest packing of equal spheres, there are also two simple packings for a 1 : 1 ratio (Figs. 1.3*e*, 1.3*f*). Arrays for 2 : 1 and 3 : 1 ratios (Figs. 1.3*g*, 1.3*h*) are readily obtained by filling vacancies in Figures 1.2*c* and 1.2*d*. A second array for the 3 : 1 ratio (Fig. 1.3*i*) answers the question posed in the preceding section. When radii are not quite the same (e.g., copper and gold atoms of radius 1.28 Å and 1.44 Å, respectively), ordered arrays tend to develop (Section 1.5). Indeed, copper–gold alloys formed at high temperature have random arrangements of atoms, but such disordered arrays become rearranged into mixtures of ordered arrays at low temperature as atoms settle down into the best fit with their neighbors.

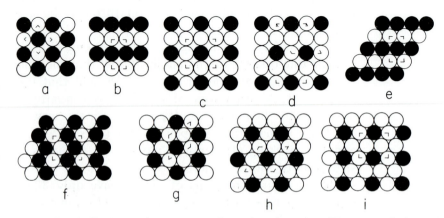

FIGURE 1.3 Packing of two types of equal circles in 2D. Corner symbols mark possible unit cells: note two choices of rectangular and oblique unit cells in *d*.

When the radius ratio is small, the large circles tend to touch each other while the small ones fit into the small holes (often called the interstices).

For square packing, Figure 1.4*a* shows a perfect fit in which all circles touch each other. This occurs only for a radius ratio of $\sqrt{2} - 1 \cong 0.41$. In triangle *pqr*, the sides have ratios $1:1:\sqrt{2}$. The length *pq* = length *qr*, and *sr* must be $\sqrt{2} - 1$. If the radius ratio increases, the larger circles lose contact with each other as the smaller circles force them apart (Fig. 1.4*b*). When the radius ratio reaches unity, the pattern matches that of Figure 1.3*a*, except for the trivial matter of a relative rotation of 45°.

Figure 1.4*c* shows how the smaller circle does not fill the hole when the radius ratio falls below 0.41. This arrangement is usually unstable in a crystal because the ions or atoms have not been pulled by the attractive forces into positions of highest stability. A rearrangement occurs to hexagonal packing of circles (Fig. 1.4*d*). Smaller circles occupy interstices and hold apart the larger circles. Even though the larger circles do not touch, this is a more stable packing than square packing because each atom is locked into position by its neighbors.

A perfect fit occurs for hexagonal packing with a radius ratio of 0.15 (Fig. 1.4*e*). To calculate this radius ratio, use triangle *pqr*, whose angles are 30°, 60°, and 90°, and whose edges are $qr:pq:pr = 1:\sqrt{3}:2$. The radius of the small circle *sr* is $2 - \sqrt{3}$. The radius ratio is $(2 - \sqrt{3})/\sqrt{3}$.

All arrays in Figure 1.4 have a $1:1$ ratio of circles, but other ratios are possible (Fig. 1.5). Half of the interstices in Figure 1.4*a* can be left empty,

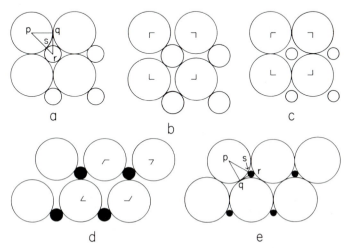

FIGURE 1.4 Packing of two types of unequal circles in 2D. Diagrams *a* and *e* show perfect fitting for square and hexagonal arrays. Diagrams *b* and *d* show arrays when the small circle is too large for a perfect fit, while *c* shows an unstable array.

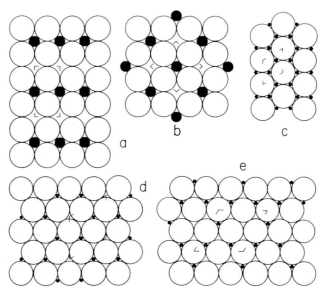

FIGURE 1.5 Square and hexagonal packing of two types of unequal circles in 2D.

giving twice as many large as small circles. There are two simple arrays: one with a rectangular unit cell (Fig. 1.5*a*) and one with a square unit cell (Fig. 1.5*b*). When a unit cell has been chosen correctly (Section 1.6), it is easy to count circles. Thus in Figure 1.5*a*, the unit cell contains two large circles and two halves of the small circles. In Figure 1.5*b*, the unit cell contains one small circle and four halves of the large circles.

Filling the empty interstices in Figure 1.4*e* gives Figure 1.5*c* with a 1:2 ratio. This array is the closest packing of two types of circles, and the circles together occupy 95.0 percent of the area. In Figure 1.4*e*, consider triangle *pqr*, whose area is $\frac{1}{2} \times pq \times qr = (pq)^2/2\sqrt{3}$. The pie-shaped slice *pqs* of the large circle has area $(pq)^2 \cdot \pi/12$. The 60° sector of the small circle has area $(sr)^2 \cdot \pi/6$, where $sr = pq \cdot (2 - \sqrt{3})/\sqrt{3}$. A similar but easier calculation for the close-packed square pattern in Figure 1.4*a* gives a combined packing efficiency of 92.0 percent.

Taken alone, the small circles in Figure 1.5*c* have the same relative positions as the circles in Figure 1.2*c*. Another hexagonal pattern (Fig. 1.5*d*) has a lower occupancy of the interstices. The unit cell has two small circles, one complete large circle, and four halves of the large circles to give a ratio of 3:2. The small circles of the array in Figure 1.5*e* have the same relative positions as the circles in Figure 1.2*d*, and the ratio is 4:3. The three small circles are immediately obvious in the unit cell. There is a complete large circle at the center of the unit cell, and four half-circles at the midpoints of the edges of the cell. The remaining large circle can be assembled from pie-shaped sectors at the corners of the cell.

1.3 Packing of Clusters

Atoms become compressed when they combine into molecules (Figs. 1.1*c*, 1.1*d*). Two atoms combine into a figure eight, which is approximately elliptical in cross section, and approximately ellipsoidal in 3D. Doublets can stand on end (Fig. 1.6*a*) to give a rectangular unit cell with one doublet per cell. This array is not packed as closely as the array in Figure 1.6*b*. Here, alternate rows of doublets click into the dimples of other rows of doublets. The unit cell is still rectangular, but it is almost twice as high and contains two doublets. Tilting of doublets produces yet another array (Fig. 1.6*c*), in which each atom touches five other atoms in addition to its partner in a doublet: indeed, this pattern is similar to the closest packing of equal circles (Fig. 1.2*b*). It is a general rule in crystal structures that the packing density increases as the number of neighbors increases, and this is true in going from Figure 1.6*a* (four neighbors) to Figure 1.6*b* (five neighbors) to Figure 1.6*c* (six neighbors). Finally, doublets can be packed in a zigzag array (Fig. 1.6*d*), in which each atom also has six neighbors. Whereas the unit cell in Figure

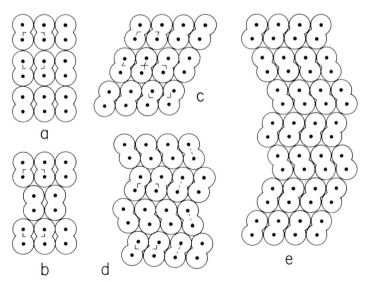

FIGURE 1.6 (*a–d*) Four regular arrays of short doublets. Unit cells are marked by corner symbols. Two possible choices are shown for *c*. Dashed lines in *d* emphasize the zigzag arrangement. (*e*) An irregular pattern.

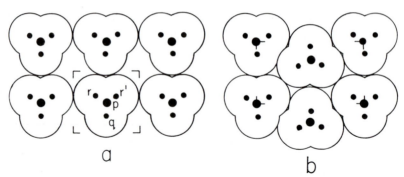

FIGURE 1.7 Two packing arrays of triplets. Actually, a molecule might have one central atom surrounded by three other atoms, as illustrated by the small circles. Corner symbols show unit cells.

1.6c is a parallelogram containing one doublet, the unit cell in Figure 1.6d is a rectangle containing two doublets. Zigzag packing of molecules is rather common in crystal structures. The shape of the unit cells in Figure 1.6 obviously depends on the elongation of the molecule and the type of packing.

Triplets do not pack well when parallel (Fig. 1.7a), but pack closely when alternate triplets are reversed (Fig. 1.7b). Quadruplets pack neatly in a square array (Fig. 1.8a), but there is a lot of empty space. The packing is closer in Figure 1.8b, but there is no elegant way of tilting the quadruplets to achieve really close packing as for the doublets in Figures 1.6c and 1.6d.

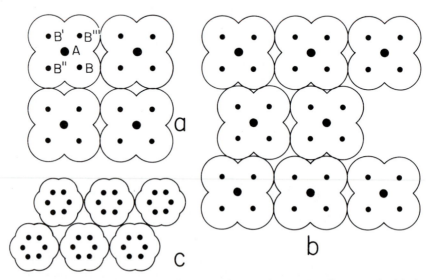

FIGURE 1.8 Packing arrays of quadruplet and hextuplet clusters. The labels in diagram a are used in Chapter 2.

Sextuplets (Fig. 1.8c) are near-circular in outline, and pack together in an hexagonal array. The reader might like to investigate how the packing efficiency varies with angle of rotation of the sextuplets. It is impossible to obtain really close packing of sextuplets in 2D. The C_6H_6 sextuplets do not lie in flat layers in the crystal structure of solid benzene, but are tilted in 3D to obtain a closer packing.

There is considerable scope for further theoretical development of the packing of clusters, but the present account illustrates the essential features.

1.4 Packing of Clusters and Circles

Many ionic structures can be described as the packing of a small spherical cation with a large anionic cluster. Calcite is a typical example in which divalent calcium cations Ca^{2+} (radius 1.0 Å) are linked to triangular CO_3^{2-} anion complexes.

As a general rule, cations are arranged as far apart as possible to minimize electrostatic repulsion between them. Of course, cations and anions are in contact to maximize the electrostatic attraction. The actual pattern depends on geometrical factors. Short anionic doublets might pack with small cations in the pattern of Figure 1.9a, which is found in the calcium carbide, CaC_2, structure. Each cation is surrounded by six components of the doublets, two from the ends of doublets, and two each from two broadside doublets. The cations are widely spaced and shielded electrostatically by the anions. Note that this pattern is merely a distortion of the NaCl pattern in Figure 1.1a. Figure 1.9b shows a possible pattern when the anion complex is more elongated. This pattern has 4-fold rotational symmetry.

The reader might like to investigate packing between cations and clusters of triangular, square, and hexagonal shapes. Perfect fits cannot be expected except at particular sizes of the packing units.

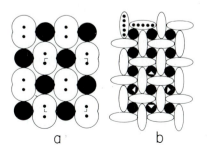

a   b

FIGURE 1.9 Packing arrays of small circles and large complexes. Corner marks show unit cells. Each doublet in *a* is formed by merging two circles. Each ellipse in *b* is a convenient approximation for the condensation of five circles, as shown in the upper left.

1.5   Packing Mistakes

Most natural crystals grow in an impure environment. Atoms or ions of various types land on growing surfaces (called ***crystal faces***). The correct species tend to "stick" to the surface, whereas the wrong ones tend to move away. The greater the difference in chemical properties of the wrong ones from the right ones, the purer is the crystal (when other things are equal). Mistakes usually increase as the temperature increases.

Atoms or ions of the wrong type can occur at random (Fig. 1.10). If the wrong species is too small, the neighbors may move toward it slightly in order to minimize disturbance in the interatomic forces. If too large, the neighbors may move away. Errors may occur merely by omission, or by insertion of interstitial species. The resulting three types of disorder are called ***substitutional disorder, omission disorder,*** and ***interstitial disorder.***

In ionic crystals, disorder may result in local charge imbalance that must be remedied elsewhere: thus replacement of Na^+ (radius 1.02 Å) by Ca^{2+} (1.00 Å) can be coupled with nearby replacement of Si^{4+} (0.26 Å) by Al^{3+} (0.39 Å), as in some aluminosilicate minerals including plagioclase feldspar. Omission of just one ion (e.g., Na^+) in the crystal structure of NaCl would produce a local electrostatic charge. Coupled omission of Na^+ and Cl^- ions leads to electrostatic neutrality: the ion pairs are called ***Schottky defects.***

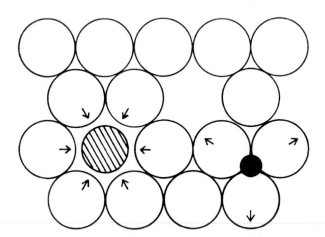

FIGURE 1.10   Three types of disorder. The hatched circle replaces an open circle (***substitutional disorder***); neighboring circles could move inward to reduce spatial misfit. ***Omission disorder*** is shown at the upper right, but potential movement of nearby circles is not shown. For ***interstitial disorder,*** the small circle occupies a new type of position between the open circles; neighboring atoms should move apart to reduce the misfit.

Bombardment by a nuclear particle can drive an atom from its proper site into an interstitial site producing a *Frenkel defect*.

Slow cooling of a disordered crystal usually allows interchange of atoms or ions to produce a more ordered pattern. The atoms or ions may either cluster together and ultimately separate out into separate crystals, or maximize their separation to give an ordered *superstructure* in a single crystal (Fig. 1.11). The intermediate stage of formation of a superstructure involves development of *domains* that grow larger with time. Each domain nucleates randomly and has no knowledge of position of *nucleation* of neighboring domains. As growth proceeds, domains expand into contact with neighboring domains. Domains that fortuitously have the same parity of choice of atomic positions (e.g., q and r in Fig. 1.11) merge into a larger domain. Domains that fortuitously have the wrong parity (e.g., domains p and q) cannot

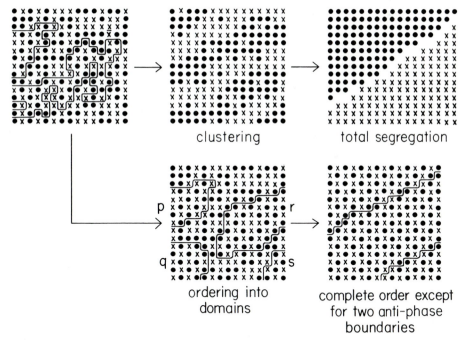

clustering   total segregation

ordering into domains

complete order except for two anti-phase boundaries

FIGURE 1.11 Increase of order producing either total segregation via clustering, or complete order except for antiphase boundaries between domains. For convenience of drawing, two types of atoms are represented by dots and crosses. The starting pattern was obtained by placing dots and crosses according to odd or even numbers from a table of random numbers. Two irregular areas turned out to have regular alternation of dots and crosses. The intermediate stages of ordering were obtained by interchanging adjacent dots and crosses semi-randomly so that domains or clusters tended to grow. In real crystal structures, geometrical strain tends to give domains elongated in a special direction.

merge, and end up with an ***antiphase domain boundary*** in which pairs of like species touch each other instead of unlike species. Domains with the wrong parity can be reversed only by movement of all atoms in the domain. The crosshatched pattern in microcline ($KAlSi_3O_8$) results from antiphase domain boundaries produced when Al and Si atoms move into ordered positions upon annealing of sanidine ($K(Al_{0.25}Si_{0.75})_4O_8$). The ($Al_{0.25}Si_{0.75}$) indicates random occupancy of one site by three times as many Si as Al atoms.

Returning to the intermediate state of ***clustering,*** there is a misfit when the two species differ in size. When the clusters are small, the species remain in contact. Ultimately they become too large for the strain to be accommodated by distortion, and the clusters break apart into separate crystals. Disordered (Na_xK_{1-x})$AlSi_3O_8$ varieties of feldspar minerals break up into x units of albite $NaAlSi_3O_8$ and ($1-x$) units of K-feldspar $KAlSi_3O_8$. Fine-scale intergrowths of albite and K-feldspar are known as perthite.

Some crystals show ***stacking disorder.*** Regular packing of doublets gave the parallel and zigzag arrays of Figure 1.6*c* and 1.6*d*. Random intergrowths of *c*-type and *d*-type fit together nicely, and an example is given in Figure 1.6*e*. This type of disorder occurs only when there are two or more choices of position for a packing unit (here, the leftward or rightward tilt of a doublet) that do not change the position of first neighbors at the contact of the packing units (here, the triangular contacts).

1.6   Unit Cell and Symmetry

A ***unit cell*** of a pattern will yield the entire pattern when translated repeatedly without rotation (Fig. 1.12*a*). The repetition produces an infinity of identical unit cells, and the pattern is regular. In order to fill space without any gaps, the unit cell must be a parallelogram in 2D, or a parallelepiped in 3D. A triangular unit cell would fill only half of the area when translated repeatedly without rotation (Fig. 1.12*b*). A hexagon will fill space by repeated translation without rotation, but is three times bigger than a 120° rhombus (Fig. 1.12*c*).

Each unit cell can be specified geometrically by the lengths of the two sides and the included angle, conventionally labeled *a, b* and γ. In vector notation, the two vectors **a** and **b** are sufficient. The entire pattern is obtained by translating the unit cell to infinity by all integral multiples of **a** and **b**. Of course, it is a waste of time to cover an entire sheet of paper with repeated unit cells, and a crystallographer usually shows only one unit cell or just a few unit cells.

How many shapes of parallelograms are possible? A preliminary answer (Fig. 1.13) can be obtained from the packing drawings: square (Fig. 1.2*a*),

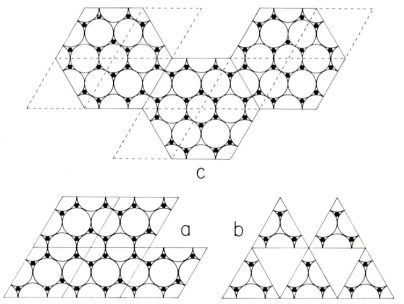

FIGURE 1.12 Diagrams illustrating why a parallelogram is chosen for the shape of a unit cell. The sphere packing can be represented by a parallelogram-shaped unit cell (*a*), of which five are shown to represent part of the infinite pattern. A triangular unit cell (*b*) cannot represent the whole pattern, because the gaps can be filled only by rotating alternate triangles. A hexagonal unit cell (*c*) can be decomposed into parallelograms when the pattern is infinite. Note that the lines are merely mathematical conveniences, and that only the circles would be real.

120° rhombus (Fig. 1.2*b*), rectangle (Figs. 1.3*b* and 1.6*a*), and oblique parallelogram (Fig. 1.6*c*). The shape of a unit cell is a consequence of the symmetry. A 4-fold rotor enforces $a = b$ and $\gamma = 90°$. A 120° rhombus is enforced by 3-fold or 6-fold symmetry. A rectangular unit cell is required by mirror symmetry. Twofold rotation symmetry does not place constraints on the shape of the unit cell. Symmetry is explored systematically in the next chapter. The edges of an oblique parallelogram can be accidentally equal, but this is not the result of symmetry except perhaps when $\gamma = 120°$, as in Figure 1.13*b*.

For a given pattern, an infinite number of unit cells can be chosen. When symmetry is present, it is customary to choose the smallest unit cell whose shape shows the special geometry resulting from the symmetry. Thus, a square is used when 4-fold symmetry is present, and not an oblique parallelogram. When the highest symmetry is 2-fold, any oblique parallelogram is possible, but usually a cell is chosen with $a < b$ and γ greater than but close to 90°.

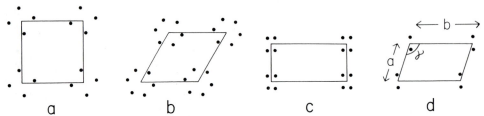

FIGURE 1.13 The four shapes of parallelograms used as unit cells in 2D patterns. The points in (*a*) obey 4-fold rotational symmetry, and require a square unit cell. Those in (*b*) have 6-fold rotational symmetry and require a 120° rhombus as unit cell. The rectangular unit cell in (*c*) has mirror lines as edges. Twofold rotors occur at the corners of the oblique cell in (*d*). Note the conventional choice of $b > a$ and $\gamma > 90°$. Even though a whole pattern can be generated from just the points in a unit cell, it is customary to show points just outside the unit cell when they help to show the symmetry.

1.7 Shape of Crystals

Whereas an ideal crystal must be infinite, a real crystal must terminate. From the geometrical viewpoint, there is no reason why the terminations should not be random and irregular. However, from the chemical viewpoint, the surface energy is smaller when the atoms and/or molecules terminate in special ways. Research on surfaces is still immature, but simple qualitative ideas are sufficient for the present purpose. The **surface energy** tends to be lower when the surface is flatter; that is, the surface should not be pockmarked by holes. Hence, surfaces tend to be planar in 3D, and can be represented by straight lines in 2D. Theoretical boundaries should be exactly planar or straight, but real crystals have various imperfections. Strong chemical bonds usually tend to lie parallel to surfaces or project into the crystal, while the surface species present only weak unsatisfied bonds to the exterior.

Square-packed circles could form 2D crystals as in Figure 1.14. The simplest boundaries are the horizontal and vertical rows (Fig. 1.14*a*), while the next simplest ones are formed by 1:1 zigzags (Fig. 1.14*b*). Actually, it is possible to invent an infinite number of boundaries in which the zigzags are formed by counting *n* circles in one direction and *m* circles in the other direction. However, the packing is neater and the surface area is smaller when *n* and *m* are small. The integers *n* and *m* can be used as indices to specify the orientation of the boundary. In Figure 1.14*c*, the upper part of the composite crystal has faces with $n/m = 2/1$.

The key features of crystal boundaries are that (i) the angles between them are determined only by the internal crystal structure, and (ii) the relative sizes of the crystal boundaries depend on the rate of growth of the crystal boundaries. All five of the hypothetical crystals in Figure 1.14 are based on square packing of circles, whose unit cell is a square. All the boundaries have orientations related to a square by varying n and m. Crystal (d) has larger (**1,0**) than (**1,1**) boundaries, while crystal (e) has larger (**1,1**) than (**1,0**) boundaries. The small boundaries in the two crystals would be wiped out by addition of circles like the dashed ones, and crystals of complex shape result only when all the faces grow at almost the same rate. The only difference between crystals (d) and (e) is the relative rate of growth of the (**1,0**) and (**1,1**) faces. When two crystals collide during growth, the crystals must be stunted at the contact. If a crystal grows outward from a wall (e.g., in a vesicle), only one end becomes developed. Very rarely, two crystals float together in their mother liquor at exactly the correct orientation and merge to form one crystal as in (c).

Although crystals of a particular chemical and structural species tend to grow with a particular shape (e.g., cube for table salt; octahedron for spinel),

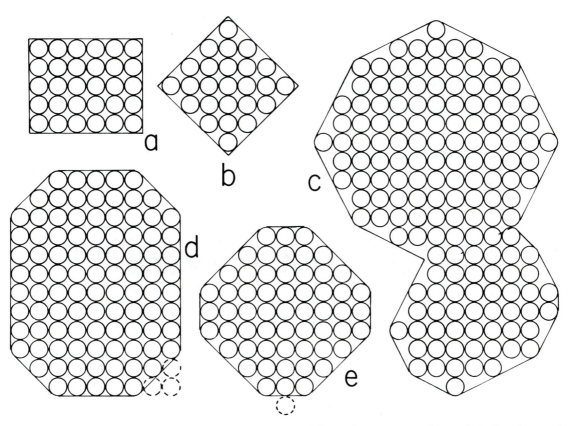

FIGURE 1.14 Some possible crystal shapes obtained from the square packing of circles. In a real crystal 1 mm across there will be approximately 5×10^6 atoms on each crystal boundary, and the lines will be a good representation of the rows of circles.

the shape may vary (but *not* the angles) for some species (e.g., feldspar). The causes of such variations are not properly understood, and several factors are probably involved: absorption of impurity atoms may hinder growth on some boundary faces; atomic bonding may change with temperature; and so on. From the viewpoint of mathematical crystallography, such variations are unimportant, and the key feature is the *constancy of angle between crystal boundaries with the same indices for all crystals of a particular chemical and structural type.* Different structural materials will have different angles between the crystal boundaries, and the angles can be related to the symmetry and shape of the unit cell.

1.8 Mathematics of Crystal Shapes

From the mathematical viewpoint, it is necessary only to consider crystals as assemblages of unit cells stacked together like bricks in a pyramid or tiles on a floor. The crystal boundaries are obtained by drawing lines that touch the corners of the unit cells, although, of course, the actual boundaries of crystals will be determined by atoms. Figure 1.15 develops a hypothetical crystal from the unit cell in Figure 1.6*a*. For convenience the unit cell is turned through 90°, and the crystal has been given symmetrical boundaries so that there are two **mirror lines** passing through the point *O*. This point was deliberately chosen not to lie at the corner of a unit cell just to show that this is unnecessary, but it is usually placed at the corner of a unit cell for convenience. A real crystal will not be exactly symmetrical, but a hypothetical equivalent can be drawn. The boundaries will remain parallel, but the relative lengths of the boundaries will differ as in the two crystals drawn at the upper and lower left of Figure 1.15.

In order to describe the crystal shape, two **axes** are drawn parallel to the edges of the unit cell, and are labeled with the same symbols used to specify the cell edges (Fig. 1.13). The two axes are actually mirror lines in the hypothetical symmetrical crystal, but not in real distorted crystals. There are five types of boundaries, each repeated by the mirror symmetry. One type is parallel to the *a*-axis, one to *b,* and the remaining three are formed by moving 1*a* and 1*b,* or 3*a* and 2*b,* or 2*a* and 1*b*. The small diagram at the upper right shows how the orientations of the boundaries can be found without drawing so many unit cells as in the main diagram. Draw just one unit cell with lengths proportional to *a* and *b*. To get the orientation of the face (3*a,* 2*b*), join the coordinates *a*/2,0 and 0,*b*/3. For the face (2*a,* 1*b*), join the coordinates *a*/1,0 and 0,*b*/2. Or in general, for face (*na,mb*), where *n* and *m* are integers, join the coordinates *a*/*m,* 0 and 0,*b*/*n*. When *n* = **0** and *m* = **1**, the coordinates are *a,*0 and 0,∞, which results in a line parallel to the *b*-axis. The angle ϵ is given by

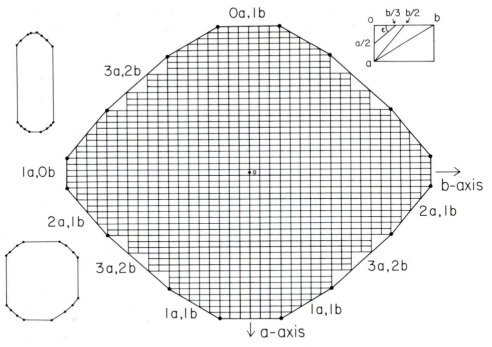

FIGURE 1.15 Possible crystal shapes for the crystal structure of Figure 1.6*a* after rotation by 90°. The large crystal in the center was constructed from an assemblage of unit cells. For clarity the junctions of the faces are marked by dots. The small diagrams at the upper and lower left contain boundaries at the same orientation as those in the central diagram, but the relative lengths of the boundaries are different giving, respectively, a crystal elongated on *a* and one that is near equant. The diagram at the upper right shows how the orientations of the boundaries can be constructed from an enlarged drawing of a unit cell.

$$\tan \epsilon = \frac{a/2}{b/3} = \frac{3a}{2b}$$

For face (*na,mb*), the angle ϵ would be given by $\tan \epsilon = na/mb$. This simple relation is the key to the shape of those 2D crystals that have a rectangular cell. For a square unit cell with $a = b$, $\tan \epsilon = n/m$.

The first studies of the shape of crystals were made before X-ray diffraction revealed the internal structure. F.C. Phillips, in his charming "Introduction to Crystallography," briefly described how N. Steno, D. Guglielmini and J.B.L. Romé Delisle developed the ***Law of Constancy of Angle*** in a particular crystalline material, and how Abbe Haüy in 1784 proposed that a crystal contains a regular stacking of a fundamental cleavage unit. Just as for the modern concept of a unit cell, this leads to the ***Law of Rational Indices***.

Whereas an X-ray crystallographer can measure the shape and size of a unit cell, an early morphological crystallographer was forced to make an arbitrary choice.

The final accepted procedure of morphological crystallography is as follows: (i) measure the *angles* between crystal boundaries, but ignore the relative sizes of the faces; (ii) look for possible symmetry; (iii) choose the most convenient axes with respect to the symmetry; (iv) choose any boundary (usually a prominent one) to obtain the ratio a/b by assigning $n = 1$ and $m = 1$; and (v) index other boundaries by calculation.

W.H. Miller (1839) assigned **Miller indices** (h,k) as follows: For any crystal, select a and b axes in 2D (and *a, b, c* in 3D). The axes need not be orthogonal, but must be parallel to actual or hypothetical edges of the crystal. Choose any boundary of the crystal as the ***parametral boundary;*** and draw a parallel line (in 2D) or plane (in 3D) that intercepts the b-axis at unit length. This parametral boundary will cut the a-axis at the distance a/b from the origin, where a/b is called the ***axial ratio*** (in 3D, there will also be c/b for the c-axis). To index any other boundary, draw its parallel through the unit point on the b-axis. The intercept on the a-axis must be a/b multiplied by a simple fraction that can be expressed as k/h, where **k** and **h** are integers. The Miller indices are **h** and **k**. Readers must remember that these are expressed in round brackets as **(hk)** in 2D or **(hkl)** in 3D. The indices are reduced to the smallest possible values: for example, $h = 2$ and $k = 3$, rather than $h = 4$ and $k = 6$.

Figure 1.6*a* was constructed using atoms of radius 1.25 Å whose spacing gave unit-cell edges $a = 2.5$ Å and $b = 4.25$ Å. The resulting crystal in Figure 1.15 has the plane (1*a*, 1*b*) at an angle of $\tan^{-1} 2.5/4.25$ to the b-axis (i.e., 30.47°, which can be checked directly on the figure with a protractor). For the original drawing of Figure 1.16*a*, the unit length (*OB*) on the b-axis was taken as 10 cm, but after reproduction for this book it is 3.3 cm. The parametral boundary (1*a*, 1*b*) cuts the a-axis at point A which is 5.9 cm from the origin *O* (or 2.0 cm after reproduction). Therefore, the axial ratio is now fixed at $5.9/10 = 0.59$. Actually it should be $2.5/4.25 = 0.5882$ by calculation. The boundary (3*a*, 2*b*) in Figure 1.15 lies at 41.4° to *b* by measurement with a protractor, and 41.42° by calculation. The corresponding line on Figure 1.16*a* meets the a-axis at *D*, which was 8.8 cm from *O* before reproduction. According to the Miller procedure, $k/h = 8.8/5.9 = 1.49$, or 1.5 within drawing error. Therefore $k = 3$ and $h = 2$, and the boundary would be given the Miller indices (23). Readers should *note that these indices are exchanged from the earlier* (3*a*, 2*b*) *description* or confusion will result. The (2*a*, 1*b*) boundary, which lies at 49.5° by measurement and 49.64° by calculation, gives point *E* at 11.8 cm from *O*, and Miller indices (12).

A morphological crystallographer without knowledge of the unit cell might have indexed the 41.42° boundary as (11), giving $a/b = 0.8824$. The

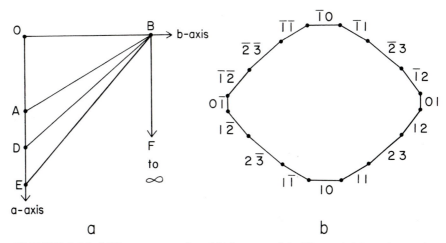

FIGURE 1.16 Miller construction (*a*) for crystal in Figure 1.15, and complete indexing of the boundaries (*b*).

30.47° boundary would become (**32**), and the 49.64° one would become (**34**).

Returning to the first indexing, which is the correct one for the unit cell chosen in Figure 1.6*a*, the boundaries on the other three quadrants of the crystal will have the same indices as those on the first quadrant except for a sign change. A boundary that cuts the negative direction of the *a*-axis is given a negative value of **h**, and one that cuts the negative *b*-axis will have negative **k**. Thus the (**11**) boundary is the prototype of three symmetric boundaries (**$\bar{1}1$**), (**$1\bar{1}$**) and (**$\bar{1}\bar{1}$**). Crystallographers place a minus sign over a numeral to save space. Figure 1.16*b* shows the complete indexing.

The small drawing at the upper right of Figure 1.15 is equivalent to Figure 1.16*a* as far as directions are concerned, and is preferred when the lengths of the unit cell are known.

The boundary parallel to the *a*-axis becomes the line *BF* in Figure 1.16*a* and indexes as (**01**). The line parallel to the *b*-axis becomes the line *OB* in Figure 1.16*a* and indexes as (**10**).

For rectangular axes, calculations are easy. Thus the angle between the boundary (**hk**) and the *b*-axis is given by $\tan^{-1}$ (**ka/hb**). For oblique axes, the Miller construction follows the same procedure, but the calculations are more complex. Figure 1.17 is based on the lower unit cell in Figure 1.6*c*, whose dimensions are *a* = 2.5 Å, *b* = 3.94, γ = 101° for an atom of radius 1.25 Å. A geometrical construction for the particular crystal used in the left-hand drawing is shown at the right. The (**10**) boundary must be parallel to the *b*-axis, and the (**01**) boundary must be parallel to the *a*-axis. The parametral plane (**11**) cuts unit lengths off the *a* and *b* axes and joins the corners *A* and *B* of the unit cell at the lower right. By measurement, the angle to the *b*-axis

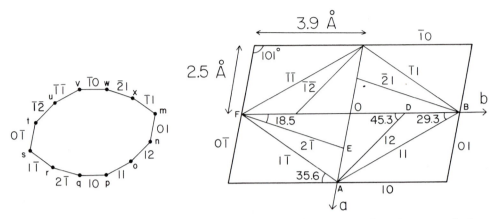

FIGURE 1.17 Construction of hypothetical crystal for a unit cell with a = 2.5 Å, b = 3.9 Å, and γ = 101°. An origin O is chosen at the center of four unit cells.

is found to be 29.3°. The (12) plane passes through points A and D, which lie at a and $b/2$, respectively, along the a and b axes. Similarly the (1$\bar{1}$) plane joins A and F, while the (2$\bar{1}$) plane joins E and F. The hypothetical crystal at the left was deliberately constructed to give equal length for all boundaries. Starting at point m, the line segment mn was drawn parallel to OA. Then no was constructed parallel to DA, op to BA, pq to BO, qr to EF, rs to AF, st to AO, tu to AB, and so on. It may be seen from the right-hand diagram that a boundary (hk) must be parallel to ($\bar{\text{h}}\bar{\text{k}}$), and from the left-hand diagram that such a pair of boundaries lies symmetrically about the center of the hypothetical crystal. Of course, a real crystal would not have boundaries of exactly equal length.

Figure 1.18 shows how the orientation of a boundary can be calculated from trigonometry. A boundary (hk) cuts off a/h and b/k, respectively, from the a- and b-axes. Using the sine formula in triangle POQ:

$$\frac{a/\text{h}}{\sin\epsilon} = \frac{b/\text{k}}{\sin(180-\gamma-\epsilon)}$$

Rewriting:

$$\frac{b\text{h}}{a\text{k}} \cdot \sin\epsilon = \sin(180-\gamma-\epsilon) = \sin(\gamma+\epsilon)$$

Expanding $\sin(\gamma+\epsilon)$ into $\sin\gamma \cos\epsilon + \cos\gamma \sin\epsilon$ results in:

$$\sin\epsilon\left[\frac{b\text{h}}{a\text{k}} - \cos\gamma\right] = \sin\gamma \cos\epsilon \quad \text{or} \quad \tan\epsilon = \frac{\sin\gamma}{[(b\text{h}/a\text{k}) - \cos\gamma]}$$

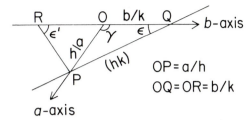

FIGURE 1.18 Trigonometric construction of boundaries (**hk**) and (**h̄k**) for an oblique unit cell with dimensions *a*, *b*, and *γ*.

For Figure 1.17, the boundary (**12**) gives:

$$\tan \epsilon = \frac{\sin 101°}{(3.9/5.0 - \cos 101°)} \quad \text{and} \quad \epsilon = 45.32°$$

For the boundary (**11**):

$$\tan \epsilon = \frac{\sin 101°}{(3.9/2.5 - \cos 101°)} \quad \text{and} \quad \epsilon = 29.28°$$

For a boundary (**h̄k**) in triangle *POR:*

$$\frac{a/h}{\sin \epsilon'} = \frac{b/k}{\sin(\gamma - \epsilon')}$$

The reader can readily obtain:

$$\tan \epsilon' = \frac{\sin \gamma}{(bh/ak + \cos \gamma)}$$

For (**2̄1**):

$$\tan \epsilon = \frac{\sin 101°}{(7.8/2.5 + \cos 101°)} \quad \text{and} \quad \epsilon = 18.53°$$

For (**1̄1**):

$$\tan \epsilon = \frac{\sin 101°}{(3.9/2.5 + \cos 101°)} \quad \text{and} \quad \epsilon = 35.64°$$

The above formulas apply directly to a 120° rhombus, and:

$$\tan\epsilon = \frac{\sin 120°}{(h/k - \cos 120°)} \text{ for an (hk) boundary}$$

Because the length of a crystal boundary is not meaningful from the viewpoint of internal structure and symmetry, it is convenient to use **circle diagrams** (Fig. 1.19). Draw a circle of suitable arbitrary radius. Imagine the crystal to lie inside the circle. From the center of the circle, draw radii that are normal to (i.e., perpendicular to) crystal boundaries. To avoid confusion, only the construction for **(23)** is shown. Mark the intersections of the radii with the circumference by dots, and label with Miller indices. Remove the crystal and construction lines. The remaining circle and dots contain all the information about the orientation of the boundaries. The positions of the *a*- and *b*-axes can be shown by their intersections with the circle. For a rectangular unit cell, the representation point for the *a*-axis coincides with that for the **(10)** boundary, and similarly for *b* and **(01)**. For an oblique cell, the representation points for *a* and **(10)** do not coincide because they are separated by $(\gamma - 90°)$.

The circle diagram is especially convenient for showing symmetry because the misleading effects caused by accidents of crystal growth have been removed. There are two mirror lines in the left-hand circle, but no mirror lines in the right-hand one.

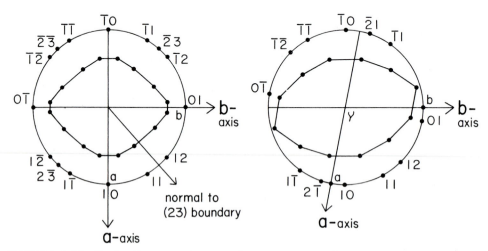

FIGURE 1.19 Circle diagrams showing the orientation of the boundaries of the inscribed crystals taken from Figure 1.15 and 1.17.

EXERCISES   (*More Difficult)

1.1   Redraw Figure 1.1a for Na^+ and Br^- ions with respective radii 1.02 and 1.96 Å. Do the Br^- ions touch, or do the Na^+ and Br^- ions touch? Check your answer by turning to Section 1.2.

1.2   Calculate the percentage of space occupied by the Na^+ and Br^- ions when in closest contact without distortion.

1.3   Count the number of atoms in each of the unit cells in Figure 1.2.

1.4* Invent a more complicated way of arranging circles of two types in a closest-packed array, and determine the smallest unit cell.

1.5   Using Be^{2+} and O^{2-} ions with respective radii of 0.27 and 1.36 Å, find out whether the Be^{2+} ions will fill the holes between O^{2-} ions in the pattern of Figure 1.4e.

1.6   Calculate the packing efficiency for the pattern in Figure 1.4a.

1.7   Redraw Figure 1.6a using circles of radius 1 cm separated by 1.5 cm in each doublet. Measure the lengths of the cell edges.

1.8* Repeat Exercise 1.7 for Figures 1.6b and 1.6c. [**Hint**. Draw doublets on several small pieces of tracing paper, and slide them over each other to get the best fit.]

1.9* Redraw Figure 1.6d using an alternation of two zigs and two zags instead of one zig and one zag.

1.10 By trial and error, find the largest circles that fit in the interstices between the clusters of Figure 1.7a, 1.7b, and 1.8c. Determine the radius ratio for each new circle to the part-circles of the cluster.

1.11* Using a table of random numbers, construct your own version of a random pattern of dots and crosses. Look for domains in which the dots and crosses alternate. If time permits, proceed to diagrams that simulate clustering and ordering into domains.

1.12 What is the unit cell content of Figure 1.12a?

1.13 Locate the 4-fold rotors in Figure 1.13a and the mirror lines in Figure 1.13c.

1.14 Construct your own crystals by varying the relative sizes of boundaries using Figure 1.14 as a guide. Measure the angles between the boundaries and the edges of the smallest unit cell. Check that they obey the tangent relation.

1.15* Sketch dumbbell molecules from Figure 1.6a into some of the outer unit cells of the central drawing of Figure 1.15. Find out how difficult it is to predict what a crystal surface will look like! There is no definitive answer to this exercise, and you might wish to discuss your ideas with other readers.

Remember that the drawings in Figure 1.15 are designed for mathematical calculations and are not meant to represent real crystals.

1.16 For a square unit cell calculate the angles between (a) the boundaries with Miller indices (**10**) and (**54**), and (b) ($\bar{4}$**3**) and (**43**).

1.17 Redraw the crystal of Figure 1.16*b* so that each boundary has the same length of 2 cm. Use the angles given in the text. An accuracy of 0.2 mm is desirable.

1.18* Redraw Figure 1.17 but with $a = 3.5$ Å instead of 2.5 Å. Measure the angles carefully from the right-hand drawing using a full-circle protractor. Check the angles by calculation.

1.19 Draw circle diagrams for regular polygons with 3 to 8 sides.

1.20* Draw the circle diagram for Exercise 1.18.

1.21 Calculate the angles between (a) the (**10**) and (**11**) boundaries, and (b) the (**01**) and (**12**) boundaries, for a unit cell that is a 120° rhombus (i.e., for hexagonal symmetry).

1.22* The use of a 120° rhombus for the unit cell of a hexagonal pattern is correct, but is psychologically difficult. The Miller–Bravais axial system uses a third axis at 120° to the *a*- and *b*-axes, and an extra index **i** with respect to this third axis. Show by simple trigonometry that $\mathbf{h} + \mathbf{k} + \mathbf{i} = 0$; for example, the boundary with Miller indices (**23**) would have Miller–Bravais indices (**23**$\bar{5}$).

ANSWERS

1.1 The radius ratio of 0.520 is greater than the critical ratio of 0.41 for square close packing, and the Na^+ and Br^- ions would be in contact in 2D.

1.2 Choose a unit cell containing two of each ion. Its area is $(5.96)^2$ Å², and it contains 2 Na^+ with area $2\pi \cdot 1.02^2$ and 2 Br^- with area $2\pi \cdot 1.96^2$. Answer is 86.4 percent.

1.3 (a) dashed, 1; dotted, 1; dot–dash, 2. (b) 1. (c) 2. (d) 3. (e) 4. (f) 6. (g) 12.

1.4 There are many possible answers. Make sure that the unit cell is a parallelogram that can give the entire pattern by repetitive translation without rotation. It must be the smallest parallelogram.

1.5 The radius ratio of 0.199 is greater than the critical value of 0.15 for hexagonal closest packing, and the Be^{2+} and O^{2-} ions would be in contact in 2D.

1.6   The same as for Figure 1.5*a,* listed in text as having a packing efficiency of 92.0 percent.

1.7   3.5 and 2.0Å.

1.8   The answer will vary a little, especially for the modification of Figure 1.6*c.* Approximate answers are 6.5 and 2.0 Å; 3.7 and 2.0 Å, ~113° (upper cell).

1.10   0.6; 0.14; 0.4.

1.12   Four large and six small circles.

1.13   Four-fold rotors occur at the corners and the center of the cell in Figure 1.13*a.* Mirror lines occur along the edges and median lines in Figure 1.13*c.*

1.16   (a) 38.66°; (b) 73.74°: between normals to boundaries.

1.21   (a) 30°; (b) 19.11°.

2

Tessellation, Lattice, General Point Symmetry, Space Group, and Crystallographic Point Symmetry in 2D

Preview This chapter considers the mathematics of 2D symmetry. A floor can be covered neatly by tiles, and 2D crystals can be represented by repetition of a unit cell. Each point in a 2D crystal displays symmetry which controls the arrangement of surrounding points. Because crystal packing is regular, atoms will lie on a lattice, which is an array of points with the same environment. There are 17 space groups, each of which describes a symmetrical arrangement of points lying on lattices. Any particular crystal has atoms that fit together according to the symmetry of one of the space groups. The size of the unit cell, the space group of symmetry elements, and the positional coordinates of each type of atom, provide a description of a crystal structure.

2.1   Tessellation

A *tessellation* covers an area with tiles. The name derives from the Roman era when mosaic decorations were popular. There are only three regular tessellations (Figs. 2.1a–2.1c). By regular is meant that every tile has the same shape, and every intersection of the tiles has the same arrangement of tiles around it; that is, the tiles are congruent, and the intersections are congruent. The three regular tessellations contain either equilateral triangles, or squares, or regular hexagons. Other regular polygons will not fill 2D space. An n-sided regular polygon has an internal angle of $(180 - 360/n)°$. For $n = 3$, 4, and 6, the angles of $60°$, $90°$, and $120°$ leave integers when divided into $360°$, thereby allowing 6, 4, and 3 tiles, respectively, to fit together at each corner of a polygon. For $n = 5$, 7, and 8, the angles of $108°$, $\sim128.6°$, and $135°$ do not divide integrally into $360°$, and leave a gap when tiles are fitted together.

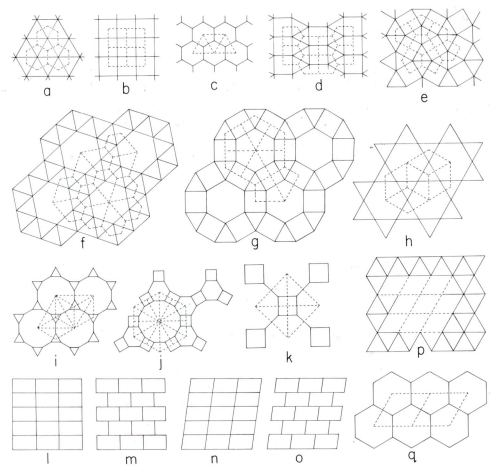

FIGURE 2.1 Tessellations (full lines) and their reciprocals (dashed lines). Drawings *p* and *q* shown special relationships between hexagonal and triangular tessellations.

Eight semiregular tessellations can be obtained from mixing two or three types of regular polygons in the right proportion (Figs. 2.1*d*–2.1*k*). In each tessellation, all intersections are congruent, but the polygons are not congruent. Four of the semi-regular tessellations can be obtained by joining centers of the spheres in Figures 1.2*d*–1.2*g*.

Every tessellation has a ***reciprocal tessellation*** or ***dual tessellation*** obtained by interchanging intersections with polygons. To obtain a reciprocal tessellation, place a point inside each polygon (the centroid is a convenient position), and join points of two adjacent polygons that share an edge. For a regular tessellation, the reciprocal tessellation is obtained by constructing the perpendicular bisector of each polygonal edge, but for an irregular tessellation this procedure may not work. The reciprocal of the square tessellation (Fig. 2.1*b*) is merely another square tessellation (dashed line) translated by half a diagonal of a square. The triangular and hexagonal tessellations are reciprocal to each other. Each reciprocal of a semiregular tessellation has congruent polygons but incongruent junctions. The polygons are not regular, and three reciprocal tessellations contain nonregular pentagons (Figs. 2.1*d*–2.1*f*).

Two-dimensional space can also be filled regularly by rectangular tiles either stacked one above the other (Fig. 2.1*l*) or staggered as in walls made of bricks (Fig. 2.1*m*). The tiles could be oblique (Figs. 2.1*n*, 2.1*o*). Returning to the regular tessellations, a pair of equilateral triangles can be combined into a 120° rhombus that fills space neatly (Fig. 2.1*p*). Alternatively, each hexagon of Figure 2.1*c* can be broken up and recombined into 120° rhombuses (Fig. 2.1*q*).

It is obvious that there are many ways of filling an area with undecorated tiles, including ones of lower symmetry than shown here (O'Keeffe and Hyde, 1980). Much greater complexity arises when the tiles are decorated. This book must concentrate on crystallography, but some readers may wish to explore the use of symmetry in art and science in the following two books: Weyl (1952), *Symmetry,* an elegant essay that covers the key points; and Shubnikov and Koptsik (1974), *Symmetry in Science and Art,* a comprehensive account with considerable amounts of advanced mathematics. Readers may also wish to examine the works of the Dutch artist Escher, as reproduced in Macgillavry (1965), *Symmetry Aspects of M.C. Escher's Periodic Drawings.* Niggli (1926, 1928) and Fischer (1968) used packed circles to derive 2D nets.

2.2 General Point Symmetry

Rotation symmetry was mentioned briefly in the sections on close packing. Systematic mathematical description of *point symmetry* is illustrated in Figure 2.2 using a three-atom molecule as the motif undergoing the symmetry operations. An *n*-fold rotor produces successive rotations of $360°/n$ about the point. Thus a 6-fold rotor produces rotations of 60°, 120°, 180°, 240°, and 300° before the individual operations add up to 360°, which returns the motif to the original position. Crystallographers try to use simple notation. A *rotation operator* is usually denoted by the appropriate numeral, for example, **6**, or by the appropriate polygon, for example, a filled hexagon. The names *diad* (for **2**), *triad* (**3**), *tetrad* (**4**), and *hexad* (**6**) are used. In order to denote no rotation symmetry, the term *monad* (or **1**) is used; obviously a rotation of $360°/1$ merely gives the original motif. A circle has infinite rotation symmetry about its center.

Reflection or mirror *symmetry* (Fig. 2.2) changes the *chirality,* just as a left hand is related to a right hand. It is impossible to turn a left-handed

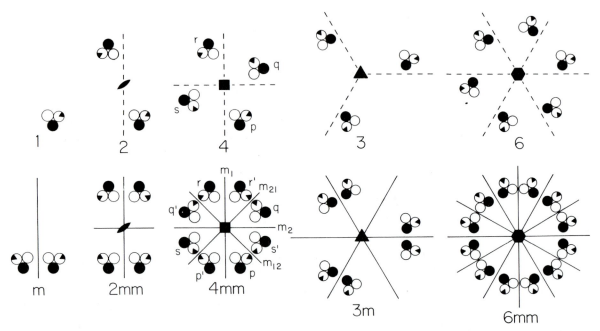

FIGURE 2.2 The 10 types of crystallographic point symmetry in 2D. Dashed lines are merely visual aids, while solid lines represent mirror symmetry.

motif into a right-handed motif by rotation or translation. The conversion can be accomplished only by imagination. A ***mirror line*** is denoted **m** and is represented by a thick line in drawings. *Chiros* means "hand" in Greek.

Mirror symmetry can be combined with rotation symmetry to give ***point groups*** **2m**, **3m**, **4m**, **6m**, and so on. Combination of a *2*-fold rotation with a mirror line produces another mirror line at 90° (Fig. 2.2) and the ***point group symbol*** **2mm** expresses all three symmetry operations. Actually, any two symbols (i.e., **2m** or **mm**) are sufficient to specify this group because the third one arises automatically. Combination of a mirror line with a tetrad axis automatically produces three more mirror lines. Thus motif *p* is rotated by the tetrad axis to give motifs *q, r,* and *s*. Mirror line $\mathbf{m}_1$ reflects motif *p* into motif *p'*, *s* into *s'*, *q* into *q'*, and *r* into *r'*. The tetrad axis turns $\mathbf{m}_1$ into $\mathbf{m}_2$, the mirror line $\mathbf{m}_{12}$ relates motif *p* with motif *s'*, and so on. The symmetry elements can be considered in any sequence, and the answer is always the same. If one motif is omitted, or one symmetry element is omitted, there is something wrong. The next section expresses the above concepts in terms of mathematical group theory, whose rules give a precise statement of these relationships. Most crystallographers prefer a pictorial representation for introduction to symmetry, but some ultimately prefer the austere beauty of an abstract description.

Returning to point group **4mm**, the mirror lines actually occur as two sets. The lines **m₁** and **m₂** are related by the rotation tetrad, as are the lines **m₁₂** and **m₂₁**, but **m₁** is not equivalent to **m₁₂** or **m₂₁**.

Combination of a mirror line with a rotation triad gives the point group **3m**. Three mirror lines are needed for completion of the group. Because these mirror lines are related by the triad, and are equivalent to each other as far as symmetry is concerned, the point group symbol is given as **3m** and not as **3mmm**. Of course, the three mirror lines occupy different positions. Point group **6mm** has two different types of mirror lines, as does **4mm**.

A point group of high symmetry can be degraded into a point group of lower symmetry. Removal of motifs *q* and *s* turns point group **4** into point group **2**, and further removal of motif *r* gives point group **1**. Thus there is a hierarchy of **subgroups**. Quartz changes from high quartz to low quartz at ~573°C with reduction of point symmetry from 6-fold to 3-fold. The point group of low quartz is a subgroup of the point group of high quartz.

Emphasis has been placed here on 2-, 3-, 4-, and 6-fold rotation. Five-fold, 7-fold, and higher multiplicity can occur around isolated points but not around points at the intersection of congruent tiles in regular tessellations. This important result is discussed thoroughly in later sections. Several plants have petals related by 5-fold rotation symmetry (e.g., *Gentiana pumila*, in Shubnikov and Koptsik, 1974, Fig. 22), and the organism *Botryllus helleborus* has 7-fold symmetry. Hence, it is necessary to distinguish between **general point symmetry** and the restricted type of **crystallographic point symmetry**.

2.3   Application of Group Theory

Detailed descriptions of **group theory** can be found in many textbooks such as Cotton (1971), *Chemical Applications of Group Theory* and Burns and Glazer (1978), *Space Groups For Solid State Scientists*. The following is merely a practical introduction.

Point group **4** contains four operations: the identity operation of rotation by any integral multiple of 360°, denoted **1** for the *monad;* an anticlockwise rotation of 90°, denoted **4**, which turns motif *p* into motif *q* (Fig. 2.2); an anticlockwise rotation of 270°, denoted **4³**, which turns *p* into *s;* and a rotation of 180°, denoted **2** for a *diad*, which turns *p* into *r.* A **4** operation followed by a **2** operation "multiply" together to produce a **4³** operation, for example, rotate 90° anticlockwise from *p* to *q* and continue another 180° to *s*, which is 270° from *p.* This is expressed in the following **multiplication table** by the intersection of the **4** row with the **2** column.

Multiplication table for point group **4**

		Second Operation			
		1	4	2	4^3
First Operation	1	1	4	2	4^3
	4	4	2	4^3	1
	2	2	4^3	1	4
	4^3	4^3	1	4	2

There is a symmetry operation at the intersection of each row and each column thereby closing the group. The point group 2 had only the operations 1 and 2 and its multiplication table is merely:

1	2
2	1

This multiplication table can be found as a subgroup in the multiplication table of point group **4**.

The reader should check the correctness of the following multiplication table for group **6**, where the anticlockwise rotations of 60°, 120°, 180°, 240°, 300°, and 360° are denoted respectively by **6, 3, 2, 3^2, 6^5**, and **1**.

Multiplication table for point group **6**

		Second Operation					
		1	6	3	2	3^2	6^5
First Operation	1	1	6	3	2	3^2	6^5
	6	6	3	2	3^2	6^5	1
	3	3	2	3^2	6^5	1	6
	2	2	3^2	6^5	1	6	3
	3^2	3^2	6^5	1	6	3	2
	6^5	6^5	1	6	3	2	3^2

The point group **4mm** has four rotation operators (4, 2, 4^3, 1) and four mirror lines, which are labeled arbitrarily as m_1, m_2, m_{12}, and m_{21} as in Figure 2.2. The multiplication table is

		Second Operation							
		1	4	2	4^3	m_1	m_2	m_{12}	m_{21}
First Operation	1	1	4	2	4^3	m_1	m_2	m_{12}	m_{21}
	4	4	2	4^3	1	m_{21}	m_{12}	m_1	m_2
	2	2	4^3	1	4	m_2	m_1	m_{21}	m_{12}
	4^3	4^3	1	4	2	m_{12}	m_{21}	m_2	m_1
	m_1	m_1	m_{12}	m_2	m_{21}	1	2	4	4^3
	m_2	m_2	m_{21}	m_1	m_{12}	2	1	4^3	4
	m_{12}	m_{12}	m_2	m_{21}	m_1	4^3	4	1	2
	m_{21}	m_{21}	m_1	m_{12}	m_2	4	4^3	2	1

Point group **4** is a subgroup of point group **4mm**, and the upper-left block of four columns and four rows is identical with the multiplication table of point group **4**. The multiplication tables for point groups **2mm, m, 2**, and **1** are also present in this table for **4mm**, and can be extracted by using the appropriate rows and columns.

2.4 Lattice

A *lattice* is an array of points with the same vectorial environment. This means that an observer could stand at any point and have exactly the same view, taking into account both distance and absolute orientation of the other points. A lattice must be infinite. The *lattice points* must be spaced regularly. In Figure 2.3, the array of points in diagram *a* is a lattice (except, of course, that the points do not go to infinity!), but those in diagram *b* are not regularly spaced and those in diagram *c* are of three kinds. Points *p* have neighbors to the north, east, south, and west, points *q* only to the north and south, and points *r* only to the east and west. Taken separately, points *p* form a lattice, as do points *q* and points *r;* hence this array is a set of three *parallel lattices.* Although each set of points is a lattice, and all three lattices are congruent except for relative translation, the composite set of points is not a single lattice. In order to describe parallel congruent lattices it is necessary to use space group symmetry, as developed in the next section.

The geometry of a lattice may vary, and the unit cell may have a shape that is either a square, or a 120° rhombus, or a rectangle, or an oblique parallelogram, as described in Section 1.6. It is always possible to choose a *primitive* unit cell for a single lattice, that is a unit cell containing only one lattice point. For simplicity, the unit cell joins four lattice points at the

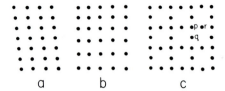

FIGURE 2.3 (*a*) Lattice points with the same vectorial environment except, of course, at the edges. (*b*) A nonlattice array of points whose spacing changes regularly away from the top left. (*c*) Three intersecting lattices.

corners of a parallelogram: of course, each lattice point is shared between four unit cells.

Looking back at the packing drawings in Chapter 1, the circles can be replaced by dots at their centers to obtain either a lattice or a set of parallel congruent lattices. The circles in Figure 1.2*a* form a primitive square lattice with a unit cell shown by the dashed lines. The circles in Figure 1.2*b* form a primitive hexagonal lattice, whose unit cell is again shown by dashed lines. Figure 1.2*c* can be represented by two parallel lattices, each of primitive hexagonal type, and so on for other patterns in Figure 1.2. Figures 1.5*a* and 1.6*a* can be represented by two parallel lattices of primitive rectangular type.

It is important to realize that placement of an atom on a lattice point makes it necessary to consider chemical character as well as mathematical position. In Figure 1.6*a*, the two lattices are occupied by the same type of atom, but in Figure 1.5*a*, the two lattices are occupied by different types of atoms. When looking at actual crystal structures always check the chemical character as well as the mathematical position of a lattice array.

Figure 1.6*b* apparently consists of four parallel lattices of primitive rectangular type and with the same chemical character. But is this the best description? Figure 2.4 shows that the circles can be represented by two parallel congruent lattices, each of which contains two points per unit cell. One point (*p*) lies at each corner of the cell (dashed lines), and the other point (*q*) lies at the exact center to give a ***non-primitive centered cell***. Points *r* and *s* generate a second centered lattice (dotted lines). Actually an oblique parallelogram (continuous line) can be selected for the unit cell, thereby allowing use of just two primitive lattices. But this description loses the convenience of a rectangular unit cell, and crystallographers prefer to use a non-primitive centered lattice when the geometry is more favorable. More fundamentally, the symmetry of the lattice is important in controlling the geometry, and crystallographers choose the smallest unit cell whose sides and angles have the simplest geometrical relations to the lattice symmetry. In order to combine two primitive lattices into a centered lattice, it is absolutely necessary that they have the same chemical character, as well as being

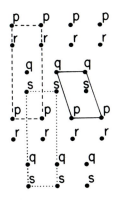

FIGURE 2.4 Lattice description of the circles in Figure 1.6*b*. The centers of the circles are represented by dots on two centered rectangular lattices, whose unit cells are outlined by dashes and dots. A primitive oblique lattice (continuous line) loses the convenience of rectangular geometry.

congruent except for the relative translation. Because the atoms in Figure 1.3*e* are of two types, they lie on two primitive, chemically distinct lattices and not on one centered lattice. The positions would allow a centered lattice, but not the chemistry.

How many 2D lattices exist? Can a square lattice be centered? Yes (Fig. 1.2*a,* dot–dash lines), but a smaller unit cell (dotted or dashed lines) can be chosen whose geometry is still square. Therefore, a centered square lattice is unnecessary, and only a primitive lattice is needed. A centered

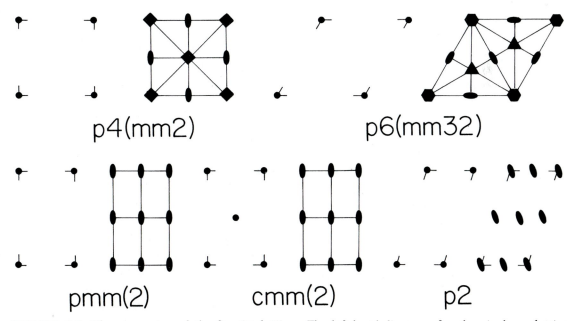

FIGURE 2.5 The symmetry of the five 2D lattices. The left-hand diagram of each pair shows lattice points and corner marks of the unit cell. The right-hand diagram shows the symmetry elements. Symmetry elements in brackets arise automatically when the unbracketed symmetry elements are used. [Note that an experienced crystallographer could add glide lines.]

oblique lattice is unnecessary because a new oblique unit cell can be chosen with only one point per cell. Addition of a point to the center of each triangle in a hexagonal lattice does not produce a lattice, because the points do not have the same vectorial environment. Hence there are five 2D lattices: primitive square, primitive hexagonal, primitive rectangular, centered rectangular, and primitive oblique.

The regular geometry of the lattice points produces symmetry, and vice versa. For clarity, Figure 2.5 shows the symmetry elements in a separate unit cell from the one that contains the lattice points. The square lattice has rotation tetrads at the corner and center of the unit cell, rotation diads halfway along the edges, and mirror lines joining all the corners and centers. The symbols **p** for primitive lattice and **4** for the tetrad are sufficient to describe the lattice, and the two types of mirror lines and the rotation diad are produced by the interaction of the lattice translations with the rotation operations of the tetrad. The hexagonal lattice is similarly related by triads, diads, and mirror lines as well as the hexads. The primitive and centered rectangular lattices have a rotation diad as well as two mirror lines.

2.5   Space Groups of Equivalent General Positions and Symmetry Elements

In Figure 2.5, symmetry elements pass through the lattice points. Moving the lattice points away from the symmetry elements produces a group of parallel congruent lattices. Turning back to Figure 2.4, the rectangular lattices marked by *p* and *r* are actually identical except for their relative position. In Figure 1.6c, the circles in each doublet are congruent except for their relative position, and they can be related by a rotation diad halfway between them. The zigzag pattern in Figure 1.6d has a curious kind of back-and-forth symmetry that is partly reflection in a mirror line and partly movement along the mirror line. Every regular 2D pattern can be described by one of seventeen **space groups**, no matter whether it is a layer in a crystal, or a tiled floor, or a sheet of wallpaper. This section develops the space groups systematically.

Figure 2.6 shows three space groups involving just rotation axes. The space group **p1** has an oblique unit cell because there are no constraints from the symmetry. There is only one lattice point because the primitive lattice has only one point and the monad axis merely relates the point to itself. The point can be placed anywhere in the unit cell. In a real crystal, there could be several atoms bonded together. If these atoms belong to space group **p1**, the symmetry places no special relationships between any of the atoms. The only restrictions are the translation vectors of the lattice, which enforce

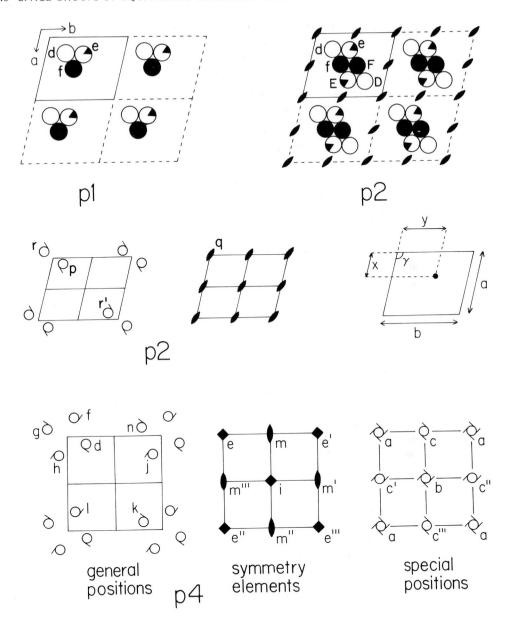

FIGURE 2.6 Space groups **p1**, **p2**, and **p4**. In the diagram at the upper left, four unit cells are shown in which a 3-atom motif is related by **p1** symmetry. At the upper right, the 3-atom motif has been rotated by the 2-fold rotor of **p2**. The two diagrams at the middle left show the relation between the two general equivalent positions (left) and the symmetry elements (right) for **p2**. Similar diagrams are shown at lower left and middle for **p4**, together with a further diagram at the right for special positions. The diagram at middle right shows how fractional coordinates x and y are determined.

repetition of the atoms to infinity. A three-atom motif *def* is shown in Figure 2.6, but only the open circle *d* is necessary to show the symmetry of the space group. Strictly speaking, a point could be shown, but crystallographers conventionally use a circle rather than a point, as in the standard *International Tables for X-ray Crystallography*. Readers should recall from the preceding section that placement of an atom on a lattice point in a real crystal structure gives a particular chemical character to that point. Thus in the diagram for **p1**, there are three parallel congruent lattices, each with a different chemical character.

For space group **p2**, the positions must occur in pairs that obey the rotation diad. To construct this space group, draw two identical oblique unit cells, one for the **equivalent positions** and one for the symmetry elements. For convenience, draw lines to quarter the cell. Pick one position anywhere in the unit cell (e.g., *p*), place a rotation diad at the corner of the unit cell for convenience (point *q*), and generate the second position (*r*) by 180° rotation, and *r'* by translation. Search for all the other symmetry elements that have been generated, in this case further diads. There are only two points per unit cell, but it is convenient to show points just outside the cell.

To help readers distinguish between rotation and mirror operations, a short line has been added to the conventional circle. For space group **p2**, there are rotation diads at the corner, center, and midpoints of edges, and this is why it is convenient to draw lines dividing the unit cell into quarters.

The diagram at the extreme upper right actually does show four unit cells in which the three-circle motif from **p1** has been operated upon by a rotation diad to give space group **p2**. From the viewpoint of symmetry, it is necessary to show only one starting position plus the other positions generated by symmetry and lattice operations. In real crystal structures, there are several types of atoms, except for the very simple structures. These atoms will be packed together closely like the patterns in Chapter 1. The three-circle motifs do not cover the whole area, and other circles would be needed to complete a close-packed pattern. Nevertheless, it is useful for teaching the principles of space group symmetry. Note that although there are six lattice positions here, there are only three chemically distinct positions. The atom on position *d* has the same bonding to atoms on positions *e* and *f*, as does the atom on *D* to atoms on *E* and *F*. The only difference is the absolute orientation in space of the *def* and *DEF* triplets. Only the relative orientation is important for chemical bonding.

The following information is needed to describe a 2D crystal structure: (i) the **cell dimensions** *a, b,* and γ, (ii) the **space group symbol,** and (iii) a list of the types of atoms and their **atomic coordinates** *x* and *y*. The diagram at the middle right shows how *x* and *y* are measured using lines parallel to the cell edges. In *International Tables for X-ray Crystallography, x* and *y* are given as fractions of the cell edge to which they are parallel. Thus the depicted position lies at x 0.38, y 0.57. Some crystallographers give x and y directly in angstrom units, such that the fractional coordinates are X/a and Y/b. This book, however, follows *International Tables for X-ray Crystallography*.

The pattern in the uppermost right can be expressed as follows:

Open circle: 0.24, 0.24; 0.76, 0.76 or $\pm$ (0.24, 0.24)
Filled circle: 0.49, 0.39; 0.51, 0.61 $\pm$ (0.49, 0.39)
Sliced circle: 0.26, 0.48; 0.74, 0.52 $\pm$ (0.26, 0.48)

Note that the equality of the first two coordinates is accidental.

A simpler way is to recognize from the space group diagrams for **p2**, that a *general position* with coordinates (x,y) is transformed by the rotation diad to $(\bar{x},\bar{y})$ where the bar is the conventional way of showing a negative co-ordinate. Because of the lattice repeats, $\bar{x} = 1 - x$, and $\bar{y} = 1 - y$, and the position actually shown in the drawing is $1 - x$, $1 - y$. Thus the second position can be omitted once the space group symbol has been given, and the following description suffices:

p2: open circle 0.24, 0.24; filled circle 0.49, 0.39; sliced circle 0.26, 0.48.

In Figure 1.6c, the doublets are packed together with **p2** symmetry. For the unit cell at the lower right, the cell dimensions would be a 2.5, b 3.9 Å, $\gamma = 101°$ for circles of radius 1.25 Å (see Section 1.8). The coordinates of the circles are x 0.16, y 0.21, and x 0.84, y 0.79, or briefly, $\pm(0.16,0.21)$.

Thus far, only *general equivalent positions* have been considered. If the starting position falls on a rotation axis, the rotation operation does not generate any further positions and the *multiplicity* decreases. In **p2**, position $(0,0)$ at the corner of the unit cell falls on a rotation diad, and is not multi-plied. Whereas a general equivalent position has multiplicity 2, this *special position* has multiplicity 1. Special positions also occur at $(0.5,0)$, at $(0,0.5)$, and at $(0.5,0.5)$.

The next space group, **p4**, has a general equivalent position with multi-plicity 4. The unit cell must be a square in order to obey the *4*-fold sym-metry. It would be helpful if the reader took a sheet of paper and followed this sequence of operations: (i) draw three identical squares and the midway lines; (ii) choose any position inside the left-hand square, but not at a corner or center, and draw a circle such as d; (iii) place a filled square to represent a tetrad at e on the middle square; (iv) generate positions f, g, and h with the tetrad; (v) use the lattice translations to give all the other positions on the left-hand drawing; (vi) look for additional symmetry elements in the middle drawing. Tetrads will be found at e', e'', and e''', and one will be found at i that relates the positions d, j, k, and l. A diad will be found at m that relates positions *d* and *n*. The remaining diads at *m'*, *m"*, and *m'''* can be located by inspection of the positions in the left-hand drawing, or can be located by using the operations of the tetrad at *i*. Readers can experiment with the sequence of operations, but the final answer will be the same if all operations have been carried out correctly. Some readers may wish to construct a multiplication table for the operations, but a purely graphical approach will be used here for space groups because most readers prefer to see the constructions visually.

The multiplicity of the general position is 4, and the coordinates are (x,y), $(\bar{y},x)$, $(\bar{x},\bar{y})$, $(y,\bar{x})$. This notation confuses most readers at first, and a specific example is needed for clarification. Position *d* has coordinates $x_1 = 0.08$, $y_1 = 0.20$. Position *f* has coordinates $x_2 = -0.20$, $y_2 = 0.08$, which is expressed algebraically as $x_2 = -y_1$ and $y_2 = x_1$, or briefly, positions *d* and *f* are expressed as (x,y), $(\bar{y},x)$. Similarly position *g* has coordinates $x_3 = -0.08$, $y_3 = -0.20$, giving $x_3 = -x_1$, $y_3 = -y_1$, or briefly, $\bar{x},\bar{y}$. Readers should now determine the coordinates for position *h*.

There are three types of special positions in **p4**. Positions *d*, *f*, *g*, and *h* can be merged into position *a* of the right-hand drawing. This position lies on a tetrad and has a **point symmetry** of 4. Its multiplicity is 1. Positions *d*, *j*, *k*, and *l* can be merged into position *b* of the right-hand drawing. This position has the same properties as position *a*, except that it lies at the center instead of the corner of the cell. A third special position is formed by merging positions *d* and *n* to give *c*. Simultaneously, positions *h* and *l* merge to give *c'*, and so on. This special position has a multiplicity of 2 and point symmetry of **2**.

For each space group in the *International Tables for X-ray Crystallography*, there are the two diagrams showing the equivalent general positions and the symmetry elements, but not a third diagram for the special positions. The latter are given in a table showing the multiplicity, a letter selected by R.W.G. Wyckoff, the point symmetry, and the coordinates. For **p4**, the table is:

Multiplicity	Wyckoff Notation	Point Symmetry	Coordinates of Equivalent Positions
4	*d*	1	$x,y; \bar{x},\bar{y}; y,\bar{x}; \bar{y},x$
2	*c*	2	$0.5,0; 0,0.5$
1	*b*	4	$0.5,0.5$
1	*a*	4	$0,0$

The general position, always with point symmetry **1**, is listed first, and the special positions are listed in order of decreasing multiplicity and increasing point symmetry. The Wyckoff symbols are listed upwards in alphabetical sequence.

Once the reader has mastered the procedure, all properties of a space group ensue just from the symbol. From **p4**, the reader deduces that the lattice is primitive, that the cell is square, and that the general equivalent positions result from operation of a rotation tetrad. Everything else follows logically and rigorously, but a single mistake can ruin everything just as in a program for an electronic calculator. Always work carefully and check every step.

The remainder of this section consists of a systematic derivation of the remaining members of the 17 space groups in 2D. The space group **pm** (Fig. 2.7) must be rectangular in order to obey the mirror operation. There is only one starting position because the lattice is primitive. The mirror line can be placed either perpendicular to the *a*-axis or to the *b*-axis. To distinguish the the two choices, space group symbols **pm1** and **p1m**, respectively, are used. For **pm1**, the mirror line is placed perpendicular to the *a*-axis, and is shown by a heavy line at *r* in the right-hand drawing. This mirror line produces position *q* from the starting position *p*. Lattice translations give the other positions. Strictly speaking, it is necessary to show only positions *p* and *q'*, but crystallographers choose to show positions in adjacent unit cells when they help to display the symmetry. Attachment of short lines to the circles helps to show the mirror symmetry, but such lines are not given in the *International Tables for X-ray Crystallography*. Instead, a comma is used to show the chirality. Position *p* has no comma, but position *q* has a comma to show that the mirror operation has changed the chirality. Positions related by the lattice translation have the same chirality, for example, *p, p', p'',* and *p'''*. Readers might ask how a point or a circle can have a chirality. Of course, they cannot, but a three-atom cluster such as in Figure 2.6 would have a chirality. A second mirror line at *s* is needed to complete the symmetry operations. Symmetry operations are spaced twice as closely as the cell repeats. The algebraic description follows:

Multiplicity	Wyckoff Notation	Point Symmetry	Coordinates of Equivalent Positions
2	*c*	1	$x,y; \bar{x},y$
1	*b*	m	$0.5,y$
1	*a*	m	$0,y$

Two mirror lines must lie perpendicular to each other, and they generate a rotation diad at their intersection. The corresponding space group **p2mm** has a general position of multiplicity 4, two types of special positions on mirror lines with multiplicity 2, and one type of special position at the inter-section of mirror lines with multiplicity 1. Readers might try to generate the algebraic description before reading the following table:

Multiplicity	Wyckoff Notation	Point Symmetry	Coordinates of Equivalent Positions
4	i	1	$x,y; \bar{x},y; \bar{x},\bar{y}; x,\bar{y}$
2	h	m	$0.5,y; 0.5,\bar{y}$
2	g	m	$0,y; 0,\bar{y}$
2	f	m	$x,0.5; \bar{x},0.5$
2	e	m	$x,0; \bar{x},0$
1	d	2mm	$0.5,0.5$
1	c	2mm	$0.5,0$
1	b	2mm	$0,0.5$
1	a	2mm	$0,0$

The Wyckoff notation is merely a convention, and the reader might choose a different sequence of positions with the same point symmetry.

It is not possible to combine three mirror lines in 2D, but it will be possi-ble to do this in 3D. Can other rectangular space groups be developed? Fig-ure 1.6d has an intriguing zigzag sequence of doublets. Each circle has the same environment as all the other circles if the absolute orientation in space is ignored. This zigzag array can be expressed by space group **pg**, where **g** stands for the new concept of *glide line*, which is denoted by a dashed line. A glide line involves a double operation. First, position p is reflected in glide line s to give imaginary position q. This position is then translated (or glided) to give the final real position r. A glide line always involves a translation of one half of the cell repeat along its length, which for glide line s is $b/2$. Space group **pg1** has a general position with multiplicity 2, and coordinates $x,y;$ $\bar{x}, 0.5 + y$. There are no special positions because the multiplicity cannot be reduced in any way. Setting a point on a glide line (e.g., at $x = 0$) still leaves two points $(0,y; 0,0.5 + y)$.

The concept of special position was invented to assist in determination of crystal structures. Suppose that the unit cell volume, density, and chemi-cal composition allowed the conclusion that a unit cell contained only one atom of a certain element in space group **p2mm**; then that atom must lie in one of the four Wyckoff positions a, b, c, or d in the table of the preceding paragraph. Because the multiplicity is not reduced when a general position is moved to a glide line, the position remains general and does not become a special position.

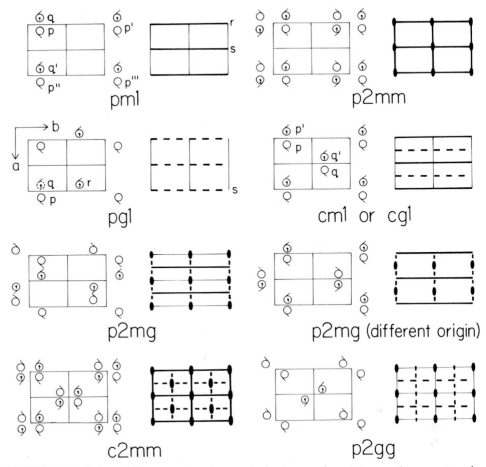

FIGURE 2.7 Pairs of diagrams showing equivalent general positions and symmetry elements for space groups with mirror or glide operations and a rectangular unit cell. All are drawn with the *a* axis pointing downward and the *b* axis pointing to the right.

The next space group **cm1** combines a centered rectangular cell with a mirror line. Two general positions x,y and $0.5 + x, 0.5 + y$ are needed for the centered cell. These positions, labeled p and q, are reflected in the mirror lines to give p' and q'. Glide lines are automatically produced to relate p to q', and q to p'. All centered lattices contain extra symmetry elements that are needed to give enough multiplications in the group table. As soon as p' was produced by reflection from p, a glide operation was automatically produced between p' and q. This interlocking relation between positions and symmetry elements is one of the elegant and beautiful features of space groups.

Space group **cm1** can also be described as space group **cg1**. From the viewpoint of symmetry, it does not matter whether a mirror line is placed perpendicular to the *a*-axis or to the *b*-axis, because it involves merely an arbitrary choice of axial labels. Morphological crystallographers usually chose $b > a$, and if this convention is still followed, a formal distinction is needed between space groups **pml** (with **m** perpendicular to *a* and a monad along *b*) and **plm** (*m* perpendicular to *b* and monad along *a*). Thus there are only three space groups with a rectangular unit cell and a mirror line and/or a glide line, namely, **pm1** (= **plm**), **pgl** (= **plg**) and **cml** (≡ **cgl** = **clm** = **clg**).

Space group **pmg** has a rotation diad, which is produced automatically when the **m** and **g** operations have been completed, and its full symbol is **p2mg**. The space group is conventionally drawn with the diad at the cell corners, which results in the mirror line lying at $x = 0.25$. If the mirror line is placed at $x = 0$, the diads move to (0.25,0) and related positions. At first sight, the pattern of equivalent general positions looks different, but the general positions can be moved to obtain exact registry of the positions and the symmetry elements. The unit cells do not match in absolute position, of course. Special positions occur on mirror lines and rotation diads.

Space group **cmm** has rotation diads (hence the full symbol **c2mm**) and glide lines. There are two types of rotation diads, one at the intersection of mirror lines, and one at the intersection of glide lines. Space group **p2gg** has diad axes at the centers of the squares formed by the glide lines.

This completes all the rectangular space groups. A brief review is desirable to reinforce a reader's understanding. A mirror operation enforces rectangular geometry of the unit cell. The rectangular lattice can then be either primitive (**p**) or centered (**c**). One edge is chosen to be perpendicular to the mirror operation and one edge parallel to the mirror operation. The mirror operation can be a pure mirror operation (**m**) or a double operation with a glide of either $a/2$ or $b/2$ (denoted **g**). From the symmetry viewpoint **pm1** is the same as **p1m,** but in a real crystal structure, the choice of axial orientation is meaningful with respect to atomic packing. The number of space groups is limited by systematic enumeration:

pml equals **plm** by relabeling axes.

pgl equals **plg** by relabeling axes.

cml equals **clm** by relabeling axes and **cgl** by the interaction between the **c** and **m** operations to give **g**.

p2mm is unique, but in a crystal structure the mirror lines have different controls on the atomic packing.

p2mg equals **p2gm** by relabeling axes.

p2gg is unique, but the glide lines are different in a crystal structure.

c2mm equals **c2gg** because the interaction between the **c** and **m** operations automatically generates the g operations.

A square lattice yields only three space groups. A rotation tetrad is needed to enforce the square shape. The simplest space group is **p4** (Fig. 2.6). Addition of a mirror line perpendicular to the *a*-axis gives space group **p4mm** (Fig. 2.8). A mirror line must occur perpendicular to the *b*-axis because of the operation of the tetrad axis. The combination of 90° rotation plus reflection about the *a*-axis produces a second set of mirror lines that lie at 45° to the first set. This second set of diagonal mirror lines gives the second **m** in the space group symbol. The first **m** includes the mirror lines perpendicular to *a* and to *b* because these are identical as far as symmetry is concerned. Glide lines are also produced in diagonal positions. For a diagonal line, the cell repeat is **a** + **b** and the glide operation is (**a** + **b**)/2. Special positions occur on tetrad and diad axes and on mirror planes as follows:

Multiplicity	Wyckoff Notation	Point Symmetry	Coordinates of Equivalent Positions
8	*g*	1	$x,y; \bar{x},\bar{y}; y,\bar{x}; \bar{y},x; \bar{x},y; x,\bar{y}; \bar{y},\bar{x}; y,x$
4	*f*	m	$x,x; \bar{x},\bar{x}; \bar{x},x; x,\bar{x}$
4	*e*	m	$x,0.5; \bar{x},0.5; 0.5,x; 0.5,\bar{x}$
4	*d*	m	$x,0; \bar{x},0; 0,x; 0,\bar{x}$
2	*c*	2mm	$0.5,0; 0,0.5$
1	*b*	4mm	$0.5,0.5$
1	*a*	4mm	$0,0$

Combination of a tetrad axis with a glide line is tricky! The obvious way is to place the glide line perpendicular to the *a*-axis, and passing through the tetrad (Fig. 2.8, bottom). This leads inexorably to a large unit cell with 16-fold multiplicity of the equivalent general positions. A new unit cell (dashed square) can be chosen that is still a square but that has only half the area. This is rotated 45° and redrawn in Figure 2.8 (middle). Glide lines occur in the small cell at $x = 0.25$, $x = 0.75$, $y = 0.25$, and $y = 0.75$ as well as mirror lines at diagonal positions and a second set of glide lines halfway between the mirror lines. The space group has the full symbol **p4gm** to show the first set of glide lines that are perpendicular to *a* (and also to *b*) and the mirror lines in diagonal positions. Special positions occur on rotation axes and on mirror lines.

Five space groups have a 120° rhombus for a unit cell. Space group **p3** (Fig. 2.9) has only triad axes, but space group **p6** has hexads, triads, and diads. Addition of a mirror line perpendicular to the *a*-axis turns **p3** into **p3m1**. The monad indicates that no symmetry is associated with the directions that are at 30° to the *a*- and *b*-axes. Glide lines are automatically pro-

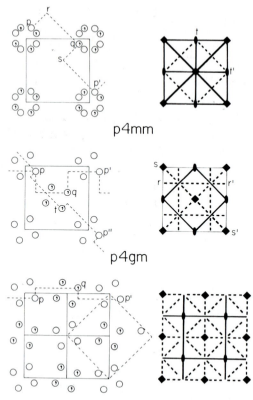

p4mm

p4gm

p4gm with glide line perpendicular to *a*
at x = 0

FIGURE 2.8   Pairs of diagrams showing equivalent general positions and symmetry elements
for tetragonal space groups. For **p4mm,** the
diagonal glide line tt' reflects point p to r and
translates it to q. Point q is reflected to s and
translated to p'. For **p4gm** (middle), glide line
rr' relates p to q and p', and glide line ss' relates
p to t and p''. In the incorrect orientation of
p4gm the glide line at $x = 0$ relates p to q and
p'. Addition of the glide line at $y = 0$ leads automatically to the other symmetry elements. It is
then necessary to choose a new cell (dashed
square) that has the same symmetry elements as
the middle diagram for **p4gm.**

duced between the mirror lines. Addition of a mirror line along the *a*-axis
turns **p3** into **p31m.** Now there is no symmetry associated with *a*, and each
mirror line is perpendicular to one of the lines that lie at 30° to the axes.
The reader should remember that a rotation axis is always placed parallel to

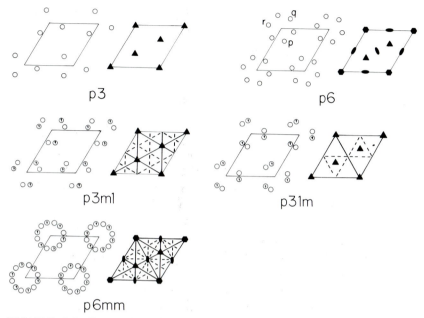

FIGURE 2.9 Pairs of diagrams showing equivalent general positions and symmetry elements for hexagonal space groups.

a line, and a mirror operator is always placed perpendicular to a line. Space group **p31m** has glide lines midway between the mirror lines as a result of the combined operation of the 120° rotation with the mirror operation. Addition of a mirror line to **p6** automatically generates a mirror line at 30°, giving the space group symbol **p6mm**. Glide lines are again produced halfway between the mirror lines.

Just as for tetragonal space groups, the coordinates for equivalent positions in a hexagonal space group may be confusing. Here is the list of positions for **p6**:

Multiplicity	Wyckoff Notation	Point Symmetry	Coordinates of Equivalent Positions
6	d	1	$x,y; \bar{y}, x - y; y - x, \bar{x}$ $\bar{x}, \bar{y}; y, y - x; x - y, x$
3	c	2	$0.5, 0; 0, 0.5; 0.5, 0.5$
2	b	3	$1/3, 2/3; 2/3, 1/3$
1	a	6	$0, 0$

To understand the coordinates of the general positions, consider the position p with $x_1 = 0.30$, $y_1 = 0.22$. The next position given algebraically as $\bar{y}, x - y$

will be $x_2 = \bar{y}_1 = -0.22$, $y_2 = x_1 - y_1 = 0.08$, and is shown as position q in Figure 2.9. The position $y - x, \bar{x}$ becomes $x_3 = -0.08$, $y_3 = -0.30$, and is shown as position r.

2.6 Assignment of Space Group to a Specific Pattern

The structure in Figure 1.8a obviously belongs to a tetragonal space group, but which one? The unit cell corners can be placed at the centers of the atomic clusters. A tetrad axis occurs there, and also at the center of the unit cell. Mirror lines occur along the edges of the unit cell and at 45° diagonal positions. The obvious space group is **p4mm**, but will all the details check out? Yes. Figure 2.8 shows atomic centers rather than the circles in Figure 1.8a. Atom A can be assigned to the following special position: multiplicity 1, Wyckoff symbol a, point symmetry **4mm**, coordinates 0,0. Atoms B, B′, B″, B‴ can be assigned to a site with $x = y = 0.22$ and multiplicity 4, Wyckoff symbol f, point symmetry **m**, coordinates x,x; $\bar{x},\bar{x}$; $\bar{x},x$; and $x,\bar{x}$. There is no general position in this atomic pattern, and, indeed, most simple crystal structures such as those of diamond, gold, and halite have atoms lying only on special positions. The square packing of circles (Fig. 1.2a) also belongs to **p4mm**, and only the special position at $x = 0$, $y = 0$ is occupied. Space group **p4** also has this special position, but the space group with the highest symmetry must be used. Figure 1.13a belongs to space group **p4**, and the four dots occupy a general position with x 0.19, y 0.03.

The pattern in Figure 1.6c illustrates a general position in space group **p2**. For the cell at the lower right, $x = 0.15$ and $y = 0.16$.

Several examples of rectangular patterns were given in Chapter 1. Figure 1.7a illustrates space group **p1m**. The unit cell is almost but not quite square. The mirror line is perpendicular to the longer axis b. Atoms p and q are in special positions on a mirror line, respectively, at $x \sim 0.5$, y 0.5 and x 0.71, y 0.5, and atoms r and r′ are in a general position with r at x=0.41 y=0.33. The pattern in Figure 1.6a belongs to space group **pmm** with two atoms per cell in a special position on a mirror plane with x=0.21, y=0. Figure 1.6b illustrates point group **cmm** with four atoms per cell in a special position on a mirror plane with x=0.11, y=0. The zigzag pattern in Figure 1.6d is the clue to the presence of glide planes. Using the outlined unit cell with a horizontal and b downward (i.e., rotated 90° to the usual orientation), the space group is **pgg**, and a general position with x=0.12, y=0.10 is occupied.

Hexagonal closest packing of circles (Fig. 1.2b) has one circle per cell in a special position with point symmetry **6mm** in space group **p6mm**. This position also occurs in **p3, p3m1, p31m,** and **p6**, but the space group with highest symmetry is used. The pattern of Figure 1.2c has circles lying on a

2-fold special position with point symmetry **3m** in space group **p6mm**. A general position in this space group is illustrated in Figure 1.2*g*. This pattern is deceiving because the *x* and *y* coordinates obey a special geometrical relationship that results in the circles lying in rings with square and hexagonal shape. This is not required by the symmetry of space group **p6mm**. Remember that the blackening of some circles merely indicates rings, and not a separate kind of atom.

Such special relations can occur in any space group. Thus the reader can imagine that the three circles in **pl** of Figure 2.6 are identical, thus giving a cluster that obeys 3-fold symmetry *when isolated*. The *overall* environment still is controlled by **pl**. It is not possible to construct a space group for **p5**, because pentad rotation axes will not fit together in space, but it is possible to put a pentagonal cluster of atoms in *general positions* of any space group.

EXERCISES   (*More Difficult)

2.1   Study the three regular tessellations in Figure 2.1. Then close the book, and draw them accurately. Construct the reciprocal tessellations. A tile edge of 1 or 2 cm is convenient.

2.2   Repeat Exercise 2.1 for the semi-regular tessellations in Figures 2.1*h* and 2.1*k*.

2.3*   Repeat Exercise 2.1 for the semi-regular tessellations in Figures 2.1*d*–2.1*g*, 2.1*i*–2.1*j*.

2.4   Locate the smallest unit cell for Figures 2.1*a*–2.1*c*, 2.1*h*, and 2.1*k*. Select a cell that has the most regular geometry and that is a parallelogram. Determine the cell dimensions by both measurement and calculation for a tile of edge 2 cm. Determine the number of tile junctions per cell.

2.5*   Repeat Exercise 2.4 for the semi-regular tessellations in Figures 2.1*d*–2.1*g*, 2.1*i*–2.1*j*.

2.6   Locate the smallest unit cell for Figures 2.1*l*–2.1*o*. Select a unit cell that has the most regular geometry and that is a parallelogram. Determine the number of tile junctions per cell.

2.7   Study the way in which the point symmetry operates in Figure 2.2. Then close the book, and construct the 10 diagrams using the same three-circle motif. Of course, the starting position is not important, but the symmetry operations must be correct. In case of difficulty, copy the motif on a piece of tracing paper, pin it over a rotation operator, and rotate by the appropriate angle. To demonstrate mirror symmetry, turn over the piece of tracing paper.

2.8* Construct the multiplication tables for the point groups, and check them against the tables given in Chapter 2.

2.9   Use the three-circle motif to show point groups **5** and **5m**. Should **5m** really be called **5mm?**

2.10 Study the five types of lattices in Figure 2.5. Then close the book and draw them. Place a point at the center of the **p4** lattice. Are all the points still congruent? Can you find a smaller unit cell which fits **p4**? Place a point at the center of one of the equilateral triangles of the unit cell of the **p6** lattice. Are the points still congruent? No, they are not. Place points at the center of both equilateral triangles of the **p6** cell. Are the points still congruent? Yes, they are. Can a smaller primitive cell be chosen? Place a point at the center of the **p2** cell. Can a smaller primitive cell be chosen? Make similar attempts to place extra points halfway along the cell edges. All attempts should fail, and the reader will end up with only five types of lattices.

2.11 Draw an oblique unit cell with $a=5$ cm, $b=7.8$ cm, $\gamma=101°$. Locate the point with coordinates $x=0.16$ and $y=0.21$, and draw a circle of radius 2.5 cm about the point. Draw a second unit cell with the same shape and orientation. Place a lens-shaped symbol at one corner to represent a rotation diad. Operate on the first point to give a second point related by the diad, and draw a second circle of radius 2.5 cm. This circle should overlap the first one. Remove the overlapping arcs to give a figure eight. Repeat this motif by the cell translations, and compare the result with Figure 1.6c.

2.12 Draw a square of edge 5 cm. Locate a point at the coordinates $x=0.18$, $y=0.31$. Draw a second square, and place a tetrad at one corner. Operate on the first point to produce three more points, and repeat them by the edge translations of the square. Join the points to produce a tessellation that almost matches that in Figure 2.1e.

2.13* Determine the space group for points exactly at the nodes of the tessellation in Figure 2.1e. Be warned that the special geometry from the squares and equilateral triangles results in local mirror symmetry that is not in the space group.

2.14 For the tessellation in Figure 2.1k, trace the nodes at which the branches intercept. Determine the position of the unit cell. Locate all the symmetry elements, using a separate diagram. Determine the space group. Be warned that the nodes lie on a special position. Determine the coordinates.

2.15* The semi-regular tessellation in Figure 2.1k can be distorted by changing the coordinates and space group. First, retain the space group found in Exercise 2.14, and change the coordinates to $x=0.15, y=0.5$. What is the shape of the tessellation produced by joining the nodes? Second, change the coordinates to $x=0.15, y=0.4$ and the space group to **p4**.

2.16* Copy Figure 1.9*b* on a sheet of tracing paper. Put five atomic centers in the correct positions in each of the elliptical molecules. Determine the space group, and draw the two conventional diagrams. Make a list of the general and special positions, and assign all atoms of Figure 1.9*b* to the appropriate positions.

2.17 Trace the centers of the circles in Figure 1.6*a.* Then change the coordinates from the special position on the mirror line at *x*=0.21, *y*=0 to *x*= 0.21, *y*=0.05 for a general position, and plot all the equivalent positions in **pmm.**

2.18 Repeat Exercise 2.17 for Figure 1.6*b* by changing the coordinates to *x*=0.11, *y*=0.05 in **cmm.**

2.19* Draw a rectangular unit cell with *a* 2.5 cm, *b* 7.5 cm. Locate the point *x*=0.12, *y*=0.10, and draw a circle of radius 1.25 cm about the point. Draw a second unit cell and place a glide line at *x* = 0.25 and another one at *y* = 0.25. Operate on the first point with these glide lines to produce three more points in the first diagram. Repeat these points by lattice translations. Draw circles about all the points, and show that the circles can be converted into close-packed doublets like those in Figure 1.6*d.*

2.20 What is the space group for Figure 1.12*a?*

2.21 What are the space groups for Figures 1.1*b*, 1.2*d*, 1.2*f*, and 1.2*g*? Give the point symmetry of the positions occupied by the circles.

ANSWERS

2.4 Figure 2.1*a*–120° rhombus, *a* = *b* = 2 cm, one junction; Figure 2.1*b*–square, *a* = *b* = 2 cm, one junction; Figure 2.1*c*–120° rhombus, *a* = *b* = $2\sqrt{3}$ cm = 3.464 cm, two junctions; Figure 2.1*h*–120° rhombus, *a* = *b* = 4 cm, three junctions; Figure 2.1*k*–square, *a* = *b* = 2 (1 + $\sqrt{2}$) cm = 4.828 cm.

2.5 Figure 2.1*d*–rectangular, *a* = 2 cm, *b* = 4(1 + $\sqrt{3}/2$) cm = 7.464 cm, four junctions; Figure 2.1*e*–square, *a* = *b* = 3.864 cm [measurement is easy; for calculation use right-angle triangle with edges 1 cm and (2 + $\sqrt{3}$) cm], four junctions (*note:* the obvious place for the corners of the cell is at the center of squares); Figure 2.1*f*–120° rhombus, *a* = *b* = 5.292 (for calculation use right-handed triangle with edges 4 and $2\sqrt{3}$ cm), six junctions (the obvious place for the cell corners is at the center of hexagons); Figure 2.1*g*–120° rhombus, *a* = *b* = 2(1 + $\sqrt{3}$) = 5.464 cm, six junctions; Figure 2.1*i*–120° rhombus, *a* = *b* = 4 + $2\sqrt{3}$ cm = 7.464 cm, six junctions; Figure 2.1*j*–120° rhombus, *a* = *b* = 6 + $2\sqrt{3}$ cm = 9.464 cm, twelve junctions.

2.6   Figure 2.1*l*—rectangular, one junction; Figure 2.1*m*—rectangular, four junctions (*note:* the cell is centered in contrast to the primitive cell for Figure 2.1*l*); Figure 2.1*n*—oblique, one junction; Figure 2.1*o*—oblique, two junctions (*note:* a centered cell is too big since the symmetry does not constrain the shape as for the rectangular cell in Figure 2.1*m*).

2.9   No, there is only one type of mirror line when the rotation axis is an odd number. When the axis is even, there are two types of mirror line.

2.13   **p4gm**. True mirror lines pass only through equilateral triangles, whereas "local" mirror symmetry can be found in the squares.

2.14   The nodes are in space group **p4mm** in a special position with point symmetry **m**. For the coordinates you should obtain $x = 0.207$, $y = 0.5$, or some equivalent set. For a semi-regular tessellation x can be calculated as $(\sqrt{2} - 1)/2$.

2.15   The first distortion gives enlarged squares and shrunken octagons. The second distortion gives tilted squares and very distorted octagons.

2.16   **p4gm**.

Multiplicity	Wyckoff Notation	Point Symmetry	Coordinates of Equivalent Positions
8	*d*	1	$x,y; y,\bar{x}; \frac{1}{2} - x, \frac{1}{2} + y; \frac{1}{2} - y, \frac{1}{2} - x$ $\bar{x},\bar{y}; \bar{y},x; \frac{1}{2} + x, \frac{1}{2} - y; \frac{1}{2} + y, \frac{1}{2} + x$
4	*c*	m	$x, \frac{1}{2} + x; \bar{x}, \frac{1}{2} - x; \frac{1}{2} + x, \bar{x}; \frac{1}{2} - x, x$
2	*b*	mm	$0.5, 0; 0, 0.5$
2	*a*	4	$0, 0; 0.5, 0.5$

The black atom lies in Wyckoff position *a*. The central atom of the molecule lies in position *b*, and the other four atoms occur in two pairs of positions of type *c*.

2.20   **p6mm**.

2.21   Figure 1.1*b*—C atoms of graphite layer are in a special position with point symmetry **3m** in **p6mm**; Figure 1.2*d*—special position with point symmetry **2mm** in **p6mm**; Figure 1.2*f*—special position with point symmetry **m** in **p6mm**; Figure 1.2*g*—general position with point symmetry **1** in **p6mm**.

3
Regular Polyhedra: Crystal Drawing

Preview This chapter uses polyhedra as the first step in learning to think about crystals in 3D. The corners of the five regular polyhedra correspond to the centers of touching atoms in some crystal structures, and the tetrahedron, cube, and octahedron form the external surface of some crystals. Various mathematical concepts (reciprocity, truncation, stellation) are used in the development of semi-regular and complex polyhedra useful in crystallography, either for filling of space or for the geometry of the adjacent atoms.

Press four spheres together and locate their centers (tennis balls are convenient). They lie at the vertices of a *regular tetrahedron* (Fig. 3.1); *vertices* is the plural of *vertex*. A tetrahedron is the simplest example of a *polyhedron*, which is the 3D analog of the 2D polygon. A tetrahedron need not be regular, but when it is regular, each *face* is an equilateral triangle. Then the four vertices are congruent to each other, and so are the four faces. Three of the six *edges* meet at each vertex to give a *valency* of 3.

Eight spheres can be placed at the vertices of a **cube**. The hole between eight touching spheres is greater than that between four touching spheres, and the eight spheres will collapse unless held carefully. The cube can be called a *regular hexahedron* because it has six regular faces (i.e., squares): *hedra* is the Greek word for a base or seat. Each of the eight vertices lies at the junction of three squares and three edges.

Six spheres pack together tightly when placed at the vertices of a *regular octahedron*. Like the regular tetrahedron, this polyhedron is composed of equilateral triangles, but four triangles meet at each vertex instead of three triangles of the regular tetrahedron.

Octahedral and tetrahedral packings of atoms are important in many crystal structures. Polyhedra are also important as external surfaces of crystals. In addition to the cube (found in common salt), octahedron (spinel and fluorite), and tetrahedron (tetrahedrite), there are many kinds of polyhedra with lower symmetry.

Systematic study of polyhedra is a part of pure mathematics, and a rigorous treatment, as in H. M. S. Coxeter, *Regular Polytopes,* 2nd ed. (1963),

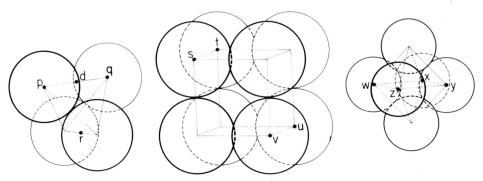

FIGURE 3.1 Close packing of spheres whose centers lie at the vertices of regular polyhedra. *Left:* four spheres at vertices of a regular tetrahedron; each sphere touches the
other three spheres. *Middle:* eight spheres at vertices of a cube; each sphere touches three
other spheres. *Right:* six spheres at vertices of a regular octahedron; each sphere touches
four other spheres. The polyhedra are drawn in clinographic projection (Figs. 3.4 and
3.5).

is rather time-consuming and forbidding. *Regular Figures* by L. Fejes Tóth
(1964) is somewhat easier. Fortunately, the important concepts can be
expounded easily by use of diagrams and models, especially if the reader will
build the latter. Most of my classes on polyhedra develop a holiday atmo-
sphere because of the sheer fun of constructing polyhedra. Such enthusiasm
can be found in H. M. Cundy and A. P. Rollett, *Mathematical Models,* 2nd
ed. (1961); A. F. Wells, *Models in Structural Inorganic Chemistry* (1970);
and A. F. Wells, *Structural Inorganic Chemistry,* 4th ed. (1975).

The five polyhedra with regular planar faces and vertices were known to
early philosophers, and the name **Platonic solids** became attached to them
probably because of the huge prestige of Plato, the Greek scholar (about
400 B.C.). Pyramids were known to the Egyptians, and prisms were a part of
earliest architecture. Mystical significance was attached to the Platonic
solids by the Pythagoreans (sixth century B.C.). Existing knowledge of
polyhedra was incorporated into the *Elements* by the Greek mathematician
Euclid (300 B.C.). The 13 **semiregular solids** are named after **Archimedes**, the
Syracuse mathematician (about 250 B.C.), whose investigations on polyhedra
are lost. Enumeration of the uniform *convex polyhedra* (including prisms
and antiprisms) was completed by Kepler, the German astronomer
(AD. 1571–1630), whose mathematical concept of harmony in the solar
system is associated partly with musical scales and partly with use of polyhedra to circumscribe planetary orbits. Convex means that there are no
re-entrant angles between faces. Reciprocals of the 13 *Archimedean polyhedra* are named after **Catalan**, the mathematician, who codified them in
1865. *Parallelohedra,* the polyhedra that fill space when parallel, were
developed by the Russian mineralogist–crystallographer Fedorov (1851–

1919). Topologic study of polyhedra developed from the publication in 1758 of *Elementa doctrinae solidorum* by Euler, the Swiss mathematician (1707–1783), and is now treated as a branch of graph theory.

3.1 Fully Regular (Platonic) Solids

Only five solids have congruent faces, each composed of a regular polygon, which meet at congruent vertices (Fig. 3.2). The regular tetrahedron, regular octahedron, and **regular icosahedron** are composed, respectively, of 4, 8, and 20 equilateral triangles. The regular hexahedron is composed

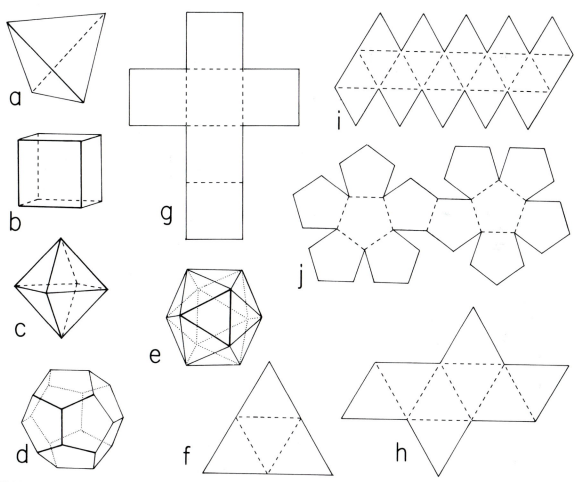

FIGURE 3.2 The five regular polyhedra. Diagrams *a, b, c, d,* and *e,* respectively, are clinographic projections of the regular tetrahedron, cube, regular octahedron, regular dodecahedron, and regular icosahedron. Diagrams *f* to *j* are corresponding templates. Cut along the solid lines, score but do not cut through the dashed lines, fold inward, and tape.

of 6 squares, and the ***regular dodecahedron*** (often seen as a calendar-cum-paperweight) of 12 pentagons. All five polyhedra can be constructed from thin cardboard by cutting out the flattened shapes in Figure 3.2, creasing or scoring along the dashed lines, folding inward, and taping the edges. These operations must be meticulous for a neat fit. Tabs and glue can be used instead of tape.

The regular tetrahedron automatically rests on a face with one vertex pointing upwards. One of the four triad rotation axes is immediately obvious in a plan (Fig. 3.3*a*). In drawings of crystal structures, a tetrahedron may point downward (3.3*b*) or be oriented (3.3*c*) with the top edge horizontal and the bottom edge also horizontal but crossed at 90°. This third orientation reveals a diad rotation axis that joins midpoints of opposing edges. Each triad axis requires two more diad axes. Note that each diad axis uses two out of the six edges.

The regular hexahedron automatically rests on a square face, thereby showing one of its three tetrad axes, but it can be oriented with either one of the four body diagonals perpendicular to the paper (Fig. 3.3*d*) or one of the

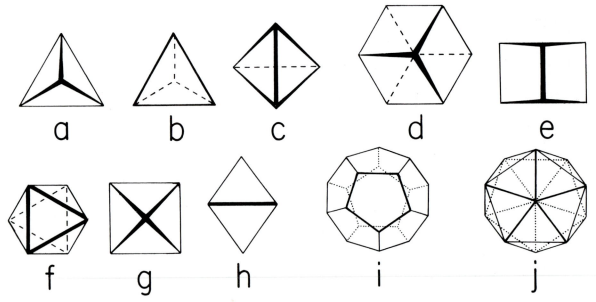

FIGURE 3.3 Different plans of the regular tetrahedron (*a–c*), cube (*d, e*) and regular octahedron (*f–h*). Plans of the regular dodecahedron (*i*) and regular icosahedron (*j*) display the 5-fold symmetry. Lines pointing diagonally upward in *a, c, d, e,* and *g* are tapered. Dashed lines in *b* and *d* are pointing diagonally downward, and those in *c* and *f* are horizontal. The thick lines in *e* and *g* hide a set of lines related by a horizontal mirror plane. The thick line in *h* represents a vertical square, and the thin lines represent two superimposed lines tilting either upward or downward. In *i*, the two pentagons in the center are horizontal, and all other lines are tilted to form 10 more pentagons. In *j*, all the radiating lines are tilted, while the lines forming the perimeter are tilted even more steeply.

face diagonals (3.3*e*). These two orientations display the triad and diad rotation axes that, respectively, join opposing vertices or midpoints of opposing edges.

The regular octahedron automatically lies on a triangular face (Fig. 3.3*f*), thereby showing one of its four triad axes, and two other orientations (3.3*g*, 3.3*h*) display a tetrad and a diad axis.

The regular dodecahedron rests on a pentagonal face, thereby revealing one of the 6 pentad axes (Figure 3.3*i*). It also has 10 triad axes. The regular icosahedron would rest on a triangular face, showing one of the 10 triad axes, but can be drawn down one of 6 pentad axes (3.3*j*).

Biologically oriented readers may wish to look up drawings of Radiolaria, an order of marine protozoa in the class Actinopoda. Siliceous skeletons of *Circoporus octahedrus, Circogonia icosahedra,* and *Circorrhegma dodecahedra* are built from a reticular network corresponding to curved faces of the octahedron, icosahedron, or dodecahedron. Complex spicules project outward from vertices like antennas from space capsules: see *Symmetry,* H. Weyl (1952) and *On Growth and Form,* new edition, D'Arcy W. Thompson (1948). Models of viruses have spheres lying at the corners of polyhedra.

3.2 Elementary Crystal Drawing

Try sketching even a simple polyhedron, and you will discover that a systematic procedure is desirable. Most artists use a perspective with disappearing points to which lines converge away from the viewer, while some distort deliberately. Crystallographers prefer to retain parallelism of corresponding edges of polyhedra, and use a ***clinographic projection,*** which produces a homogeneous distortion.

Take an origin and three mutually perpendicular axes *a, b,* and *c.* Unit lengths on the axes are represented in the clinographic projection by an ***axial cross*** (Fig. 3.4) with the *c* axis vertical, the *b* axis pointing to the right and slightly downward, and the *a* axis pointing almost to the reader, but strongly foreshortened to about eight o'clock. This right-handed arrangement of axes is used conventionally by crystallographers, and corresponds to the thumb (*b*) and first (*c*) and second (*a*) fingers of the right hand: left-handed axes would be related by a mirror operation, and correspond to the thumb (*b*) and first (*c*) and second (*a*) fingers of the left hand. The amount and direction of foreshortening are a matter of choice. Following C.F. Naumann (1797–1873), a German crystallographer–mineralogist–geologist, draw a horizontal line *Op* of convenient length from an origin *O.* Construct *qr* where $Oq = Op/3$ and $qr = Op/9$ and perpendicular to *Op.* The line *Or* represents the +*a* axis, and the backward extension to *r'* the −*a* axis. Construct *spt* perpendicular to *Op* with $ps = Op/27$ and $pt = Oq = Op/3$. Check

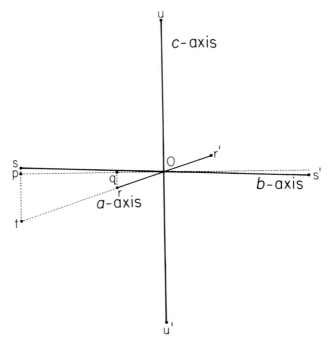

FIGURE 3.4 Construction of axes for a clinographic projection.

that point *t* is on the extension of *Or.* Line *Os* represents the – *b* axis, and the right-hand extension *Os'* the +*b* axis. The +*c* axis extends upward to *u*, where *Ou* = *Ot*, and angle *pOu* is 90°: the – *c* axis extends downward to *u'*. This construction gives convenient tilts to the *a* and *b* axes. Distances in the *bc* plane are distorted only slightly, but those projecting in the *a* direction are shortened about threefold by the perspective.

To draw the regular octahedron, join points at unit lengths on the axes (Fig. 3.5*a*). All the triangular faces appear distorted from their equilateral shape in 3D, but the brain can learn to accommodate for the distortion.

To draw the cube, construct parallels as in Fig. 3.5*b*). The resulting points lie at coordinates $(1a, 1b, 1c)$, $(1a, \bar{1}b, 1c)$, $(1a, \bar{1}b, \bar{1}c)$, $(1a, 1b, \bar{1}c)$, $(\bar{1}a, 1b, 1c)$, $(\bar{1}a, \bar{1}b, 1c)$, $(\bar{1}a, \bar{1}b, \bar{1}c)$, and $(\bar{1}a, 1b, \bar{1}c)$, where the bar represents a negative length.

To draw a regular tetrahedron, join four out of eight vertices of the cube. The two possible choices yield the positive (3.5*c*) and negative (3.5*d*) orientations. Of course, such a distinction is meaningful only when axial directions have been specified.

Construction of the regular dodecahedron and regular icosahedron requires use of the ***golden ratio***, $(\sqrt{5} + 1)/2$, which results from the fivefold symmetry. It is beyond the scope of this book to follow the use of the golden ratio, the pentagram, and the Fibonacci series (Leonardo of Pisa,

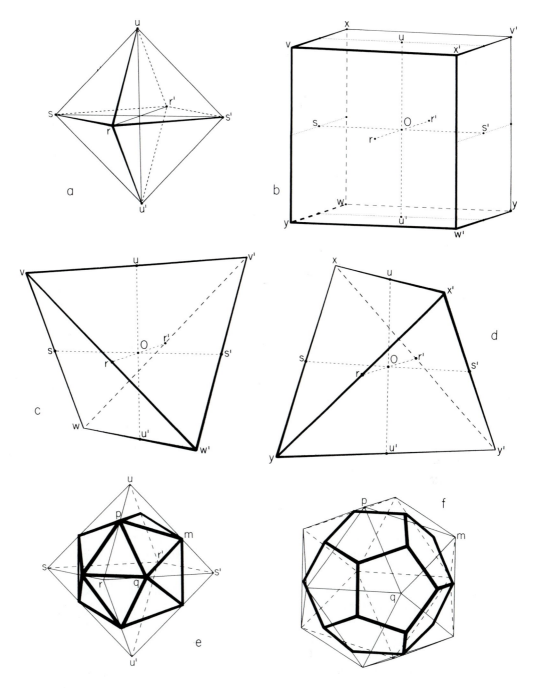

FIGURE 3.5 Construction of clinographic projections of regular polyhedra: octahedron (*a*), cube (*b*), regular tetrahedron in positive (*c*) and negative (*d*) orientations, regular icosahedron (*e*), and regular dodecahedron (*f*). The icosahedron differs in orientation from Figure 3.2*e*. In *e*, the length ratios *rp/pu, um/ms'*, and *s'q/qr* are $(\sqrt{5} + 1)/2$. The points *r*, *r'*, *s*, *s'*, *u*, and *u'* correspond to points in Figure 3.4.

thirteenth century) in art, architecture, biological structure, and sorcery. To construct the regular icosahedron, draw a regular octahedron. Locate point *p,* which divides edge *ur* in the golden ratio (3.5*d*). Similarly locate points *q* and *m* so that the triad axis is obeyed for face *urs'*. Locate points on all other edges to obey all four triad axes of the octahedron. Join each point to its five nearest neighbors. The golden ratio results in all triangles being equilateral, whereas any other ratio results in 12 isoceles triangles and 8 equilateral triangles. To construct the regular dodecahedron (3.5*f*), join midpoints of adjacent faces of the regular icosahedron.

3.3   Schlegel Diagram and Euler Relation

Topology is a branch of mathematics useful to crystallographers. The topological properties of a polyhedron can be shown in 2D by the **Schlegel diagram.** Place a polyhedron with one face in contact with a sheet of paper. Imagine that the polyhedron is collapsed neatly into the chosen face so that all faces and vertices retain their spatial relations. A tetrahedron rests on a triangular face, and the opposing vertex is collapsed onto this face (Fig. 3.6*a*) so that the other three faces become isosceles triangles. The resulting Schlegel diagram is identical to Figure 3.3*a*. The cube rests on a square face, and the opposing face shrinks into a smaller square during the collapse (3.6*c*), while each of the other four faces becomes a trapezium (3.6*b*). The octahedron rests on an equilateral triangle, and the opposing face shrinks to a smaller equilateral triangle, while the other faces become isosceles triangles (3.6*d*). From the topologic viewpoint, the original polyhedron need not be regular because only the relative arrangement of edges, vertices, and faces is important. However, it is helpful to draw a Schlegel diagram in its most symmetrical shape. Try to sketch the Schlegel diagrams of the regular dodecahedron and icosahedron before studying Figure 3.6*e* and 3.6*f*.

Mathematicians have enjoyed inventing new proofs of the **Euler relation** for a polyhedron, $F + V = E + 2$, where E, F, and V are the number of edges, faces, and vertices, respectively. The simplest demonstration uses the Schlegel diagram. Consider the cube (Figs. 3.6*g*–3.6*l*). The first diagram (3.6*g*) has an edge and a face marked with dots. Because E and F are on opposite sides of the Euler relation, removal of an edge and a face (3.6*h*) modifies the equation to $(F - 1) + V = (E - 1) + 2$. Repeating this procedure three times removes all the outer edges and faces, leaving a square with four spikes (3.6*k*). Each spike contains one vertex and one edge, and simultaneous removal merely reduces each side of the Euler relation by one to give $F + (V - 1) = (E - 1) + 2$. The final diagram (3.6*l*) is merely a square. Removal of four vertices and four edges leaves two faces, which are the square in the final

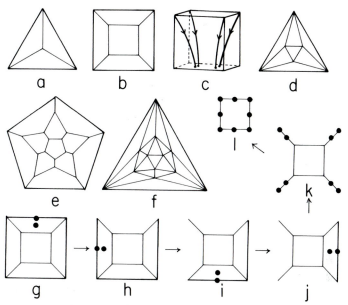

FIGURE 3.6 Schlegel diagrams of Platonic solids. (*a*) Tetrahedron, (*b*) cube, (*c*) method of construction for cube, (*d*) octahedron, (*e*) (pentagonal)-dodecahedron, (*f*) icosahedron. The sequence *g* → *l* illustrates a demonstration of Euler's relation: dots show items removed to give the next diagram of the sequence.

diagram, and the square onto which the cube was projected to give the Schlegel diagram. Hence these two elements provide the justification for the number 2 in the Euler relation.

⬇ (This procedure works for all *simply connected* polyhedra. All polyhedra used in crystallography are simply connected, but mathematicians also consider bizarre *multiply connected* polyhedra such as a wedding ring with a ⬆ faceted instead of a rounded surface.)

3.4 Reciprocal Polyhedron (Dual)

Crystallographers profit greatly from the concept of *reciprocity*. In Chapter 1, a boundary line of a 2D crystal was replaced by a perpendicular line radiating outward from the center of the crystal (Fig. 1.19), and ultimately by a dot on a circle. In 3D, a crystal face can be replaced by a line normal to the face. The faces of one polyhedron can be exchanged reciprocally with the vertices of a second polyhedron, and vice versa. Place a dot at the center of each face of a cube, and join the six dots to produce a regular octahedron

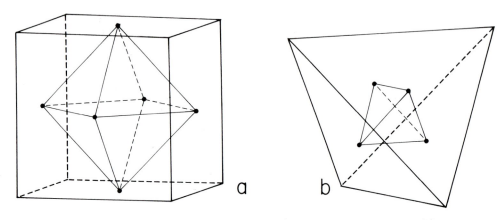

FIGURE 3.7   Construction of duals of cube (*a*) and tetrahedron (*b*).

(Fig. 3.7*a*). Each vertex of the octahedron is related to a face of the cube. Thus the regular octahedron is the **reciprocal polyhedron**, that is, the ***dual*** of the cube, and vice versa. The dual of a regular tetrahedron in the positive orientation is a tetrahedron in the negative orientation (3.7*b*), and the tetrahedron is its own dual. In Figure 3.7, the polyhedra are fully regular, but from the viewpoint of topology this is quite unnecessary. The dual of any distorted cube is an octahedron and vice versa. The regular icosahedron and regular dodecahedron are dual to each other (Fig. 3.5*f*).

3.5   Vertex and Edge Truncation

Symmetrical ***truncation of vertices*** (Fig. 3.8*a*) converts a regular tetrahedron into a ***truncated tetrahedron***. Each triangular face is converted into a hexagon. Small slices leave a hexagon with three long sides and three short ones. Further slicing can give a regular hexagon, and the resulting polyhedron (heavy lines) is the ***regular truncated tetrahedron***. It is not a Platonic fully regular solid, because it contains two types of faces (regular hexagon and equilateral triangle), but its vertices are congruent. A convenient notation (the **Schläfli symbol**) contains the sequence of polygons around the vertex— that is, 3.6.6, or briefly, 3.6^2. Further slicing of the regular truncated tetrahedron results ultimately in conversion of each hexagon into an equilateral triangle and formation of a regular octahedron (dotted lines). Continued slicing (not shown) could convert the original tetrahedron, which is in a positive orientation, into a smaller tetrahedron in the negative orientation.

Vertex truncation of the cube can turn each square face into a regular octagon. The resulting ***regular truncated cube*** (3.8*b*) has two types of faces (octagon and equilateral triangle) and congruent vertices (3.8.8, or briefly, 3.8^2). Further truncation converts each octagon into a square, but the resulting polyhedron is not a cube but a cubo-octahedron, or briefly, ***cuboctahedron,*** with congruent vertices at which squares alternate with equilateral triangles to give the symbol 3.4.3.4, or briefly, $(3.4)^2$.

Vertex truncation of a regular octahedron produces a square at each vertex. This square grows as the original triangular face is cut back to a regular hexagon, at which stage the ***regular truncated octahedron*** is produced ($3.8c$). The congruent vertices have the symbol 4.6.6, or 4.6^2. Further

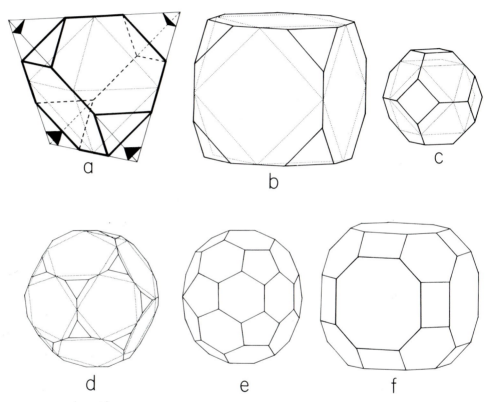

FIGURE 3.8 Vertex truncation of Platonic solids. In *a*, black triangles show small slices from the vertices of a regular tetrahedron in positive orientation. Further truncation gives the regular truncated tetrahedron (heavy lines), and even further truncation gives the regular octahedron (dotted lines). Progressive truncation of a cube (*b*) gives the regular truncated cube (heavy lines) and the cuboctahedron (dotted lines). For clarity, only the front faces are shown. Progressive truncation of a regular octahedron produces first the regular truncated octahedron (heavy lines) and then the cuboctahedron (dotted lines). Diagrams *d* and *e* show the regular truncated dodecahedron and regular truncated icosahedron (continuous line). Further truncation of the regular dodecahedron gives the icosidodecahedron (dotted line). Truncation of the cuboctahedron (*f*) gives a nonregular truncated cuboctahedron that can be converted into the great rhombicuboctahedron by a symmetrical deformation.

truncation yields the cuboctahedron when the hexagons are reduced to equilateral triangles.

Vertex truncation of the regular dodecahedron produces the **regular truncated dodecahedron** (3.8*d*) with symbol 3.10.10. A triangular face grows at each vertex until each regular pentagon is converted into a regular 10-sided polygon (decagon). Further truncation gives the **icosidodecahedron** (dotted line). Truncation of the regular icosahedron yields the **regular truncated icosahedron** (3.8*e*) when a pentagon at each vertex grows

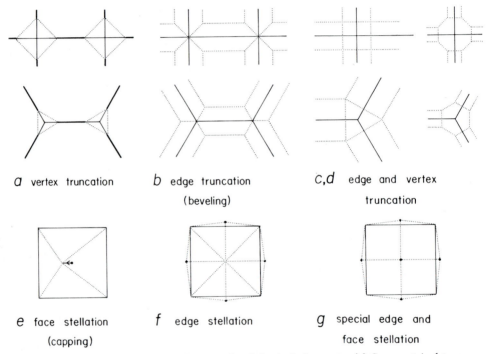

a vertex truncation

b edge truncation
(beveling)

c,d edge and vertex
truncation

e face stellation
(capping)

f edge stellation

g special edge and
face stellation

FIGURE 3.9   Truncation and stellation of polyhedral elements. (*a*) Symmetrical truncation of two adjacent vertices, the upper ones with four edges (continuous line) meeting at each vertex, and the lower ones with three edges. New edges are shown by dotted line, and surviving parts of the original edges by the heavy line. (*b*) Edge truncation without any truncation of the original vertices except that produced by the edge truncation. All original edges (continuous line) are removed. (*c, d*) Two types of combined edge and vertex truncation. For the right-hand pair of diagrams, the vertex truncation is deeper than for the left-hand pair, thereby converting the square or triangle into an octagon or hexagon. (*e*) Face stellation. The midpoint (dot) of the face is moved perpendicularly outward, and the sideways displacement results merely from convenience of drawing. (*f*) Edge stellation. The center of each edge should be moved outward symmetrically with respect to the two faces that meet at the edge. The diagonals of the square do not move, and become valleys between the adjacent triangular faces. (*g*) Special edge and face stellation. If the midpoint of the square face is moved outward by the correct amount, each pair of triangular faces loses the re-entrant angle at the valley and merges into a kite-shaped quadrilateral.

large enough to convert equilateral triangles to regular hexagons and thereby produce the vertex symbol of 5.6^2. Further truncation gives the icosidodecahedron.

Symmetrical vertex truncation of the cuboctahedron produces a rectangle at each vertex (3.8*f*), with the longer edges crossing the original square faces. When the equilateral triangle has been cut down to a regular hexagon, the square has been cut down to a nonregular octagon. Symmetrical deformation of this truncated cuboctahedron can give a symmetric polyhedron, often loosely called the ***truncated cuboctahedron,*** after Kepler, but perhaps better called the ***great rhombicuboctahedron.*** In Figure 3.8*f*, the rectangles have sides in the ratio $\sqrt{2}:1$, and the great rhombicuboctahedron is produced by turning the rectangles into squares while retaining the regularity of the hexagons and the directions of the edges.

Figure 3.9*a* is a schematic depiction of truncation of a vertex. A new face (dotted line) causes removal of parts of original edges (thin line), leaving the heavy lines. Simple edge truncation (Fig. 3.9*b*) produces beveled edges, as is commonly used in carpentry. New edges radiate from the vertex, which is cut back as beveling proceeds. A further set of edges is produced in the surfaces of the surviving faces. Simple edge truncation is of little interest in crystallography, but combined edge and vertex truncation leads to useful polyhedra. Two symmetrical products result depending on the relative rates of edge and vertex truncations. In Figure 3.9*c*, the 3- and 4-valent vertices are replaced, respectively, by a triangular or a square face. In Figure 3.9*d*, the vertices are truncated more rapidly than the edges to produce either a hexagonal or an octagonal face.

The ***small rhombicuboctahedron*** (Fig. 3.10*a*) is produced by combined

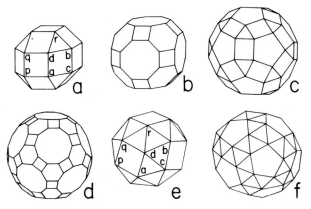

FIGURE 3.10 Clinographic projections of six Archimedean polyhedra. (*a, b*) Small and great rhombicuboctahedron. (*c, d*) Small and great rhombicosidodecahedron. (*e, f*) Snub cube and snub dodecahedron, both in the right-handed configuration.

edge and vertex truncation of a cube as in Figure 3.9*c*, and the great rhombi-bicuboctahedron (Fig. 3.10*b*) by further vertex truncation as in Figure 3.9*d*. Alternatively, these two polyhedra may be obtained by combined edge and vertex truncation of an octahedron. Combined edge and vertex truncation of either the regular dodecahedron or the regular icosahedron will give either the **small** or **great rhombicosidodecahedron** (Figs. 3.10*c*, 3.10*d*) depending on the relative depths of the cuts.

3.6   Semi-regular Polyhedra: Stellation

Whereas the fully regular solids have only one type of regular polygon and one type of vertex, the semiregular **Archimedean polyhedra** have more than one type of regular polygon. Congruency of the vertices is retained. Eleven of the Archimedean polyhedra were obtained in the preceding section by vertex and edge truncation. The remaining two Archimedean polyhedra are obtained by distortion of other Archimedean polyhedra. To obtain the **snub cube** (Fig. 3.10*e*) from the small rhombicuboctahedron (Fig. 3.10*a*), turn square *adbc* into two equilateral triangles *acd* and *bdc* by moving *c* and *d* toward each other. Square *adqp* and equilateral triangle *dbr* retain their shape but are rotated. Symmetrical changes for the other polygons complete the transformation. The **snub dodecahedron** (Fig. 3.10*f*) can be obtained from the small rhombicosidodecahedron (Fig. 3.10*c*) by a similar transformation of each square into two equilateral triangles. Strictly speaking, there are 15 rather than 13 Archimedean polyhedra, since the snub cube and the snub dodecahedron each occur as a pair of mirror images. To obtain the mirror image of Figure 3.10*e*, distort Figure 3.10*a* so that the new triangles are *abd* and *acb*. The mirror image will be left-handed in contrast to the right-hand nature of Figure 3.10*e,* and the pair of polyhedra are called *enantiomorphs.*

Also belonging to the group of semi-regular polyhedra with congruent vertices and more than one type of regular polygon is the infinite series of **regular prisms** with square sides and **regular antiprisms** with equilateral triangles as sides. To a crystallographer, a **prism** is an assemblage of faces, all of whose edges are parallel (Fig. 3.11*a*). The faces cannot enclose space by themselves, and it is necessary to close the top and bottom to produce a polyhedron. When the top and bottom are regular polygons, and the sides are squares, the resulting **regular closed prism** obeys the conditions for a semi-regular polyhedron. In common usage, the noun *prism* includes the top and bottom faces, and the adjective *closed* will be omitted when the context is clear. A **trigonal prism** results when the ends are equilateral triangles (3.11*b*). Square ends result in a cube, which is a Platonic solid, of course. An infinite number of prisms results as the top and bottom polygons have more

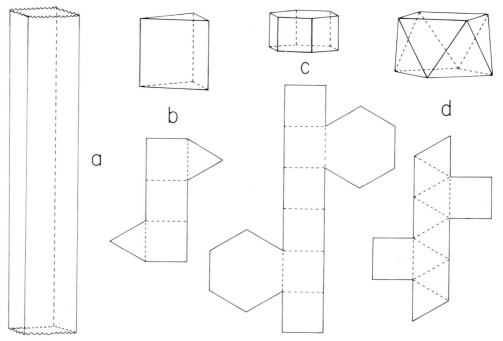

FIGURE 3.11 Clinographic projections and templates of prisms. (*a*) Diagram illustrating parallel edges of prism; the edges and faces should be infinite, and are terminated by the jagged lines for convenience. (*b*) Trigonal prism. (*c*) Hexagonal prism. (*d*) Square antiprism.

and more sides. Hexagonal ends result in a ***hexagonal prism*** (3.11*c*), and so on, ultimately giving a cylinder when the polygonal caps develop an infinite number of edges, and become circles.

An ***antiprism*** has the top and bottom faces rotated with respect to each other (3.11*d*). The vertices alternate, and are linked to give equilateral triangles, which are tilted and point alternately upward and downward. The ***square antiprism*** is important in the packing of atoms. Eight balls are packed more tightly when placed at the vertices of a square antiprism than when placed at the corners of a cube, as in Figure 3.1*b*. There is an infinite series of antiprisms, but the regular triangular antiprism is actually a regular octahedron.

A second set of semi-regular polyhedra is obtained by constructing duals of the Archimedean polyhedra, prisms and antiprisms. These new polyhedra have congruent faces and incongruent vertices. The dual of a regular prism is a ***Catalan bipyramid,*** that is, a double pyramid with a horizontal plane of symmetry specified by the prefix *bi-* (Fig. 3.12*a*). The octahedron, of course, can be regarded as a special case of a tetragonal bipyramid. Trigonal and hexagonal bipyramids will turn out to be important for the shapes of hexagonal crystals, including quartz.

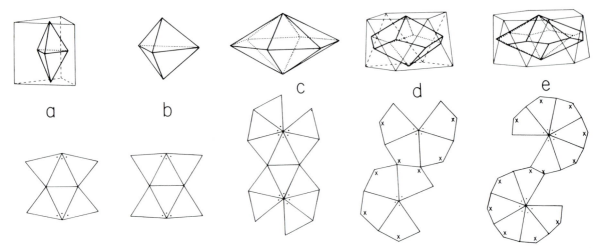

FIGURE 3.12   Clinographic projections and templates of bipyramids and trapezohedra. (*a*) Construction of the Catalan trigonal bipyramid as the dual of the closed trigonal prism. (*b*) Trigonal bipyramid shortened twofold along the vertical axis with respect to the dual of the regular closed prism. (*c*) Hexagonal bipyramid, which is dual of haxagonal prism. (*d*) Construction of the Catalan tetragonal trapezohedron as the dual of the square antiprism. (*e*) Hexagonal trapezohedron.

For the topologic definition of the dual (i.e., interchange of faces and vertices), there is no control over the detailed shape of the dual. For Platonic and certain other solids, the simplest way to construct the dual is to join the midpoints of adjacent faces, but in some polyhedra of low symmetry this procedure gives a solid with either nonplanar faces or the wrong topology. For elegance and simplicity, it is usual to draw a dual in its most symmetric shape.

Joining the midpoint of the faces of an Archimedean trigonal prism gives a trigonal bipyramid (Fig. 3.12*a*) whose faces are isosceles triangles with angles 51.32°, 64.34°, and 64.34°. Doubling the horizontal over the vertical dimension (3.12*b*) changes the angles to 57.42°, 61.29°, and 61.29°. Both trigonal bipyramids are topologic duals of a trigonal prism. The shape of the trigonal bipyramid could be adjusted so that all five vertices lie on the surface of a sphere. When all six faces are equilateral triangles, the trigonal bipyramid becomes a member of the **deltahedra,** a group of polyhedra all of whose faces are equilateral triangles (Section 3.9). A hexagonal bipyramid (3.12*c*) cannot be converted into a deltahedron, because 6 × 60° = 360°.

The topologic dual of an antiprism is a **Catalan trapezohedron** (3.12*d*). Joining the midpoints of the faces of a regular antiprism gives a special shape to the kite-shaped quadrilaterals that compose the trapezohedron. The shape is different if the vertices are moved to the surface of a sphere. A **trapezohedron** is characterized by the zigzag pattern of edges across its waist, plus the top and bottom vertices, and will be considered further in the section on crystal morphology. Just as for bipyramids, there is an infinite series of trapezohedra, of which the tetragonal and hexagonal varieties are given in Figures 3.12*d* and 3.12*e*.

Of the duals of the 13 Archimedean polyhedra (*sensu stricto*), the following are of special interest to crystallographers: **rhombic dodecahedron** (dual of cuboctahedron), **triakis tetrahedron** (from truncated tetrahedron), **tetrakis hexahedron** (from truncated octahedron), **triakis octahedron** (from truncated cube), **trapezoidal icositetrahedron** (from small rhombicuboctahedron), **hexakis octahedron** (from great rhombicuboctahedron) and **pentagonal icositetrahedron** (from snub cube). All seven polyhedra (Fig. 3.13) are possible surfaces of isometric crystals.

The rhombic dodecahedron has 12 congruent faces, each of which is a rhombus (i.e., a parallelogram with equal edges) with alternate angles of 109.47° and 70.53°. It must be clearly distinguished from the regular (or pentagonal) dodecahedron, whose faces are regular pentagons with edges meeting at 108°. Figure 3.13*a* shows the easiest way to construct the rhombic dodecahedron in clinographic projection. Draw a cube using positions *s*–*y'* from Figure 3.5*b*. Locate midpoints *A–D*, and draw four lines from *u'* halfway toward *A–D*. Draw parallels to complete the polyhedral edges. Joining the midpoints of faces of a cuboctahedron (Fig. 3.8*b*) does not give the rhombic dodecahedron (Fig. 3.13*b*). A flat trigonal pyramid (continuous heavy line) is generated above each triangular face of the cuboctahedron, and it is necessary to move each pyramid top inward (arrows) to obtain the rhombic dodecahedron (dotted line).

Joining the midpoints of faces of a cuboctahedron actually gives a triakis octahedron (Fig. 3.13*b*) when lines are added in the valleys between the trigonal pyramids. These valleys outline an octahedron, and the prefix *triakis* means that each octahedral face has been turned into three faces. The triakis octahedron is the dual of the truncated cube, and is most easily envisaged as the product of *face stellation* of the octahedron.

To stellate a face symmetrically (Fig. 3.9*e*), pull out its midpoint away from the center of a polyhedron. The original face is now replaced by a pyramid with as many faces as there were edges around the original face. Thus, a square face turns into a square pyramid, and the adjective *tetrakis* would be used. This operation of face stellation develops the following semi-regular solids: triakis octahedron (Fig. 3.13*c*), triakis tetrahedron (Fig. 3.13*d*), tetrakis hexahedron (3.13*e*) and hexakis octahedron (3.13*f*). Face stellation is also known as *capping,* when only one face or some faces are stellated.

For *edge stellation* (Fig. 3.9*f*), the midpoint of an original face remains fixed as the midpoint of each edge is raised. A re-entrant angle appears at each face diagonal. Face stellation can occur simultaneously with edge stellation. If there is a special relation between the distances moved by the

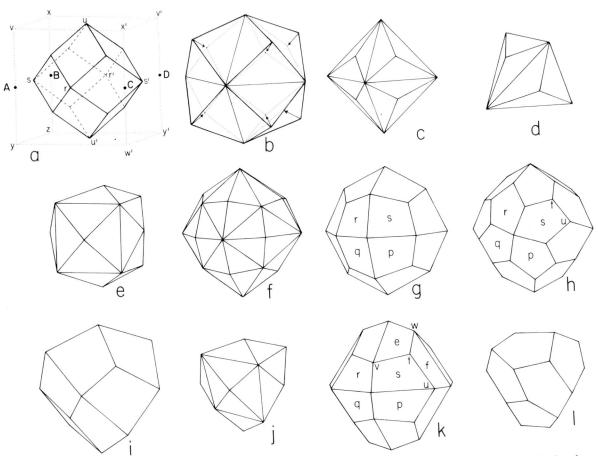

FIGURE 3.13 Duals of Archimedean polyhedra. In these clinographic projections, the rhombic dodecahedron *a* has a unique shape, but all others except *b* have a shape chosen to give a good perspective and to obey a particular set of Miller indices for the faces of an isometric crystal (Chapter 6). (*b*) Triakis octahedron (heavy line) with vertices at centroids of faces of a cuboctahedron. Arrows show conversion into the rhombic dodecahedron. (*c*) Triakis octahedron with Miller indices {**332**}. (*d*) Triakis tetrahedron {**211**}. (*e*) Tetrakis hexahedron {**210**}. (*f*) Hexakis octahedron {**321**}. (*g*) Trapezoidal icositetrahedron {**211**}. (*h*) Pentagonal icositetrahedron {**321**}. (*i*) Trapezoidal dodecahedron {**221**}. (*j*) Hexakis tetrahedron {**321**}. (*k*) Dyakis dodecahedron {**321**}. (*l*) Tetrahedral pentagonal dodecahedron {**321**}.

midpoints of a face and its edges, the re-entrant angles disappear. Thus in Figure 3.9*g*, the original square face is replaced by four kite-shaped quadrilaterals that meet to form half of a trapezohedron. This special relation is found (Fig. 3.13*g*) in the trapezoidal icositetrahedron (i.e., 24 trapezoidal faces), which results from special edge and face stellation of either the cube or octahedron. It also occurs (3.13*i*) in the trapezoidal dodecahedron (i.e., 12 trapezoidal faces), which is similarly developed from a tetrahedron.

When such a special relation does not occur, combined edge and face stella-tion looks like Figure 3.9*f* in projection onto the face, but the midpoint of the face has moved. The hexakis tetrahedron (3.13*j*) is thereby produced from a tetrahedron, and the hexakis octahedron (3.13*f*) from an octahedron. The hexakis octahedron could also be called the octakis hexahedron because it can be developed from the cube by combined edge and face stellation. Furthermore, the hexakis octahedron can be obtained by face stellation of a rhombic dodecahedron. Crystallographers commonly shorten names by re-moving the *akis* or *aki* or *kis* to give ***tristetrahedron, tetrahexahedron, hex-octahedron*** and ***hextetrahedron.***

The pentagonal icositetrahedron (i.e., 24 pentagons) is the dual (Fig. 3.13*h*) of the snub cube, and is obtained from the trapezoidal icositetra-hedron (3.13*g*) by rotating a group of four faces such as *p, q, r,* and *s* about a 4-fold axis, together with similar rotations for equivalent groups of four faces. New edges like *tu* are produced, thereby turning each kite-shaped quadrilateral into a pentagon. The ***dyakis dodecahedron*** (3.13*k*) is derived from the trapezoidal icositetrahedron by rotating *e, f,* and *s* about the triad axis with slight lengthening of the edges *vt, tu,* and *tw* so that faces *p, q, r,* and *s* are still related by two mirror planes. Another way of obtaining the dyakis dodecahedron, which is also called the ***didodecahedron*** or ***diploid,*** is by asymmetric edge and face stellation of an octahedron.

The ***trapezoidal dodecahedron*** (Fig. 3.13*i*), also called a ***deltoid dodecahedron*** or a ***deltohedron,*** is a distortion of the rhombic dodeca-hedron, whereas the ***tetrahedral pentagonal dodecahedron*** (3.13*l*), also called a ***tetartoid,*** can be regarded as a distortion of the tristetrahedron.

The hierarchy of polyhedra with increasing distortion and decreasing symmetry is developed further in Chapter 6.

3.7 Space-Filling Polyhedra

Probably all readers stacked cubic blocks together during childhood, and know that a honeycomb contains hexagonal prisms. How many other polyhedra can fill space by themselves?

In 2D, any parallelogram can fill space by regular translation without change of orientation (Chapter 2), and the equivalent in 3D is a ***par-allelepiped*** whose six faces are parallelograms. The geometry of the paral-lelepiped depends on the symmetry of the packing (Fig. 3.14). When the parallelepiped is a cube, there are tetrad rotation axes perpendicular to the square faces, triads along body diagonals, diads joining midpoints of op-posing edges, and two sets of mirror planes. When the parallelepiped is a tetragonal prism, there is a tetrad axis along the unique axis. A rectangular parallelepiped with unequal perpendicular edges (sometimes known as a

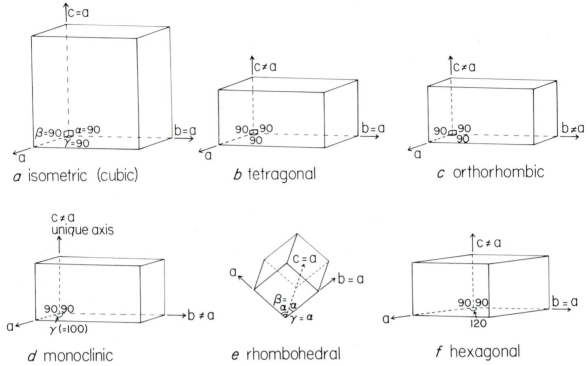

FIGURE 3.14 Clinographic projections of space-filling parallelepipeds. (*a*) Cube. (*b*) Tetragonal prism. (*c*) Rectangular parallelepiped. (*d*) Monoclinic parallelepiped. (*e*) Rhombohedron. (*f*) 120° (hexagonal) parallelepiped. Axes *a, b,* and *c* are placed along three intersecting edges, and angles α, β, and γ are conventionally assigned so that α is between the *b* and *c* axes, β between *c* and *a,* and γ between *a* and *b*. Each parallelepiped is used as a unit cell for a crystal system or subsystem with the listed name. The triclinic parallelepiped with *a* ≠ *b* ≠ *c* and α ≠ β ≠ γ is not shown. Note the spatial distortion caused by the clinographic projection: thus the top and bottom faces of the tetragonal parallelepiped are squares.

cuboid) has three perpendicular diad axes. A monoclinic parallelepiped has one oblique and two 90° angles between the edges, as is required by one diad axis. Three nonequivalent oblique edges are found in a triclinic parallelepiped that has no rotation symmetry, except for the formal monad symmetry. Two more special shapes are associated with 3-fold and 6-fold symmetry. A ***rhombohedron,*** which has three equal edges and three equal oblique angles, is consistent with a single triad axis down the unique body diagonal. A hexagonal prism will fill space, and can be decomposed into three prisms with 120° rhombuses on the top and bottom, and identical rectangles as the four side faces. Strictly speaking, the horizontal edges of close-packed 120° prisms do not obey 3-fold rotation symmetry, but the vertices obey 6-fold rotation symmetry. Just as in 2D, the above parallelepipeds will be used as unit cells for 3D crystal structures, and the symmetry of the polyhedron will be used to classify crystals into *systems.*

All these parallelepipeds can be derived from the cube by homogeneous deformation. Although homogeneous deformation is easy to understand qualitatively, it requires a matrix for rigorous mathematical description (*affine transformation*). The simplest statements are that straight lines remain straight, that parallel lines remain parallel, and that the amount of deformation depends on the orientation of the line. Thus, a sphere can transform into an ellipsoid but not into an egg-shaped body. The rhombohedron results from extension or compression along a body diagonal. The tetragonal prism results from extension or compression along a tetrad axis, and so on.

Can other types of polyhedra, which are not parallelepipeds, pack together to completely fill space by simple translation without change of orientation? Yes. The space-filling polyhedra are known as *parallelohedra,* or *Fedorov solids,* after their enumerator. In addition to ones based on the cube and the hexagonal prism, there are others based on the cuboctahedron, the rhombic dodecahedron, and the *elongated rhombic dodecahedron,* making five in all (Fig. 3.15). For each prototype, an infinite series of polyhedra can be developed by homogeneous deformation. Those polyhedra that result from deformation of the cube have already been listed, and they are infinite in number because the ratios of the edge lengths can change continuously. The hexagonal prism can have any ratio of vertical to horizontal dimension. It can be tilted like the leaning tower of Pisa so long as the parallel edges remain parallel. Packing of the truncated octahedron is particularly interesting because it is the basis of the silicate framework of the sodalite family of minerals (Section 8.6.7). The silicon and aluminum atoms lie at the vertices, and oxygen atoms lie near the midpoints of the edges. The elongated rhombic dodecahedron is actually a nonspecial combination of a tetragonal prism and a tetragonal dipyramid, and the truncated octahedron can be regarded as a special combination of a cube and an octahedron.

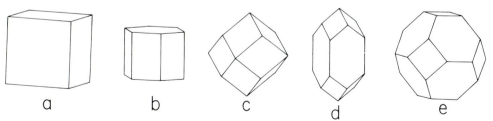

FIGURE 3.15 Clinographic projections of Fedorov space-filling polyhedra. (*a*) Cube. (*b*) Hexagonal prism. (*c*) Rhombic dodecahedron. (*d*) Elongated rhombic dodecahedron. (*e*) Truncated octahedron.

3.8   Summary of Properties of Regular and Semi-regular Polyhedra

Table 3.1 summarizes the properties of the regular and semi-regular polyhedra. The *face symbol* lists the type of polygon (triangle 3, quadrilateral 4, pentagon 5, etc.) and the number of polygons as a superscript. Thus, the cuboctahedron has a face symbol of $(3^8 4^6)$, using round brackets. The Schläfli symbol specifies the sequence of polygons around the vertex. The number of polygons meeting at a vertex must equal the number of edges meeting at that vertex (i.e., the valency v). Often the *vertex symbol* is sufficient. This lists the valency of the vertex, and the number of vertices of that

Table 3.1   Properties of regular and semiregular polyhedra.

Name	Face Symbol	Vertex Symbol	Schläfli Symbol	Face Angle(s)	Dihedral Angle(s)	Subtended Angle
Platonic polyhedra						
Tetrahedron	3^4	3^4	3.3.3	60°	70.53°	109.47°
Cube	4^6	3^8	4.4.4	90°	90°	70.53°
Octahedron	3^8	4^6	3.3.3.3	60°	109.47°	90°
Dodecahedron	5^{12}	3^{20}	5.5.5	108°	116.57°	41.82°
Icosahedron	3^{20}	5^{12}	3.3.3.3.3	60°	138.18°	63.43°
Archimedean polyhedra						
Truncated tetrahedron	$3^4 6^4$	3^{12}	3.6.6	60°:120°	70.53°:109.47°	50.48°
Truncated cube	$3^8 8^6$	3^{24}	3.8.8	60°:135°	90°:125.26°	32.65°
Truncated octahedron	$4^6 6^8$	3^{24}	4.6.6	90°:120°	109.47°:125.27°	36.87°
Truncated dodecahedron	$3^{20} 10^{12}$	3^{60}	3.10.10	60°:144°	116.57°:142.62°	19.40°
Truncated icosahedron	$5^{12} 6^{20}$	3^{60}	5.6.6	108°:120°	138.19°:142.62°	23.28°
Cuboctahedron	$3^8 4^6$	4^{12}	3.4.3.4	60°:90°	125.26°	60°
Icosidodecahedron	$3^{20} 5^{12}$	4^{30}	3.5.3.5	60°:108°	142.62°	36°
Small rhombicuboctahedron	$3^8 4^{18}$	4^{24}	3.4.4.4	90°:120°	135°:144.74°	41.89°
Great rhombicuboctahedron	$4^{12} 6^8 8^6$	3^{48}	4.6.8	90°:120°:135°	125.26°:135°:144.74°	24.92°
Small rhombicosidodecahedron	$3^{20} 4^{30} 5^{12}$	4^{60}	3.4.5.4	60°:90°:108°	148.28°:159.09°	25.87°
Great rhombicosidodecahedron	$4^{30} 6^{20} 10^{12}$	3^{120}	4.6.10	90°:120°:144°	142.62°:148.28°:159.09°	15.11°
Snub cube (d and î types)	$3^{32} 4^6$	5^{24}	3.3.3.3.4	60°:90°	142.98°:153.23°	43.67°
Snub dodecahedron (d and î)	$3^{80} 5^{12}$	5^{60}	3.3.3.3.5	60°:108°	152.93°:164.18°	26.82°
Archimedean prism	$4^n n^2$	3^{2n}	4.4.n	90°:(180-360/n)	90°:180-360/n	$\tan^{-1}[\sin(180/n)]$
Archimedean antiprism	$3^{2n} n^2$	4^{2n}	3.3.3.n	60°:(180-360/n)	–	$\tan^{-1}[2\sin(90/n)]$
Catalan polyhedra						
Triakis tetrahedron	3^{12}	$3^4 6^4$	3^3 and 3^6	33.56°:112.89°	129.52°	70.53°:109.47°
Triakis octahedron	3^{24}	$3^8 8^6$	3^3 and 3^8	31.37°:117.25°	147.35°	54.74°:90°
Tetrakis hexahedron	3^{24}	$4^6 6^8$	3^4 and 3^6	48.19°:83.62°	143.13°	54.73°:70.53°
Triakis icosahedron	3^{60}	$3^{20} 10^{12}$	3^3 and 3^{10}	30.47°:119.05°	160.60°	37.38°:63.43°
Pentakis dodecahedron	3^{60}	$5^{12} 6^{20}$	3^5 and 3^6	55.69°:68.62°	156.72°	37.38°:41.81°
Rhombic dodecahedron	4^{12}	$3^8 4^6$	4^3 and 4^4	70.53°:109.47°	120°	54.74°
Rhombic triacontahedron	4^{30}	$3^{20} 5^{12}$	3^3 and 4^5	63.43°:116.57°	144°	37.38°
Trapezoidal icositetrahedron	4^{24}	$3^8 4^{18}$	4^3 and 4^4	81.58°:115.26°	138.11°	35.27°:45°
Hexakis octahedron	3^{48}	$4^{12} 6^8 8^6$	$3^4, 3^6$ and 3^8	37.77°:55.02°:87.20°	155.08°	35.27°:45°:54.73°
Trapezoidal hexecontahedron	4^{60}	$3^{20} 4^{30} 5^{12}$	$4^3, 4^4$ and 4^5	67.78°:86.97°:118.27°	154.13°	20.91°:31.72°
Hexakis icosahedron	3^{120}	$4^{30} 6^{20} 10^{12}$	$3^4, 3^6$ and 3^{10}	32.77°:58.24°:88.99°	164.89°	20.91°:31.72°:37.38°
Pentagonal icositetrahedron	5^{24}	$3^{32} 4^6$	5^3 and 5^4	80.75°:114.81°	136.33°	26.77°:37.02°
Pentagonal hexecontahedron	5^{60}	$3^{80} 5^{12}$	5^3 and 5^5	67.47°:118.13°	153.18°	15.82°:27.07°
Catalan bipyramid	3^{2n}	$4^n n^2$	3^4 and 3^n	$\cos^{-1}[\sin^2(\pi/n)]$	–	–
Catalan trapezohedron	4^{2n}	$3^{2n} n^2$	4^3 and 4^n	$\cos^{-1}[0.5-\cos(\pi/n)]$	–	–

type as a superscript. Thus the rhombic dodecahedron has a vertex symbol of $[3^84^6]$, using square brackets. A dual has a face symbol that matches the vertex symbol of its antidual: thus, the face symbol of the cuboctahedron matches the vertex symbol of its dual, the rhombic dodecahedron, and vice versa. The number of edges is obtained by adding the superscripts in the face and vertex symbols and subtracting 2 as in Euler's relation. Each polygon can be specified by its *face angles* (e.g., three 60° angles for an equilateral triangle). Two faces intersect at the ***dihedral angle,*** which is expressed as the supplement of the angle between the face normals. The two vertices at the end of each edge form a triangle with the centroid of the polyhedron. The angle at the centroid is called the ***subtended angle.***

All geometrical properties of the regular solids are fixed, but those of some of the semiregular solids are not. Thus for a triakis tetrahedron, the degree of stellation can change with consequent change of shape of the isosceles triangles. A unique geometry is obtained for the semiregular solids by placing all vertices on the surface of a sphere. Further details for Archimedean and Catalan polyhedra are given in Cundy and Rollett (1961).

3.9   Other Polyhedra Useful to Crystallographers

 Systematic enumeration of topologically different *convex* polyhedra yielded the following number $N(n)$ of polyhedra with n vertices: $N(4) = 1, N(5) = 2, N(6) = 7, N(7) = 34, N(8) = 257$ (Britton and Dunitz, 1973). Although one of the polyhedra of interest to crystallographers has 28 vertices, it is obvious that systematic enumeration of all polyhedra with $N < 28$ is impractical. Fortunately, most useful polyhedra have considerable symmetry, and can be classed into simple groups.

Systematic enumeration begins with the tetrahedron, which is the only polyhedron with four vertices. Capping one of the faces generates the first of the two polyhedra with five vertices. This ***monocapped tetrahedron*** becomes a trigonal bipyramid if all the triangles are congruent and isosceles. A polyhedron composed only of triangles, no matter what shape, is called a ***triangulated polyhedron.*** If the triangles are equilateral, the polyhedron is placed in the group of ***deltahedra*** (Fig. 3.16), of which only eight representatives exist. The regular tetrahedron and trigonal bipyramid are the first two representatives of the deltahedra. Returning to the topology of the monocapped tetrahedron, whose face and vertex symbols are $[3^6](3^24^3)$, removal of one edge converts two adjacent triangular faces into a quadrilateral to give the second topologic type of polyhedron with five vertices. This polyhedron with symbols $[3^44^1](3^44^1)$ is a self-dual, and is a square pyramid in its most symmetric shape (Fig. 3.17a).

Systematic enumeration continues by discovery for each n of all polyhedra composed entirely of triangular faces (not necessarily equilateral).

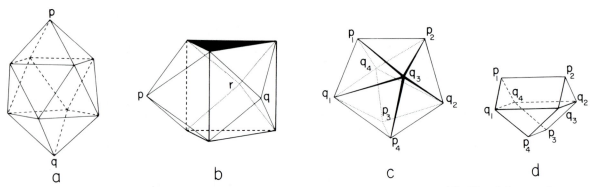

FIGURE 3.16 Three convex deltahedra (*a–c*) and the disphenoidal octahedron (*d*). The 4.4-capped square antiprism (*a*) is constructed from Figure 3.11*d* by stellating the two square faces to give vertices *p* and *q*. The 4.4.4-capped trigonal prism (*b*) is constructed from the Archimedean trigonal prism by stellating the three square faces to give vertices *p*, *q*, and *r*. The triangulated dodecahedron (*c*) is also known as the bisdisphenoid because its eight vertices lie at the vertices of two sphenoids (denoted by *p* and *q*, respectively). The vertices of the disphenoidal octahedron are labeled to show the relation to the triangulated dodecahedron. Diagram *a* is an accurate clinographic projection, whereas diagrams *b* to *d* are slightly distorted.

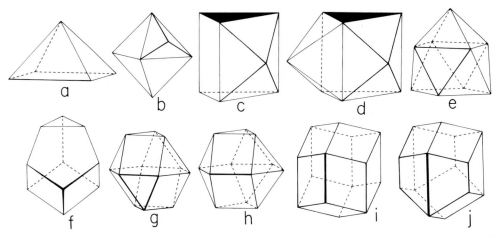

FIGURE 3.17 Ten polyhedra: (*a*) Square pyramid. (*b*) Monocapped octahedron. (*c*) Monocapped trigonal prism. (*d*) Bicapped trigonal prism. (*e*) Monocapped square antiprism composed of 1 square and 12 equilateral triangles. (*f*) Dual of the monocapped octahedron oriented with the triad axis vertical. (*g*) Cuboctahedron oriented with one of the four triad axes vertical. (*h*) "Twinned cuboctahedron" oriented with the triad axis vertical and a mirror plane horizontal. (*i*) Dual of the cuboctahedron in *g*. This is a rhombic dodecahedron oriented with one of the four triad axes vertical. (*j*) Dual of the "twinned cuboctahedron." Note how the vertical rhombuses of the rhombic dodecahedron have turned into trapeziums.

Edges are then removed one at a time to produce a quadrilateral face, a pentagonal face, or two quadrilateral faces, and so on, until all possible polygons are used. Of the seven polyhedra with six vertices, the octahedron (the third deltahedron when fully regular), the pentagonal pyramid $[3^5 5^1]$ $(3^5 5^1)$, and the trigonal prism $[3^2 4^3](3^6)$ have the highest symmetry. For seven vertices, the hexagonal pyramid $[3^6 6^1](3^6 6^1)$, the pentagonal bipyramid $[3^{10}](4^5 5^2)$, the *monocapped octahedron* $[3^{10}](3^{14} 3^5 3^3)$ in Figure 3.17b, and the *monocapped trigonal prism* $[3^6 4^2](3^2 4^5)$ in Figure 3.17c are the most interesting. The pentagonal bipyramid is the fourth deltahedron when its faces are equilateral.

For eight vertices, the cube, square antiprism, *bicapped trigonal prism* $[3^{10} 4^1](4^6 5^2)$ in Figure 3.17d, *triangulated dodecahedron* $[3^{12}](4^4 5^4)$ in Figure 3.16c, and *disphenoidal octahedron* $[3^4 4^4](3^4 4^4)$ in Figure 3.16d are useful to crystallographers. The triangulated dodecahedron is the fifth deltahedron. Vertices p_1 to p_4 lie at the vertices of a tetrahedron pulled out along a diad axis, and vertices q_1 to q_4 form a compressed tetrahedron. Such distorted tetrahedra are known as *tetragonal sphenoids* (Chapter 6), and an alternate name for the triangulated dodecahedron is *bisdisphenoid*. The triangulated dodecahedron can be turned into the *disphenoidal octahedron* by removal of edges $p_1 q_3$, $p_2 q_4$, $p_3 q_1$, and $p_4 q_2$, and by conversion of $q_1 q_3 q_2 q_4$ into a square.

For nine vertices, the *tricapped trigonal prism* $[3^{14}](4^3 5^6)$ in Figure 3.16b is the sixth deltahedron when the faces are equilateral. Also present is the *monocapped square antiprism* in Figure 3.17e. The capping is always assumed to be on the face or faces with the most edges (e.g., square in preference to triangle).

For 10 vertices, the *bicapped square antiprism* $[3^{16}](4^2 5^8)$ in Figure 3.16a gives the seventh deltahedron when the faces are equilateral. The pentagonal antiprism $[3^{10} 5^2](4^{10})$ also has ten vertices, as does the dual of the monocapped octahedron $[3^{14} 4^3 5^3](3^{10})$ in Figure 3.17f. An ugly 18-hedron with 11 vertices (unillustrated) occurs in borane anions.

The regular icosahedron $[3^{20}](5^{12})$ with 12 vertices completes the set of eight deltahedra. Also useful are the hexagonal antiprism $[3^{12} 6^2](4^{12})$, the cuboctahedron $[3^8 4^6](4^{12})$, and the anticuboctahedron, or *"twinned cuboctahedron,"* also $[3^8 4^6](4^{12})$. The latter two polyhedra differ only in the linkage across the waist (Figs. 3.17g, 3.17h). Also interesting is the unnamed octahedron $[4^4 5^4](3^{12})$ in Figure 3.18a, which is the dual of the triangulated dodecahedron. It can be obtained from the cube by splitting two opposing faces into pairs of squares rotated in projection by 90°, and can be converted into the disphenoidal octahedron by merging A to B, C to D, E to F, and G to H.

The dual of the tricapped trigonal prism, $[4^3 5^6](3^{14})$ in Figure 3.18b,

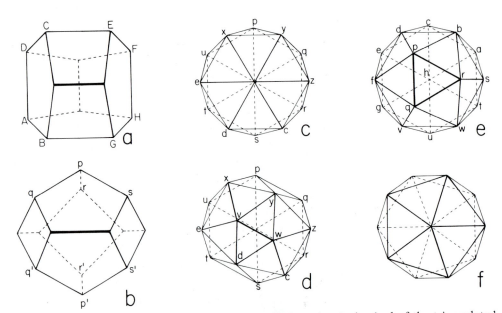

FIGURE 3.18   Six polyhedra. (*a*) Plan of octahedron that is the dual of the triangulated dodecahedron in Figure 3.16*c*. It cannot be constructed from squares and regular pentagons. Glue together four pentagons and then measure the gaps for linking trapeziums. (*b*) Elevation of 14-hedron that is the dual of the tricapped trigonal prism. The most symmetric shape has a triad axis joining vertices *p* and *p'*, and relating vertices *q, r*, and *s*. A mirror plane relates primed and unprimed vertices. (*c*) Plan down the hexad axis of a bicapped hexagonal antiprism in its regular configuration. To make a cardboard model, construct a hexagonal antiprism from 2 hexagons and 12 equilateral triangles. Cap each hexagon with 6 isosceles triangles. All vertices lie on the surface of a sphere when each isosceles triangle has angles 48.26° and 65.87° (twice). In the plan, each isosceles triangle is foreshortened to look like an equilateral triangle, and each equilateral triangle from the waist of the antiprism is foreshortened to an isosceles triangle! (*d*) Plan of unnamed 26-hedron, which can be constructed as follows. Using 12 equilateral triangles, construct the "waist" of an hexagonal antiprism. Using 6 isosceles triangles (48.26°, 65.87° twice), cap the base of the antiprism. Instead of capping the hexagonal top of the antiprism with 6 isosceles triangles, it is necessary to use 8 triangles that meet to give 2 new vertices *v* and *w*. To accommodate these vertices, vertices *d* and *y* are pulled inward with respect to the positions for the bicapped hexagonal antiprism. A model can be completed by using equilateral triangles for *exv* and *cwz*, isosceles triangles (51½°, 64¼° twice) for *dvw* and *yvw*, and isosceles triangles (70°, 55° twice) for *evd, yxv, zwy* and *dcw*. These angles are not uniquely fixed. (*e*) Tetracapped truncated tetrahedron viewed in plan down a triad axis. First, make a truncated tetrahedron from four hexagons and four equilateral triangles. Each hexagonal face is then capped with six isosceles triangles. When each such triangle has angles 51.32° and 64.34° (twice), all vertices lie on a sphere. Triangles *pqr, cde, ast*, and *guv* are the equilateral triangles of the truncated tetrahedron. Vertices *b, f, h*, and *w* are new vertices produced by capping hexagonal faces *cdprsa, edpqvg, catuge*, and *stuvqr*. (*f*) Regular icosahedron viewed in plan down a pentad axis. Note that it is the regular version of a bicapped pentagonal antiprism.

has 14 vertices. More symmetrical are the triakis octahedron, the tetrakis hexahedron, the rhombic dodecahedron $[4^{12}](3^8 4^6)$, which is the dual (Fig. 3.17*i*) of the cuboctahedron, and an unnamed 14-hedron $[4^{12}](3^8 4^6)$, which is the dual (3.17*j*) of the "twinned cuboctahedron."

The *bicapped hexagonal antiprism* $[3^{24}](5^{12} 6^2)$ in Figure 3.18*c* can have equilateral triangles around the waist, but must have isosceles triangles in the caps. All 14 vertices will lie on the surface of a sphere when the isosceles triangles have angles of 48.26° and 65.87° (twice). Less symmetrical, but still needed by crystallographers, is the unnamed 26-hedron $[3^{26}](5^{12} 6^3)$ in Figure 3.18*d,* which is constructed by first capping the lower hexagon of a hexagonal antiprism, and then "roofing" the upper hexagon with a ridge *vw*. The *tetracapped truncated tetrahedron* $[3^{28}](5^{12} 6^4)$ is called a *Friauf polyhedron* (Fig. 3.18*e*). All four hexagonal faces are capped with isosceles triangles, whose angles must be 51.32° and 64.34° (twice) for all vertices to lie on a sphere; the original triangles remain equilateral.

Close-packed atoms in crystal structures lie at the vertices of triangulated polyhedra. The tetrahedron and octahedron (Fig. 3.1) are so common that they can be used for classification of crystal structures. In complex alloys with atoms of different sizes, the regular icosahedron (Fig. 3.18*f*) and the three polyhedra of the preceding paragraph are important.

Each triangulated polyhedron is reciprocal to a **trivalent polyhedron**. In a clathrate compound, the cage consists of an infinite array of atoms each bonded to four neighbors. The chemical bonds can be represented geometrically by the edges of space-filling trivalent polyhedra. The dual (Fig. 3.19*a*) of the bicapped square antiprism has the symbol $[4^2 5^8](3^{16})$. An 11-hedron (3.19*b*) has the symbol $[4^2 5^8 6^1](3^{18})$. The regular (pentagonal) dodecahedron $[5^{12}](3^{20})$ is also trivalent. Three 14-hedra are trivalent. The polyhedron with symbol $[5^{12} 6^2](3^{24})$ consists of opposing hexagons linked by a waist of 12 pentagons alternatively pointing up and down (3.19*c*); it is usually called the *tetrakaidecahedron* in papers on clathrates.

This 14-hedron will not fill space by itself, but three other 14-hedra will do so. The truncated octahedron $[4^6 6^8](3^{14})$ will fill space without change of orientation, and is a parallelohedron (Fig. 3.19*d*). The two other 14-hedra, $[4^4 5^4 6^6](3^{13})$ and $[4^2 5^8 6^4](3^{14})$, will fill space only by changing the orientation, but not the shape, of adjacent polyhedra; they are *space-filling solids* but are not parallelohedra. Figures 3.19*e* and 3.19*f* show how the truncated octahedron can be transformed into the $[4^2 5^8 6^4]$ type of 14-hedron by way of the $[4^4 5^4 6^6]$ type. See A.F. Wells, *Three-Dimensional Nets and Polyhedra* (1977), pages 139 to 141, for further details.

A 15-hedron $[5^{12} 6^3](3^{26})$ has 3-fold rotational symmetry (3.19*g*), a 16-hedron $[5^{12} 6^4](3^{28})$ has isometric symmetry with four intersecting triad axes (3.19*h*), a 17-hedron $[4^3 5^9 6^2 7^3](3^{28})$ is unusual because of the presence of heptagons (3.19*i*), and an 18-hedron $[5^{12} 6^6](3^{32})$ is obtained by

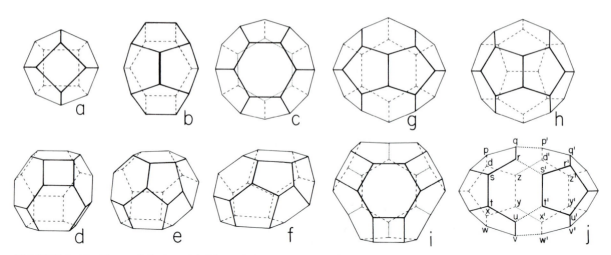

FIGURE 3.19   Ten polyhedra: (*a*) Plan of the dual of the bicapped square antiprism. A rotation tetrad and vertical mirror planes are present. (*b*) Plan of an 11-hedron. (*c*) Plan of a 14-hedron down the hexad axis. Vertical mirror planes are also visible. (*d*) Perspective drawing of truncated octahedron with a triad axis lying up–down in the plane of the paper. (*e, f*) Corresponding views of the $[4^2 5^8 6^4]$ and $[4^4 5^4 6^6]$ types of 14-hedra showing how the truncated octahedron can be progressively modified. (*g*) Plan of a 15-hedron. A triad axis lies east–west in the plane of the paper. Note how most faces match those of the 14-hedron in *c*. (*h*) Plan of a 16-hedron with isometric symmetry. A triad axis passes through the center of each hexagon and a vertex between three pentagons. Note how most faces match those of the 15-hedron in *g* to the left. (*i*) Plan of a 17-hedron with a vertical triad axis. Two edges of each heptagon and three pentagons are almost collinear in projection, but the intervening vertex can be located from the intersection with the dotted line. (*j*) Plan of an 18-hedron. Removal of the edges marked with dots together with some flexing allows primed and unprimed vertices to merge. The resulting figure is a regular (pentagonal) dodecahedron viewed with a triad axis east–west in the plane of the paper. Some polygons are nonplanar in the drawing, which represents a star-tubing model with equal distances between adjacent vertices. To obtain flat polygons it is necessary to have unequal edges.

splitting a regular dodecahedron along a zigzag waist of 12 edges, and inserting 6 new edges to give a new waist of six hexagons (3.19*j*).

Trivalent polyhedra can be constructed rapidly using three spikes of tetrahedral plastic stars and plastic tubing. Curvature in the tubing takes care of any angular misfits. Polyhedra with higher valency can be constructed by choice of appropriate spikes from multivalent plastic stars.

The following papers illustrate the use of polyhedra by crystal chemists: Evans (1971), Frank and Casper (1958, 1959), Jeffrey and McMullan (1967), King (1969, 1970, 1972), King and Rouvray (1977), Nyman and Andersson (1979), Samson (1958, 1965, 1968), Wade (1976), Chabot et al. (1981), Hellner and Koch (1981), and Nyman and Hyde (1981).

EXERCISES   (* More Difficult)

3.1 Construct the five Platonic solids from templates in Figures 3.2*f*–*j*. Further exercises will rely on accurate construction of the regular tetrahedron, cube, and regular octahedron. Manila folders are recommended, or an equivalent grade of thin cardboard. Because polyhedra will be packed together in later exercises, all students should use the same scale. An edge length of 3 cm is recommended. College instructors may wish to make large models out of stiff cardboard. Internal struts are desirable for large models. An accuracy of better than 0.2 mm is needed for good closure of models.

3.2 Count the number of faces, vertices, and edges of the Platonic solids. Make a table, and check that Euler's relation is obeyed.

3.3 Place four, six, or eight spheres in contact at the vertices of a tetrahedron, octahedron, or cube. Table-tennis or lawn-tennis balls are convenient. Enclose three spheres by books, and show that a fourth sphere will fit neatly into the dimple, thereby giving a tetrahedron. Remove the upper sphere, and hold three spheres above the first three to give an octahedron. Enclose four spheres by books to give a square pattern. Hold four other spheres directly above to give the cubic arrangement. If the top spheres are not held in position, one may drop into the dimple above the bottom four spheres. These five spheres lie at the vertices of a square pyramid, which is half of an octahedron. College instructors may wish to build permanent models from large polystyrene balls used for Christmas decorations.

3.4* Glue together balls lying at the vertices of a regular icosahedron to provide a model for atomic clusters found in alloys or globular clusters in viruses. [**Hint.** glue together several groups of three balls. After the glue has dried, connect with single balls. Hold balls in position with modeling clay or books.]

3.5 Place the appropriate polyhedron over each diagram in Figure 3.3, and identify corresponding vertices. Close the book and reconstruct Figure 3.3 using the polyhedral models as a direct guide. Then put away the models, and try to imagine them floating in space. Reconstruct Figure 3.3 again without looking at models. Such a visual feat is very difficult at first, but many crystallographers achieve this ability after prolonged practice. Do not despair if you cannot do it. These exercises are designed to give you plenty of opportunity to develop your ability. Ultimate success will be very rewarding.

3.6 Construct axes for a clinographic projection (Fig. 3.4). Construct clinographic projections of the tetrahedron, cube, and octahedron (Fig. 3.5).

3.7 Make Schlegel diagrams of the five Platonic solids, and check with Figure 3.6. Only the relative positions of vertices matter, and you may get different shapes of the polygons.

3.8 Mark each center of the cube with a dot. Imagine that the dots are joined to give a regular octahedron (Fig. 3.7a). Repeat with an octahedron, and construct a clinographic projection in which dots at the center of the faces of an octahedron are joined to give a cube. [**Hint.** to obtain the center of a triangle, join each vertex to the midpoint of an opposing edge, and take the intersection of the three lines.] Repeat for a tetrahedron in the negative orientation to obtain a tetrahedron in a positive orientation.

3.9 Check the Euler relation for an octahedron by stripping away the Schlegel diagram as in Figures 3.6*g* to 3.6*l* for the cube.

3.10 Draw faint pencil lines on the surface of models of the Platonic solids to show how semiregular solids can be produced by symmetrical truncation of the vertices (Fig. 3.8).

3.11 Construct clinographic projections to illustrate truncation of the tetrahedron, cube, and octahedron. Check against Figure 3.8.

3.12* Draw Schlegel diagrams to illustrate vertex truncation of Platonic solids.

3.13 Make models of at least two Archimedean polyhedra using regular polygons cut from thin cardboard. Students in a class could produce a communal set. A single reader might wish to build the truncated octahedron, the truncated tetrahedron and the cuboctahedron, and ignore the more complex models. *The two enantiomorphic varieties of the snub cube are instructive, but time-consuming to build.

3.14 Construct models of the Archimedean trigonal prism, hexagonal prism, and square antiprism, using the templates in Figure 3.12. Show that eight spheres pack more closely at the vertices of an Archimedean square antiprism than for a cube.

3.15 Construct models of the trigonal bipyramid, hexagonal bipyramid, tetragonal trapezohedron, and hexagonal trapezohedron using the templates in Figure 3.12.

3.16 Construct a clinographic projection of a rhombic dodecahedron as in Figure 3.13*a.*

3.17** Construct models of Catalan polyhedra using cardboard polygons with face angles from Table 3.1.

3.18 Construct models of the seven space-filling parallelepipeds (Fig. 3.14). The tetragonal prism requires two congruent squares (suggested size 3 cm) and four congruent rectangles (suggested size 2 X 3 cm). Tetragonal prisms can be assembled together to fill space in *parallel* orientation. The orthorhombic parallelepiped requires two each of three rectangles. Suggested sizes are 2 × 3, 3 × 4 and 2 × 4 cm. The monoclinic parallelepiped requires two each of two rectangles (suggested sizes 2 × 3 and 3 × 4 cm) and two parallelograms (suggested size 2 × 4 cm with arbitrary angles of 70° and 110°). The rhombohedron is constructed from six rhombuses (suggested size 3 × 3 cm with arbitrary angles of 75° and 105°). The hexagonal prism is constructed from two rhombuses (suggested size 3 × 3 cm with mandatory angles of 60° and 120°) and four rectangles (suggested size 3 × 2 cm). A triclinic parallelepiped (not illustrated) can be constructed from two each of three parallelograms of arbitrary shape except for matching of edge lengths (suggested size 2 × 3 cm and 65°, 115° angles; 2 × 4 cm and 85°, 95° angles; 3 × 4 cm and 75°, 105° angles).

3.19* Show that models of the five Fedorov solids (Fig. 3.15) will stack together in parallel orientation. Models of the cube and hexagonal prism should have been built earlier. The truncated octahedron may have been built in Exercise 3.13*. If not, build it now. The rhombic dodecahedron can be built from twelve rhombuses with equal edges (use 3 cm) and angles of 70.53° and 109.47°. Note that four angles of 70.53° meet at one type of vertex, and three angles of 109.47° at the second type. To construct the elongated rhombic dodecahedron, assemble the four upper faces using rhombuses like those in the rhombic dodecahedron. Do the same for the four lower faces. Cut out four hexagons with angles 109.47°, 125.26°, 125.26°, 109.47°, 125.26°, and 125.26°. Edges next to each 109.47° angle must be 3 cm, but those between the 125.26° angles are arbitrary (suggested value is 4 cm).

3.20* Construct the eight deltahedra from equilateral triangles of edge length 3 cm. Construction of the tetrahedron, trigonal bipyramid, octahedron, and pentagonal bipyramid is straightforward. The triangulated dodecahedron (Fig. 3.16c) can be constructed easily as follows: assemble the pentagonal pyramid with apex at q_3; repeat for the pentagonal pyramid with apex at q_4; fasten together at edges $q_1 p_1$, $p_1 p_2$, and $p_2 q_2$; fasten together two triangles at edge $p_3 p_4$; complete the model by inserting this double triangle between free edges of the pentagonal pyramids. To make the tricapped trigonal prism (Fig. 3.16b), construct a trigonal prism from two triangles and three squares; then make three square pyramids, and fasten one to each square face of the trigonal prism. To construct the bicapped square antiprism (Fig. 3.16a), assemble a square antiprism from two squares and eight triangles, and then fit square pyramids to each square. The eighth deltahedron is the regular icosahedron (Exercise 3.1).

3.21* Construct the polyhedra in Figure 3.17. Figures 3.17a, c, d, e, g, and h can be constructed from equilateral triangles and squares. For the mono-capped octahedron, first construct an octahedron, and then add a pyramid.

The faces of the pyramid are isosceles triangles, and angles of 117.25° and 31.37° (twice) will result in all vertices lying on a sphere. The rhombic dodecahedron (3.17*i*) was constructed in Exercise 3.18. To construct Figure 3.17*j*, fasten together three rhombuses with edge length 3 cm to give the upper three faces, and similarly for the lower three faces. Each of the six faces around the waist is a trapezium with angles and edges of 109.47°, 2 cm, 109.47°, 3 cm, 70.53°, 4 cm, 70.53°, and 3 cm in sequence.

3.22** Construct the polyhedra in Figure 3.18 using the hints given in the figure legend.

3.23** Make models of the polyhedra in Figure 3.19. Figures 3.19*a, c*, and *d* (already constructed in Exercise 3.13) can be assembled from regular polygons. The easiest way to make all these models is to use tetrahedral stars (available commercially in either metal or plastic) for the vertices and tubing for the edges. One prong of each star is unused and should project outward. The tubing should be flexible to accommodate the angular misfit. A more tedious way is to glue or weld rods.

4

Trigonometry and Stereographic Projection

Preview This chapter develops the mathematical tools necessary for calculation of distances and angles. A stereographic projection is used to show 3D orientations on a sheet of paper, and spherical trigonometry is used for calculations. The relations between lines and planes are needed for calculations on lattices and crystal shapes. Because crystallographers may choose different unit cells for the same crystal structure, it is necessary to use a square matrix for transformation between them.

4.1 3D Coordinate Geometry

Interatomic distances and bond angles are important parameters in crystal structures, and it is often necessary to calculate them from the fractional coordinates of the atoms.

In 3D, there are seven possible shapes of the parallelohedra that can be stacked together to fill space (Fig. 3.14). In a crystal structure, a parallelohedron is chosen as a unit cell just as in 2D, and three edges are used for the coordinate axes *a*, *b*, and *c* (Fig. 4.1). The interaxial angles are labeled α, β, and γ. The first Greek letter (alpha) expresses the angle opposite the first Latin letter (*a*), that is, the one between the axes labeled with the second and third Latin letters (*b* and *c*). Similarly β is placed between *a* and *c*, and γ between *a* and *b*. In triclinic crystals, all parameters are nonequivalent but in crystals with higher symmetry there are special geometrical relations (Fig. 3.14):

triclinic	$a \neq b \neq c;$	$\alpha \neq \beta \neq \gamma$
monoclinic	$a \neq b \neq c;$	$\alpha = \beta = 90°, \gamma \neq 90°$
orthorhombic	$a \neq b \neq c;$	$\alpha = \beta = \gamma = 90°$
tetragonal	$a = b \neq c;$	$\alpha = \beta = \gamma = 90°$
isometric	$a = b = c;$	$\alpha = \beta = \gamma = 90°$

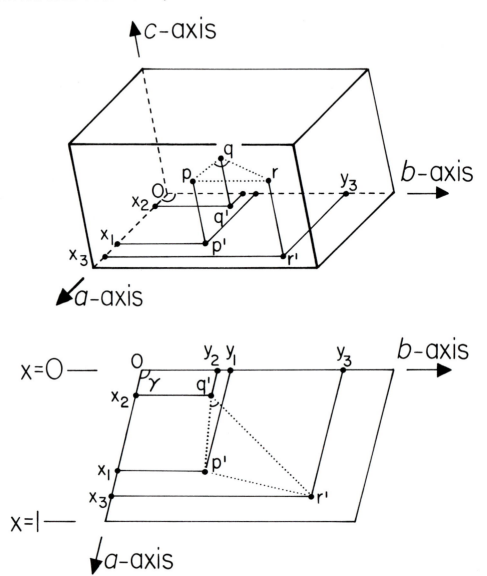

FIGURE 4.1   How to plot atomic coordinates in a perspective drawing (upper diagram) and a projection (lower diagram) of a unit cell. For the projection down the c-axis, draw a parallelogram with sides of length proportional to a and b, inclined at γ to each other. A point p' is plotted at fractional coordinates x_1 and y_1, and absolute coordinates $a.x_1$ and $b.y_1$. In the perspective drawing, the point p has fractional coordinates x_1, y_1, and z_1, and absolute coordinates $a.x_1$, $b.y_1$, and $c.z_1$. Point p' in this drawing is obtained by constructing parallel lines from points on the a and b axes, and the line pp' is parallel to the c-axis with distance $pp' = c.z_1$. The line pp' is not perpendicular to the plane of the a- and b-axes unless the c-axis is orthogonal to the a- and b-axes. Distance in the upper diagram are distorted by the perspective, but those in the lower diagram are undistorted.

rhombohedral	$a = b = c;$	$\alpha = \beta = \gamma \neq 90°$
hexagonal	$a = b \neq c;$	$\alpha = \beta = 90°, \gamma = 120°$

In all unit cells, a position is described by fractional coordinates x, y, and z (Fig. 4.1). Note that two cell edges can be accidentally equal within experimental error.

For rectangular axes (i.e., for orthorhombic, tetragonal, and isometric crystals), distance and angle calculations are straightforward. Consider three positions p, q, and r with fractional coordinates x_1, y_1, z_1; x_2, y_2, z_2; x_3, y_3, z_3. Just as in 2D the Pythagoras equation gives the distance between two points.

$$\mathbf{(pq)}^2 = a^2(x_2 - x_1)^2 + b^2(y_2 - y_1)^2 + c^2(z_2 - z_1)^2$$

$$\mathbf{(qr)}^2 = a^2(x_3 - x_2)^2 + b^2(y_3 - y_2)^2 + c^2(z_3 - z_2)^2$$

$$\mathbf{(rp)}^2 = a^2(x_1 - x_3)^2 + b^2(y_1 - y_3)^2 + c^2(z_1 - z_3)^2$$

The angle $\angle\mathbf{pqr}$ is given by

$$\mathbf{(rp)}^2 = \mathbf{(pq)}^2 + \mathbf{(qr)}^2 - 2\mathbf{pq} \cdot \mathbf{qr} \cdot \cos \angle\mathbf{pqr}$$

For monoclinic axes, the oblique angle γ between the a and b axes requires an extra term, as is readily deduced from the dot product in vector multiplication.

$$\mathbf{(pq)}^2 = a^2(x_2 - x_1)^2 + b^2(y_2 - y_1)^2 + c^2(z_2 - z_1)^2$$
$$+ 2ab(x_2 - x_1)(y_2 - y_1)\cos\gamma$$

Care must be taken that an error is not made with the sign of the additional term. In many scientific papers, monoclinic axes are chosen with $\alpha = \gamma = 90°$ and $\beta \neq 90°$. For this choice of axes,

$$\mathbf{(pq)}^2 = a^2(x_2 - x_1)^2 + b^2(y_2 - y_1)^2 + c^2(z_2 - z_1)^2$$
$$+ 2ca(z_2 - z_1)(x_2 - x_1)\cos\beta$$

Distances in a hexagonal unit cell can be obtained by placing $\gamma = 120°$ in the first equation for a monoclinic unit cell, and replacing b with a.

For triclinic axes, there are six terms.

$$(\mathbf{pq})^2 = a^2 (x_2 - x_1)^2 + b^2 (y_2 - y_1)^2 + c^2 (z_2 - z_1)^2$$
$$+ 2ab(x_2 - x_1)(y_2 - y_1) \cos \gamma$$
$$+ 2bc(y_2 - y_1)(z_2 - z_1) \cos \alpha$$
$$+ 2ca(z_2 - z_1)(x_2 - x_1) \cos \beta.$$

For rhombohedral axes, this equation simplifies by replacing b and c with a, and β and γ with α.

A direction can be specified uniquely by drawing a line through the origin of the three axes (Fig. 4.2), and by listing the cosine of each angle between the line and an axis. The three **direction cosines** denoted l, m, and n, have a special relationship when the axes a, b, and c are rectangular. Measure a unit distance along the line. The coordinates of the point are l, m, n, and the Pythagoras equation requires that $l^2 + m^2 + n^2 = 1$. Again for rectangular axes only, the angle between two lines with direction cosines l_1, m_1, n_1 and l_2, m_2, n_2 is given by $\cos^{-1} (l_1 l_2 + m_1 m_2 + n_1 n_2)$. This relation can be proved by tedious rewriting of the earlier expression for $\angle pqr$.

For nonrectangular geometry, the equations for the angle between lines are often treated better by spherical trigonometry than by coordinate geometry.

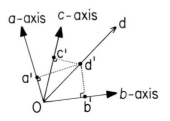

FIGURE 4.2   Direction cosines. Take three axes a, b, and c, shown here at an unusual orientation for convenience of drawing. To specify a general direction, draw a line Od through origin O. Mark off unit length on Od to give point d'. Drop perpendicular lines to give points a', b', and c'. Because Oa' is the cosine of the angle between Oa and Od, it is the direction cosine l. Similarly Ob' is m and Oc' is n.

4.2   Spherical Trigonometry

A **spherical triangle** joins three points on the surface of a sphere (Fig. 4.3). Each point represents a line drawn from the center of the sphere. Each side of the triangle is the *curved* intersection with the sphere of the plane defined by two of the points and the center of the sphere. This intersection is the shortest distance *on the surface of the sphere* between the two points, but is longer than the *straight* line joining the points. The side of the triangle is a segment of a **great circle,** which is defined as the intersection between a sphere and a plane passing through its center. A **small circle** is defined as the intersection between a sphere and a plane that cuts the sphere off-center. Whereas the side of a planar triangle is expressed as a linear distance, the side

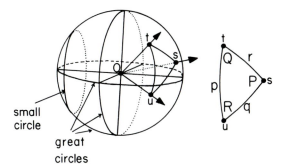

FIGURE 4.3   Spherical trigonometry. The left-hand diagram is a near-clinographic projection of a sphere. Points *s, t,* and *u* lie on the surface of the sphere and the arcs joining them are the intersections of planes *Ost, Otu,* and *Ous* with the sphere. The right-hand diagram shows the spherical triangle with labeling of the sides and angles. The sphere has unit radius and side *p* = radian value of angle *tÔu; q* = radian value of angle *sÔu; r* = radian value of angle *sÔt.*

of a spherical triangle is expressed by the distance measured along the segment of the great circle. For a sphere of unit radius, the distance along this arc is equal to the radian value of the angle subtended by the arc at the center of the sphere. Each angle of a spherical triangle is measured on the surface of the sphere between the two sides just as they meet at the corner of the triangle. Specifically, the angle lies between the two tangents formed at the intersection of the two curved sides. Each spherical triangle will be specified by three angles labeled *P, Q,* and *R* and three sides labeled *p, q,* and *r.* The angle *P* is chosen to lie opposite the side *p; Q* opposite *q,* and *R* opposite *r.* Each side is actually expressed as the angle subtended at the center of the sphere, and the reader must not become confused when a spherical triangle is described by six angles.

A globe of the world is helpful in understanding spherical trigonometry. All lines of longitude are great circles, but only one line of latitude (the equator) is a great circle. The other lines of latitude are small circles, whose radius decreases as a pole is approached. A spherical triangle must not have small circles as sides. Only a great circle gives the minimum distance between two points on the surface of a sphere, and there is an infinite number of small circles passing through two points, except for two points that are directly opposite each other, like the North and South poles. The angles of a spherical triangle add up to more than 180°: thus, a triangle composed of the North Pole, 0°W 0°N, and 90°W 0°N has three angles of 90° giving a total of 270°. The larger are the sides of the triangle, the larger is the sum of the angles. The area of a spherical triangle on a sphere of unit radius is $(P + Q + R - \pi)$ radians.

In 2D, the sides and angles of a triangle are related by the sine equation $p/\sin(180 - P) = q/\sin(180 - Q) = r/\sin(180 - R)$, or briefly, $p/\sin P = q/\sin Q = r/\sin R$, where **p, q,** and **r** are sides and **P, Q,** and **R** are the respective opposing angles. In 3D, the corresponding relation is $\sin p/\sin P = \sin q/\sin Q = \sin r/\sin R$. In 2D, the cosine equation is $p^2 = q^2 + r^2 - 2qr \cos P$, and in 3D there are two corresponding equations:

$$\cos p = \cos q \cdot \cos r + \sin q \cdot \sin r \cdot \cos P$$

$$\cos(180 - P) = \cos(180 - Q) \cos(180 - R)$$
$$+ \sin(180 - Q) \sin(180 - R) \cos(180 - p)$$

The symbols of the first cosine equation can be rotated to give

$$\cos q = \cos r \cdot \cos p + \sin r \cdot \sin p \cdot \cos Q$$

$$\cos r = \cos p \cdot \cos q + \sin p \cdot \sin q \cdot \cos R$$

Because $\cos(180 - P) = -\cos P$ and $\sin(180 - Q) = +\sin Q$, and so on, the second cosine equation can be written:

$$\cos P = -\cos Q \cdot \cos R + \sin Q \cdot \sin R \cdot \cos p$$

$$\cos Q = -\cos R \cdot \cos P + \sin R \cdot \sin P \cdot \cos q$$

$$\cos R = -\cos P \cdot \cos Q + \sin P \cdot \sin Q \cdot \cos r$$

A spherical triangle with one or more sides or angles equal to 90° is known as a ***Napierian triangle,*** after John Napier. Because $\cos 90° = 0$ and $\sin 90° = 1$, the cosine equations are simplified. It is customary to use a special procedure for remembering the equations for a right-angle triangle. Write down the angles and sides of the triangle in a hexagon:

If the 90° angle was an internal angle (i.e., P, Q, or R) remove it; leave the adjacent angles alone; and replace the remaining three angles by their complements (i.e., subtract them from 90°). Thus, for $R = 90°$, write

$$q \perp p$$

```
        q ⊥ p
       /      \
    90-P      90-Q
       \      /
        90-r
```

If the 90° angle was a side (i.e., p, q, or r), subtract all the angles from 180°, and then follow the above steps. Thus for $r = 90°$, write down

```
       180-p                    p-90
      /      \                 /    \
  180-R     180-Q         R-90       \
    |         |            |          180-Q
  180-q     180-r        q-90        /
      \      /                 \    /
       180-P                    180-P
```

All the Napierian equations can then be obtained from:

sine of any component = product of tangents of adjacent components or

product of cosines of nonadjacent components

Thus $\sin(90 - P) = \tan q \cdot \tan(90 - r) = \cos p \cdot \cos(90 - Q)$ for the first example, and $\sin(R - 90) = \tan(q - 90) \cdot \tan(p - 90) = \cos(180 - P) \cdot \cos(180 - Q)$ for the second.

    To use these formulas for a spherical triangle, note that three out of the six elements are sufficient to define a triangle, and that the other three elements can then be calculated one at a time. Examples are given in the next section.

    All the above equations are symmetrical with respect to the pairs p and $(180 - P)$, q and $(180 - Q)$, and r and $(180 - R)$, as will be made clear in Section 4.7.

4.3   Stereographic and Gnomonic Projections

Initially most readers will profit from tracing spherical triangles on the surface of a globe, but will rapidly ask for a simple way of projecting a sphere onto a flat sheet of paper. Geographers use many types of projection, but none is fully satisfactory, because distortion is inevitable. Crystallographers normally use the **stereographic projection** because it condenses all the angular information inside a single circle (Fig. 4.4*b*). In the perspective drawing of Figure 4.4*a*, a desired direction is shown by the line *Od*, which joins the center *O* of a sphere to a point *d* on its surface. A horizontal plane passing through the sphere is selected for the site of the stereographic projection. This plane intersects the sphere in a great circle $ab\bar{a}\bar{b}$. The line perpendicular to the plane and the great circle cuts the sphere at *c* and $\bar{c}$, which are known as the **poles** of the great circle, just as the North and South poles are related

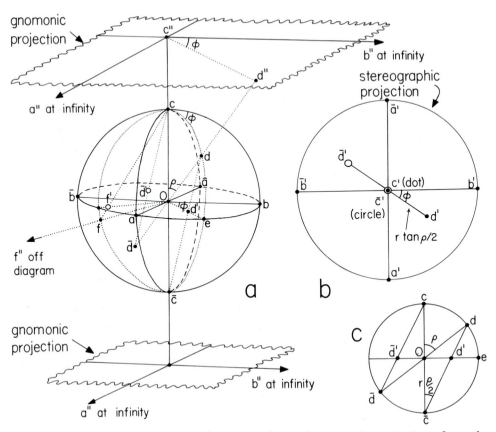

FIGURE 4.4   The construction of stereographic and gnomonic projections from the spherical projection. (*a*) Near-clinographic projection of sphere oriented on orthogonal directions $a O \bar{a}$, $b O \bar{b}$, and $c O \bar{c}$. The direction $\bar{d} O d$ gives a stereographic projection at d' and a gnomonic projection at d''. The opposite direction $d O \bar{d}$ gives a stereographic projection at $\bar{d}'$, and the direction Of gives stereo-point f'. The direction $\bar{b} O b$ gives stereo-point b' and gnomonic point b'' at infinity as indicated by arrows in the two planes of the gnomonic projection. The gnomonic point f'' would be off the diagram also. (*b*) Stereographic projection laid flat on the paper. r is the radius of the circle. Stereo-points c' (dot) and $\bar{c}'$ (circle) coincide at the center of the diagram. The distance $c'd'$ is $r \tan \rho/2$ where ρ is the angle $c \hat{O} d$ in diagram (*a*). (*c*) Diagram showing how the position of d' is related to ρ.

to the equator of the Earth. If d is above the plane of the stereographic projection, join d and $\bar{c}$ with a straight line. The point d' where this line passes through the central plane is the stereographic projection of the direction Od. Every upward-pointing direction will intersect the upper half of the sphere in a point, and every such point will produce a point on the plane of the stereographic projection; mathematicians would say that there is a unique relation between the spherical (*d*) and stereographic (*d'*) projections with a one-to-one ***mapping*** between them. In order to show a downward-pointing direction, use the north pole instead of the south pole, and find the intersection of *fc* with the plane of the stereographic projection. The resulting point *f'* must be distinguished with a symbol to show that it represents a downward rather than an upward direction. Most crystallographers follow the *International Tables* in using a dot for an upward direction and a circle for a downward direction. The opposite direction to *Od* is represented by a circle $\bar{d}'$ that lies in the plane of the stereographic projection symmetrically to *d'* on the other side of the center *O*. The perspective drawing of the stereographic projection is an ellipse (Fig. 4.4*a*), but becomes a circle when laid flat on the paper in Figure 4.4*b*.

For rectangular crystallographic axes, it is customary to place the *c*-axis upward, *b* to the right, and *a* toward the reader, but this convention may be disregarded when a special view is required of a crystal. The radius (*r*) of the spherical and stereographic projections is purely arbitrary, and 10 cm is convenient for most purposes. A direction is usually expressed by the ***polar coordinates*** ϕ and ρ, where ϕ is the angle between the great circles *cb* and *cd* and ρ is the angle $\angle cOd$. The lines *c'b'* and *c'd'* in the stereographic projection (Fig. 4.4*b*) represent segments of great circles in the spherical projection, and the angle ϕ is exactly the angle that would be measured on the surface of the sphere between the great circles as they meet at point *c*. The angle ρ, however, is represented by the distance *c'd'*, where $c'd' = r \tan \rho/2$. This relation is derived in Figure 4.4*c*. Take the section $cdec\bar{d}$ of the sphere. In any circle, an arc (like *cd*) subtends only half the angle at the circumference (e.g., at $\bar{c}$) as at the center (e.g., at *O*). In the triangle $Od'\bar{c}$, $Od' = r \tan \rho/2$.

The ***gnomonic projection*** is popular with some crystallographers because it allows assignment of Miller indices of crystal faces merely by direct measurement. Take a plane perpendicular to *Oc* at a convenient distance from *O*. Project *Od* to *d''* to obtain the gnomonic projection of direction *Od*. Unlike the stereographic projection, which is confined to a circle, the gnomonic projection goes to infinity. Directions in the *Oab* plane can be represented by arrows pointing to infinity, but this is awkward. Directions pointing downward (e.g., *Of*) can be projected onto another plane symmetrically placed below the sphere, or can be represented by the backward direction on the upper gnomonic projection if a special symbol is used to distinguish the projected point from a point representing an upward direction. Whereas all small and great circles on a sphere project as circles (or straight lines for infinite radius) on a stereographic projection, some do not project as straight lines on a gnomonic projection.

4.4   Stereographic Projection of Simple Polyhedra

Place a cube (Fig. 4.5*a*) with its six faces perpendicular to the directions *Oa*, *Ob*, and *Oc* in Figure 4.4. Each face of the cube is represented by a line perpendicular to the face, and the stereo points (Fig. 4.5*b*) lie at a', $\bar{a}'$, b', $\bar{b}'$, c', and $\bar{c}'$ of Figure 4.4*b*). The points a', b', $\bar{a}'$, and $\bar{b}'$, readily demonstrate the operations of the tetrad axis along *Oc*, but the operations of the tetrad axes along *Oa* and *Ob* may require careful study by the reader.

Now place a regular octahedron with its tetrad axes in the same orientation as for the cube (Fig. 4.5*c*). The eight face normals lie parallel to the triad axes, and must lie at equal angles to *Oa*, *Ob*, and *Oc*. Using the equation for the direction cosines in an orthogonal system of axes ($l^2 + m^2 + n^2 = 1$) and putting $l = m = n$ results in $l = \sqrt{1/3} \simeq \cos 54.74°$. From symmetry, the eight stereo points in Figure 4.5*d* plot at the combinations of $\phi = 45°$, $135°$, $225°$ and $315°$, and $\rho = 54.74$ and $125.26°$. Each stero point lies at a distance of $r \cdot \tan 27.37°$ from the center. The resulting four pairs of dot and

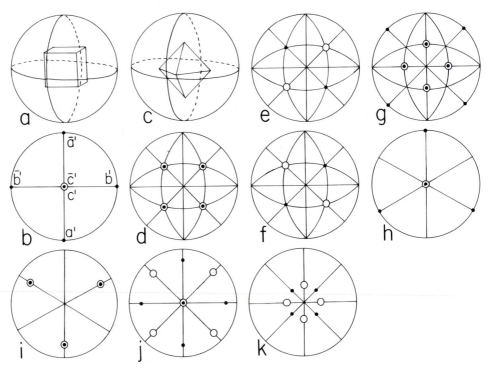

FIGURE 4.5   Stereographic projections of polyhedra. (*a* and *b*) Perspective drawing and stereogram for cube. (*c* and *d*) Perspective drawing and stereogram for regular octahedron. (*e* and *f*) Stereograms of regular tetrahedron in positive and negative orientations. (*g*,h,i,j, and k.) Stereograms of rhombic dodecahedron, closed trigonal prism, trigonal bipyramid, square antiprism, and symmetrical tetragonal trapezohedron.

circle (Fig. 4.5*d*) neatly demonstrate the tetrad axes and mirror planes of symmetry, but the operations of the triad axes require careful thought. Actually, the stereo points for the regular octahedron can be obtained simply by finding the intersection of circles. With center *b'* in Figure 4.5*b*, draw a circle passing through *a'* and *ā'*, and repeat for circles related by the tetrad axis along *Oc*.

Because the regular tetrahedron has four faces parallel to four out of the eight faces of the regular octahedron when it is in the positive and negative orientations (Fig. 3.5*c*, 3.5*d*), it is easy to plot the corresponding stereo points (Figs. 4.5*e*, 4.5*f*). In both orientations, the diad symmetry axes, which join the midpoints of opposing edges, lie parallel to *Oa, Ob,* and *Oc.*

Each face of the rhombic dodecahedron (Fig. 3.13*a*) is parallel to one axis and cuts off equal intercepts from the other two axes: hence the normal to each face lies at one of the 12 permutations of 45°, 45°, and 90° to the axes when account is taken of positive and negative directions. Four faces give stereo points in the perimeter (Fig. 4.5*g*), and eight faces give four pairs of circled dots at $\rho = 45°$ on the great circles radiating out from the center to the stereo points for the *Oa* and *Ob* lines. These pairs also lie on the circles drawn in Figure 4.5*d*.

The closed trigonal prism projects as a dot and circle at the center of the stereographic projection (commonly called a ***stereogram***) and three equally spaced dots on the perimeter (Fig. 4.5*h*). All closed prisms will have a dot and a circle at the center, and the other stereo points on the perimeter, as the reader can demonstrate using the closed hexagonal prism (Fig. 3.11*c*) as an example.

The trigonal bipyramid (Fig. 3.12*a*) has a horizontal plane of symmetry, and its six face normals must project as three sets of circled dots obeying the vertical triad axis; but what is the distance from the center? Because the particular trigonal bipyramid in Figure 3.12*a* is the dual of an Archimedean trigonal prism, the angle ρ is determined by the regular shape of the polygons. Straightforward calculation using coordinate geometry gives

$$\rho = \cot^{-1} [1/(\cos 30°) - \cos 30°] \simeq 73.90°.$$

This angle can be checked from the polyhedron constructed from the template in Figure 3.12. In the stereogram (Fig. 4.5*i*), the stereo points lie at $r \cdot \tan 36.95°$ from the center. The trigonal bipyramid in Figure 3.12*b* was constructed by doubling the horizontal dimension over Figure 3.12*a*, and its value of ρ is $\cot^{-1} [2 \times (\cot 73.90°)] = 60.00°$.

The Archimedean square antiprism (Fig. 3.11*d*) gives a circled dot for the top and bottom faces and a set of four dots and four circles for the inclined faces. The latter must alternate regularly around a small circle (Fig. 4.5*j*). Again it is necessary to calculate the angle ρ for the inclined face normals. From 3D coordinate geometry, ρ is $\mathrm{cosec}^{-1}\,[(\sqrt{2}-1)/\sqrt{3}] \simeq 76.16°$, which can be checked from the model.

The Catalan tetragonal trapezohedron (Fig. 3.12*d*) is the dual of the square antiprism, and it projects as four dots and four circles alternating regularly on a small circle with $\rho \simeq 34.86$ (Fig. 4.5*k*).

Construction of the stereograms for the regular icosahedron and regular dodecahedron requires care. For the regular dodecahedron, examine the plan in Figure 3.3*i*. The top and bottom faces will give a circled dot at the center of a stereogram, and the other 10 faces will give alternate dots and circles equally spaced on a small circle because of the symmetry (Fig. 4.6*a*). From Table 3.1, the angle between adjacent faces is $\simeq 116.57°$, and the small circle has $\rho \simeq 63.43°$. Actually, it is easy to calculate this angle using spherical trigonometry. Draw the great circle *gdefg'*. Because of the 3-fold symmetry, the triangle *cdf* has equal sides and equal angles. The triangle *cef* has $\angle fce = 36°$ and $\angle cfe = 72°$ because of the 5-fold symmetry, and $\angle cef = 90°$ from the mirror symmetry. The side *cf* can be calculated from this Napierian triangle as follows:

$$\sin(90 - fc) = (\tan 18°) \cdot (\tan 54°),\ \text{giving}\ cf \simeq 63.435°.$$

For the regular icosahedron oriented as in Figure 3.3*j*, the 20 faces will give alternating dots and circles on two small circles (Fig. 4.6*b*). Draw the

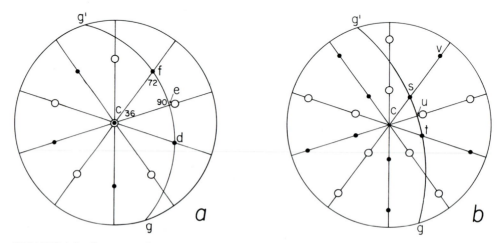

FIGURE 4.6 Stereographic projections of regular dodecahedron (left) and regular icosahedron (right). The great circles through *d* and *f* (left) and *s* and *t* (right) must pass through *g* and *g'*, because of symmetry. Point *e* is halfway between *f* and *d* (left). Point *u* is halfway between *s* and *t* (right). Angles *scu, cus,* and *usc* are 36, 90, and 60°, respectively.

great circle between s and t. The triangle cus has $\angle scu = 36°$, $\angle cus = 90°$, and $\angle csu = 60°$ from the symmetry elements. The side cs is calculated from the Napierian triangle as follows:

$$
\begin{array}{cc}
\quad 90° & \quad us \perp cu \\
us \diagup \quad \diagdown cu & \diagup \qquad \diagdown \\
| \qquad \quad | & 30° \qquad \quad 54° \\
60° \quad \; 36° & \diagdown \qquad \diagup \\
\diagdown sc \diagup & 90\text{-}sc
\end{array}
$$

$\sin(90 - sc) = (\tan 30°) \cdot (\tan 54°)$, giving $cs \simeq 37.377°$. Side $sv = 2 \times$ (side us) by symmetry and $\sin 54° = (\cos 30°) \cdot (\cos us)$, giving $us \simeq 20.905°$, and $sv \simeq 41.810°$. Consequently, v has $\rho \simeq 79.187°$.

4.5   Geometry of Stereographic Projection

Looking at Figure 4.4, imagine that c and $\bar{c}$ are the North and South poles of the Earth, and that a is at 0°N 90°W and b is at 0°N 0°W. The lines of longitude project as radiating lines from the center of the stereographic projection (Fig. 4.7a), and the lines of latitude project as concentric circles to give a ***polar net***. To obtain the radius of the circles, multiply the radius of the stereographic projection by $\tan[(90 - \epsilon)/2]$, where ϵ is the latitude. The spacing of the circles increases as ϵ decreases, and it is obvious that areas on the surface of a sphere are distorted when projected stereographically.

The polar net is useful for plotting directions for which ρ $(= 90 - \epsilon)$ and ϕ (= the westerly longitude) have been measured, as in two-circle goniometry of crystals. For other purposes, the ***meridian*** (or ***Wulff***) ***net*** is suitable.

Reorient the Earth so that in Figure 4.4a, the North and South poles are at b and $\bar{b}$, respectively, and the Greenwich meridian (longitude 0°W) is along $\bar{b}cb$. In the stereographic projection (Fig. 4.7b), the Greenwich meridian is a straight line between $\bar{b}'$ and b' in Figure 4.4b, and the equator is a straight line between a' and $\bar{a}'$. All lines of longitude project as great circles that radiate out from b to $\bar{b}$ and pass through an appropriate point on the equator. This point lies at $r \tan \eta/2$ from the center, where η is the longitude. All lines of latitude (except for the equator) are small circles that are not concentric. Each circle passes through three points with polar coordinates in Figure 4.4a: $\rho = 90°$, $\phi = 90 - \epsilon°$; $\rho = 90°$, $\phi = \epsilon - 90°$; $\rho = \epsilon°$, $\phi = 0°$. In the stereographic projection, these points project as radial coordinates: r, $90 - \epsilon°$; r, $\epsilon - 90°$; $r \tan \epsilon/2$, $0°$. The Wulff net contains great circles and small circles of all possible shapes, and allows easy construction of the great circle that passes through any two stereo points, and measurement of the angle between them.

The easiest way to learn is to follow an example. The polar and Wulff nets (Fig. 4.7) were constructed from first principles, and show lines of latitude and longitude only for 10° intervals. Commercial nets usually con-

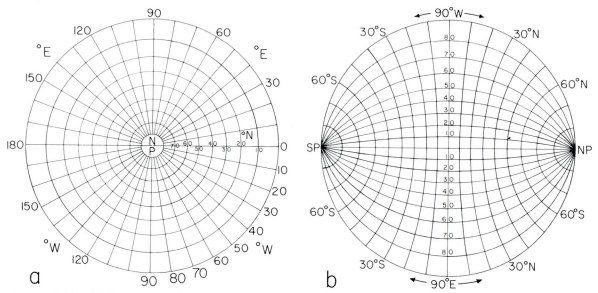

FIGURE 4.7 (*a*) Polar net with lines of longitude and circles of latitude every 10°. The north pole is at the center, and the equator is at the perimeter. (*b*) Meridian (Wulff) net with lines of longitude radiating out as great circles from the north and south poles, and lines of latitude occurring as small circles.

tain 2° intervals, thereby allowing positions to be plotted with an accuracy up to 0.5° if care is taken. It is possible to accomplish all stereographic plots with a Wulff net. Remove Figure 4.8 from the book and either glue it or tape its edges to a piece of bristol board or other stiff material with a flat surface. Carefully prick a hole through the exact center with a sharp compass point or needle. Widen the back side of the hole, and insert a thumb tack from the back side so that it projects through the exact center of the Wulff net. Select a thumb tack whose point is symmetrical, and avoid a tack with a burr. Cut a piece of tracing paper and put one or two pieces of transparent tape about 4 mm across at the center on both sides. Press the tracing paper over the thumb tack so that the thumb tack passes through the tape. Check that the tracing paper rotates accurately around the thumb tack, and handle it gently so that the tack does not rip out a hole. Using a sharp pencil and a ruler, draw the equator and Greenwich meridian to give a reference.

To plot the poles of the regular octahedron (as in Fig. 4.5*d*), rotate the tracing paper 45° about the starting position, and count off 54.74° from the center along the equator and Greenwich meridian. Any stereo point can be plotted in this way by rotating an angle ϕ and counting off an angle ρ (Fig. 4.9). Referring to Figure 4.4*b*, the rotation provides the straight line on which d' lies, and counting off the angle ρ automatically corresponds to calculating the distance $r \tan \rho/2$. The reader should profit by plotting the stereo points in Figures 4.5*g*, 4.5*i*, 4.5*j* and 4.5*k* using the angles given in the preceding section.

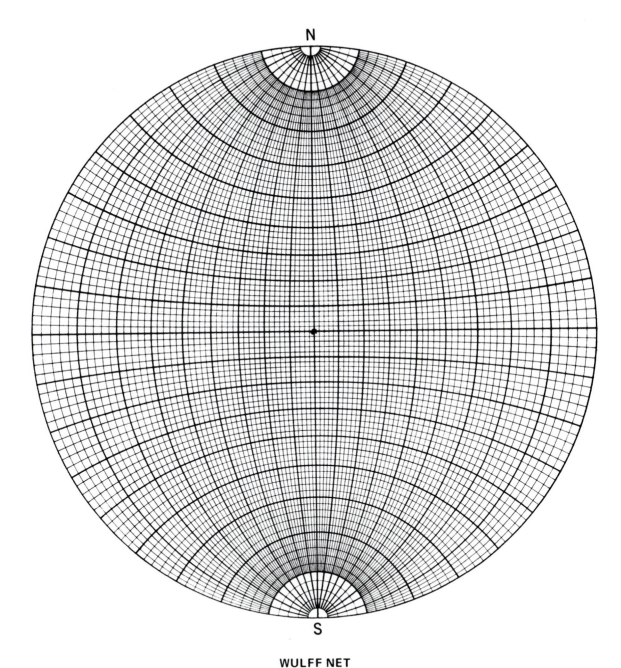

WULFF NET

FIGURE 4.8 Wulff net with 2° intervals.

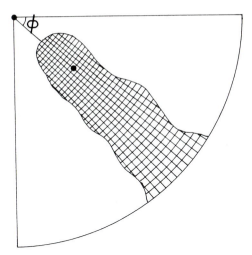

FIGURE 4.9 Use of Wulff net to plot a direction specified by polar coordinates ρ and ϕ. The point lies at $\phi\ 41°\ \rho\ 38°$. Only a small portion of the Wulff net is shown.

Suppose now that two stereo points have been plotted, and it is necessary to draw the great circle between them, and to measure the angle between them. Rotate the tracing paper until the two stereo points lie on a great circle of the Wulff net. Trace the great circle with a pencil, and measure the angle by counting off the number of small circles that cross the great circle. Of course, it is usually necessary to interpolate between circles. This procedure is illustrated in Figure 4.10. Both points are dots, and represent directions above the projection. What happens if one point is above (i.e., a dot) and one is below (i.e., a circle)? Construct the **opposite** of the second point (Fig. 4.11), and rotate about the Wulff net until the first point and the

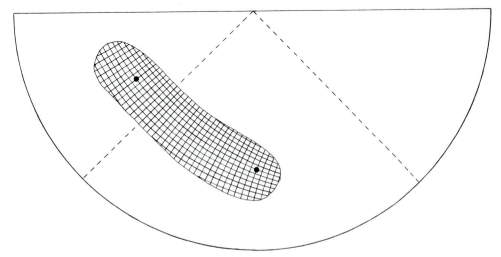

FIGURE 4.10 Use of Wulff net to plot a great circle between two directions pointing upward. Only part of the Wulff net is shown.

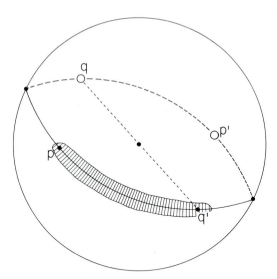

FIGURE 4.11 Use of Wulff net to plot a great circle between two directions *p* and *q*, one pointing up and one pointing down. Construct *q'*, the opposite of *q*. Rotate the Wulff net to find the great circle that passes through *p* and *q'*. Draw in the symmetrical circle (dashed line), and check that it goes accurately through *q* and *p'* (the opposite of *p*).

opposite lie on the great circle. Trace the great circle to the perimeter, and continue on the symmetric circle using a dashed line until the second point is reached. If both points are below, merely rotate the Wulff net until they lie on a great circle, and trace over the great circle with a dashed line. The great circle may project as a straight line instead of a curve.

 A third procedure involves construction of a point that lies at specific angles from two other points (Fig. 4.12*a*). Suppose that points *p* and *q* are already plotted, and that point *r* lies at 40° to *p* and 30° to *q*. Construct a small circle of 40° about *p* and one of 30° about *q*, and place *r* at the intersection. Actually there are two intersections, and a third piece of information is required to resolve the ambiguity. To construct a small circle, rotate the tracing paper over the Wulff net until the equator passes through the point. Measure off the required angle in both directions on the equator, and in both directions on the great circle that crosses at 90°. The first two points uniquely define the small circle since the center of the circle is halfway between them. The second two points provide a check on the accuracy of the construction. Because of the areal distortion of the stereographic projection, the center of the small circle is displaced from the starting point, except for a small circle about the center of the stereographic projection. This procedure is valid when all four points lie above the projection, or all four lie below. But what procedure is to be used when it is necessary to count over the perimeter and around the other side? Locate points u, v, v', v'', and v''', and draw a circle that passes between v', v, and v''. The circle intersects the perimeter in w and w'. Draw the circle that passes through w, w', and v'''. The required small circle consists of the arcs $wv'vv''w'$ and $wv'''w'$.

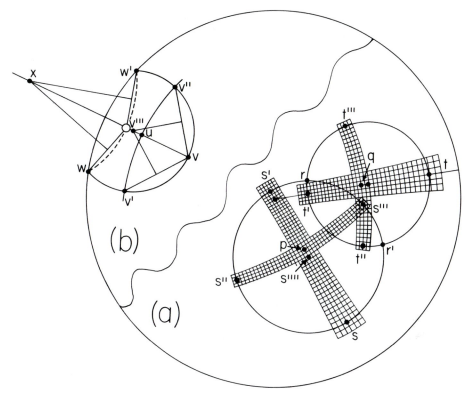

FIGURE 4.12 (*a*) Construction of the poles r and r' that lie at 40° to p and 30° to q. Rotate the tracing paper so that the equator passes through p (see Figure 4.7*b* for location of equator). Measure off 40° in each direction to give s and s'. Find the midpoint (s'''') of s and s', and draw a circle of radius $ss'/2$ about s''''. Measure off 40° along the great circle that crosses the equator at 90° at p. Check that the two points s'' and s''' lie on the circle. Repeat the procedure for q. The center of the circle is the unlabeled dot just to the right of q. (*b*) Construction of a small circle that crosses the perimeter. Locate u the center of the small circle, and rotate the tracing paper over the Wulff net so that two great circles pass through u. Count across the small circles (not shown to avoid overlap) to obtain v, v', v'', and v'''. The first three points represent upward directions, whereas the fourth one (labeled with a circle) represents a downward direction. Draw the circle through v', v, and v'' after locating the center, which is slightly displaced from u. To locate this center, construct the perpendicular bisectors of vv' and vv''. Points w and w' are obtained from the intersection of the circle with the perimeter. Draw the circle dashed line) between w, v''', and w' after locating the center x from the intersection of the perpendicular bisectors of wv''' and $v'''w'$.

Structural petrologists commonly use the Lambert equal-area projection. This is not used by crystallographers, because it does not give simple geometrical constructions. Small and great circles in the stereographic projection are distorted circles in the equal-area projection. The stereographic projection preserves the angles between circles (note that all circles cross at 90° in the Wulff net), but the equal-area projection does not.

All constructions in a stereographic projection can be made with pencil, ruler, and compass, as described in Phillips (1971) and McKie and McKie (1974). Use of a Wulff net is faster and sufficiently accurate if a 10-cm radius is used.

4.6 Stereographic Projection of Axes of Unit Cells

Isometric, tetragonal, and orthorhombic unit cells have orthogonal axes, which plot at a', b', and c' on Figure 4.4b. A monoclinic unit cell with $\gamma \neq$

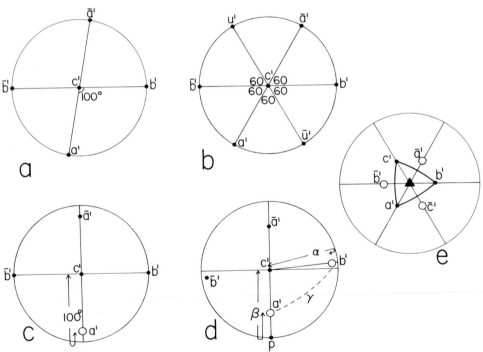

FIGURE 4.13 Stereograms of nonorthogonal axes. (*a*) Monoclinic with γ 100°. (*b*) Hexagonal. (*c*) Monoclinic with β 100°. (*d*) Triclinic with α 93°, β 116°, and γ 91°. The angle γ is measured along the great circle joining $a'b'$. The angle β is measured from c' to p and then underneath to a'; α would be measured on a great circle that goes to the perimeter and then turns underneath to b'. (*e*) Rhombohedral. The interaxial angle α would be measured between a' and b', or b' and c', or c' and a' along the arcs of the great circles (heavy lines).

90° is represented by Figure 4.13*a*, where γ is chosen arbitrarily to be 100°, c' is in the center, and b' is to the right. The stereo point a' is displaced 10° clockwise from the position of a' for orthogonal axes. Figure 4.13*b* shows hexagonal axes for which $\gamma = 120°$; u' is the stereo point for the auxiliary u-axis at 120° to the a and b axes. If the monoclinic axes are chosen with $\beta \neq 90°$, c' is placed in the center, b' to the right on the perimeter, and a' is underneath at $\phi = 90°$ and $\rho = \beta$ (Fig. 4.13*c*). For triclinic axes, it is customary to place c' in the center, and a' at $\phi = 90°$ and $\rho = \beta$ (Fig. 4.13*d*). The stereo point b' cannot be at the right on the perimeter, unless α and γ were accidentally equal exactly to 90°. To obtain the position of b', it is necessary to construct small circles of angle α about c' and angle γ about a'. Figure 4.13*d* is drawn with the angles for a metamorphic anorthite: $\alpha = 93.12°$, $\beta = 115.91°$, $\gamma = 91.26°$. Rhombohedral axes are best displayed with the projection of the triad axis at the center, and the three axes pointing symmetrically upward (Fig. 4.13*e*).

4.7   Reciprocal Geometry

A crystal face can be represented directly by a parallel plane that passes through a sphere. The corresponding stereographic projection is a great circle. It is simpler to represent the face reciprocally by its *face normal,* which is represented stereographically by a pole of the great circle.

A unit cell can be represented either by three nonparallel edges, as in the preceding section, or by three nonparallel faces. These faces are represented stereographically by three great circles. Thus the face that is parallel to the axes a and c is represented by the great circle *pcqa* in Figure 4.14, the face parallel to the c and b axes by the great circle *scrb*, and the face parallel to the a and b axes by the great circle *tabuāb̄*. The first two great circles actually project as straight lines.

It is easier to represent each face by its face normal, and the resulting three directions are called the *reciprocal axes* and labeled with stars. The $a^\star$ axis is perpendicular to the b and c axes; $b^\star$ to c and $a;$ and $c^\star$ to a and b. Merely by reversing the above procedure, a is perpendicular to $b^\star$ and $c^\star$, and so on. The reciprocal angle $\alpha^\star$ is between the $b^\star$ and $c^\star$ axes; $\beta^\star$ between $c^\star$ and $a^\star$; and $\gamma^\star$ between $a^\star$ and $b^\star$.

In Figure 4.14, $a^\star$ must lie in the perimeter because it is at 90° from c, which is at the center. Furthermore, $a^\star$ must be the pole of the great circle *sbcrb*. Similarly, $b^\star$ must lie in the perimeter at the pole of the great circle *pācqa*. The angle $\gamma^\star$ is measured off directly on the perimeter, and is the difference in ϕ values of $a^\star$ and $b^\star$. The $c^\star$ axis is represented by the pole of the great circle *tb̄aubat*. The angle $\beta^\star$ is measured on the great circle between $a^\star$ and $c^\star$, while $\alpha^\star$ is the angle between $b^\star$ and $c^\star$.

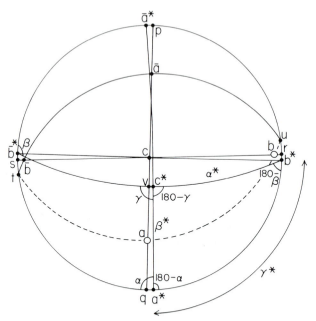

FIGURE 4.14 Stereogram of direct and reciprocal axes for a triclinic mineral: anorthite α 93.12; β 115.91; γ 91.26, α* 85.92; β* 63.97; γ* 87.08°.

The next step is to show that the internal angles of the spherical triangle $a^{\star}b^{\star}c^{\star}$ are $180 - \alpha$, $180 - \beta$, and $180 - \gamma$. In triangle $qvb^{\star}$, the angle $\angle vb^{\star}q$ must equal the side qv because sides $vb^{\star}$ and $qb^{\star}$ are 90°. This can be seen from Figure 4.4a, where angle $\angle bOe$ (which equals side eb) is equal to the angle ϕ between the great circles that meet at c. The direction q is perpendicular to c in Figure 4.14, and v is perpendicular to a. Hence $vq = 180 - \beta$. Similarly, the internal angle $\angle c^{\star}a^{\star}b^{\star}$ is $180 - \alpha$ and $\angle a^{\star}c^{\star}b^{\star}$ is $180 - \gamma$.

Using the equations from Section 4.2,

$$\sin\alpha^{\star}/\sin(180 - \alpha) = \sin\beta^{\star}/\sin(180 - \beta) = \sin\gamma^{\star}/\sin(180 - \gamma)$$

$$\cos\alpha^{\star} = \cos\beta^{\star}\ \cos\gamma^{\star} + \sin\beta^{\star}\ \sin\gamma^{\star}\ \cos(180 - \alpha)$$

$$\cos\beta^{\star} = \cos\gamma^{\star}\ \cos\alpha^{\star} + \sin\gamma^{\star}\ \sin\alpha^{\star}\ \cos(180 - \beta)$$

$$\cos\gamma^{\star} = \cos\alpha^{\star}\ \cos\beta^{\star} + \sin\alpha^{\star}\ \sin\beta^{\star}\ \cos(180 - \gamma)$$

To calculate α, β, and γ when $\alpha^{\star}$, $\beta^{\star}$, and $\gamma^{\star}$ have been measured for a crystal, use the following equations:

$$\cos\alpha = (\cos\beta^{\star}\ \cos\gamma^{\star} - \cos\alpha^{\star})/\sin\beta^{\star}\ \sin\gamma^{\star}$$

$$\cos\beta = (\cos\gamma^\star \ \cos\alpha^\star - \cos\beta^\star)/\sin\gamma^\star \ \sin\alpha^\star$$

$$\cos\gamma = (\cos\alpha^\star \ \cos\beta^\star - \cos\gamma^\star)/\sin\alpha^\star \ \sin\beta^\star$$

All these expressions are related by cyclic rotation of α, β, and γ and $\alpha^\star$, $\beta^\star$, and $\gamma^\star$. Finally, it is merely a matter of choice concerning which set of axes is labeled with stars. Hence, switching starred and unstarred axes in Figure 4.14 yields

$$\cos\alpha^\star = (\cos\beta \ \cos\gamma - \cos\alpha)/\sin\beta \ \sin\gamma$$

and so on.

But why bother with reciprocal axes? The answer is that measurements of the angles between crystal faces lead to the reciprocal axes $a^\star$, $b^\star$, and $c^\star$, and it is necessary to calculate the position of the direct axes a, b, and c. Furthermore, X-ray diffraction also gives the position of the reciprocal axes. Only for orthogonal axes do the direct and reciprocal axes coincide. For monoclinic axes, the equations can be readily obtained from the triclinic equations by setting the appropriate angles to $90°$.

Finally, in Section 4.2 the equations for spherical triangles were stated to be symmetrical with respect to p and $180 - P$, q and $180 - Q$, and r and $180 - R$. This arises from the relationship between direct and reciprocal geometry, as illustrated by the relationship between the sides and angles of the triangles abc and $a^\star b^\star c^\star$ in Figure 4.14.

4.8 Relation Between Lattice Directions and Planes and Crystal Edges and Faces

A line passing through lattice nodes has an angular orientation that can be specified by indices **U**, **V**, and **W**. Take any lattice node as origin O. For indices **UVW**, the line passes through the origin O and a point with co-ordinates Ua, Vb, and Wc, where **U**, **V**, and **W** are any integers. Figure 4.15 shows various *lattice directions* selected merely for convenience of drawing. A direction $\overline{\mathbf{U}}\overline{\mathbf{V}}\overline{\mathbf{W}}$ is merely the opposite to **UVW**. The edges of a unit cell are denoted by [100], [010], and [001], which directions are parallel, respectively, to the a, b, and c axes. A square bracket is always used to denote a lattice direction. The face diagonals of a unit cell are denoted by [110], [011], and [101], and the body diagonals by [111], [$\bar{1}\bar{1}$1], [$\bar{1}$11], [1$\bar{1}$1], or by an opposite set of symbols, for example, [11$\bar{1}$] for [$\bar{1}\bar{1}$1]. Indices [**UVW**] are independent of the shape and symmetry of a lattice, but the geometry of the lattice is needed to calculate the orientation of a line from [**UVW**].

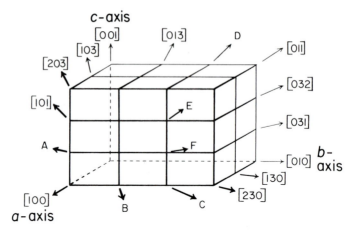

FIGURE 4.15 Lattice directions. An arbitrary stack of 2 × 3 × 3 unit cells is shown. Directions A to F are used in Exercise 4.15.

By analogy with Figures 1.14 and 1.15, any plane passing through lattice nodes is a potential *face* of a crystal, and the ***Miller indices*** (**hk**) in 2D are extended to (**hkl**) in 3D. Choose any lattice node of a unit cell of a lattice as the origin of a set of three right-handed axes parallel to the edges *a, b,* and *c* (Fig. 4.16). Join the coordinates $(a/h,0,0)$, $(0,b/k,0)$, and $(0,0,c/l)$ to obtain a ***lattice plane*** (**hkl**). Caution: boldface one and el look similar.

This construction is sufficient to give the orientation of a crystal face. The area and shape of the face, of course, depend on how the crystal grows. Round brackets are used to distinguish a crystal face (**hkl**) from a lattice

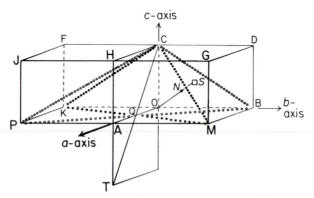

FIGURE 4.16 Lattice planes. The open and filled circles show planes (211) and (2$\bar{1}$1), respectively. *CQT* is the lattice direction [$\bar{1}$02], and *TQC* is the direction [10$\bar{2}$]. *ON* is the perpendicular to plane *BCQ* and the reciprocal point *S* is conventionally shown by a square.

direction [**UVW**]. The intersection of two faces gives an *edge,* and each edge is parallel to a lattice direction. Thus, in Figure 4.16, the intersection between lattice planes *BCPQ* and *MCKQ* is the line *CQ,* which projects through node *T.* Faces with parallel intersections (edges) belong to a **zone.**

For a lattice direction (or edge) to lie in a lattice plane (or face), the indices are related. The equation of a plane passing through $(a/\mathbf{h},0,0)$, $(0,b/\mathbf{k},0)$, and $(0,0,c/\mathbf{l})$ is:

$$\frac{\mathbf{h}x}{a} + \frac{\mathbf{k}y}{b} + \frac{\mathbf{l}z}{c} = 1$$

For convenience, take a parallel plane passing through the origin at:

$$\frac{\mathbf{h}x}{a} + \frac{\mathbf{k}y}{b} + \frac{\mathbf{l}z}{c} = 0$$

A lattice direction [**UVW**] passes through the origin with equation:

$$\frac{x}{\mathbf{U}a} = \frac{y}{\mathbf{V}b} = \frac{z}{\mathbf{W}c}$$

For the lattice direction to lie in the plane, substitute $y = \mathbf{V}bx/\mathbf{U}a$ and $z = \mathbf{W}cx/\mathbf{U}a,$ and clear the fractions to obtain the condition:

$$\mathbf{h}\mathbf{U} + \mathbf{k}\mathbf{V} + \mathbf{l}\mathbf{W} = 0$$

In morphological crystallography this condition is known as the **Weiss zone law,** named after the German mineralogist C.S. Weiss (1780–1856).

In Figure 4.16, planes *BCQ* and *CKQ* have indices (211) and $(2\bar{1}1)$, respectively, and the zone law can be used to obtain the indices [**UVW**] for *CQ*:

$$2\mathbf{U} + \mathbf{V} + \mathbf{W} = 0 \quad \text{and} \quad 2\mathbf{U} - \mathbf{V} + \mathbf{W} = 0$$

whence $\mathbf{U} = \bar{1}$, $\mathbf{V} = 0$, and $\mathbf{W} = 2$, by inspection.

In general, the indices for the lattice direction lying in two planes $(\mathbf{h}_1\mathbf{k}_1\mathbf{l}_1)$ and $(\mathbf{h}_2\mathbf{k}_2\mathbf{l}_2)$ are given by solving the two simultaneous equations:

$$\mathbf{h}_1\mathbf{U} + \mathbf{k}_1\mathbf{V} + \mathbf{l}_1\mathbf{W} = 0$$

$$\mathbf{h}_2\mathbf{U} + \mathbf{k}_2\mathbf{V} + \mathbf{l}_2\mathbf{W} = 0$$

giving

$$U = k_1 l_2 - k_2 l_1 ; V = l_1 h_2 - l_2 h_1 ; W = h_1 k_2 - h_2 k_1.$$

For (211) and (2$\bar{1}$1), [20$\bar{4}$] is obtained, which reduces to [10$\bar{2}$].

The above equations have been concerned only with orientation and not with absolute position. In a lattice there must be an infinite set of planes (**hkl**) because each node must be treated the same as any other node. In Figure 4.16, a plane (111) is obtained at *ABC* and a parallel plane lies at *POF*. Each plane, of course, should be of infinite extent, but it is impossible to draw an infinite set of parallel planes. Furthermore, **h**, **k**, and **l** can independently assume integral values from minus infinity to plus infinity to give a triply infinite set of planes. Fortunately, it is possible to give the essential information about each set of planes, namely, orientation and spacing, by a single point. Choose any point as origin, say *O*, and draw a perpendicular to a set of planes (**hkl**). Thus for (211), the perpendicular line passes through point *N* in plane *BCQ*. The distance *ON* is the **spacing** of the planes, and is denoted *d*(**hkl**). Locate point *S* on the extension of line *ON*, where *OS* = λ/*d*(**hkl**). Point *S* is called the **reciprocal point** for the set of planes (**hkl**), and its vector from point *O* is denoted **d***(**hkl**). The parameter λ is arbitrary, but once chosen must stay constant. The reciprocal points for all the integral values of **h**, **k**, and **l** fall on the nodes of a lattice, which provides an easy geometrical solution to X-ray diffraction.

4.9   Construction of Stereograms for Crystal Faces

The orientation, but not the absolute position, of a face of a crystal can be plotted uniquely as a point on a stereogram from knowledge of the cell dimensions and the Miller indices.

Consider a face (**hkl**) cutting off *a*/**h**, *b*/**k**, and *c*/**l** from the axes (Figure 4.17*a*) of a triclinic crystal of anorthite, which were already plotted in the stereogram of Figure 4.14. The dimensions of the pseudocell are $a = 0.817$ nm, $b = 1.287$, $c = 0.708$. Figure 4.17*b* shows the stereo points for (111), ($\bar{1}$11), (1$\bar{1}$1), and ($\bar{1}\bar{1}$1) obtained by the following procedure. For generality, consider face (**hkl**) in Figure 4.17*a*. Plane trigonometry yields:

$$a\mathbf{k}/b\mathbf{h} = \sin \phi_{ab}/\sin \phi_{ba}$$

$$c\mathbf{k}/b\mathbf{l} = \sin \phi_{cb}/\sin \phi_{bc}$$

$$c\mathbf{h}/a\mathbf{l} = \sin \phi_{ca}/\sin \phi_{ac}$$

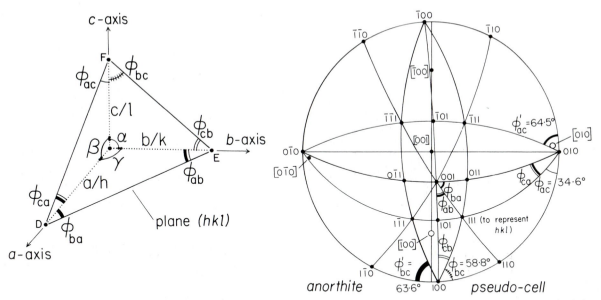

FIGURE 4.17 Angular relations for a triclinic crystal, shown in real space (left) and a stereogram (right). For the stereogram, lattice directions are shown in square brackets [*UVW*]. Round brackets are omitted around the *hkl* triplets for face-normals.

where the six angles are given suffixes in a special way. These six angles become the internal angles between great circles in the stereogram. Thus the *b* axis can be represented by the plane to which it is perpendicular, and this plane is represented by the great circle containing the poles (**001**), (**h0l**), and (**100**). The line *DE* is the intersection between the planes (**hkl**) and (**001**), and it is perpendicular to the great circle containing the poles of these planes. Similarly, the other five angles can be located using the suffixes to ϕ as a guide.

Consider now the specific example of (**111**) in anorthite. Calculation of ϕ_{bc} and ϕ_{ac} allows plotting of the (**111**) pole from two great circles. Because $\phi_{bc} + \phi_{cb} = 180 - \alpha$, the middle equation becomes:

$$\sin(180 - \alpha - \phi_{bc})/\sin \phi_{bc} = c\mathbf{k}/b\mathbf{l}$$

The parameters for anorthite give the following for $\mathbf{h = k = 1}$:

$$\sin(86.88° - \phi_{bc})/\sin \phi_{bc} = 0.5501$$

Using $\sin(P - Q) = \sin P \cos Q - \cos P \sin Q$, and rearranging, gives:

$$\sin \phi_{bc}/\cos \phi_{bc} = \tan \phi_{bc} = \sin 86.88°/(0.5501 + \cos 86.88°)$$

whence $\phi_{bc} = 58.81°$. A similar calculation gives:

$$\tan \phi_{ac} = \sin 64.09°/(0.8666 + \cos 64.09°)$$

whence $\phi_{ac} = 34.61°$.

The pole $(\bar{h}kl)$ lies on the same great circle as (100), (hkl) and $(0kl)$, and ϕ_{bc} is still applicable. However, it is necessary to replace ϕ_{ac} by ϕ_{ac}'. For $(\bar{1}11)$,

$$\tan \phi_{ac}' = \sin 64.09°/(0.8666 - \cos 64.09°)$$

whence $\phi_{ac}' = 64.47°$. Similarly, pole $(h\bar{k}l)$ has the same ϕ_{ac} as (hkl), while ϕ_{bc} is replaced by ϕ_{bc}'. For $(1\bar{1}1)$,

$$\tan \phi_{bc}' = \sin 86.88°/(0.5501 - \cos 86.88°)$$

whence $\phi_{bc}' = 63.60°$. Pole $(\bar{1}\bar{1}1)$ is plotted with ϕ_{ac}' and ϕ_{bc}'.

For an orthorhombic crystal (Fig. 4.18a), all the interaxial angles are set to 90°. Hence, $ak/bh = \tan \phi_{ab}$; $ck/bl = \tan \phi_{cb}$; $ch/al = \tan \phi_{ca}$. Furthermore, $\phi_{ab} = (100) \wedge (hk0)$; $\phi_{cb} = (001) \wedge (0kl)$; $\phi_{ca} = (001) \wedge (h0l)$. The *olivine* used for Figure 4.18a has $a = 4.76$ Å, $b = 10.20$ Å, $c = 5.98$ Å, and the above three angles for $h = k = l = 1$ are 25.02°, 30.38°, and 51.48°.

Special care is needed for a monoclinic crystal. The unit cell of a *sanidine* was chosen with $a = 8.5642$ Å, $b = 13.030$ Å, $c = 7.1749$ Å, and $\beta = 115.994°$, and the principal poles are plotted in Figure 4.18c. An alternative choice of unit cell with $a = 8.5642$ Å, $b = 7.1749$ Å, $c = 13.030$ Å, and $\gamma = 115.994°$, is shown in Figure 4.18b. In diagram 4.18b, the c axis is represented by the pole $[001]$ at the center, and the poles $[100]$ and $[010]$ lie in the circumference inclined at γ. The poles (001) and $[001]$ coincide, but (100) lies at 90° to $[001]$ and $[010]$ and is therefore at $(\gamma - 90°)$ to $[100]$. Modification of the triclinic equations yields:

$$ak/bh = \sin \phi_{ab}/\sin \phi_{ba} = \sin(100) \wedge (hk0)/\sin(hk0) \wedge (010)$$

$$ck/bl = \sin \phi_{cb}/\sin \phi_{bc} = \tan \phi_{cb}$$

$$ch/al = \sin \phi_{ca}/\sin \phi_{ac} = \tan \phi_{ca}$$

The first equation can be solved for ϕ_{ab} using $\phi_{ab} + \phi_{ba} = 180° - \gamma$. For $h = k = l = 1$, the angles are $\phi_{ab} = 28.85°$, $\phi_{cb} = 56.68°$, and $\phi_{ca} = 61.16°$. Because of the monoclinic symmetry, $\phi_{bc}' = \phi_{bc}$. However, ϕ_{ab}' is 49.96°, and is not equal to ϕ_{ab}.

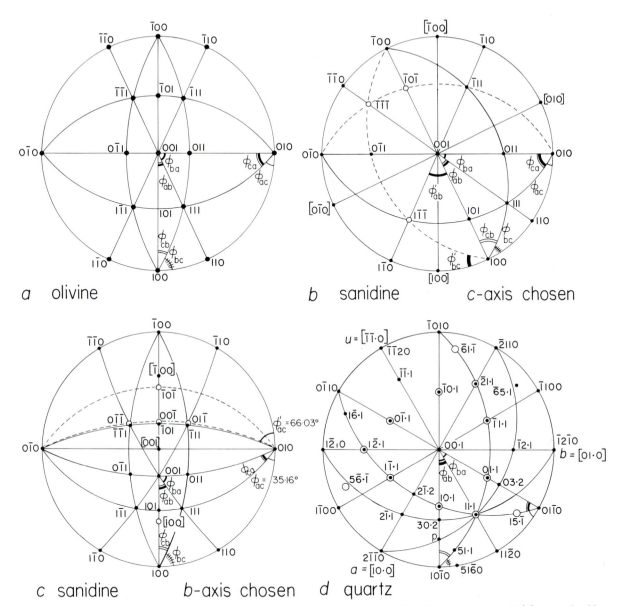

FIGURE 4.18 Angular relations for orthorhombic (*a*), monoclinic (*b, c*), and hexagonal (*d*) crystals. Unbracketed triplets are (**hkl**) for face normals. A horizontal mirror plane for olivine would produce a circle around each dot in the interior. Not all the symmetry-related poles are shown in *b* through *d*.

Diagram 4.18*c* shows the conventional choice of axes for sanidine. The pole [**001**] for the *c* axis is placed at the center and (**010**) = [**010**] at the right. The pole (**100**) is at the bottom, and [**100**] lies at (β – 90°) from it. The triclinic equations yield the following for **h = k = l = 1**:

$$\phi_{ab} = \tan^{-1}(a/b) = 33.32°$$

$$\phi_{cb} = \tan^{-1}(c/b) = 28.84°$$

$$\phi_{ac} = 35.16°; \quad \phi_{ac}' = 66.03°$$

The pole (111) can be plotted with a Wulff net using ϕ_{cb} and ϕ_{ac}. From the Napierian triangle (100), (001), (hk0) in which (100) ∧ (001) = 180 − β, the reader can show that tan(100) ∧ (hk0) = (tan ϕ_{ab}) · (sin β), whence (100) ∧ (110) = 30.57°. Sanidine has pseudocubic geometry as shown by angles near to 30° and 60°. Also note that the poles for ($\bar{1}$01) and (00$\bar{1}$) almost overlap.

To plot a stereogram for a tetragonal crystal, use the equations for an orthorhombic crystal with $a = b$, whence (100) ∧ (110) = 45°. Similarly, a hexagonal crystal can be treated as a monoclinic crystal with $a = b$, $\gamma = 120°$. However, the special angle of 120° has led to use of four instead of three axes, and careful consideration is needed.

Diagram 4.18*d* shows some principal directions for quartz (a = 4.91 Å, c = 5.41 Å). The simplest procedure is to use a and b axes at 120° to each other and at 90° to a vertical c axis. All mathematical equations and indices are then the same as those for diagram 4.18*b*, except for change of γ to 120° and equality of the a and b repeats. However, this procedure does not draw attention to the hexagonal symmetry, and it is customary to use an auxiliary axis u at 120° to a and b and 90° to c. Instead of indices (hkl) for a face, Miller–Bravais indices (hkil) are used. The first three indices allow easy recognition of faces related by the hexagonal symmetry. Thus (11.1), (2$\bar{1}$.1), (1$\bar{2}$.1), ($\bar{1}\bar{1}$.1), ($\bar{2}$1.1), and ($\bar{1}$2.1) are related by the six rotation operations of a hexad axis, which is more easily recognized from (11$\bar{2}$1), (2$\bar{1}\bar{1}$1), (1$\bar{2}$11), and so on. From simple trigonometry it can be shown that **i** = (**h** + **k**). It is merely a matter of personal choice whether three or four symbols are used, but it is wise to use a dot in the third position if the third symbol is omitted. For a lattice direction and corresponding zone axis, it is mandatory to use only three symbols because [**UVW**] corresponds to a coordinate point. Thus the a, b, and u axes must be represented by [**100**], [**010**], and [**$\bar{1}\bar{1}$0**]. A dot is used in diagram 4.18*d*, but it must not be replaced by a number.

In diagram 4.18*d*, the a, b, and u axes are oriented in the usual way with b to the right. Face normals in the circumference are given all four Miller–Bravais symbols, whereas those inside have a dot in place of **i**. The ϕ_{ab} and ϕ_{ba} angles for (11.1) must be equal at 30°. For (11.1), $\phi_{cb} = \tan^{-1}(c/b)$ = 42.23°, and $\phi_{ca} = \phi_{cb}$. Alternatively, simple coordinate geometry gives:

$$(11\bar{2}0) \wedge (11\bar{2}1) = \tan^{-1}(a/2c) = 24.41°$$

$$(10\bar{1}0) \wedge (10\bar{1}1) = \tan^{-1}(a\sin 60°/c) = 38.17°$$

To illustrate the use of zones, what is the face at the intersection of the great circles between $(01\bar{1}0)$ and $(01\bar{1}1)$, and $(11\bar{2}1)$ and $(\bar{1}2\bar{1}1)$? The first two faces give a zone symbol $(010) \times (011) \to [100]$, and the second two faces give $(111) \times (\bar{1}21) \to [\bar{1}\bar{2}3]$. Then $[100] \times [\bar{1}\bar{2}3] \to (0\bar{3}\bar{2}) \equiv (032)$ or $(0\bar{3}\bar{2})$. This face can be obtained more rapidly by the adding rule for zones. The faces (11.1) and $(\bar{1}2.1)$ can add to give (03.2), which is also obtained from $2(00.1)$ and $3(01.0)$. Similarly, (10.1) can be obtained from the zones containing (10.0) and (00.1), and $(1\bar{1}.1)$ and (11.1). Rare quartz crystals show the six faces of the hexagonal trapezohedron indexed as $(51\bar{6}1)$, $(5\bar{6}1\bar{1})$, $(1\bar{6}51)$, $(\bar{6}15\bar{1})$, $(\bar{6}511)$, and $(156\bar{1})$. The face (51.1) must lie between (10.0) and (11.1), and between (51.0) and (00.1). Hence, it can be plotted after calculation of ϕ_{ab} for $\mathbf{h}=5$ and $\mathbf{k}=1$.

4.10 Unit-Cell Transformations

The vectors $\mathbf{a}_1$, $\mathbf{b}_1$, $\mathbf{c}_1$ and $\mathbf{a}_2$, $\mathbf{b}_2$, $\mathbf{c}_2$ of two different unit cells of the same lattice are related by

$$\mathbf{a}_2 = s_{11}\mathbf{a}_1 + s_{12}\mathbf{b}_1 + s_{13}\mathbf{c}_1 \qquad \mathbf{a}_1 = t_{11}\mathbf{a}_2 + t_{12}\mathbf{b}_2 + t_{13}\mathbf{c}_2$$
$$\mathbf{b}_2 = s_{21}\mathbf{a}_1 + s_{22}\mathbf{b}_1 + s_{23}\mathbf{c}_1 \quad \text{and} \quad \mathbf{b}_1 = t_{21}\mathbf{a}_2 + t_{22}\mathbf{b}_2 + t_{23}\mathbf{c}_2$$
$$\mathbf{c}_2 = s_{31}\mathbf{a}_1 + s_{32}\mathbf{b}_1 + s_{33}\mathbf{c}_1 \qquad \mathbf{c}_1 = t_{31}\mathbf{a}_2 + t_{32}\mathbf{b}_2 + t_{33}\mathbf{c}_2$$

The set of nine coefficients s_{11} to s_{33} forms a 3×3 square matrix that is a second-order tensor denoted briefly as s_{ij} where i and j can independently assume the values of 1, 2, and 3. Similarly t_{ij} is a second-order tensor, and t_{ij} is the inverse matrix of s_{ij}, and vice versa.

Consider a monoclinic lattice (Fig. 4.19) in which the first investigator

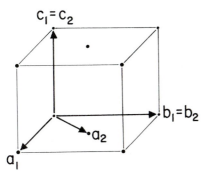

FIGURE 4.19 Two different choices of axes for a monoclinic lattice.

decided to use a unit cell with two lattice points per unit cell, one lying in the center of the *ab* face defined by the other point. A second investigator chose a unit cell of half the volume for which the *b* and *c* axes were retained, and the *a*- axis moved to a face diagonal. Then

$$a_2 = \tfrac{1}{2}a_1 + \tfrac{1}{2}b_1 + 0c_1 \qquad\qquad a_1 = 2a_2 - 1b_2 + 0c_2$$

$$b_2 = 0a_1 + 1b_1 + 0c_1 \quad \text{and} \quad b_1 = 0a_2 + 1b_2 + 0c_2$$

$$c_2 = 0a_1 + 0b_1 + 1c_1 \qquad\qquad c_1 = 0a_2 + 0b_2 + 1c_2$$

These may be expressed by matrices I and II:

	a_1	b_1	c_1
a_2	$\tfrac{1}{2}$	$\tfrac{1}{2}$	0
b_2	0	1	0
c_2	0	0	1

and

	a_2	b_2	c_2
a_1	2	-1	0
b_1	0	1	0
c_1	0	0	1

Miller indices transform the same way as the unit-cell vectors. Thus $(312)_1 \rightarrow (212)_2$:

$$(3 \times \tfrac{1}{2}) + (1 \times \tfrac{1}{2}) + (2 \times 0); (3 \times 0) + (1 \times 1) + (2 \times 0); (3 \times 0)$$
$$+ (1 \times 0) + (2 \times 1)$$

This can be checked in reverse:

$$(2 \times 2) + (1 \times -1) + (2 \times 0); (2 \times 0) + (1 \times 1) + (2 \times 0); (3 \times 0)$$
$$+ (1 \times 0) + (2 \times 1)$$

Fractional coordinates *xyz* and indices **UVW**, however, are transformed after reflecting the matrices about the north–west to south–east diagonal:

	U_1 x_1	V_1 y_1	W_1 z_1
$U_2\ x_2$	$\tfrac{1}{2}$	0	0
$V_2\ y_2$	$\tfrac{1}{2}$	1	0
$W_2\ z_2$	0	0	1

	U_2 x_2	V_2 y_2	W_2 z_2
$U_1\ x_1$	2	0	0
$V_1\ y_1$	-1	1	0
$W_1\ z_1$	0	0	1

Thus $(\frac{1}{4}, \frac{1}{2}, \frac{3}{4})_1 \rightarrow (\frac{1}{8}, \frac{5}{8}, \frac{3}{4})_2$:

$$(\frac{1}{4} \times \frac{1}{2}) + (\frac{1}{2} \times 0) + (\frac{3}{4} \times 0); (\frac{1}{4} \times \frac{1}{2}) + (\frac{1}{2} \times 1)$$
$$+ (\frac{3}{4} \times 0); (\frac{1}{4} \times 0) + (\frac{1}{2} \times 0) + (\frac{3}{4} \times 1)$$

and $[312]_2 \rightarrow [6\bar{2}2]_1$:

$$(3 \times 2) + (1 \times 0) + (2 \times 0); (3 \times -1) + (1 \times 1) + (2 \times 0); (3 \times 0)$$
$$+ (1 \times 0) + (2 \times 1)$$

Fractional coordinates are left unchanged, but $[6\bar{2}2]$ can be reduced to $[3\bar{1}1]$.

Further examples of unit-cell transformations are given in *International Tables for X-ray Crystallography,* Volume 1, Section 2.5. The transformation between rhombohedral and hexagonal unit cells is particularly interesting.

EXERCISES   (* More Difficult)

4.1 Iron occurs in meteorites as kamacite, which has an isometric unit cell with $a = 2.86$ Å and two atoms at fractional coordinates $(0,0,0)$ and $(\frac{1}{2},\frac{1}{2},\frac{1}{2})$. Draw a plan of the unit cell. Calculate the closest approach of two iron atoms. How does your answer compare with twice the atomic radius of an iron atom in Table 1.1?

4.2* Rutile has a tetragonal unit cell with $a = 4.59$ Å, $c = 2.96$ Å. Titanium atoms occur at $(0,0,0,)$ and $(\frac{1}{2},\frac{1}{2},\frac{1}{2})$. Oxygen atoms occur at $(0.305, 0.305,0)$, $(0.805,0.195,\frac{1}{2})$, $(0.695,0.695,0)$, $(0.195,0.805,\frac{1}{2})$. Draw a plan of the unit cell down the c-axis. Calculate distances from the Ti atom at $(\frac{1}{2},\frac{1}{2},\frac{1}{2})$ to the six oxygen atoms that surround it. Then calculate distances between the oxygen atoms and show that they lie near the vertices of a regular octahedron. Compare distances with ionic radii in Table 1.1. Calculate O–Ti–O angles in the octahedron.

4.3* Niccolite has a hexagonal unit cell with $a = 3.60$ Å, $c = 5.01$ Å. Nickel atoms occur at $(0,0,0)$ and $(0,0,\frac{1}{2})$ and arsenic atoms at $(\frac{1}{3},\frac{2}{3},\frac{1}{4})$ and $(\frac{2}{3},\frac{1}{3},\frac{3}{4})$. Draw a plan of the unit cell down the c axis. Show that each arsenic atom is surrounded by six nickel atoms at the corners of a trigonal prism, and that each nickel atom is surrounded by six arsenic atoms at the corners of an octahedron. Calculate the Ni–As distance and the edges of the polyhedra. Show that each nickel atom is almost as close to two nickel atoms as the six arsenic neighbors.

4.4* Marcasite is orthorhombic with $a = 4.44$ Å, $b = 5.41$ Å, $c = 3.38$ Å. Iron atoms occur at $(0,0,0,)$ and $(\frac{1}{2},\frac{1}{2},\frac{1}{2})$ and sulfur atoms at $(0.20, 0.378,0)$, $(0.80,0.622,0)$, $(0.30,0.878,\frac{1}{2})$, and $(0.70,0.122,\frac{1}{2})$. Draw a plan of the unit cell down the c axis. Show that sulfur atoms occur in pairs, and calculate the S–S distance. Show that each iron atom is surrounded by six S atoms at the corners of a distorted octahedron. Calculate the Fe–S distances and the edges of the octahedron.

4.5* A particular variety of sanidine feldspar is monoclinic with $a = 8.5642$ Å, $b = 13.0300$ Å, $c = 7.1749$ Å, and $\beta = 115.994°$. The T_1 site is occupied randomly by 75 percent Si and 25 percent Al atoms at $(0.0097, 0.1850,0.2233)$, and is surrounded by four oxygen atoms O_A $(0,0.1472,0)$, O_B $(0.8273,0.1469,0.2253)$, O_C $(0.0347,0.3100,0.2579)$, and O_D $(0.1793, 0.1269,0.4024)$. Calculate T–O and O–O distances, and O–T–O angles. Show that the four oxygen atoms form a nearly regular tetrahedron about the T_1 atom.

4.6* Low microcline, the ordered polymorph of sanidine, is triclinic with $a = 8.560$ Å, $b = 12.964$ Å, $c = 7.215$ Å, $\alpha = 90.65°$, $\beta = 115.83°$, $\gamma = 87.70°$. The T_1 site is occupied by Al at $(0.0104,0.1875,0.2169)$ and is surrounded by oxygen atoms O_A $(0.0007,0.1448,0.9831)$, O_B $(0.8202,0.1476,0.2205)$, O_C $(0.0352,0.3203,0.2514)$, and O_D $(0.1911,0.1229,0.4053)$. Calculate T–O and O–O distances, and O–T–O angles. Show that the four oxygen atoms form a nearly regular tetrahedron about the T_1 atom. Show that the mean T_1–O distance has increased from high sanidine to low microcline, and check with interionic distances expected from Table 1.1. [Note that each oxygen ion is bonded to two tetrahedral ions in feldspar minerals.]

4.7 Construct the stereograms in Figure 4.5.

4.8 Calculate the angle between face normals of a rhombic dodecahedron using first the equation $\cos^{-1}(l_1 l_2 + m_1 m_2 + n_1 n_2)$ and then an equation involving a Napierian triangle. Check the angle from a model.

4.9 Construct a polar net as in Figure 4.7a using the same radius as in Figure 4.8. A special compass may be needed to construct great circles of large radius. To find the center of a circle that passes through three points, p, q, and r, draw the perpendicular bisectors of pq, qr, and pr.

4.10 Construct a meridian net as in Figure 4.7b, using the same radius as in Figure 4.8.

4.11 Using tracing paper and either your polar net or the commercial meridian net, plot the following cities: Chicago 42°N 88°W, Buenos Aires 35°S 60°W, London 52°N 0°W, Sydney 150°E 35°S, Calcutta 22°N 88°E. Place the North Pole in the center, and the Greenwich meridian up–down. Use a dot for the northern hemisphere and a circle for the southern one. Rotate the tracing paper over the meridian net, and draw great circles between the cities. [Remember that a great circle between a city in the upper hemisphere and a city in the southern hemisphere must pass through the opposites of the two cities.] By counting across small circles, measure the angular distances from Chicago to London, and from Chicago to Sydney. Using the radius of the Earth (3957 miles), calculate great-circle distances in kilometers.

4.12 If you did not gain enough confidence with Exercise 4.11, replot the above cities with the North Pole to the extreme right and the Greenwich meridian horizontal through the middle of the stereogram.

4.13 In a particular hexagonal bipyramid, the angle between adjacent face normals, one each from the upper and lower groups, is 38°. Draw a stereographic projection of all the face normals. Measure the angle between any pair of adjacent upper (or lower) faces. Calculate the angle using Napierian geometry.

4.14 Calculate the angles α^*, β^*, and γ^* for anorthite using $\alpha = 93.12°$, $\beta = 115.91°$, $\gamma = 91.26°$.

4.15 Determine the symbols for lattice directions A–F in Figure 4.15.

4.16 Determine the indices for planes *ABC, BDHA, AHFK, CJPO, BDJP* in Figure 4.16.

4.17 What lattice direction lies in planes (**213**) and (**312**)? What plane contains lattice directions [**231**] and [**321**]?

4.18 Determine the ϕ angles for (**234**) in olivine, and plot the pole on a stereogram using a Wulff net. Check that the three great circles intersect within drawing error. Further exercises on plotting stereograms are given in Chapter 6.

4.19 Instead of centering the $a_1 b_1$ face in Figure 4.19, place a point at the body-center. Choose a_2 so that its arrowhead is now at the new point. Determine the matrices for axial transformations, and transform $(312)_1$, $(¼, ½, ¾)_1$ and $[312]_2$.

ANSWERS

4.1 The plan is given in Figure 4.20*a*. The closest approach is 2.48 Å between an atom at a corner and one at a body center. This distance compares with the atomic radius of 1.26 Å from Table 1.1.

4.2 The plan in Figure 4.20*b* shows Ti with a dot and oxygen with a circle. $pr = pt = 1.98$ Å, $pq = ps = 1.95$ Å. The atoms q and s are repeated at $z = 1$, giving q' and s', which also lie at 1.95 Å to p. Thus p has six oxygens as neighbors. $qq' = ss' = 2.96$ Å. $qs = q's' = 2.53$ Å. $qt = qr = rs = st = q't = q'r$

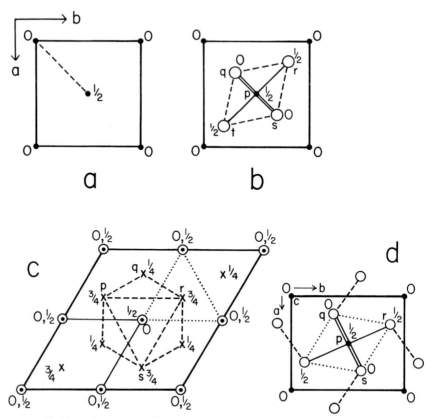

FIGURE 4.20 Answers to Exercises 4.1, 4.2, 4.3, and 4.4.

$= rs' = s't = 2.78$ Å. $\angle qpr = 90°$, $\angle qpq' = 98.92°$, $\angle qps = 81.08°$. The 12 distances are approximately equal. The octahedron is viewed down a diad axis. [The distortion in the octahedron can be explained by the repulsion between the Ti ions. The octahedra share edges of type qs to form columns along c, and the repulsion between Ti ions causes lengthening of qq' and ss' and shortening of qs and $q's'$.] From Table 1.1, the ionic radius of Ti^{4+} in six coordination (i.e., bonded to six neighbors) is 0.60, and the ionic radius of O^{2-} with three neighbors is 1.36 Å. The sum of 1.96 Å lies between the calculated distances of 1.98 and 1.95 Å. [The columns of octahedra are cross-linked by sharing of vertices to form an infinite framework of linked octahedra.]

4.3 Four unit cells are shown in Figure 4.20c. Nickel atoms are shown by circled dots at height 0 and $\frac{1}{2}$, and arsenic atoms by crosses. The Ni–As distance of 2.43 Å compares with 2.49 Å, the sum of atomic radii from Table 1.1. A trigonal prism of nickel atoms is shown by dotted lines: the vertical edges are 2.505 Å long and the horizontal edges are 3.60 Å. An octahedron of arsenic atoms is shown by dashed lines: $pq = 3.25$ Å; $pr = 3.60$ Å. The octahedron is flattened down the c axis. Each Ni atom has a Ni atom directly above and directly below at $c/2 = 2.505$ Å, which is only slightly longer than Ni—As $= 2.43$ Å. [The octahedra share faces of type prs to form columns along c, and share edges of type $pq = qr$ with octahedra of adjacent columns. The trigonal prisms share edges with each other, leaving half the volume empty because of the alternation of arsenic atoms at $z = \frac{1}{4}$ and $\frac{3}{4}$.]

4.4 The plan (Fig. 4.20d) shows Fe atoms by dots and S_2 doublets by circles linked by dashed lines. Six S atoms form an octahedron about the iron atom at the center. $pr = 2.23$ Å, $pq = 2.25$ Å, $qs = 2.97$ Å, $qr = 3.22$ Å, $rs = 3.11$ Å, $qq' = 3.38$ Å (q is at $z = 1$). [The octahedra share horizontal edges of type qs to form columns parallel to c. Adjacent columns are displaced by $c/2$, and share vertices of the octahedra. Note that the shared edges are shorter than unshared edges.]

4.5 $T_1 - O_A = 1.643$ Å, $T_1 - O_B = 1.645$ Å, $T_1 - O_C = 1.647$ Å, $T_1 - O_D = 1.643$ Å, $O_A - O_B = 2.626$ Å, $O_A - O_C = 2.744$ Å, $O_A - O_D = 2.623$ Å, $O_B - O_C = 2.713$ Å, $O_B - O_D = 2.723$ Å, $O_C - O_D = 2.679$ Å. The angles $\angle ATB = 106.0°$, $\angle ATC = 113.0°$, $\angle ATD = 105.9°$, $\angle BTC = 110.0°$, $\angle BTD = 111.8°$, and $\angle CTD = 109.0°$ are close to the angle of $109.47°$ for a regular tetrahedron.

4.6 $T_1 - O_A = 1.738$ Å, $T_1 - O_B = 1.739$ Å, $Ti - O_C = 1.745$ Å, $T_1 - O_D = 1.741$ Å, $O_A - O_B = 2.761$ Å, $O_A - O_C = 2.910$ Å, $O_A - O_D = 2.770$ Å, $O_B - O_C = 2.900$ Å, $O_B - O_D = 2.863$ Å, $O_C - O_D = 2.839$ Å, $\angle ATB = 105.2°$, $\angle ATC = 113.3°$, $\angle ATD = 105.5°$, $\angle BTC = 112.7°$, $\angle BTD = 110.7°$, $\angle CTD = 109.1°$. The mean $T_1 - O$ distance in low microcline (1.741 Å) is greater than that (1.645 Å) in high sanidine. Predicted values from Table 1.1 are 1.74 Å for Al–O and 1.61 Å for Si–O, giving 1.64_2 Å for random occupancy of 75 percent Si and 25 percent Al.

4.8 Refer to Figures 3.14a and 4.5g. Each face lies at 45° to two axes and 90° to the third. Face normal s in Figure 4.5g has direction cosines with respect to the usual orientation of axes a, b, and c: cos 45°, cos 45°, cos 90°. Face normal t, pointing upward, has direction cosines cos 45°, cos 90°, cos 45°. Hence the angle st is 60°. Napierian triangle sat has $\angle sat = 90°$ and $as = at = 45°$. Let st be the required angle, and S and T the angles at s and t $\sin(90 - st) = (\cos 45°) \cdot (\cos 45°)$, giving $st = 60°$.

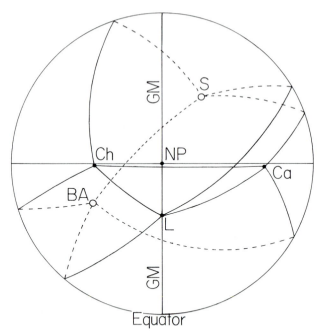

FIGURE 4.21   Answer to Exercise 4.11.

4.11   See the stereogram in Figure 4.21. Graphical measurement gave Chicago–London = $57\frac{1}{2}°$ and Chicago–Sydney = $134\frac{1}{2}°$. One degree = 3957 × 1.609 × $\pi/180$ km.   Chicago–London $\doteq$ 6390   km,   Chicago–Sydney $\doteq$ 14,950 km. Students usually obtain answers within $\frac{1}{2}$ to 1° of the above angles, corresponding to a maximum variation of about 100 km. [**Note**. exact values can be calculated using spherical trigonometry.]

4.12   See the stereogram in Figure 4.22.

4.13   See the stereogram in Figure 4.23. The measured angle is $56\frac{1}{2}°$. In Napierian triangle *pqr*, *pq* = 71°, ∠*pqr* = 30° by symmetry, and ∠*prq* = 90°. sin *pr* = [cos(90 − 71)°][cos(90 − 30)°]. *pr* = 28.21°, and the required angle is 56.42°.

4.14 α* = 85.92°, β* = 63.97°, γ* = 87.08°.

4.15   A[201], B[210], C[110], D[023], E[111], F[221].

4.16   *ABC* (111), *BDHA* (110), *AHFK* (1$\bar{1}$0), *CJPO* same as *BDHA*, *BDJP* (210).

4.17   [$\bar{1}5\bar{1}$], (11$\bar{5}$).

4.18 ϕ_{ab} = 34.99°, ϕ_{cb} = 23.74°, ϕ_{ca} = 32.14°.

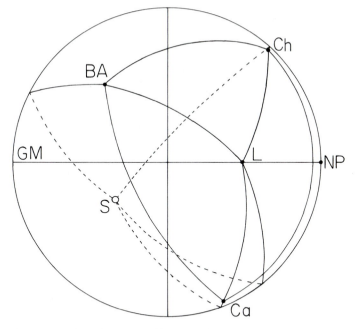

FIGURE 4.22   Answer to Exercise 4.12.

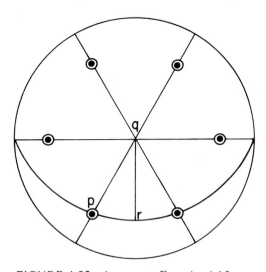

FIGURE 4.23   Answer to Exercise 4.13.

4.19

→a_1	b_1	c_1

	→a_1	b_1	c_1
a_2	½	½	½
b_2	0	1	0
c_2	0	0	1

	→a_2	b_2	c_2
a_1	2	−1	−1
b_1	0	1	0
c_1	0	0	1

$(312)_2$, fortuitously the same indices; $(\frac{1}{8},\frac{5}{8},\frac{7}{8})_2$; $[6\,\bar{2}\,\bar{1}]_2$.

5
Packing in Three Dimensions

Preview   The concepts of close versus open packing and homogeneous versus heterogeneous packing were explored for 2D in Chapter 1. Some features of 3D packing are now presented with emphasis on sphere packing and polyhedra. Readers should take note of the symmetry elements and lattices in preparation for a systematic enumeration of point groups and lattices in Chapter 6. Following a description of space groups in Chapter 7, important crystal structures are described in Chapters 8 and 9 from the viewpoint of both packing and symmetry.

As you study this chapter, you will almost follow the early history of discovery of crystal structures from the simple prediction of ordered patterns of close-packed spheres in the 1890s by W. Barlow, to the discovery of simple structures by X-ray diffraction 20 years later by W.L. Bragg, and to the increasing sophistication of ideas on packing in the 1920s, (e.g., Bragg, L. Pauling).

Although this is not a book on chemical bonding, you should bear in mind that, (a) ideal metallic bonding merely involves the closest packing of spheres, (b) covalent bonding requires special values of interatomic angles (e.g., the tetrahedral angle for sp^3 hybridization of carbon in diamond), (c) ideal ionic bonding involves minimization of the electrostatic potential and, (d) most bonds are hybrid.

For ionic and even some covalent crystals, Pauling's rules tend to be obeyed and are useful in understanding the arrangement of the anions and cations. Slightly rewritten, they are:

1. Each cation is surrounded by anions whose centers constitute a coordination polyhedron (the consequences of anions being larger than cations, Table 1.1). The cation-anion distance is the sum of the ionic radii, and the coordination number decreases as the radius ratio of cation to anion decreases (the consequence of close packing of spheres assumed to be hard).

2. The sum of valency bonds from adjacent cations is equal to the charge on an anion (e.g., Mg^{2+} surrounded by six O^{2-} generates a

valency bond of 2/6, and Si^{4+} surrounded by four O^{2-} generates a valency bond of 4/4; in olivine, each O^{2-} is bonded to one Si^{4+} and three Mg^{2+} so that $(1 \times 4/4) + (3 \times 2/6) = 2$.

3. Coordination polyhedra around cations tend not to share edges and especially faces if the cation is highly charged, or small, or both; any shared edge is shortened (the consequence of increase of electrostatic repulsion between cations). Furthermore, cations of high charge and small size lie as far apart as possible.

5.1   Packing of Nearest Neighbors

Because the strongest chemical bonding is between nearest neighbors, consider first how spheres of one type can surround a sphere of a second type. Figure 3.1 shows how four, six, and eight spheres can touch. A smaller sphere (not shown; denoted B) of the appropriate radius could fit neatly into the hole between each set of larger close-packed spheres (denoted A). If A and B were atoms, the A atoms would be called the *first neighbors* of the B atom, and the polyhedron formed by joining the centers of adjacent A atoms would be called the *coordination polyhedron* of the B atom. The number of first neighbors is called the *coordination number*. Tetrahedral, cubic, and octahedral coordinations correspond, respectively, to four, eight, and six first neighbors. Figure 5.1 shows sections in which the A and B spheres are in contact. For the octahedron (5.1c), the A spheres touch along each side of a square, and the A and B spheres along each diagonal: thus $2r_A \cdot \sqrt{2} = 2(r_A + r_B)$. The *radius ratio* for a perfect fit is $(\sqrt{2} - 1) \simeq 0.414$. For the cube (5.1b), the A spheres touch along each edge, and the A and B

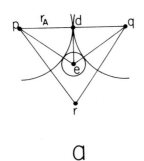
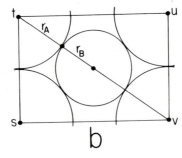
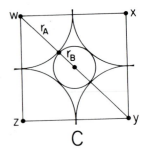

FIGURE 5.1   Diagrams for calculation of radius ratio for perfect fit in tetrahedral (*a*), cubic (*b*), and octahedral (*c*) coordination. Points *p* to *z* correspond to the points in Figure 3.1. In (*a*), angles $p\hat{e}d = q\hat{e}d = d\hat{p}r = d\hat{q}r = 54.735°$. Spheres are in contact along *pe* and *qe* but not along *re*. In (*b*), $uv = 2r_A$, $tu = 2r_A \cdot \sqrt{2}$, and $tv = 2(r_A + r_B)$. In (*c*), $wx = xy = 2r_A$ and $wy = 2(r_A + r_B)$.

spheres along each body diagonal. The section is a rectangle with edges $2r_A$ and $2r_A \cdot \sqrt{2}$, giving the Pythagorean equation:

$$(2r_A)^2 + (2r_A \cdot \sqrt{2})^2 = [2(r_A + r_B)]^2$$

whence

$$r_B/r_A = \sqrt{3} - 1 \cong 0.732$$

For the tetrahedron (5.1a), the radius ratio for perfect fit is obtained rapidly from the subtended angle $\angle peq$ of 109.47° (Table 3.1; calculated in Section 4.4): thus $r_B/r_A \cong (\text{cosec } 54.735° - 1) \cong 0.225$.

Table 5.1 lists the coordination number and ideal radius ratio for the common types of coordination polyhedra. Calculations for the Archimedean trigonal prism, Archimedean square antiprism, and regular icosahedron are straightforward. The cuboctahedron and "twinned cuboctahedron" are the coordination polyhedra obtained in the cubic closest packing and hexagonal closest packing of equal spheres (next section). Geometrical data for the monocapped octahedron, triangulated dodecahedron, and tricapped trigonal prism are from Wells (1975). Coordination numbers greater than 12 occur in certain alloys in which smaller atoms cluster around a larger central atom. Hard spheres of the same radius will not fit exactly with all their neighbors

Table 5.1 Ideal radius ratio for coordination polyhedra

Coordination Number	Coordination Polyhedron	Ideal Radius Ratio
4	Regular tetrahedron	0.225
6	Regular octahedron	0.414
	Archimedean trigonal prism	0.528
7	Monocapped octahedron	0.592
8	Archimedean square antiprism	0.645
	Triangulated dodecahedron	0.668
	Cube	0.732
9	Tricapped trigonal prism	0.732
12	Regular icosahedron	0.902
	Cuboctahedron	1
	"Twinned" cuboctahedron	1
14	Bicapped hexagonal antiprism	Irregular, ~1.2
15	(See Figure 3.19g)	Irregular, ~1.2
16	Tetracapped truncated tetrahedron	Irregular, ~1.3

in such clusters, and two or more types of A atoms are needed (Frank and Kasper, 1958), for which a crude average of the radius ratios is given in Table 5.1.

Only the coordination polyhedra with high symmetry are found in simple crystal structures: indeed, the tetrahedron and octahedron are of overwhelming importance.

Coordination polyhedra with many triangular faces provide high packing efficiency, as exemplified by the octahedron and icosahedron. The square antiprism is more efficient than the cube for coordination number 8, but the triangulated dodecahedron composed entirely of triangular faces is not quite as efficient as the square antiprism, which has two square faces.

5.2   Infinite Packing of Equal Spheres

Perhaps the most obvious kind of packing is to place a sphere at each corner of a cube (Fig. 3.1b). Regular repetition of the cube as a unit cell results in the center of each sphere lying at a node of a primitive cubic lattice (Fig. 6.1) that extends to infinity in all three directions. This cubic lattice is the analogue of the square lattice in 2D (Fig. 2.5). The cubic 3D lattice is *primitive* because there is only one lattice point per unit cell (each of the eight corners, of course, is shared between eight unit cells). The cell dimension a is determined by the diameter of each close-packed sphere.

The unit cell is shown in Figure 5.2a as a projection onto one face. This 2D projection consists of a square with the center of a sphere at each corner. By convention, a point at zero height is often shown without a symbol for the height. A point at height zero must be repeated with a point at fractional height of unity, and, indeed, for all possible integers because of the lattice repeat. The second diagram in Figure 5.2a is a stereogram for the first neighbors that lie along the directions of the face normals of a cube. The co-ordination polyhedron is an octahedron, which is the dual of a cube. The coordination number is 6. The spheres of radius r touch each other at the center of each edge of the unit cell. Each sphere has volume $4\pi r^3/3$, the cell volume is $(2r)^3$, and the packing efficiency is therefore $\pi/6 \simeq 52.4$ percent. Obviously, simple cubic packing is rather open, and, indeed, balls will not rest together in simple cubic packing unless glued or held together.

Body-centered cubic packing (Exercise 4.1, Fig. 4.20a, Fig. 5.2b) is more efficient, but still does not have closest packing. The unit cell remains isometric, but now has two spheres centered at $(0,0,0)$ and $(\frac{1}{2},\frac{1}{2},\frac{1}{2})$. Do the two spheres fall on a single lattice? Yes, because their centers have the same vectorial environment that is shown in the stereogram for the first neighbors. Each sphere is in contact with eight neighbors that lie in octahedral directions. The coordination polyhedron is a cube, which is the dual

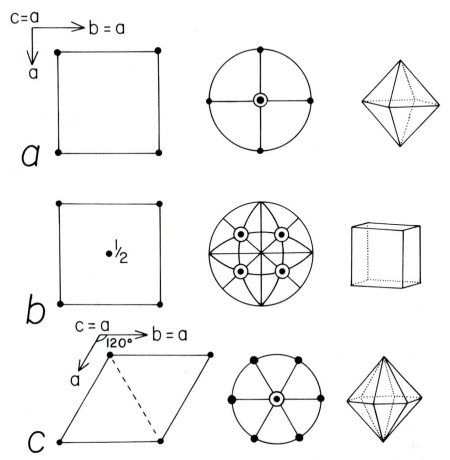

FIGURE 5.2 The close-packing of equal spheres. (*a*) Primitive cubic close-packing with plan of unit cell, stereogram of directions of nearest neighbors, and clinographic drawing of the coordination polyhedron. (*b*) body-centered cubic close-packing. (*c*) Primitive hexagonal close-packing.

of the octahedron. Because the spheres contact each other along body diagonals, $\sqrt{3} \cdot a = 4r$. The spheres do not touch along cell edges, and it is difficult to make a model by gluing balls together unless the balls are wedged apart until the glue has dried. From the cell volume $(4r/\sqrt{3})^3$ and the volume of two spheres $(8\pi r^3/3)$, the packing efficiency is $\sqrt{3} \cdot \pi/8 \backsimeq 68.0$ percent. Note that the packing efficiency increases as the coordination number increases from 6 to 8. Can the efficiency be increased?

First, try a primitive hexagonal cell to take advantage of the closest packing of circles in 2D. Each sphere has six first neighbors at 60° in the horizontal plane (Fig. 5.2*c*) and one above and one below, to give a total of eight. The coordination polyhedron is a hexagonal bipyramid. For spheres in contact, the cell dimensions are $a = c = 2r$, and the packing efficiency is

$4\pi r^3/3$ divided by $8r^3 \cdot \sin 60° = \pi/3\sqrt{3} \simeq 60.5$ percent. So this type of packing has a lower efficiency than body-centered cubic packing, even though the coordination numbers are equal. It is difficult to build a model of **simple hexagonal packing** because the second layer of spheres slips off the top of the first layer of spheres unless held until the glue dries. Why fight this tendency? Let the second layer of spheres (labeled *B*) drop neatly into the dimples formed by the first layer (labeled *A*) of close-packed spheres (Fig. 5.3*a*). Automatically, another set of dimples appears at the top of the second layer of spheres, and it is easy to continue building. Obviously, this is a dense form of packing, and, indeed, it is the closest packing of equal spheres in an infinite array.

But wait, are there two choices of relative position of adjacent layers? Yes, there are. Take any layer and denote its spheres by the nodes of a hexagonal 2D lattice labeled *A,* as in Figure 5.3*a.* A second layer of spheres can lie in the dimples labeled *B,* as in Figure 5.3*a,* or in the dimples labeled *C.* Simultaneous occupancy of *B* and *C* is impossible for spheres of one size. The choices of *B* or *C* are formally different but are not topologically different because the sequence *AB* can be turned into *AC* by rotation of 180° about any node. Since the absolute orientation in space is irrelevant, and it is only the topologic relation that is meaningful, let us choose *AB.* A third layer can go into either of the lattice positions *C* or *A,* but raised vertically again. Because the first two layers have been fixed into an absolute orientation, there is now a topologic distinction between the two possible positions *A* and *C* of the third layer. The simplest pattern is obviously ...*ABABAB*..., or briefly, *AḂ,* where the superscript dots denote the repeat unit. The unit cell is hexagonal (Fig. 5.3*b*), and contains two spheres at $(0,0,0)$ and $(\frac{2}{3},\frac{1}{3},\frac{1}{2})$. An alternative choice of $(0,0,0)$ and $(\frac{1}{3},\frac{2}{3},\frac{1}{2})$ merely corresponds to the 180° rotation described above. This type of closest packing is known as **hexagonal closest packing.** The next simplest pattern ...*ABCABC*..., or briefly, *AḂĊ,* can be described on a hexagonal unit cell (Fig. 5.3*c*), three layers high, with spheres centered on $(0,0,0)$, $(\frac{2}{3},\frac{1}{3},\frac{1}{3})$ and $(\frac{1}{3},\frac{2}{3},\frac{2}{3})$. But the symmetry is actually isometric, and the unit cell (Fig. 5.3*d*) is centered on all faces with spheres at $(0,0,0)$, $(0,\frac{1}{2},\frac{1}{2})$, $(\frac{1}{2},0,\frac{1}{2})$ and $(\frac{1}{2},\frac{1}{2},0)$. This pattern is known as **cubic closest packing,** or sometimes as **face-centered cubic packing.** The acronyms bcc, hcp, and ccp are often used for body-centered cubic packing, hexagonal closest packing, and cubic closest packing, respectively. An infinite number of closest-packed sequences is possible, including *AḂAĊ* and a completely disordered type. The sequences are known as **polytypes** because they involve merely different ways of arranging identical layers along the perpendicular direction.

The relationship between the hexagonal and cubic unit cells of cubic closest packing is best seen with a model of glued balls, as in Figure 13 of

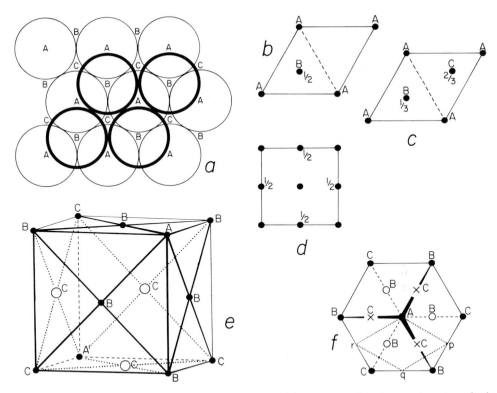

FIGURE 5.3 Closest-packing of equal spheres. (*a*) Diagram showing a closest-packed layer (thin line) with spheres centered on *A* positions, and a superimposed layer (thick line) with spheres centered on *B* positions. Positions of type *C* are unoccupied. (*b*) Hexagonal closest-packing with a sphere in position *A* at (0,0,0) and an identical sphere in position *B* at (0,0,½) of the hexagonal unit cell, which is two layers deep. (*c*) Cubic closest-packing depicted with a hexagonal unit cell three layers deep with identical spheres occupying positions *A* (0,0,0), *B* (⅔,⅓,⅓), and *C* (⅓,⅔,⅔) for the usual axial orientation. (*d*) Cubic closest-packing depicted with a cubic unit cell in which identical atoms occupy the corners and the center of all faces of the unit cell. (*e*) Clinographic drawing of the cubic unit cell showing the relationship to the hexagonal closest-packed layers. Take the *c*-repeat of the hexagonal unit cell along the body diagonal between the corners labeled *A* and *A'*. The six nodes labeled *B* represent a *B* layer, and the six nodes labeled *C* represent a *C* layer. Face diagonals are drawn for the three exposed faces, whose centers are occupied by atoms of type *B,* and for the three hidden faces whose centers are occupied by atoms of type *C* (circles). (*f*) Ditto for cube viewed down a triad axis as in Figure 3.3*d*. Atoms at the corners are shown by dots, those at the center of exposed faces by circles, and those at the center of hidden faces by crosses. The dotted lines outline a hexagonal cell, and *p, q,* and *r* show the position of *A* atoms from adjacent cubic cells.

Crystal Structure of Minerals by Bragg et al. (1965) or in Figure 4.14*a* of Wells (1975). Each reader is advised to build a model with $\dot{A}B\dot{C}$ stacking of either cork or table-tennis balls, and to identify a face-centered isometric cell. Figure 5.3*e* is a perspective drawing in which each lattice point is labeled appropriately with *A*, *B*, or *C*. Figure 5.3*f* is a plan down a triad axis. Dotted lines show the position of a hexagonal cell. Corners *p*, *q*, and *r* are occupied by Type *A* atoms from adjacent cubic cells, as can be deduced by using lattice translations.

All polytypes of closest packing of equal spheres have 12 coordination for each sphere, but there are two types of coordination polyhedra (Fig. 5.4). When viewed perpendicular to the *A*, *B*, and *C* layers, each sphere has six neighbors from its own horizontal layer, plus three each from the two adjacent layers. If an A layer is sandwiched between two *B* layers as in hexagonal closest packing, the stereogram (5.4*b*) shows three circled dots inside the primitive. But if an *A* layer is sandwiched between a *B* and a *C* layer as in cubic closest packing, the three dots alternate with the three circles (5.4*d*). For the latter arrangement, the coordination polyhedron is a cuboctahedron, shown in Figure 5.4*a* by a parallel projection down one of the four triad axes. For hexagonal closest packing, the polyhedron is a "twinned cuboctahedron," which also has eight triangular and six square faces, but has only

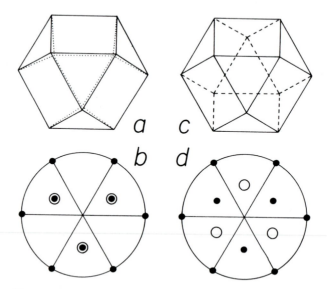

FIGURE 5.4 Coordination polyhedron and stereogram of directions to nearest neighbors of hexagonal and cubic closest-packings. (*a* and *b*) Hexagonal. (*c* and *d*) Cubic. In (*b*) and (*d*) the poles lie at 54.74 and 90° from the center.

one triad axis. Further drawings of these polyhedra were given in Figures 3.17g, and *h*.

The packing efficiency is easily calculated from the cubic unit cell, which contains four spheres of combined volume $16\pi r^3/3$. Adjacent spheres are in contact along a face diagonal (Fig. 5.3e), and $\sqrt{2} \cdot a = 4r$, giving a cell volume of $a^3 = 16\sqrt{2}r^3$. The resulting packing efficiency of ~74.0 percent is greater than the 68 percent for body-centered cubic packing.

The regular icosahedron cannot be used as the coordination polyhedron for the closest packing of *equal* spheres, because the radius ratio is 0.902 instead of unity (Table 5.1). Furthermore, even if two sizes of spheres are used, it is not possible to produce an infinite array of closest-packed spheres all of which lie on two lattices. Icosahedral clusters are important in alloys, but such clusters of near-equal spheres do not pack together as efficiently as hcp and ccp packing of equal spheres when the whole crystal is considered.

There are many complex structures in which spheres touch each other to form a continuous linkage in 3D, but in which the density of packing is low. Only the **diamond** structure is given here, but a wide range of packing densities can be obtained, as illustrated by the oxygen atoms in zeolite minerals. In diamond (Fig. 5.5, right), C atoms lie at (0,0,0), (½,½,0), (0,½,½), (½,0,½), (¼,¼,¼), (¾,¾,¼), (¾,¼,¾), and (¼,¾,¾). Each C atom has four neighbors that lie at the corners of a tetrahedron. The tetrahedra alternate from a positive to a negative orientation. The packing efficiency is only one half of that for body-centered cubic packing, as can be deduced by taking a 2 × 2 × 2 block of unit cells of bcc and removing half the positions. Construction of a ball model is difficult because there are no close-packed layers. The diamond structure is stable only because the C–C

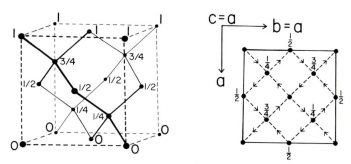

FIGURE 5.5 The diamond structure shown in clinographic projection (left) and plan (right). The *C* atoms are linked into a 4-connected net in the left-hand diagram, and the unit cell is shown by dashed lines. The directions of the branches of the net are shown by dashed lines and arrows in the right-hand drawing.

bonds adopt a tetrahedral arrangement from hybridization of the molecular orbitals. Indeed, diamond is best regarded as an infinite molecule of composition C_∞, and spheres used in model building should be flattened in the tetrahedral bond directions. The interatomic bonds can be regarded as the branches of a 3D net in which four branches meet at each atomic node (Fig. 5.5, left).

5.3   Infinite Packing of Near-Equal Spheres

Silver and gold atoms have similar atomic radii (1.44 Å) and form an alloy with random occupancy of ccp sites at all temperatures and all Ag/Au ratios. Copper and gold atoms have rather different atomic radii (1.28 Å and 1.44 Å) but still form a random ccp alloy for all Cu/Au ratios at elevated temperatures for which there is enough thermal driving force for disorder. At lower temperatures, ordered structures become stable, and Figure 5.6 shows the ***tetragonal structure for Cu_1Au_1*** and the ***primitive cubic structure for Cu_3Au_1***.

In the tetragonal structure, layers perpendicular to the c-axis are alternately occupied by Cu and Au atoms. In order to fit together in the horizontal directions, the gold atoms must be squeezed with respect to the copper atoms. Upon heating, the Cu and Au atoms occupy the sites at random, giving a face-centered cubic cell with $a \simeq 3.9$ Å. Upon cooling, the tetragonal structure can nucleate randomly to give domains (cf. Fig. 1.11). Domains that meet in phase will merge to give one domain in which the layers have alternate Cu and Au atoms. Domains that meet out of phase cannot merge, because two adjacent layers are composed entirely of either Cu or Au atoms. Such antiphase domains produce an antiphase intergrowth with boundaries perpendicular to the c axis.

Retention of the cubic axial directions for the tetragonal unit cell (con-

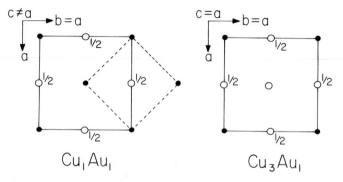

FIGURE 5.6   Superstructures in the Cu,Au chemical system. Dots and circles respectively represent gold and copper atoms.

tinuous lines in Fig. 5.6) gives an unnecessarily large cell with centering of the upper and lower faces. The smaller uncentered cell (dashed lines) with a=2.81 Å, c=3.70 Å is sufficient.

The ordered Cu_3Au_1 alloy remains cubic, but the symmetry has dropped from face-centered to primitive cubic. The atoms at the center of the faces (circles in Fig. 5.6) are no longer related to the atoms at the cell corners, and the structure must now be described as the interpenetration of four separate lattices, each of primitive type. The disordered variety of Cu_3Au_1 alloy has statistical face-centered symmetry, and nucleation can occur upon cooling with four choices of origin for the primitive unit cell, thereby producing a complex antiphase texture.

A crystal structure in which one set of symmetry-related positions is occupied by a regular sequence of two or more different types of atoms is known as a *superstructure*. Removal of the distinction between the two or more types of atoms gives the *parent* or *degenerate* structure, from which the superstructure can be regarded as a *derivative* structure. Thus the ccp structure with one set of symmetry-related positions is the parent of the Cu_1Au_1 and Cu_3Au_1 ordered superstructures. All three structures are based on one set of positions that have the same vectorial environment until atoms are placed on them in an ordered array. Only the ccp structure has the highest possible symmetry, which is face-centered cubic, while the other two structures have lower symmetry. The misleading term *superlattice* was used instead of superstructure in the old literature, and still persists.

The reader may like to invent a superstructure for the bcc structure before proceeding to the next paragraph.

Figure 5.7a shows the obvious superstructure for bcc. The site at the center of the cell is occupied by an atom different from the atom at the corner. This superstructure is found in various metallic phases including CuZn alloy (*β-brass*). Heating causes the Cu and Zn atoms to occupy the sites at random to give a degenerate parent structure. The β-brass structure has the same atomic positions as the *CsCl structure,* in which the sites are occupied by Cs and Cl atoms, or Cs^+ and Cl^- ions for the ionic model. Heating does not cause the Cs and Cl atoms to occupy randomly one set of sites, and the bcc structure is only a parent structure in a mathematical sense. Because the Cs and Cl atoms lie on two separate lattices, each with primitive cubic symmetry, the CsCl structure should not be called a body-centered cubic structure. The *Fe$_3$Al superstructure* (Fig. 5.7b) of the parent bcc structure has a cubic unit cell whose edge is doubled. The atoms lie on four interpenetrating lattices, each with face-centered symmetry. The *NaTl superstructure* (Fig. 5.7c) also has a cubic unit cell with doubled edge, and consists of four interpenetrating lattices, each with face-centered symmetry. All atoms have 8-fold coordination. In the NaTl superstructure, each atom is surrounded by four like and four unlike atoms, whereas in the Fe_3Al superstructure each Al

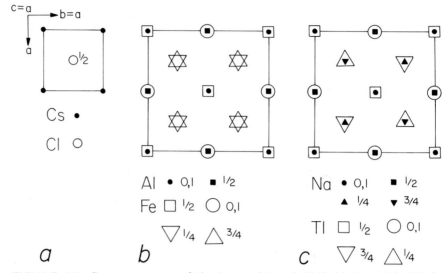

FIGURE 5.7 Superstructures of the *bcc* packing. CsCl (left), Fe₃Al (middle), and NaTl (right). The Cl atom is at height ½ at the body center of the Cs atoms. For the Fe₃Al and NaTl structures, the height is shown by the shape of the symbol and the atom type by the open or closed nature.

atom is surrounded by eight Fe atoms and each Fe atom by either four Al and four Fe atoms or by eight Fe atoms.

The ***NaCl structure*** type (Fig. 5.8, left) can be envisaged geometrically as a superstructure of the primitive cubic type of packing. Take a 2 × 2 × 2 block of unit cells with a site at each corner, and place the two types of atoms so that each atom is surrounded by six atoms of the other type dis-

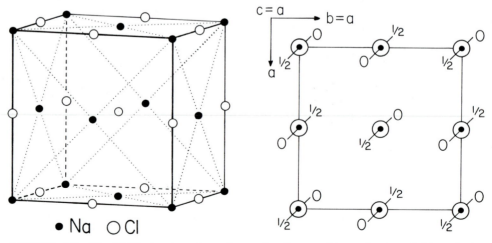

FIGURE 5.8 The NaCl structure type shown in clinographic projection (left) and plan (right). It does not matter which atom is assigned to which symbol, as indicated by deliberate exchange of symbols between left and right.

placed by one of the edge vectors of the original cell. The new cell has twice the cell dimensions of the old cell, and the atoms lie on two interpenetrating lattices, each with face-centered symmetry. Thus in Figure 5.8 (left), the dots occupy all the corners and face centers of the new cell, while the circles lie on a similar lattice displaced by half an edge of the new cell. Nonoccurrence of the NaCl structure type for any pair of atoms that have almost the same radius is to be expected because the packing efficiency for a radius ratio of unity is so low (~52%). In the next section, the NaCl structure type will be derived by the filling of small holes between large spheres that touch each other.

Using the diamond structure as a parent, regular alternation of B and N atoms produces the derivative structure (Fig. 5.9) found in **boron nitride**.

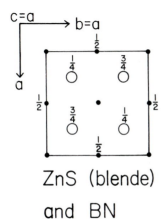

ZnS (blende) and BN

FIGURE 5.9 The ZnS (blende) structure type. It does not matter which atom is assigned to which symbol.

In this superstructure the B atoms are placed at $(0,0,0)$, $(0,\frac{1}{2},\frac{1}{2})$, $(\frac{1}{2},0,\frac{1}{2})$ and $(\frac{1}{2},\frac{1}{2},0)$ and the N atoms at $(\frac{1}{4},\frac{1}{4},\frac{1}{4})$, $(\frac{1}{4},\frac{3}{4},\frac{3}{4})$, $(\frac{3}{4},\frac{1}{4},\frac{3}{4})$, and $(\frac{3}{4},\frac{3}{4},\frac{1}{4})$. Each B atom touches four N atoms at the corners of a regular tetrahedron and vice versa. Heating does not disorder the B and N atoms. The B and N atoms fall on separate face-centered cubic lattices. In the next section, the same structure type will be found by filling small holes between large atoms, as exemplified by the mineral **blende** of ZnS composition, for which the structure type is usually named. The mineral **chalcopyrite**, **$CuFeS_2$**, is a double derivative of diamond. Sulfur atoms maintain the positions of blende, and each S atom is next to two Fe and two Cu atoms in an ordered array with tetragonal symmetry.

5.4 Ideal Void-Filling of Close-Packed Structures

In primitive cubic packing (Fig. 5.2a), there is a void at the body center of each unit cell. The largest sphere that can just fit into the void without forcing apart the eight spheres at the corners of the cell has a radius 0.732

times that of the original spheres, as calculated in Section 5.1. The atomic coordinates of the *positions* in the resulting structure are exactly the same as for the bcc arrangement of equal spheres, but two types of atoms are now used, for example, type A at $(0,0,0)$ and type B at $(\frac{1}{2},\frac{1}{2},\frac{1}{2})$, in contrast to type A at both $(0,0,0)$ and $(\frac{1}{2},\frac{1}{2},\frac{1}{2})$. Of course, the new structure was already derived as a derivative of CsCl type in the preceding section. It is obvious that interpretation of the CsCl structure type depends on the radius ratio of the atoms (or ions). If the smaller atom has a radius ratio of exactly $(\sqrt{3}-1)$ with respect to the larger atom, the smaller atoms just fill the voids between the larger atoms that are just in contact. If the radius ratio increases above $(\sqrt{3}-1)$, the larger atoms are forced apart. For the exact fit, the two types of spheres collectively occupy the following fraction of the volume: $\pi/6[1 + (\sqrt{3}-1)^3] \simeq 72.9$ percent.

In simple hexagonal packing, six spheres lie at the vertices of an Archimedean trigonal prism (Fig. 5.2c). The void is fitted exactly by a sphere with radius ratio ~0.528 with respect to the original spheres. There are twice as many voids as original spheres (Fig. 5.10a), thereby providing the positions for the crystal structure of **aluminum boride,** AlB_2 (Wyckoff, 1963, p. 362). This structure (Fig. 5.10b) provides a salutary warning about simple interpretation. First, the dimensions of the hexagonal cell ($a=3.01$ Å; $c=3.26$ Å) are not equal as would be required for an ideal structure. Second, the nearest neighbors of a boron atom, each of which fills a void, are not aluminum atoms but 3 boron atoms that lie in an equilateral triangle at 1.73 Å (i.e., $3.01/\sqrt{3}$). The next-nearest neighbors of each boron atom are 6 aluminum atoms at the vertices of an Archimedean trigonal prism with B−Al = 2.37 Å. Each Al atom has 12 B atoms at the vertices of a hexagonal prism as nearest neighbors. On a hard-sphere model, the boron and aluminum atoms would be represented, respectively, by spheres of radius 0.86_5 Å and 1.50_5 Å, whose radius ratio of 0.575 is close to but greater than the ideal value of 0.528 (cf. Table 1.1). The AlB_2 structure can be regarded as the regular alternation of hexagonal nets of B atoms and triangular nets of Al atoms.

Two other structure types utilize only half of the available voids. For clarity, label the spheres A and the voids b and c in Figure 5.10a (following Wells, 1975). The structure (Wyckoff, 1963, p. 150) of **tungsten carbide, WC,** has a tungsten atom at each A position and a carbon atom at each b position (Fig. 5.10c). The unit cell is hexagonal with a (2.91 Å) slightly greater than c (2.84 Å). Each C atom is surrounded by a trigonal prism of W atoms, and each W atom is similarly surrounded by a trigonal prism of C atoms.

The next simplest possibility is the important structure of the mineral **niccolite, NiAs,** already used for Exercise 4.3. In Figures 4.20c and 5.10d, the Ni atoms occupy the A positions and the As atoms occupy alternately the b and c positions in successive layers along the c axis. Thus the atom labeled

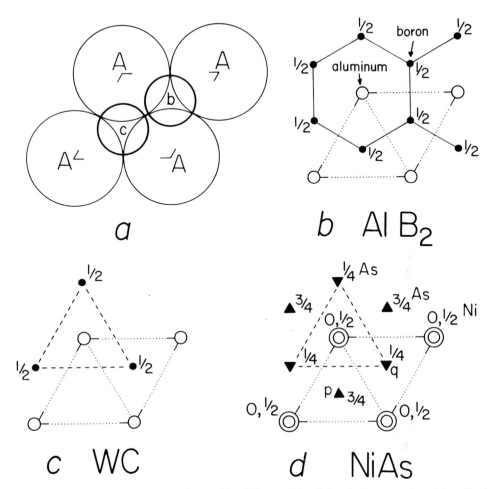

FIGURE 5.10 Structure types formed by filling voids of simple hexagonal packing. Each type has a hexagonal unit cell shown by the corner marks. (*a*) Two positions for voids (labeled *b* and *c*) between layers of spheres (labeled *A*) in simple hexagonal packing. The spheres are shown with the radius for a perfect fit. If the *A* spheres are placed at the corners of the hexagonal unit cell (i.e., at height 0,1 for the projection of their centers), the *b* and *c* spheres are centered at height $\frac{1}{2}$. (*b*) In the AlB$_2$ structure, the boron atoms occupy nodes of a hexagonal net (continuous line) at height $\frac{1}{2}$ and the aluminum atoms lie at the nodes of a triangular net (dotted line) repeated at heights 0,1, etc. (*c*) In the WC structure, both atoms lie at the nodes of triangular nets. (*d*) In the NiAs structure, the As atoms (triangle) occupy *b* and *c* sites from alternate layers (inverted triangle *q* at height $\frac{1}{4}$; upright triangle *p* at height $\frac{3}{4}$), and the Ni atoms (circle) occupy *A* sites at height 0 and $\frac{1}{2}$. Each type of atom lies on a triangular net. (See Figure 4.20*c* for coordination polyhedra.)

q is in a type b position, and the atom labeled p occupies a type c position. This alternation causes doubling of the repeat distance down the c axis. Each As atom lies at the center of a trigonal prism of Ni atoms, and each Ni atom at the center of an octahedron of As atoms. Interpretation of the structure is complicated by the existence of a short Ni–Ni distance of 2.50_5 Å along c, which is only slightly longer than the Ni–As distance of 2.43 Å. Thus it is perhaps better to interpret the coordination of Ni as 6 As + 2 Ni rather than just 6 As, and to assume that there is some approach to metallic bonding.

Readers are again cautioned about the danger of assuming that a specific structure (e.g., that of niccolite, NiAs) has the atoms touching as in the ideal structure type. The formal atomic radii of As (1.21 Å) and Ni (1.25 Å) in Table 1.1 are almost equal, thereby ruling out the filling by As atoms of the holes between close-packed Ni atoms. For a perfect fit of As atoms into the holes of type b and c between Ni atoms touching each other in A positions, the radius of As would need to be 0.66 Å. From the chemical viewpoint, arsenic is regarded as a semimetal, and a simple interpretation of the crystal structure is not expected. Some readers may have already noticed that the NiAs structure type can also be obtained by putting large spheres at positions of type p and q in Figure 5.10d and small spheres at positions of type A. If the atomic radii were not known, and only the atomic centers had been located, it would not be possible to determine which atoms would be in contact.

5.5   Ideal Void-Filling of Closest-Packed Structures: Binary Compounds

In both the hexagonal ($\dot{A}\dot{B}$) and cubic ($\dot{A}\dot{B}\dot{C}$) varieties of closest packing there are two types of holes. Four spheres at the vertices of a tetrahedron yield a hole into which a sphere can fit exactly when its radius ratio is $\simeq 0.225$. Similarly, six spheres outline an octahedral hole suitable for a sphere with radius ratio $\simeq 0.414$.

Referring to Figure 5.3a, there is a tetrahedral hole between 3 spheres of type A and 1 sphere of type B, or between 3 A and 1 C, 3 B and 1 A, 3 B and 1 C, 3 C and 1 A, or 3 C and 1 B. In hexagonal closest packing, the only possible tetrahedral voids are of types 3 A and 1 B, and of 1 A and 3 B. There is one tetrahedral void above each sphere, and one below each sphere. In cubic closest packing, there are also two tetrahedral voids for each sphere, but the letter combinations change from layer to layer.

Figures 5.11a, 5.11b show the positions of the tetrahedral voids in the unit cells. In the cubic unit cell of cubic closest packing, the coordinates of the eight voids are $(\pm\frac{1}{4},\pm\frac{1}{4},\pm\frac{1}{4})$. To show this, consider the A' atom at the origin of the unit cell in Figure 5.3e, and the three C atoms at $(\frac{1}{2},0,\frac{1}{2})$, $(\frac{1}{2},\frac{1}{2},0)$ and $(0,\frac{1}{2},\frac{1}{2})$. These four atoms outline a tetrahedron whose center

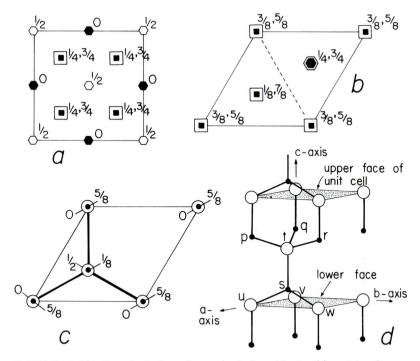

FIGURE 5.11 Tetrahedral and octahedral voids in (*a*) cubic closest-packing and (*b*) hexagonal closest-packing. The tetrahedral voids are shown by filled and open squares, and the octahedral voids by filled and open hexagons. The wurtzite structure has S atoms (circle) overlapping in projection down the hexagonal *c*-axis (diagram *c*) with Zn atoms (dot). In the clinographic projection (*d*), the tetrahedral linkages of an S atom (*s*) and a Zn atom (*t*) are shown.

is at $(\frac{1}{4},\frac{1}{4},\frac{1}{4})$, and the seven other tetrahedral voids follow by symmetry. In the hexagonal unit cell of hexagonal closest packing, the tetrahedral voids lie at $(0,0,u)$, $(0,0,\bar{u})$, $(\frac{2}{3},\frac{1}{3},\frac{1}{2}+u)$, and $(\frac{2}{3},\frac{1}{3},\frac{1}{2}-u)$, where u is $\frac{3}{8}$ (Fig. 5.11*b*). The void at $(\frac{2}{3},\frac{1}{3},\frac{7}{8})$ lies above the sphere at $(\frac{2}{3},\frac{1}{3},\frac{1}{2})$ and below the three spheres at $(0,0,1)$, $(1,0,1)$, and $(0,1,1)$.

In Figure 5.3*a*, each position *C* lies between three *A* spheres of a lower layer and three *B* spheres of an upper layer. These six spheres occupy the corners of a regular octahedron (cf. Fig. 3.3*d*) and enclose an octahedrally coordinated void. In hexagonal closest packing, the octahedral voids are centered on $(\frac{1}{3},\frac{2}{3},\frac{1}{4})$ and $(\frac{1}{3},\frac{2}{3},\frac{3}{4})$ as demonstrated in Figure 5.11*b*. To obtain the first position, take *A* positions at $(0,0,0)$, $(0,1,0)$, and $(1,1,0)$ and *B* positions at $(\frac{2}{3},\frac{1}{3},\frac{1}{2})$ and $(\frac{2}{3},\frac{4}{3},\frac{1}{2})$, and calculate the centroids. In projection, the voids lie over the *C* positions. For cubic closest packing, a representative void can be found from examination of Figures 5.3*e* and *f*. Take the centroid of the three *B* positions at $(\frac{1}{2},\frac{1}{2},1)$, $(1,\frac{1}{2},\frac{1}{2})$ and $(\frac{1}{2},1,\frac{1}{2})$ and the three *C* positions at ($\frac{1}{2}$,0,$\frac{1}{2}$), ($\frac{1}{2}$,$\frac{1}{2}$,0), and (0,$\frac{1}{2}$,$\frac{1}{2}$). The resulting center of an octahedral void ($\frac{1}{2}$,$\frac{1}{2}$,$\frac{1}{2}$) lies at the body center of the cubic cell, and is repeated by the F-centered symmetry to (0,0,$\frac{1}{2}$), ($\frac{1}{2}$,0,0) and (0,$\frac{1}{2}$,0), as in Figure 5.11*a*. The body center of the cubic cell corresponds to the position (0,0,$\frac{1}{2}$) of the hexagonal unit cell (Fig. 5.3*c*), and the *B* and *C* positions must have octahedral voids lying at *c*/2 above them to give positions ($\frac{2}{3}$,$\frac{1}{3}$,$\frac{5}{6}$) and ($\frac{1}{3}$,$\frac{2}{3}$,$\frac{1}{6}$) in the hexagonal cell.

Many structure types are obtained by filling different fractions of the tetrahedral and octahedral voids, as summarized in Wells (1975, Tables 4.5 and 4.6). Only the simpler examples are described here. Occupation of all the tetrahedral sites in ccp gives the **antifluorite** structure type, typified by Li_2O (Fig. 5.12*b*). The large oxygen ions (radius ~1.4 Å) occupy the ccp positions, and the small lithium ions (~0.6 Å) occupy all the tetrahedral voids. The radius ratio of ~0.4 is higher than the ideal value of 0.225 and, naively, the oxygen ions can be imagined to be held apart by the lithium ions. Every edge of a tetrahedron is shared with an adjacent tetrahedron, but faces are not shared. Occupancy of all the tetrahedral sites in hcp results in sharing of faces: thus occupancy of tetrahedral sites at (0,0,$\frac{3}{8}$) and (0,0,$\frac{5}{8}$) in Figure 5.11*b* would involve sharing of a horizontal face at height $\frac{1}{2}$. Face sharing is only rarely observed in ionic structures because of electro-

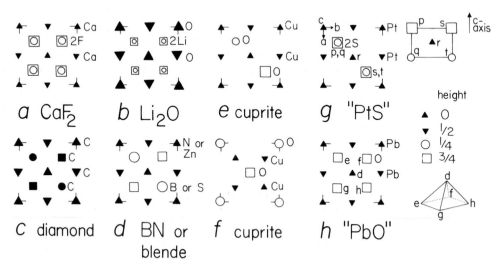

FIGURE 5.12 Structures related to fluorite (CaF_2) and antifluorite (Li_2O). The shape of the symbol shows the height (extreme middle right) and the relative sizes of the symbols in each plan refer to the relative sizes of the atoms. Thus in CaF_2 the circle and square for *F* are larger than the triangles for Ca. All symbols are the same size for diamond. The cuprite structure is shown with two choices of origin. See the text discussion for a description of the distortion and coordination in the "PtS" and "PbO" structure types.

static repulsion between nearby cations, and, indeed, no structure has been found with occupancy of all tetrahedral sites in hcp.

Regular occupancy of half the tetrahedral sites yields the **wurtzite** structure type for hcp (Figs. 5.11c, 5.11d) and **blende** type for ccp (Fig. 5.12d). Both structure types are named after minerals of composition ZnS. The blende structure was already encountered as a derivative structure of the parent diamond structure (Fig. 5.9). Because the metallic and covalent radii of Zn and S atoms are 1.37 and 1.04 Å, the blende and wurtzite structures cannot be explained by the simple fitting of small Zn atoms into holes between large S atoms. Even for an ionic model, the radius ratio of $\sim 0.6/1.8 = 0.3$ is too large for a perfect fit. The first neighbors of blende and wurtzite have an identical tetrahedral configuration, but the second neighbors are different. In wurtzite (Fig. 5.11d), three S atoms (*u*, *v*, and *w*) and three Zn atoms (*p*, *q*, and *r*) lie at the vertices of a trigonal prism. To go from a piece of wurtzite to a piece of blende, rotate the *p*, *q*, and *r* triplet 180° around the line *st* and leave *u*, *v*, and *w* alone. The easiest way to build the structures of blende and wurtzite is to use plastic tetrahedral stars and spaghetti links (see the exercises). Readers will find that in addition to building blende and wurtzite, it is possible to envisage an infinite series of polytypes in which the S and Zn atoms have different sequences of the *A*, *B*, and *C* positions for closest packing. The C atoms of the mineral **lonsdaleite** occupy all the sites of the wurtzite structure.

Numerous structure types can be invented by first reducing the fraction of occupied tetrahedral sites, and then by developing ordered arrays to give a series of derivatives from each parent. For brevity, only the simpler members of the **fluorite family** are given here (Fig. 5.12). The mineral **fluorite**, CaF_2, has atomic coordinates exactly the same as for the Li_2O structure, but the anions and cations are exchanged (Figs. 5.12a, 5.12b). In the Li_2O structure, the small Li^+ cations occupy all the tetrahedral interstices between the large O^{2-} anions. In fluorite, the F^- anions (1.3 Å) are larger than the Ca^{2+} cations (1.0 Å), but the anions are now twice as abundant as the cations. Hence, the Li^+ cations are replaced by the F^- anions and the O^{2-} anions by the Ca^{2+} cations in going from the antifluorite to the fluorite structure. This reversal results in a change of packing so that each Ca^{2+} cation is surrounded by eight F^- anions at the corner of a cube. For a hard-sphere model, the radius ratio of $1.1/1.3 \simeq 0.77$ is close to the ideal value of 0.732 for cubic coordination, and the F^- anions can be imagined to be slightly pushed apart by the Ca^{2+} cations.

The diamond structure (Fig. 5.12c) is obtained from the fluorite structure by removing half the fluorine atoms and eliminating the distinction between the remaining Ca and F atoms. Retention of the distinction gives the blende or boron nitride structure. Removal of three quarters of the F atoms from fluorite gives the atomic sites for **cuprite, Cu₂O** (Fig. 5.12*f*). This structure is usually described with the oxygen atoms at the cell corners (Fig. 5.12*e*), which merely requires a translation of $(\frac{1}{4},\frac{1}{4},\frac{1}{4})$ from diagram 5.12*f*. The oxygen atoms occupy the $(0,0,0)$ and $(\frac{1}{2},\frac{1}{2},\frac{1}{2})$ positions of a body-centered cubic lattice, and the copper atoms lie at $(\frac{1}{4},\frac{1}{4},\frac{1}{4})$, $(\frac{3}{4},\frac{3}{4},\frac{1}{4})$, $(\frac{1}{4},\frac{3}{4},\frac{3}{4})$, and $(\frac{3}{4},\frac{1}{4},\frac{3}{4})$. Each oxygen atom is surrounded by four copper atoms at the corners of a regular tetrahedron, and each copper atom lies midway between two oxygen atoms. This structure must be explained in terms of directed bonding.

The **"PtS" structure** is a second way of selecting half of the positions occupied by F atoms in fluorite. Each Pt atom (e.g., *r*) has four S atoms as first neighbors. These four S atoms (*p, q, s, t*) lie at the corners of a rectangle with edge ratio of $\sqrt{2}$ if the unit cell remains cubic as for the small diagram at the upper right of Figure 5.12. Actually, the mineral **cooperite, PtS**, is tetragonal and the *c* repeat expands somewhat over the *a* repeat so that the *p, q, s,* and *t* atoms form a rectangle with edge ratio of 1.24 instead of 1.414. This approach toward square coordination results from a strong tendency for Pt atoms to form a square planar complex with electronegative atoms. Sulfur atoms are in tetrahedral coordination, but expansion of *c* over *a* causes vertical elongation of the tetrahedron.

A third way of selecting half the positions gives the **"PbO" structure type** in which all four oxygen atoms are at height $\frac{3}{4}$. In this ideal arrangement, each Pb atom (e.g., *d* at height 1) has four oxygen atoms (*e, f, g, h*) as first neighbors such that the Pb atom lies at the apex of a square pyramid (lower right, Fig. 5.12). All structures belonging to the "PbO" family are distorted considerably from this ideal structure (Wells, 1975, p. 218).

Many structure types result from filling all or some of the octahedral interstices in hcp and ccp. For hcp, the niccolite structure results when all the octahedral holes are filled (Fig. 5.13). Spheres of types *A* and *B* (Fig. 5.3*b*) are placed in a hexagonal cell at $(0,0,0)$ and $(\frac{2}{3},\frac{1}{3},\frac{1}{2})$. Octahedral interstices are centered on $(\frac{1}{3},\frac{2}{3},\frac{1}{4})$ and $(\frac{1}{3},\frac{2}{3},\frac{3}{4})$, as in Figure 5.11*b*. Spheres of appropriate size to fit in the octahedral interstices overlap in Figure 5.13, and are labeled *c*. These positions match those already given in Figure 5.10*d*, as can be demonstrated by use of the right-hand diagram in Figure 5.13. Add $\frac{1}{4}$ to the *z* coordinate. The inverted triangle representing the center of the *A* sphere in Figure 5.13 moves to height $\frac{1}{4}$, which matches the height of the *b* sphere in Figure 5.10*a*. Similarly, the upright triangle for the *B* sphere moves to $\frac{3}{4}$, which matches the height of the *c* sphere in Figure 5.10*a*. The two *c* spheres of Figure 5.13 correspondingly match the *A* spheres of Figure 5.10*a*. A horizontal translation of the unit cell completes

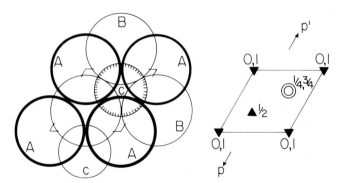

FIGURE 5.13 A diagram showing how the positions of the NiAs structure are obtained from filling octahedral holes of type *c* in hexagonal closest packing of *A* and *B* spheres. The sphere centers are shown in the right-hand diagram with symbols chosen to match those in Figure 5.10*d*.

the identity of the positions in Figures 5.10*a* and 5.13. Consequently, the remarkable result is obtained that positions of the NiAs structure can be derived either by filling all the octahedral holes of type *c* in spheres packed $\dot{A}\dot{B}$ or by filling half the trigonal-prismatic holes of types *b* and *c* in spheres packed $\dot{A}\dot{A}$. This is yet another warning that the atomic (or ionic) radii must be known independently before attempting to choose which atoms (or ions) are in contact in some but not all of the simpler types of close-packed structures.

Occupancy of all the octahedral interstices in ccp gives the *halite (NaCl)* structure type, already derived as a derivative structure of primitive cubic packing (Fig. 5.8). For NaCl itself, the radius ratio of ~0.55 is higher than the ideal value of 0.414 for filling of an octahedral hole, and Na$^+$ ions hold apart the Cl$^-$ ions in a hard-sphere model. The mineral periclase, MgO, has this structure and may be important in the deep mantle of the Earth along with perovskite, MgSiO$_3$, and perhaps fluorite-related structures.

Out of the many structure types described by Wells (1975, Table 4.6 et seq.), the following were selected because of their importance to mineralogy and high-pressure chemistry.

The *corundum* structure of Al$_2$O$_3$ and the derivative structure of *ilmenite*, Fe^{2+}Ti^{4+}O$_3$, are obtained by occupying two thirds of the octahedral holes in hcp. Indeed, the corundum structure can be imagined theoretically as a derivative of the niccolite structure type such that three Ni atoms are replaced by two Al atoms and a void, while the As atoms are replaced by O atoms. The radius ratios of $0.53/1.36 \simeq 0.39$ for Al^{3+} and O^{2-} ions in corundum and $0.78/1.36 \simeq 0.57$ for Fe^{3+} and O^{2-} ions in isostructural *hematite* bracket the ideal ratio of 0.41 for octahedral coordination. Placement of the cations in two thirds of the octahedral voids is achieved in an elegant array that harks back to Figure 1.2. Look first at the crosses that represent oxygen atoms in Figure 5.14 (left). The inclined crosses are at the corners of the unit cell in hexagonal closest packing, and occur as A-type layers, like those in Figure 1.2*b*, at $z = 0$, 1, 2, and so on. The horizontal crosses represent the B-type layers, and also have the same triangular pattern as in Figure 1.2*b*, but displaced from the A layers. In Figure 1.2*c*, one third of the circles were removed, and the octahedral cations in corundum use the resulting hexagonal pattern. The dots show occupation of octahedral interstices at $z = \frac{1}{4}$, and the dashed lines link the occupied sites. In the next layer of interstices at $z = \frac{3}{4}$, only those marked with open squares are occupied, and these are linked by zigzag lines. So far, there has been no need to make a choice of which voids to occupy and which to leave empty once the decision was made to use the hexagonal net of Figure 1.2*c*. But the next layer of octahedral interstices at $z = \frac{5}{4}$ presents two topologically different choices. The one that actually occurs in corundum (open circles linked by dotted lines) gives the highest symmetry and maximizes the distance between unoccupied interstices. From the ionic viewpoint, the latter phrase is equivalent to obtaining the best local balance of charge and a minimum for the electrostatic potential energy.

To build a model of corundum, use two sets of balls with radii nearly proportional to 0.414. Glue together a complete layer of larger balls to represent oxygen atoms at height 0. Place small balls in two thirds of the dimples, using the pattern of dots in Figure 5.14. Make three more such double layers, and superimpose them in the appropriate relative orientation.

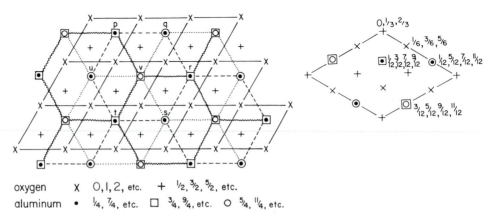

oxygen   X   0,1,2, etc.   + $\frac{1}{2}, \frac{3}{2}, \frac{5}{2}$, etc.

aluminum   • $\frac{1}{4}, \frac{7}{4}$, etc.   □ $\frac{3}{4}, \frac{9}{4}$, etc.   ○ $\frac{5}{4}, \frac{11}{4}$, etc.

FIGURE 5.14 Projection of the crystal structure of corundum onto the plane of the closest-packed layers. The heights of the left-hand drawing are given as fractions of the spacing of the layers of oxygen atoms. The coordinates for a hexagonal unit cell are shown in the right-hand drawing.

If the double layers are not glued together, it is easier to examine the topologic and geometric relations. Show that the Al atoms occur in pairs whose coordination polyhedra share a horizontal face. This results in close approach of highly charged cations, which is not favorable in a *simple* ionic model. Ilmenite, $FeTiO_3$, has the corundum structure as its pseudostructure. For each pair of octahedra that share faces, an Fe atom (or Fe^{2+} ion, in the ionic model) and a Ti atom (or Ti^{4+} ion) each occupies one of the octahedra: there is no evidence for two Ti^{4+} ions facing each other across a shared face.

Whereas the hcp structure repeats after each two layers ($\dot{A}\dot{B}$), the corundum structure has a 6-layer repeat of the AB layers: $\dot{A}deBefAfdBdeAefBfd\dot{A}$. The small letters d, e, and f represent the octahedral voids. In the first layer of interstices, d and e are occupied and f is empty; in the next layer, e and f are occupied and d is empty, and so on. The smallest unit cell is rhombohedral with vertices that project onto Figure 5.14 at v (highest), p, r, and t (next highest), q, s, and u (next highest), and a position hidden by v (lowest). As with most close-packed structures with partial occupancy of interstices, there is considerable distortion from the ideal geometry, but the rhombohedral symmetry is retained. The smallest hexagonal cell is shown in Figure 5.14 (right) with coordinates from the left-hand diagram divided by 3 to allow for the 6-layer repeat.

Wells (1975, Table 4.6) lists six structure types in which half of the octahedral interstices of hcp are occupied. The ideal pattern for the *α-PbO₂ structure* (Fig. 5.15) is easily remembered because of the simple zigzag pattern of occupied interstices. The triangles represent oxygen atoms at heights 0, 1 (upright), and ½ (reversed). Dots represent occupied interstices at height ¼ giving 120° chains like *pqrstuv*, while circles represent unoccupied interstices, which also lie in 120° chains. At height ¾ the dots and circles represent unoccupied and occupied interstices, respectively. The zigzag

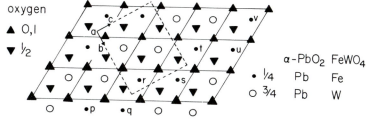

FIGURE 5.15 Projection of the crystal structure of α-PbO₂ onto the plane of the closest-packed layers. The oxygen atoms (triangles) occupy the *A* and *B* sites of *hcp*, and the lead atoms occupy half of the octahedral interstices with a characteristic zigzag sequence. The orthorhombic unit cell is shown by dashed lines. This drawing is idealized, and the accurate atomic coordinates are given in Chapter 9.

chains are not compatible with hexagonal symmetry, and the orthorhombic symmetry of the α-PbO$_2$ structure is analyzed in Chapter 9, using the unit cell shown by dashed lines. Considerable distortion occurs because of cation–cation repulsion, and the Pb^{4+} ions are slightly too big for the octahedral interstices between close-packed O^{2-} ions. The **wolframite** structure type can be derived from the α-PbO$_2$ structure by (i) replacing Pb atoms at height ¼ by Fe and Pb atoms at height ¾ by W, and (ii) distorting the structure considerably so that each Fe atom retains octahedral coordination while each W atom has four nearest neighbors at 1.8 Å and two neighbors displaced to 2.2 Å.

The third structure type with half-occupancy of octahedral interstices is found in **diaspore, α-AlO(OH)**, and **goethite, FeO(OH)**. Ignoring the hydrogen and idealizing the atomic positions produces the projection in Figure 5.16. Whereas the occupied interstices fall in zigzag chains in α-PbO$_2$, they fall in double bands in diaspore and goethite. The unit cell is again orthorhombic, and structural details are given in Chapter 9.

Readers may wonder if any structure with half-occupancy of octahedral interstices of hcp has the occupied interstices lying in single rows, as in Figure 5.17a. Indeed, this pattern leads to the **CaCl$_2$ structure** and the related structures of **rutile (TiO$_2$)** and **cassiterite (SnO$_2$)**, but considerable distortions are needed. Following Wells (1975, Fig. 4.21), the niccolite structure is first projected along the line pp' in Figure 5.13 to give Figure 5.17b. The larger circles represent the A and B spheres, but have only one quarter of the radius required for close packing. The smaller circles represent the spheres in the octahedral interstices, and the lines show the edges of octa-

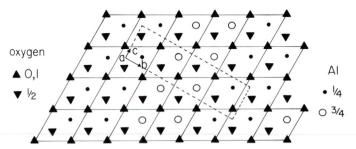

oxygen

▲ 0,l

▼ ½

Al

• ¼

○ ¾

FIGURE 5.16 Projection of the crystal structure of diaspore onto the plane of the closest-packed layers. The oxygen atoms (triangles) occupy the A and B sites of *hcp,* and the Al atoms occupy half of the octahedral interstices with a characteristic double-band sequence readily visible from the contrast between dot and circle. The orthorhombic unit cell is shown by dashed lines. This drawing is idealized, and the accurate atomic coordinates are given in Chapter 9.

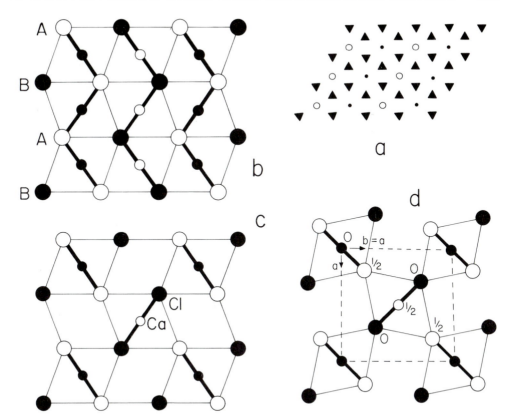

FIGURE 5.17 Derivation of the CaCl$_2$ and rutile structure types from the niccolite structure type. (a) Diagram showing how half the octahedral sites are occupied in the CaCl$_2$ and rutile structures (cf. Figures 5.15 and 5.16). (b) Niccolite structure viewed parallel to the closest-packed layers along the line pp' in Figure 5.13. Large spheres in A and B sites occupy alternate layers. Small spheres in octahedral sites lie halfway between the A and B layers. (c) Idealized CaCl$_2$ structure obtained by emptying half the octahedral sites from the niccolite structure. (d) Actual rutile structure projected down the tetrad axis onto the ab plane. Titanium atoms occur at $(0,0,0)$ and $(\frac{1}{2},\frac{1}{2},\frac{1}{2})$ and oxygen atoms at $\pm(u,u,\frac{1}{2})$ and $\pm(\frac{1}{2}-u,\frac{1}{2}+u,0)$ where u is 0.30. Atoms at height 0 are shown by filled symbols and those at height $\frac{1}{2}$ by open symbols. This distinction is also used for diagrams (b) and (c).

hedra oriented as in Figure 3.3h. Removal of alternate rows of octahedral-coordinated spheres as in Figure 5.17a gives the arrangement in Figure 5.17c, which is an idealization of the CaCl$_2$ structure. Rotation of the remaining rows of octahedra gives the rutile structure (Fig. 5.17d) viewed down the c axis of its tetragonal unit cell. This structure was already used in Exercise 4.2, and will be encountered again as an example of edge sharing of octahedra (Chapter 9).

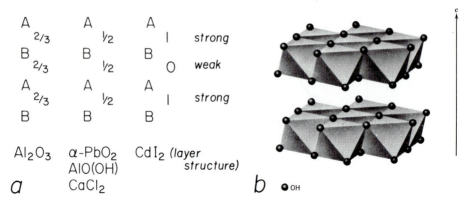

A
  2/3
B   A   A
  2/3   1/2
A   B   B
  2/3   1/2   O   weak
B   A   A
           1/2 | strong
           B   B

Al_2O_3 α-PbO_2 CdI_2 (layer
            AlO(OH)   structure)
            $CaCl_2$

a   b   ● OH

FIGURE 5.18 Comparison of structure types with occupancy of octahedral sites in hexagonal closest-packing. (a) Sequence of layers showing the fraction of occupied octahedral sites between the A and B layers of closest-packed spheres. In the CdI_2 structure, alternate layers of octahedral sites are either completely occupied or completely unoccupied, leading to strong or weak bonding. (b) Polyhedral drawing of the crystal structure of brucite, $Mg(OH)_2$, which has the same topology as the CdI_2 structure. The OH^- ions are represented by spheres about one-sixth the correct size for close packing. The center of each octahedron is occupied by an Mg^{2+} ion. From Hurlbut and Klein (1977, Figure 8.19). Reprinted by permission. (See Chapter 9 for details.)

The last of the six structure types obtained by selecting half of the octahedral sites from the niccolite structure is that of CdI_2 and the mineral *brucite, Mg(OH)$_2$*. Figure 5.18a compares the fractions of occupied octahedral sites in the corundum, α-PbO_2, diaspore, $CaCl_2$ (and rutile), and CdI_2 structures. Whereas all octahedral layers have two-thirds occupancy in corundum and one-half occupancy in the next three structure types, the octahedral layers in the *CdI_2 structure* type have alternate full and zero occupancy. This gives a layer-type structure in which there is strong chemical bonding between Cd and I across the occupied octahedral layers and only weak bonding between I atoms across the unoccupied layers. Figure 5.18b shows how the $Mg(OH)_6$ octahedra share all nonhorizontal edges to give tightly bonded layers held together only weakly by OH–OH interaction. This drawing also applies to CdI_2 when modified for change of interatomic distances.

Turning now to ccp structures, occupancy of all the octahedral sites gives the halite structure as already described. Occupancy of half the octahedral sites leads to the *$CdCl_2$ structure* type (analogous to CdI_2 with regular alternation of octahedral layers with either complete or zero occupancy) and to the structures of atacamite and anatase, which are too complex to describe here.

5.6   Ideal Void-Filling of Close-Packed Structures: Spinel and Olivine

 Although elementary students may be forced to omit this section because of time, most readers should attempt some time or other to master the spinel and olivine structures, which are the key to so many important materials. In the spinel family of materials, typified by the mineral *spinel* of composition $MgAl_2O_4$, the unit cell is cubic with a twice as large as for the unit cell of cubic closest packing of equal spheres. Thus for $MgAl_2O_4$, the a dimension of ~8 Å corresponds to twice the cell edge for closest-packed oxygen anions of radius ~1.4 Å. This unit cell with 8-fold greater volume contains 32 oxygen atoms, 8 tetrahedrally coordinated atoms (denoted T), and 16 octahedrally coordinated atoms (denoted M), to give a formula $T_8M_{16}O_{32}$, or TM_2O_4 when simplified. Figure 5.11*a* acts as a reminder that there are as many possible M sites $(0,0,\frac{1}{2};$ etc.) as O sites $(0,0,0;$ etc.) and twice as many T sites $(\frac{1}{4},\frac{1}{4},\frac{1}{4};$ etc.). Hence in spinel, one half of the M sites are occupied and one eighth of the T sites.

For an ionic model, the cations will tend to occupy those M and T sites that have the greatest separation: specifically, the coordination polyhedra should tend to share vertices in preference to edges and particularly to faces. The actual choice of sites in spinel is elegant, but easily misunderstood. For the doubled cell, a T site at $(\frac{1}{4},\frac{1}{4},\frac{1}{4})$ becomes $(\frac{1}{8},\frac{1}{8},\frac{1}{8})$ and all O, T, and M sites can be expressed by multiples of $\frac{1}{8}$. The literature on spinel minerals is confused by different choices of origin for the doubled cell; Figure 5.19 follows the choice in the Mineralogical Society of America *Short Course Notes on Oxide Minerals,* Volume 3, 1976. In Figure 5.19*a*, the unit cells for ccp are outlined by dotted lines, and the spinel cell by continuous lines joining *f, g, d,* and *h.* An occupied M site is shown by an open hexagon with inscribed height multiplied by 8. Occupied T sites are shown by open triangles. Unoccupied M and T sites are not shown. Oxygen atoms are also not shown, but would lie at $(0,0,\frac{2}{8})$, $(0,0,\frac{6}{8})$, $(\frac{1}{2},0,\frac{2}{8})$, $(\frac{1}{2},0,\frac{6}{8})$, and so on.

It is easy to remember the pattern. Just note that the heights of the M sites go $\frac{0}{8}$, $\frac{2}{8}$, $\frac{4}{8}$, $\frac{6}{8}$, $\frac{8}{8}$, $(=\frac{0}{8})$, and so on in east–west and north–south directions. This steady trend is actually along a face diagonal, since a horizontal coordinate simultaneously changes by $\frac{2}{8}$ in sympathy with the change in the vertical coordinate. This trend along a face diagonal can also be seen in the constancy of height at $\frac{4}{8}$ for M sites running NW–SE between *f* and *d.* Of course, all these trends are related by the triad rotation axes that lie along the body diagonals of the cubic unit cell. How about the T atoms? Consider the four M atoms at *p, q, r,* and *s.* In projection at *t,* there are four possible tetrahedral sites at height $\frac{1}{8}$, $\frac{3}{8}$, $\frac{5}{8}$, and $\frac{7}{8}$ for the double cell of ccp. The site at height $\frac{1}{8}$ is the furthest away from the M atoms, and,

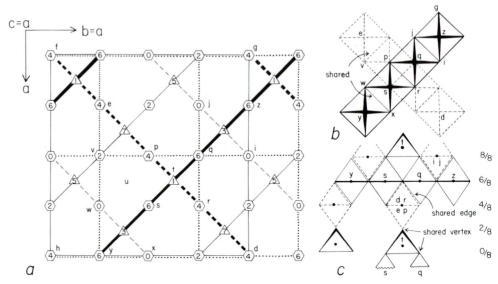

FIGURE 5.19 Occupancy of octahedral (*M*) and tetrahedral (*T*) sites in spinel. (*a*) Four complete cells and two half-cells of cubic closest-packing are shown by dotted lines, and the spinel cell by the continuous lines between *f, g, d,* and *h.* Heights of *M* sites (hexagon)and *T* sites (triangle) are given in one-eighth of the edge of the spinel cell. Diagonal lines link centers of edge-shared octahedra. (*b* and *c*) The relation between two lines of edge-shared octahedra in plan *b* and elevation *c.* In the plan, all of the octahedra have the same orientation as in Figure 3.3*g* (but rotated 45°). In the elevation, the octahedra have the same orientation as in Figure 3.3*h,* and the tetrahedra are viewed parallel to one edge.

indeed, it is the one that is occupied. There are tetrahedral sites in projection at *u* between M atoms at *w, v, p,* and *s,* but because these M atoms are at heights $^0/_8$, $^2/_8$, $^4/_8$, and $^6/_8$, all tetrahedral sites projecting onto *u* are closer to M atoms than the *t* site. So the apparently complex features of the spinel structure can be attributed to maximization of the separation of the M and T atoms, and this can be explained on the ionic model by minimization of the electrostatic repulsion between cations.

Diagonal lines in Figure 5.19*a* join M atoms with the same height. This equality of height results in edge sharing of the coordination polyhedra. Consider the M atom at height $^6/_8$ on *s.* It is coordinated to two oxygens at $^4/_8$ and $^8/_8$ projecting onto *s,* and to four oxygens at $^6/_8$ projecting onto *p, r, x,* and *w.* The resulting octahedron (Fig. 5.19*b*) is projected down a tetrad axis. The M atom at *q* has a similar octahedron of oxygen atoms around it, and the two octahedra share an edge composed of oxygen atoms at height $^6/_8$ on *p* and *r.* An infinite chain of edge-linked octahedra runs along the diagonal direction *ysqz* (Figs. 5.19*b,* and *c*), and similarly for all the other face-diagonal directions. The octahedron centered at height $^4/_8$ on *r* shares inclined edges with octahedra centered at height $\frac{6}{8}$ on *s* and *q* (Fig. 5.19*c*). Each octahedron shares one edge with each of six adjacent octahedra (e.g., *s* at $\frac{6}{8}$ shares with octahedra whose centers project onto *p, q, r, x, y,* and *w*). The T atom at height $\frac{1}{8}$ on *t* is surrounded by a tetrahedron of oxygen atoms at height $\frac{2}{8}$ on *p* and *r* and $\frac{0}{8}$ on *s* and *q*. This tetrahedron is viewed down an edge in Figure 5.19*c*, and each vertex is shared with a vertex from octahedra centered on the *p, q, r,* and *s* sites. In summary, each octahedron shares half its edges with adjacent octahedra to form a 3D framework of linked octahedra, and this framework is cross-linked by isolated tetrahedra that share all vertices with adjacent octahedra. But it must be emphasized that this is a mathematical abstraction, and that the fundamental basis is the cubic closest packing of oxygen atoms, whose interstices are partly occupied by M and T atoms. This is seen best by gluing spheres together in layers that lie perpendicular to a triad axis.

For clarity, consider first Figures 5.20*a* and *b*. A hexagonal layer of closest-packed spheres is viewed in projection as the large circles in mutual

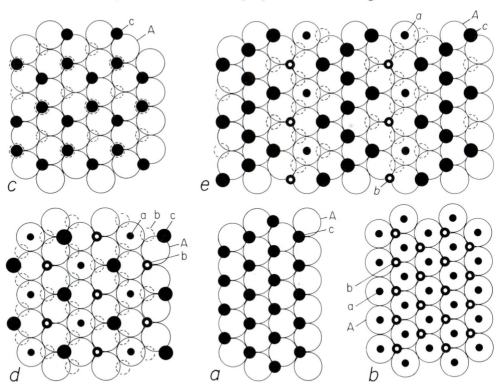

FIGURE 5.20 Plans of closest-packed layers (*A*) and occupied sites of octahedral (large filled and dashed circles) and tetrahedral types (small filled and heavy-line circles) in positions *a, b,* and *c*. Diagrams (*a*) and (*b*) show all possible octahedral and tetrahedral sites. Diagrams (*c*), (*d*), and (*e*) show, respectively, the occupied sites in corundum, spinel, and olivine structure types.

contact. In (Fig. 5.20*a*), half of the dimples are occupied by spheres (filled circles in projection) just big enough to fill octahedral holes. Both types of spheres are centered on the nodes of a triangular net. Label the spheres *A* and *c* as in Figure 5.13. Consider the two layers as an open-faced sandwich. To get the *c* spheres in octahedral coordination, the spheres from the bottom of the next sandwich must be in *B* positions so that each *c* sphere has 3 *A* spheres underneath and 3 *B* spheres above. To obtain the **periclase** structure stack the layers . . . *AcBaCbA* . . ., where the large spheres *A*, *B*, and *C* are in cubic closest packing and the small spheres occupy half of the octahedral sites. To obtain the **corundum** structure (Fig. 5.14), remove one third of the small spheres (Fig. 5.20*c*) and stack the sandwiches . . . *Ac'Bc″Ac'″Bc'Ac″Bc'″A* . . ., where *c'*, *c″*, and *c'″* are the three different ways of retaining two thirds of the small spheres, as in Figure 5.14. The large spheres, of course, are in hexagonal closest packing.

For tetrahedral sites (Fig. 5.20*b*), an open-faced sandwich can be constructed in which each small sphere lies either in a dimple (small circle in heavy line) or on the top of a large sphere (filled circle). Label the large spheres with an *A*. The second set of small spheres must be labeled with an *a* because the spheres rest directly over the *A* spheres. The first set of small spheres is labeled *b*. In order to complete the tetrahedral coordination of the *b* spheres, the next open-faced sandwich must be positioned so that its large spheres are directly over the *b* spheres, and hence must be in the *B* position. Similarly, the second set of small spheres was deliberately chosen to lie over the *A* spheres, and could not be placed in *c* positions that would lead to octahedral coordination. Stacking of the open-faced sandwiches would give the sequence *AbaBpq*, where *p* and *q* are not yet determined. For hexagonal closest packing of the large spheres, the sequence would be *AbaBabA* and for cubic closest packing it would be *AbaBcbCacA*. This assumes full occupancy of the tetrahedral sites.

To construct the spinel structure from spheres, first make an open-faced sandwich (Fig. 5.20*d*) in which only one quarter of the octahedral dimples (type *c*) are occupied in a triangular array. Now place the tetrahedrally coordinated spheres in one quarter of the *a* and one quarter of the *b* sites, noting that the occupied sites are at the greatest possible distance from the octahedrally coordinated spheres. Indeed, this pattern is easy to remember for it is highly symmetrical and would minimize the cation repulsion for ionic forces. Unfortunately, an open-faced sandwich is not enough for the spinel structure. Underneath the sandwich, place octahedrally coordinated cations in *b* positions (dashed circles in Fig. 5.20*d*) such that three quarters of the sites are occupied. Again it is easy to remember which sites are unoccupied because it is obvious from the viewpoint of cation repulsion. The unoccupied sites are those directly under the *b* sites occupied by tetrahedrally coordinated spheres. Note that the remaining three quarters of the sites occupy the nodes of a kagomé net (Fig. 1.2*d*). To complete the spinel structure, first place large spheres in *C* positions *below* the two-faced sandwich to give the combined sequence:

$$C \quad b \quad A \quad b \quad c \quad a$$
$$\tfrac{3}{4}\,M \qquad\qquad \tfrac{1}{4}\,T \quad \tfrac{1}{4}\,M \quad \tfrac{1}{4}\,T$$

Layers *C* and *A* have full occupancy of oxygen atoms, but the other layers are occupied only fractionally. The multiple sandwiches can be stacked in threes to give the complete repeat down each body diagonal of the cubic unit cell:

$$C \quad b \quad A \quad b \quad c \quad a \quad B \quad a \quad C \quad a \quad b \quad c$$
$$\tfrac{3}{4}\,M \qquad\qquad \tfrac{1}{4}\,T \quad \tfrac{1}{4}\,M \quad \tfrac{1}{4}\,T \qquad\qquad \tfrac{3}{4}\,M \qquad\qquad \tfrac{1}{4}\,T \quad \tfrac{1}{4}\,M \quad \tfrac{1}{4}\,T$$

$$A \quad c \quad B \quad c \quad a \quad b \quad C$$
$$\tfrac{3}{4}\,M \qquad\qquad \tfrac{1}{4}\,T \quad \tfrac{1}{4}\,M \quad \tfrac{1}{4}\,T$$

Most readers will have found this treatment rather exhausting. Have courage—thorough study will demonstrate the essential simplicity behind the patterns! I find the spinel structure to be a beautiful and elegant solution to the packing of an MT_2O_4 compound. It seems so obvious when you comprehend it. Just imagine the thrill when W. L. Bragg and S. Nishikawa independently solved the spinel structure from measurement of the X-ray diffraction pattern in 1915. What a contrast to the tragic unhappiness caused by the carnage of the unnecessary First World War!

Returning to the cool elegance of the spinel structure, there are three subtypes. The obvious way to assign two types of atom X and Y to the TM_2O_4 subunit is to put 1 X into the T site and 2 Y into the twofold M site to get **normal spinel**. The second obvious way is to assign the atoms X and Y completely at random to the T and M sites so that the T site is occupied statistically by $(\tfrac{1}{3}\,X + \tfrac{2}{3}\,Y)$ to give **fully disordered spinel**. The third way is not so obvious, and is approached best from a crystal-chemical direction. Consider true spinel with one Mg^{2+} and two Al^{3+} ions per four oxygens. In normal spinel, the Mg^{2+} would be tetrahedral and the Al^{3+} would be octahedral. But this presents a packing problem because the Mg^{2+} ion in 4-coordination has a radius of 0.58 Å (Table 1.1), and the Al^{3+} ion in 6-coordination has a radius of 0.53 Å. For *ideal* cubic-closest packing, the octahedral hole should be larger than the tetrahedral hole. The Mg^{2+} ion would fit better in 6-coordination (radius 0.72 Å) and the Al^{3+} ion in 4-

coordination (0.39 Å), but the atomic ratio of Mg/Al is the opposite of the population ratio of the M and T sites. From the packing viewpoint, the best answer is to fill the T sites with Al^{3+} and the M sites with a one-to-one random mixture of Mg^{2+} and Al^{3+}. The resulting third type of spinel with the schematic formula $Y_T(X + Y)_M O_4$ is known as ***inverse spinel.*** For several reasons, the crystal chemistry of spinels is very complex, and intermediate distributions occur between the three ideal end members. The oxygen atom moves slightly from the ideal position to change the shape of the coordination polyhedra. Furthermore, superstructures and noncubic distorted varieties are known, but all have the essential topology of the spinel structure.

For hexagonal closest packing, the ***olivine*** structure (Mg_2SiO_4) also has MT_2O_4 character. This structure is analyzed in detail in Chapter 9, but the packing of the spheres is conveniently discussed here. In Figure 5.20e, the M atoms occupy octahedral sites in a zigzag band, commonly described in the mineralogical literature as a serrated chain. Between the zigzag bands, T atoms form a zigzag chain with occupancy of alternate positions in dimples and above oxygen spheres. The open-faced sandwich can be denoted:

$$A \qquad b \qquad c \qquad a$$
$$\tfrac{1}{8} \text{ T} \quad \tfrac{1}{2} \text{ M} \quad \tfrac{1}{8} \text{ T}$$

Dashed circles show the positions of the M atoms of the open-faced sandwich that lies below, and the positions of the T atoms can be readily deduced. The full repeat for the olivine structure is:

$$A \qquad b \qquad c \qquad a \qquad B \qquad a \qquad c \qquad b \qquad A$$
$$\tfrac{1}{8} \text{ T} \quad \tfrac{1}{2} \text{ M} \quad \tfrac{1}{8} \text{ T} \qquad\qquad \tfrac{1}{8} \text{ T} \quad \tfrac{1}{2} \text{ M} \quad \tfrac{1}{8} \text{ T}$$

The olivine structure is not as elegant topologically as the spinel structure, and the complex distortions are described in Chapter 9.

There are other structure types with partial occupancy of both tetrahedral and octahedral interstices in hcp and ccp. Readers should consult Moore and Smith (1970) for drawings of the β-Mg_2SiO_4, manganostibite, staurolite, and kyanite structures, which are too complex to discuss here. Incidentally, kyanite is the high-pressure polymorph of Al_2SiO_5, as expected theoretically from its denser packing than for sillimanite and andalusite, while the β-Mg_2SiO_4 structure is produced when olivine transforms at high pressure, as probably occurs in the Earth's upper mantle. The β-Mg_2SiO_4 structure does not have the topology of spinel, but is often described incorrectly as a distorted spinel variety.

5.7 Open Packing of Tetrahedrally Coordinated Frameworks

Analogous to the 2D open packing of circles (Figs. 1.2c–1.2g) is the 3D open packing of spheres. In tetrahedrally coordinated frameworks, four spheres touch each other to enclose a tetrahedral site. Each sphere belongs to two tetrahedral clusters, thereby allowing continuity throughout the 3D framework. The corresponding crystal structures are stabilized by occupation of each tetrahedral site by a small atom (or ion). Thus in *tridymite,* SiO_2, each tetrahedral hole between four oxygen atoms (or anions on the ionic model) is occupied by a silicon atom (or cation). Take the open packing (Fig. 1.2d) in which the centers of the circles lie on the nodes of a kagomé net. Convert each circle into a sphere, and place a sphere of identical size in each alternate dimple, leaving unoccupied the intervening dimples. Each set of four adjacent spheres encloses a tetrahedral void into which a small sphere is placed. Most readers will profit from construction of such an open-faced sandwich from balls of appropriate size. Actually, it is unnecessary to insert the small balls, because they are hidden by the large balls. Open-faced sandwiches can now be stacked vertically to give an infinite framework (of course, the model is only a finite piece!). Each upward-projecting sphere will click into a downward-facing dimple of the next sandwich.

This simple method of construction yields the structure of tridymite, in which the Si atoms occupy all the tetrahedral voids between the oxygen atoms. Denote the occupied positions of the kagomé layer (Fig. 5.21a) by A_1, A_2, and A_3, and the unoccupied position by A_4. Denote the upper spheres by B_4 and the corresponding tetrahedral sites by b_4. The other possible oxygen and tetrahedral positions (C_4 and c_4) lie underneath the sandwich (dashed circles). Layers are stacked in the following sequence in tridymite:

$$\ldots (A_1 A_2 A_3) \quad b_4 \quad B_4 \quad b_4 \quad (A_1 A_2 A_3) \quad c_4 \quad C_4 \quad c_4 \quad (A_1 A_2 A_3) \ldots$$

$$\text{O} \qquad \text{Si} \quad \text{O} \quad \text{Si} \qquad \text{O} \qquad \text{Si} \quad \text{O} \quad \text{Si} \qquad \text{O}$$

From the topologic viewpoint, the simplest description of the tridymite structure is to consider only the b_4 and c_4 sites of the tetrahedrally bonded Si atoms. Join each site to the four nearest sites to produce a *4-connected 3D net.* Each atom lies at a node of the net, and each node is connected to the next node by a *branch.* The net is 4-connected because each node has four branches. Figure 5.21b is a clinographic drawing of part of the tridymite net in which the relation to the hexagonal unit cell (dotted line) is shown. The next simplest description utilizes vertex sharing of tetrahedra (Fig. 5.21c). Join the center of each oxygen atom to the centers of the six adjacent oxygen atoms, thereby generating the edges of tetrahedra. Each tetra-

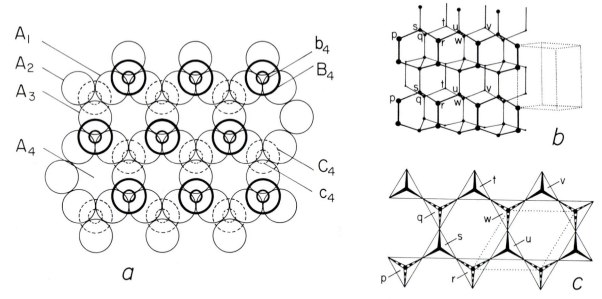

FIGURE 5.21   Ideal crystal structure of tridymite. (*a*) Plan of hexagonal layer of atoms of oxygen (large circles) and silicon (small circles). (*b*) Clinographic projection of 3D 4-connected net, whose nodes are occupied by the Si atoms. The relative size of the circles illustrates the perspective. (*c*) Plan of the tetrahedral framework with alternate tetrahedra pointing upward (heavy line) and downward (dashed line). The centers of the tetrahedra are labeled with the same symbols as the corresponding nodes of the 4-connected net.

hedron shares each of its four vertices with a neighboring tetrahedron to yield a ***tetrahedral framework***. The edges of the tetrahedra can be regarded as the branches of a 6-connected 3D net, and the oxygen atoms can be regarded as lying on the nodes. However, this description is too ponderous, and the 4-connected 3D net formed from the sites of the Si atoms is sufficient and easier to draw.

Is it possible to invent other tetrahedral frameworks? Indeed it is, and there is an infinite number. Fortunately, all the frameworks found in natural and synthetic materials can be assembled from simple building units, as illustrated for zeolite structures by Meier and Olson (1978). Systematic enumeration of 4-connected 3D nets from simple units is in progress (Smith, 1977; 1978; 1979; Smith and Bennett, 1981), and the following treatment selects only a few simple nets that illustrate the stacking principles. In most crystal structures, various atoms and molecules occupy some or all of the void space between the atoms of the 4-connected 3D net, thereby preventing the net from twisting and collapsing about the voids. Tridymite, when pure or nearly pure SiO_2, has the ideal geometry of Figure 5.21 at high temperature, but at low temperature the thermal vibrations are too small to maintain the maximum volume of the void space. The Si—O—Si bond angles are reduced as the oxygen atoms move into the empty regions (Dollase and

Baur, 1976). *Kalsilite,* $KAlSiO_4$, has the same framework as tridymite but with regular alternation of Al^{3+} and Si^{4+} ions on the tetrahedral nodes. The lower charge on the framework allows one K^+ ion to enter the structure for each Al^{3+}, and this K^+ ion occupies a void and props apart the tetrahedral framework even at low temperature (Perrotta and Smith, 1965). Kalsilite and related structures are known as *stuffed frameworks,* following M.J. Buerger (1954). The following treatment ignores all the complexities caused by atomic ordering on nodes, by collapse around voids, and by stuffing.

Returning to tridymite, the reader may wish to check whether the open-faced sandwiches can be linked in any other ways. Instead of using the sequence $(A_1 A_2 A_3)$ $b_4 B_4 b_4$ $(A_1 A_2 A_3)$, it is possible to use the sequence $(A_1 A_2 A_3)$ $b_4 B_4 b_4$ $(C_1 C_2 C_3)$, which corresponds to rotation of the tetrahedron $B_4 b_4$ $(A_1 A_2 A_3)$ by $180°$ about the $B_4 b_4$ line to give the tetrahedron $B_4 b_4$ $(C_1 C_2 C_3)$, as shown in Figures 5.22a and b. Cyclic manipulation of the symbols gives the regular sequence:

$$\ldots (A_1 A_2 A_3)\, b_4 B_4 b_4\, (C_1 C_2 C_3)\, a_4 A_4 a_4\, (B_1 B_2 B_3)\, c_4 C_4 c_4\, (A_1 A_2 A_3) \ldots$$

of *cristobalite.* By analogy with the $\dot{A}\ddot{B}$ and $\dot{A}B\dot{C}$ stacking of closest-packed spheres, readers might guess that cristobalite has isometric symmetry. Indeed, the Si atoms of cristobalite occupy the same relative positions as the C atoms of *diamond,* and the oxygen atoms lie nearly midway between adjacent Si atoms (Fig. 522b). Appropriate stacking of the open-faced sandwiches of close-packed spheres will yield the cristobalite structure. Whereas tridymite has straight nonintersecting channels parallel to the c-axis of its hexagonal cell, cristobalite has zigzag intersecting channels parallel to the four face diagonals of the cubic unit cell. Stacking faults are typical of natural and synthetic varieties of tridymite and cristobalite, especially of the disordered variety of silica known as CT-silica. An infinite number of regular stacking sequences can be invented. All these structures are *polytypes* in which parallel open-faced sandwiches are merely cross-linked in different ways. *Carnegieite,* $NaAlSiO_4$, is a stuffed derivative of cristobalite, and is a polymorph of high-Na *nepheline,* $NaAlSiO_4$, which is a stuffed derivative of tridymite. Polytypism has not been observed between these stuffed derivatives.

The *sodalite* framework (Fig. 5.23a) is obtained by using the edges of close-packed truncated octahedra as branches. Fedorov described the *truncated octahedron* as one of the five parallelohedra that will fill space (Fig. 3.15e). To aid visualization, Figure 5.23 is displayed as a stereo pair. Some fortunate readers will be able to "swivel" their eyes independently, especially if a card is placed vertically between the two views. However, most readers will need to use a stereo viewer adjusted to their eye separation. In sodalite, $Na_6 Al_6 Si_6 O_{24} \cdot 2\, NaCl$, the tetrahedral nodes are occupied alternately by

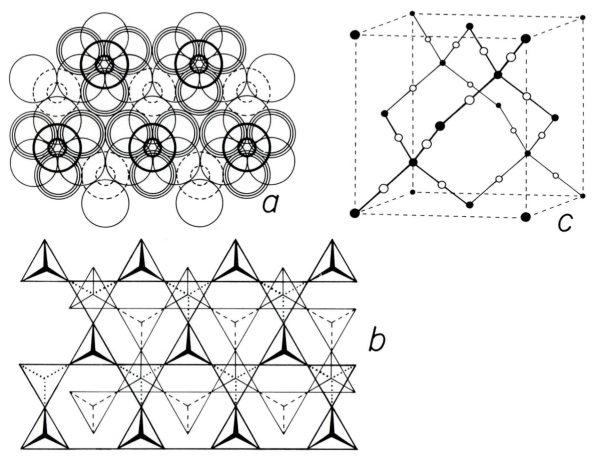

FIGURE 5.22   Ideal crystal structure of cristobalite. (*a*) Plan of hexagonal layers of atoms of oxygen (large circles) and silicon (small circles). With increasing height, the oxygen atoms are shown by a dashed line, thin line, thick line, and triple line. (See Figure 5.21*a* for a clearer depiction of the lower three layers of oxygen atoms.) (*b*) Plan of tetrahedral framework. With increasing depth, the tetrahedra are distinguished by thin dashed line, thin continuous line, dotted line, and thick tapered line. (*c*) Clinographic projection of cubic unit cell. The oxygen atoms (open circle) are placed at the centers of the branches connecting the silicon atoms (filled circle), but in the real structure each oxygen atom is statistically distributed in three positions displaced about 0.5 Å in a plane perpendicular to the Si-Si branch. Four atoms lie in projection along edges of the unit cell (dashed line), but this is accidental. The size of the circles decrease with depth of perspective.

Al^{3+} and Si^{4+} ions, while an O^{2-} ion lies near to but not quite at the midpoint of each branch. A sodium ion (not shown) lies near the center of each hexagon, and a chlorine ion lies at the center of each truncated octahedron (commonly known to zeolite scientists as the **sodalite unit**). Thus, sodalite can be regarded as a stuffed derivative of a hypothetical open framework of composition TO_2. No material is known with the structure of the unstuffed framework.

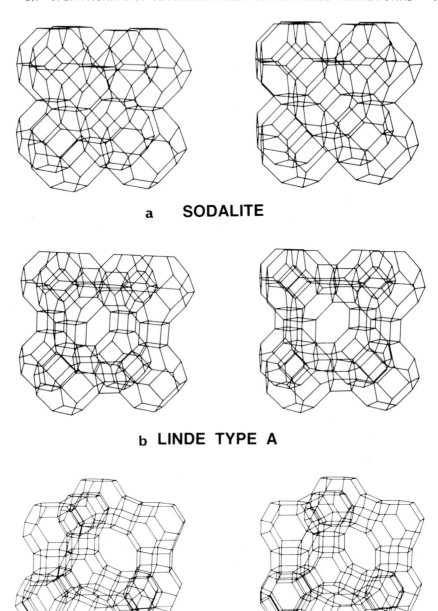

a   **SODALITE**

b **LINDE TYPE A**

c **FAUJASITE**

FIGURE 5.23   Stereo-pairs of the tetrahedral nodes and branches of (*a*) sodalite, (*b*) Linde A zeolite, and (*c*) faujasite. From "Atlas of Zeolite Structure
Types", W.M. Meier and D.H. Olson (1978). Published by the Structure Commission of the International Zeolite Association. Reprinted by permission.

The framework of the ***Linde-A zeolite,*** $Na_{12}Al_{12}Si_{12}O_{48} \cdot 27\,H_2O$, is obtained by replacing each 4-ring of the sodalite framework by a cube (Fig. 5.23*b*). The unit cell is still cubic, but it now contains twice as many tetrahedra. Whereas the sodalite framework is body centered with a truncated octahedron at the body center as well as the corners of the unit cell, the Linde-A framework is primitive. A truncated octahedron remains at each corner of the unit cell, but the body center is now occupied by another Archimedean solid, the ***great rhombicuboctahedron*** (Fig. 3.10*b*). An 8-ring is generated around the center of each face of the unit cell. When the zeolite is heated, the water molecules can escape through the 8-rings, ultimately leaving behind a dehydrated material of composition $Na_{12}Al_{12}Si_{12}O_{48}$, which is used industrially as a molecular sieve and ion exchanger. Molecules small enough to pass through an 8-ring can be adsorbed, whereas larger ones are kept out. Two sodium ions, Na^+, can be replaced by one divalent ion, for example, Ca^{2+}, and so on for other ions.

Replacement of each hexagon of the sodalite framework by a hexagonal prism gives the framework of the mineral *faujasite,* which has a series of synthetic analogs, of which the ***Linde-X zeolite*** ($Na_{96}Al_{96}Si_{96}O_{384} \cdot {\sim}240\,H_2O$) is a convenient representative (Fig. 5.23*c*). The unit cell is still cubic, but only a small piece of the framework is shown in a view down a triad axis. Access to the 3D system of intersecting channels is restricted by the relatively large 12-rings, which allow passage of molecules up to ~7 Å in diameter. The remarkable adsorption and ion-exchange properties of the synthetic faujasite-type zeolites have been exploited commercially, particularly in catalysis; in lectures on the use of such catalysts in the petroleum industry, I have used the joking title "How Archimedes saves the petroleum industry a billion dollars a year!"

 Whereas all three of the sodalite, Linde-A, and faujasite frameworks can be obtained by placing the tetrahedral nodes at the vertices of close-packed regular and semiregular convex polyhedra, most 4-connected 3D nets are best considered in terms of linkage of polygons (or rings). *Cancrinite,* typified by $Na_6Al_6Si_6O_{24} \cdot CaCO_3 \cdot 2\,H_2O$, has its 12 tetrahedral ions Al^{3+} and Si^{4+} at the nodes of horizontal parallel 6-rings cross-linked by tilted 4-rings (Fig. 5.24*a*). These 4-rings can be considered mathematically to condense into vertical zigzag chains, but these chains are merely arbitrary components of the framework. It is necessary to remind oneself often that the framework is merely a mathematical concept, and that the oxygen atoms pack together to form an open array of spheres. From Figure 5.24*a*, it may be seen that the centers of the 6-rings of cancrinite match the *B* and *C* positions of closest-packed spheres, when viewed in projection down the vertical axis of the hexagonal cell. Indeed, there is an infinite family of 3D nets related to the cancrinite net (Smith, 1963), of which the following are illustrative:

$\dot{B}\dot{C}(\equiv\dot{A}\dot{B}\equiv\dot{A}\dot{C})$ Cancrinite

$\dot{A}B\dot{C}$ Sodalite

$\dot{A}BA\dot{C}$ Synthetic zeolite Losod

$\dot{A}AB\dot{B}$ Gmelinite

$\dot{A}ABBC\dot{C}$ Chabazite

$\dot{A}A\dot{B}$ Offretite

$\dot{A}ABAA\dot{C}$ Erionite

Relabeling converts $\dot{B}\dot{C}$ into $\dot{A}\dot{B}$ or $\dot{A}\dot{C}$, just as for hcp of equal spheres. The relation of cancrinite to sodalite can be seen by looking down any one of the four triad axes of the cubic unit cell of sodalite, and, indeed, cancrinite and sodalite are related in an analogous way to hcp and ccp. Whereas cancrinite ($\dot{A}\dot{B}$) has single 6-rings cross-linked by tilted 4-rings, *gmelinite* ($\dot{A}AB\dot{B}$) has double 6-rings. Each pair of 6-rings is joined by vertical 4-rings to give a hexagonal prism, and the prisms are then cross-linked by tilted 4-rings (Fig. 5.24b). *Chabazite* ($\sim Ca_6 Al_{12} Si_{24} O_{72} \cdot 40 H_2O$) is similar to gmelinite ($\sim Na_8 Al_8 Si_{16} O_{48} \cdot 24 H_2O$) except that the hexagonal prisms now are stacked $AABBCC$ instead of $AABB$ (Fig. 5.24c). Gmelinite has a hexagonal unit cell, and is remarkable for the wide nonintersecting channels restricted by 12-rings. The different stacking of chabazite results in the hexagonal prisms lying at the corners of a rhombohedral unit cell with $\alpha \sim 95°$. A large cavity lies at the center of each unit cell, and access between the cavities is given by 8-rings. Two more zeolites, *offretite* ($\dot{A}A\dot{B}$: Fig. 5.24d) and *erionite* ($\dot{A}ABAA\dot{C}$: Fig. 5.24e) were confused for many years until the topological difference was recognized.

Merlinoite has its tetrahedral nodes at the vertices of octagonal prisms joined by isolated squares (Fig. 5.25a). The prisms lie at the corners and body center of a tetragonal unit cell. Another possible description involves recognition of vertical chains shaped like crankshafts with alternate vertical and tilted links. Such crankshaft-shaped chains also occur in *gismondine* (Fig. 5.25b) and in other frameworks, including that of *feldspar* (Chapter 8). The framework of *quartz* (also described in Chapter 8) contains cross-linked helices. A bifurcated chain occurs in the frameworks of the *natrolite* family of zeolites. This chain can be cross-linked in three different ways, one of which is in *edingtonite* (Fig. 5.25c). The chain extends up–down in the plane of the paper, and the nodes lie at the corners of twisted 4-rings. Finally to illustrate the complex nature of some of the tetrahedral frameworks found in zeolites, Figures 5.25d and 5.25e show, respectively, the linkages in *dachiardite* and *heulandite*. These frameworks contain 5-rings sharing branches, and readers may take comfort from the knowledge that even an expert on zeolite structures finds it difficult to build models of such complex frameworks.

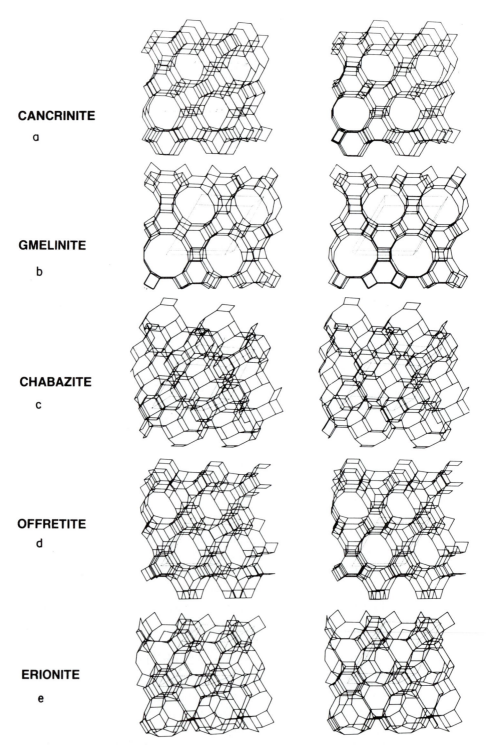

CANCRINITE

a

GMELINITE

b

CHABAZITE

c

OFFRETITE

d

ERIONITE

e

FIGURE 5.24 Stereo-pairs of the tetrahedral nodes and branches of (*a*) cancrinite, (*b*) gmelinite, (*c*) chabazite, (*d*) offretite, and (*e*) erionite. From Meier and Olson (1978). Published by the Structure Commission of the International Zeolite Association. Reprinted by permission.

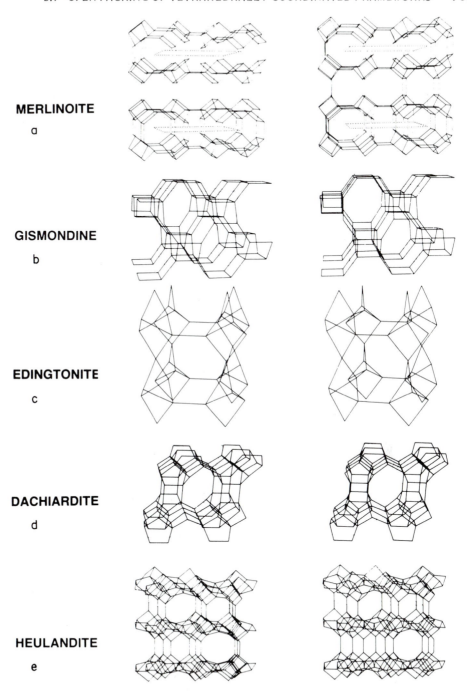

MERLINOITE

a

GISMONDINE

b

EDINGTONITE

c

DACHIARDITE

d

HEULANDITE

e

FIGURE 5.25 Stereo-pairs of the tetrahedral nodes and branches of (*a*) merlino-ite, (*b*) gismondine, (*c*) edingtonite, (*d*) dachiardite, and (*e*) heulandite. From Meier and Olson (1978). Published by the Structure Commission of the International Zeolite Association. Reprinted by permission.

The water molecules of **clathrate** structures (Jeffrey and McMullan, 1967; Wells, 1975) fall at the nodes of 4-connected nets, most of whose branches outline space-filling arrays of polyhedra. In one type of clathrate, the water molecules occupy the tetrahedral nodes of the sodalite type of structure (Fig. 5.23*a*), and a large molecule occupies the center of the trun-

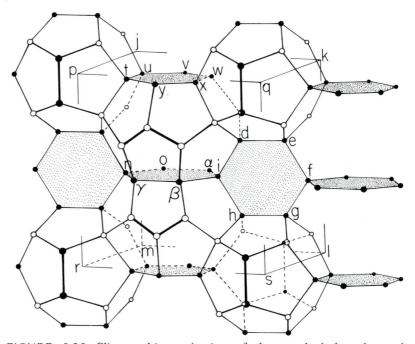

FIGURE 5.26 Clinographic projection of the tetrahedral nodes and branches of the Type I gas hydrate structure. A pentagonal dodecahedron is drawn at the corners *p, q, r,* and *s* of the isometric unit cell. To avoid confusion the dodecahedra are omitted from the four rear corners of the cell, *j, k, l,* and *m*. Adjacent pentagonal dodecahedra are connected by a hexagon (stippled). Thus the pentagonal dodecahedra centered about *p* and *q* are linked by the hexagon *tuvwxy* centered about the midpoint of the cell edge *pq*, and the dodecahedra centered about *q* and *s* are linked by the hexagon *defghi*, centered about the midpoint of the cell edge *qs*. The triad rotation axis along each body diagonal will generate a third type of symmetry-related hexagon (not shown) centered on the cell edges *pj, qk, sl,* and *rm*. Hexagons *tuvwxy* and *noαiβγ* are interchanged with each other to generate the opposing faces of 14-hedra, usually called tetra-kaidecahedra (Figure 3.19*c*). Adjacent 14-hedra are rotated 180° and share faces to form infinite **columns** parallel to cell edge *pr*. Similar columns (not shown) are formed parallel to *pj* and *pq*. Only the frontal faces of six polyhedra are shown, except for the pentagonal dodecahedron centered on *s*, for which the hidden faces are outlined by dashed lines. The sizes of the circles and width of polyhedral edges decrease away from the reader. Filled circles belong to a hexagon whereas open circles do not.

cated octahedron. In most other clathrates, the polyhedra consist of **pentagonal dodecahedra** and one or more polyhedra with 12 pentagonal faces and 2, 3, or 4 hexagonal faces (Fig. 3.19). The **Type I gas hydrate** structure (Fig. 5.26) has a pentagonal dodecahedron at each corner of a cubic unit cell and a second pentagonal dodecahedron in a different angular orientation at the center of the cell. These dodecahedra are cross-linked by single branches to outline six 14-hedra, each lying at the center of a face of the cubic unit cell. There are 46 tetrahedral nodes occupied by H_2O molecules, and in the structure of **argon hydrate** an argon atom lies at the center of each polyhedron to give a formula of 8 $Ar \cdot 46\ H_2O$. Chlorine molecules are too large to enter the pentagonal dodecahedra, and occupancy of just the six 14-hedra yields the formula of 6 $Cl_2 \cdot 46\ H_2O$. Other structure types of clathrates are too complex to give here.

To conclude this section, let us emphasize that the 3D tetrahedral frameworks can be constructed by gluing together balls, but that these balls are far from being in closest packing. Natural and synthetic materials are stabilized by the chemical bonding, which usually involves nonframework as well as framework atoms. Although this section was introduced from the viewpoint of packing of spheres, it was convenient mathematically to proceed to emphasis on the topology of 4-connected 3D nets whose nodes are occupied by the tetrahedrally coordinated spheres. Because there is an infinite number of 4-connected 3D nets, it is impossible to give a comprehensive treatment. Readers are referred to *Three-Dimensional Nets and Polyhedra* by A.F. Wells (1977) for details of many other 3D nets. Nodes need not be 4-connected, and some useful 3D nets with different connec- tivities are given in the next section. Low-density nets and sphere packings are particularly interesting (Fischer, 1976; Brunner, 1979).

5.8 Some Nontetrahedral 3D Nets; Nets Related to Space-Filling Polyhedra

The fundamental concepts of **nontetrahedral 3D nets** can be illustrated with some of the structures already described in this chapter. Connect the centers of adjacent spheres in simple cubic packing (Fig. 5.2*a*) to obtain the 6-connected net of Figure 5.27*a*. Each sphere center lies at a **node**, and each node lies at the intersection of six **branches**. For simple cubic packing, the unit cell is a cube, and each branch of the resulting net is an edge of a unit cell. From the topologic viewpoint, a net need not be in its most symmetric shape. Only the **connectivity** of the net is important. The branches need not have the same length, and the angles between branches need not be the same. In crystals, the underlying nets usually occur in the most symmetric shape, or closely thereto.

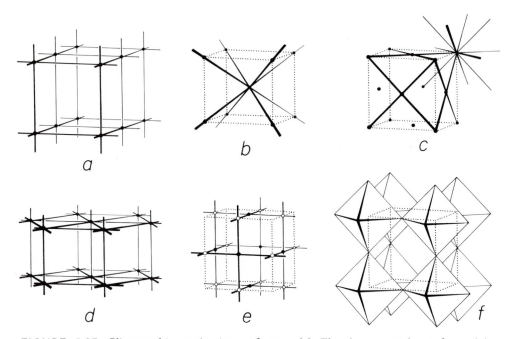

FIGURE 5.27 Clinographic projections of nets. (*a*) The 6-connected net formed by joining nearest neighbors in simple cubic packing. Two unit cells are shown. (*b*) The 8-connected net formed by joining nearest nodes of a body-centered cubic lattice. The unit cell is outlined by dotted lines. Only the node at the body center is depicted with all 8 branches. (*c*) The 12-connected net formed by joining nearest nodes of a face-centered cubic lattice. Only the node at the upper right is depicted with all 12 branches. Nodes at the center of the three exposed faces are depicted with 4 out of the 12 branches. (*d*) The 8-connected net formed by joining nearest nodes of a hexagonal lattice. (*e*) The 4-connected net formed by connecting adjacent atoms in the NbO structure, whose cubic unit cell is shown by dotted lines. Each Nb atom (filled circle) is bonded to four oxygen atoms (open circle) at the corners of a square, and *vice versa*. (*f*) The 8-connected net formed by joining adjacent oxygen atoms in the ReO_3 structure. A rhenium atom lies at each corner of the cubic unit cell (dotted line), and is at the center of an octahedron of oxygen atoms.

For the body-centered packing of spheres (Fig. 5.2*b*), the corresponding net is 8-connected (Fig. 5.27*b*); for face-centered cubic packing, the net is 12-connected (Fig. 5.27*c*). A 12-connected net would also be found for hexagonal closest packing of spheres (not illustrated). The net formed by connecting the C atoms of **diamond** (Fig. 5.5, left) is 4-connected, and is identical with the net that would link the Si atoms of **cristobalite** (Fig. 5.22). The net for simple hexagonal packing of spheres (Fig. 5.2*c*) is 8-connected (Fig. 5.27*d*). Bonds between the Nb and O atoms in the **NbO structure** form a 4-connected net in which four branches form a planar cross at each node (Fig. 5.27*e*).

In the **ReO₃ structure,** a rhenium atom lies at each corner of the cubic unit cell and an oxygen atom at the midpoint of each edge. Each Re atom lies at the center of an octahedron of oxygen atoms, and the edges of the octahedra form a 6-connected net (Fig. 5.27*f*). This net corresponds to an octahedral framework in which all vertices are shared between two octahedra. It is the simplest of the octahedral frameworks, and is 8-connected. A simpler description, analogous to that used for the tetrahedral frameworks, is to connect the Re atoms to give the 6-connected net already illustrated in Figure 5.27*a*. The positions of the O atoms are then obtained from the midpoint of each branch.

The branches of the ReO₃ 8-connected net can be considered as the edges of a space-filling array of octahedra and cuboctahedra. The number of ways in which Platonic and Archimedean solids can be packed together to fill space was determined by Andreini in 1907 (Wells, 1975). In addition to the ReO₃ type, the Linde-A type with cubes, truncated octahedra, and great rhombicuboctahedra was already described (Fig. 5.23*b*). A further type is obtained from the framework of boron atoms in the **CaB₆ structure** (Fig. 5.28*a*). Instead of placing an oxygen atom at the midpoint of each edge, as in the ReO₃ structure, pairs of boron atoms are placed at coordinates $\pm(u,0,0)$, $\pm(0,u,0)$, and $\pm(0,0,u)$, where $u = 1/(2 + \sqrt{2})$. The branches of the resulting 5-connected net generate the edges of a space-filling array of octahedra and truncated cubes. The B atoms of the **UB₁₂ structure** lie at the nodes of

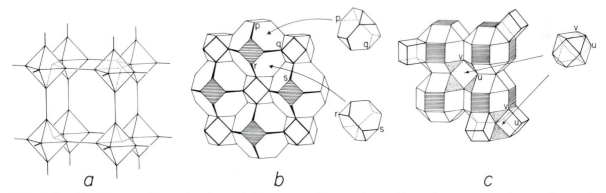

FIGURE 5.28 Clinographic projections of three nets whose nodes and branches correspond to vertices and edges of close-packed regular and semiregular polyhedra. (*a*) Space-filling array of octahedra and truncated cubes. (*b*) Array of truncated octahedra in parallel orientation, all of whose square faces are connected by cuboctahedra. Four square faces are hatched to enhance the perspective, but unhatched faces are crystallographically equivalent. The close-packing of the truncated octahedra and cuboctahedra generates voids which are filled by truncated tetrahedra, as illustrated by the arrows and labeled vertices. (*c*) Array of rhombicuboctahedra sharing faces perpendicular to the tetrad axes. A cube is attached to each face lying perpendicular to the diad axes. Each resulting void is occupied by a cuboctahedron, as exemplified by the arrows and matching vertices.

another 5-connected net whose branches correspond to the edges of the space-filling array of truncated tetrahedra, truncated octahedra, and cubocta-hedra (Fig. 5.28*b*). Another space-filling array is generated from cubes, cuboctahedra, and small rhombicuboctahedra (Fig. 5.28*c*).

5.9   Perovskite Structures

The mineral **perovskite,** $CaTiO_3$, gives its name to a remarkable family of close-packed crystal structures. The ideal variant of the perovskite structures is obtained by adding a third type of sphere at the body center of the unit cell of the ReO_3 structure just described in the preceding paragraph (Fig. 5.29*a*). The new sphere (denoted P) is coordinated to the 12 first neighbors (R) at the midpoints of the edges, and the spheres of type R also form octahedra around the Q type of spheres at the corners of the cell. When account is taken of the sharing of sites with adjacent unit cells, the cell content is PQR_3. In order to obtain a perfect fit the P and R spheres should be exactly the same size, and, indeed, when the distinction between the P and R spheres is removed, the P and R spheres collectively generate cubic closest packing of equal spheres. This can be seen algebraically as follows. In perovskite, the atomic coordinates are:

$$P \; \tfrac{1}{2},\tfrac{1}{2},\tfrac{1}{2} \quad Q \; 0,0,0 \quad R \; 0,0,\tfrac{1}{2}; \;\; 0,\tfrac{1}{2},0; \;\; \tfrac{1}{2},0,0$$

Displacement of the origin of the unit cell by $\tfrac{1}{2},0,0$ gives new coordinates:

$$P \; 0,\tfrac{1}{2},\tfrac{1}{2} \quad Q \; \tfrac{1}{2},0,0 \quad R \; \tfrac{1}{2},0,\tfrac{1}{2}; \;\; \tfrac{1}{2},\tfrac{1}{2},0; \;\; 0,0,0$$

These coordinates for P and R collectively yield the face-centered cubic cell of cubic closest packing.

For the mineral perovskite, Ca = P, Ti = Q, and O = R. When envisaged as an ionic compound, this means that Ca^{2+} and O^{2-} ions are collectively in the same positions as cubic closest packing. In most perovskite-type structures, the atoms or ions do not fit the exact requirements for close packing of spheres. The calcium cation is too small for 12-coordination to oxygen anions, and at room temperature the oxygen anions collapse around the calcium cation with distortion to orthorhombic symmetry. Above 900°C, increased thermal vibration stabilizes the ideal structure. There is an enormous literature on the structural details in the perovskite family. Particularly important commercially is the ferroelectric behavior of low-temperature varieties of $BaTiO_3$. Readers are referred to Wyckoff (1960), Wells (1975), and Megaw (1957) for an introduction to the burgeoning literature.

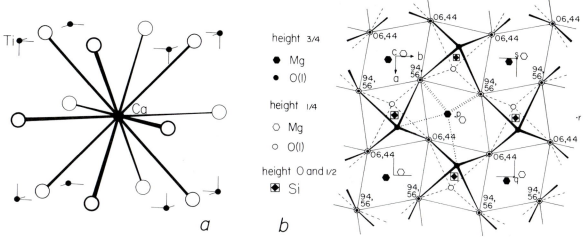

FIGURE 5.29 Perovskite structures. (*a*) Clinographic projection of cubic unit cell of ideal perovskite, as typified by the structure of $CaTiO_3$ at high temperature. (*b*) Plan down the *c*-axis of the orthorhombic unit cell of the $MgSiO_3$ structure. The octahedra of oxygen atoms (circles) around the Si atoms (squares) are rotated with respect to the ideal arrangement in Figure 5.27*f*. Bonds to the six nearest oxygen atoms from one of the Mg atoms (hexagon) are shown by dotted lines. The corners of a cubic pseudo-cell project onto *p*, *q*, *r*, and *s*. Along the *c*-axis, the repeat distance would be halved. Heights at ¼ and ¾ are shown by symbols. Heights of O(2) atoms are given to the nearest hundredth.

As an example of the distortions in perovskite structures consider the structure of the $MgSiO_3$ variety. This material is stable only at high pressure, and is believed to be the major component of the Earth's lower mantle. After pressure quenching from 300 kbar (3×10^{10} Pa), the crystal structure (Fig. 5.29*b*) has an orthorhombic unit cell with $a = 4.78$ Å, $b = 4.93$ Å, $c = 6.90$ Å (Yagi, Mao, and Bell, 1978). In projection down the *c*-axis, the Si atoms (squares) overlap in pairs at heights 0 and ½. Oxygen atoms of type O(1) lie at heights ¼ and ¾, while oxygen atoms of type O(2) lie at heights 0.056, 0.444, 0.556, and 0.944. Each Si atom is surrounded by two O(1) and four O(2) to give a tilted octahedron depicted by dashed and continuous lines. Because of the horizontal displacement of the O(1) atoms at heights ¼ and ¾, the octahedra tilt alternately in the *ac* plane, but this is not obvious in the *c*-axis projection because of exact overlap of octahedra centered on Si atoms at heights 0 and ½. Magnesium atoms (hexagons) lie at heights ¼ and ¾. Tilting of the octahedra allows 6 of the oxygen atoms to approach each Mg atom (dotted lines) more closely than the remaining 6 out of the 12 neighbors for an ideal perovskite structure [Mg–O(1) = 2.10 Å; Mg–O(1)′ = 2.12 Å; Mg–O(2) = 2.06 Å (2-fold multiplicity); Mg–O(2)′ = 2.20 Å (2-fold); Mg–O(2)″ = 2.47 Å (2-fold); Mg–O(1)″ = 2.75 Å; Mg–O(1)‴ = 2.95 Å; Mg–O(2)‴ = 3.16 Å (2-fold)].

The experimental error is about 0.10 Å for two standard deviations. It is not possible to draw a sharp cutoff when discussing the bonding between the Mg and O ions, but it is obvious that the angular tilting of the SiO_6 octahedra allows the coordination of the Mg ion to drop from 12-fold to a value more suitable for the small ionic radius of Mg^{2+}. Only the six shortest distances are shown in Figure 5.29b. The orthorhombic cell has four times the volume of the cubic cell of ideal perovskite, as explained in the figure legend. The space group is *Pbnm* (see Chapter 7 for explanation), and the atomic coordinates are developed from these starting positions:

	x	y	z		x	y	z
Mg	0.974	0.063	¼	O(1)	0.096	0.477	¼
Si	0	½	0	O(2)	0.696	0.291	0.056

The $MgSiO_3$ variety of perovskite has a particularly large distortion from the ideal structure, and other varieties have a small distortion. Hexagonal, tetragonal and monoclinic varieties are known, as well as superstructures.

5.10   Alloy Structures

Pearson (1972) makes great use of nets and polyhedra in his thorough treatment *The Crystal Chemistry and Physics of Metals and Alloys*. It is quite impossible to do justice here to the remarkable range of structures of alloys, which range from simple superstructures (such as the ones in Figs. 5.6 and 5.7) to incredible structures involving the triangulated polyhedra listed in Chapter 3 (Samson, 1968). Just as a small taste, consider the *$MgCu_2$ structure,* which belongs to the family named after F. Laves. The cubic cell has $a = 7.05$ Å.

First place the Cu atoms in the averaged positions of the oxygen atoms of *cristobalite* (Fig. 5.22c). These positions are halfway between those of the C atoms in *diamond*. Figure 5.30a shows the positions projected onto one face of the cubic unit cell. Recalling that the coordinates in *diamond* are:

$$0,0,0; \; ½,½,0; \; 0,½,½; \; ½,0,½; \; ¼,¼,¾; \; ¾,¾,¾; \; ¼,¾,¼; \; ¾,¼,¼$$

the coordinates of the 16 Cu atoms are:

$$⅛,⅛,⅞; \; ⅛,⅜,⅝; \; ⅛,⅝,⅜; \; ⅛,⅞,⅛;$$
$$⅜,⅛,⅝; \; ⅜,⅜,⅞; \; ⅜,⅝,⅛; \; ⅜,⅞,⅜;$$
$$⅝,⅛,⅜; \; ⅝,⅜,⅛; \; ⅝,⅝,⅞; \; ⅝,⅞,⅝;$$
$$⅞,⅛,⅛; \; ⅞,⅜,⅜; \; ⅞,⅝,⅝; \; ⅞,⅞,⅞$$

As for the oxygen atoms of cristobalite, the Cu atoms lie at the vertices of a tetrahedral framework. In Figure 5.30*a*, each tetrahedron projects like Figure 3.3*c*. There are two types of orientation. Between the tetrahedra are polyhedral voids shaped like truncated tetrahedra. Thus the Cu atoms shown in Figure 5.30*b* outline a ***truncated tetrahedron*** viewed down a diad rotation axis. Place an Mg atom at the center ($\frac{1}{2},\frac{1}{2},\frac{1}{2}$) of the truncated tetrahedron. Complete the $MgCu_2$ structure by placing seven more Mg atoms at:

$$0,0,\tfrac{1}{2}; \; 0,\tfrac{1}{2},0; \; \tfrac{1}{2},0,0; \; \tfrac{1}{4},\tfrac{1}{4},\tfrac{1}{4}; \; \tfrac{1}{4},\tfrac{3}{4},\tfrac{3}{4}; \; \tfrac{3}{4},\tfrac{1}{4},\tfrac{3}{4}; \; \tfrac{3}{4},\tfrac{3}{4},\tfrac{1}{4}$$

Each Mg atom is surrounded by 12 Cu atoms at 2.92 Å at the vertices of the truncated tetrahedron, but it is also close to 4 Mg atoms at 3.05 Å. Each of these Mg atoms (filled circles) lies on a line projecting outward through the center of one of the four hexagonal faces of the truncated tetrahedron. When all 16 nearest neighbors are considered, the coordination polyhedron becomes the ***tetracapped truncated tetrahedron*** (Fig. 3.18*e*), which is one of the ***Friauf triangulated polyhedra***. Addition of the dotted lines in Figure 5.30*b* generates this triangulated polyhedron. Each Cu atom is surrounded by 6 Cu atoms at 2.49 Å and 6 Mg atoms at 3.05 Å. Although these distances are so different, the 6 Cu atoms do not form a polyhedron that excludes the 6 Mg atoms. All 12 atoms must be chemically bonded to the central atom. Taken together, the 6 Cu and 6 Mg atoms lie at the vertices of another triangulated polyhedron, which is a strongly distorted ***icosahedron*** (Fig. 5.30*c*).

An alternative description of this structure uses ***triangular and kagomé nets*** stacked perpendicular to a triad axis of the cubic unit cell. Pearson (1972) used α, β, and γ to denote the three possible ways of orienting a kagomé layer with respect to *A, B,* and *C* positions of a triangular layer (Fig. 5.30*d*). Using this notation, the $MgCu_2$ structure is expressed as:

Layer Number	1			2	3			4	5			6
Net Type	*A*	*B*	*C*	α	*B*	*C*	*A*	β	*C*	*A*	*B*	γ
Atomic Content	Mg	Cu	Mg	3 Cu	Mg	Cu	Mg	3 Cu	Mg	Cu	Mg	3 Cu

Figure 5.30*e* shows the hexagonal unit cell of the $MgCu_2$ structure, as obtained by viewing the cubic unit cell down any one of the four triad axes. Atomic positions are shown in projection by the layer number. The Mg and Cu atoms are distinguished by using tilted and upright symbols, respectively.

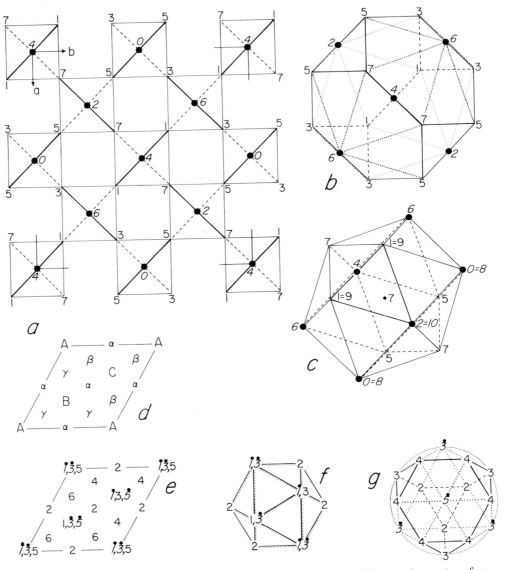

FIGURE 5.30 Projections of the crystal structure of MgCu$_2$. (*a*) Projection onto a face of the cubic unit cell. All heights are expressed in eighths of the cell repeat. The Cu atoms lie at heights $\frac{1}{8}$, $\frac{3}{8}$, $\frac{5}{8}$, and $\frac{7}{8}$ at the nodes of a tetrahedral framework. Each tetrahedron is represented by a square with two diagonals as in Figure 3.3*c*. The Mg atoms (filled circle) lie at heights $\frac{0}{8}$, $\frac{2}{8}$, $\frac{4}{8}$, and $\frac{6}{8}$. (*b*) Nearest neighbors and coordination polyhedra around a magnesium atom at ($\frac{1}{2}$, $\frac{1}{2}$, $\frac{4}{8}$) in the center of diagram (*a*). The 12 nearest neighbors are Cu atoms lying at the vertices of a truncated tetrahedron (continuous and dashed lines), viewed down a 2-fold axis [actually an inverse tetrad axis] instead of being drawn in clinographic projection as in Figure 3.8*a*. Four Mg atoms (filled dots) are the next nearest neighbors. Dotted lines show how the four hexagonal faces of the truncated tetrahedron are capped to give the tetracapped truncated tetrahedron when all 16 atoms

For ease of viewing, a dot is placed over the symbol for each Mg atom. Thus the A position of layer one is represented by the number 1 at $x = y = 0$ in Figure 5.30e. Because the site is occupied by Mg, the numeral 1 is tilted, and marked with a dot. The full list of hexagonal coordinates is:

Mg $0,0,0$; $\frac{1}{3},\frac{2}{3},0$; $0,0,\frac{2}{6}$; $\frac{2}{3},\frac{1}{3},\frac{2}{6}$; $\frac{1}{3},\frac{2}{3},\frac{4}{6}$; $\frac{2}{3},\frac{1}{3},\frac{4}{6}$

Cu $\frac{2}{3},\frac{1}{3},0$; $\frac{1}{3},\frac{2}{3},\frac{2}{6}$; $0,0,\frac{4}{6}$; $0,\frac{1}{2},\frac{1}{6}$; $\frac{1}{2},0,\frac{1}{6}$; $\frac{1}{2},\frac{1}{2},\frac{1}{6}$;

$\frac{1}{6},\frac{1}{3},\frac{3}{6}$; $\frac{1}{6},\frac{5}{6},\frac{3}{6}$; $\frac{2}{3},\frac{5}{6},\frac{3}{6}$; $\frac{1}{3},\frac{1}{6},\frac{5}{6}$; $\frac{5}{6},\frac{1}{6},\frac{5}{6}$; $\frac{5}{6},\frac{2}{3},\frac{5}{6}$

Note that layer one is assigned height $z = 0$, layer two $z = \frac{1}{6}$, and so on.

Figures 5.30f and g, respectively, show the hexagonal projection of the coordination polyhedra around a copper and a magnesium atom. Overlap of many edges may cause difficulty in visualizing the two polyhedra, especially the tetracapped truncated tetrahedron.

EXERCISES   (* More Difficult)

Note. Models can be built either by gluing spheres together or by linking balls by spokes, or plastic stars by tubing. Several kinds of material are available commercially, but it is possible to manage with discarded tennis, golf, or polystyrene balls if money is short.

5.1 Build models of primitive cubic packing, body-centered cubic packing, primitive hexagonal packing, and of hexagonal and cubic closest packing of spheres. If spheres are used, it will be necessary to hold at least the first layer in position before gluing. One way is to cut a piece of plywood or stiff card-board and coat it with moulding clay. Press the first layer of spheres into the clay at appropriate distances. Alternatively, drill holes into the plywood into which the spheres will rest. Ball-and-spoke models can be built either

---

are used for the coordination polyhedron. (c) Nearest neighbors and coordination poly-hedron around a copper atom at $(\frac{5}{8}, \frac{5}{8}, \frac{7}{8})$. To generate the polyhedron, atoms at heights $\frac{1}{8}$ and $\frac{2}{8}$ are replaced by lattice-related atoms at heights $\frac{9}{8}$ and $\frac{10}{8}$. The 6 Mg atoms (large filled circles) lie at greater distances than the 6 Cu atoms (unmarked vertices) from the central Cu atom (small filled circle), and the icosahedron is nonregular. Many lines overlap in projection. (d) The nomenclature used by Pearson (1972, Fig. J-4) for tri-angular and kagomé nets. (e) The projection of hexagonal unit cell of $MgCu_2$ with heights expressed in one-sixth of cell repeat. Copper atoms are denoted by upright symbols, and Mg atoms are denoted by sloping ones. (f) A nonregular icosahedron of 6 Cu and 6 Mg atoms around a Cu atom at $(\frac{1}{2}, \frac{1}{2}, \frac{2}{6})$. Most edges overlap in projection. (g) Truncated tetrahedron (continuous and dashed lines) and tetracapped truncated tetrahedron (addi-tional dotted lines) about a magnesium atom at $(\frac{2}{3}, \frac{1}{3}, \frac{5}{16})$.

from commercial plastic stars and tubing or from drilled balls and rods. Balls can be drilled either with the aid of a special jig, or can be purchased with holes drilled into cubic, octahedral, and dodecahedral directions.

5.2 What will be the edge lengths of the unit cells of the above models if made of spheres of radius 1.4 Å representing oxygen atoms?

5.3 Put away the models, and draw sphere centers in unit cells for the above kind of packing. Draw both the hexagonal and cubic unit cells for cubic closest packing.

5.4 Assuming that the larger spheres in the NaCl structure type touch each other as well as the smaller spheres, what is the percentage of empty space?

5.5 Draw stereograms showing the first and then the second nearest neighbors for the NaCl and CsCl structure types.

5.6 Construct a ball-and-spoke model of diamond. Label alternate nodes to obtain a model for ZnS (blende).

5.7* Construct a ball-and-spoke model of chalcopyrite in which the tetrahedral nodes are either colored or labeled to distinguish the Cu, Fe, and S atoms. The easiest way is to examine a perspective drawing of chalcopyrite in a standard monograph, for example, *Sulfide Mineralogy,* edited by P.H. Ribbe (1974). A more difficult way is to use just the following atomic coordinates in a tetragonal unit cell with $a = 5.25$ Å, $c = 10.32$ Å:

$$Fe \quad 0,0,0; \ \tfrac{1}{2},\tfrac{1}{2},\tfrac{1}{2}; \ \tfrac{1}{2},0,\tfrac{3}{4}; \ 0,\tfrac{1}{2},\tfrac{1}{4}$$

$$Cu \quad 0,0,\tfrac{1}{2}; \ \tfrac{1}{2},\tfrac{1}{2},0; \ \tfrac{1}{2},0,\tfrac{1}{4}; \ 0,\tfrac{1}{2},\tfrac{3}{4}$$

$$S \quad \tfrac{1}{4},\tfrac{1}{4},\tfrac{1}{8}; \ \tfrac{1}{4},\tfrac{1}{4},\tfrac{5}{8}; \ \tfrac{3}{4},\tfrac{3}{4},\tfrac{1}{8}; \ \tfrac{3}{4},\tfrac{3}{4},\tfrac{5}{8};$$

$$\tfrac{1}{4},\tfrac{3}{4},\tfrac{3}{8}; \ \tfrac{1}{4},\tfrac{3}{4},\tfrac{7}{8}; \ \tfrac{3}{4},\tfrac{1}{4},\tfrac{3}{8}; \ \tfrac{3}{4},\tfrac{1}{4},\tfrac{7}{8}$$

Note that the c repeat is doubled over that for blende because of the regular alternation of Cu and Fe at $c/2$. In each 6-ring of the 4-connected net, alternate nodes are occupied by S and the remaining three nodes are occupied either by 2 Fe and 1 Cu or by 2 Cu and 1 Fe. ** Can you invent other ways of placing Fe and Cu in a *regular* array on the nodes?

5.8 Build sphere models or ball-and-spoke models of CsCl and NaCl structure types.

5.9* Build sphere models of the Cu_1Au_1 and Cu_3Au_1 superstructures of ccp. Use balls of two colors.

5.10* Build sphere models or ball-and-spoke models of the Fe_3Al and NaTl superstructures of the CsCl structure type.

5.11* Build sphere models of the AlB_2, WC, and NiAs structure types using two sizes of balls to approximate occupancy of small voids between larger spheres in simple hexagonal packing (not hcp).

5.12* Build sphere models of fluorite and antifluorite (Li_2O) structure types using balls of appropriate sizes. Build a ball-and-spoke model using nodes of two colors, and observe that this single model represents both the fluorite and antifluorite structures. Compare the ball-and-spoke model with that of diamond (Exercise 5.6). Build a ball-and-spoke model of cuprite, and observe the much smaller number of branches than in the fluorite model. **Demonstrate that a sphere model of cuprite is extremely unstable unless very strong glue or locking pins are used.

5.13* Build a packing model of corundum using layers like those in Figure 5.20*c*. A model for ilmenite can be built from balls of different colors.

5.14* Build a packing model of spinel using layers like those in Figure 5.20*d*. **Fasten together cardboard octahedra and tetrahedra to demonstrate the sharing of edges as illustrated in the right-hand diagrams of Figure 5.19.

5.15* Build a sphere model of the α-PbO_2 structure. **Either label appropriate balls of the α-PbO_2 model, or build a new model of the idealized $FeWO_4$ structure.

5.16** Build a sphere model of the diaspore structure.

5.17** Using octahedra, build models of the $CaCl_2$ and rutile structures.

5.18** Using octahedra, build a layer of the CdI_2 structure. ***Using rods, fasten together layers of octahedra as in Figure 5.18*b*.

5.19 Using balls, build a layer of the tridymite and cristobalite structures (Fig. 5.21*a*). Stack layers vertically to give the tridymite structure. Build a ball-and-spoke model to illustrate the 4-connected net whose nodes lie at the positions of the Si atoms in tridymite.

5.20* Build a ball-and-spoke model of cristobalite.

5.21* Build ball-and-spoke models of the 4-connected nets in the sodalite, Linde-A, and faujasite structures.

5.22* Build models from cardboard polyhedra of the sodalite, Linde-A, and faujasite structures.

5.23* Build ball-and-spoke models of the 4-connected nets in the cancrinite, gmelinite, and chabazite structures.

5.24** Build ball-and-spoke models of the 4-connected nets in the structures of offretite, erionite, and synthetic zeolite Losod.

5.25** Build ball-and-spoke models of merlinoite, gismondine, and edingtonite.

5.26*** Show that there are only two other ways of linking together the chains in edingtonite, and check your answer with Smith (1965).

5.27*** Build ball-and-spoke models of dachiardite and heulandite.

5.28** Build a ball-and-spoke model of the 4-connected net in Type I clathrates. Alternatively, pack together cardboard models of dodecahedra and 14-hedra. ***Build models of other types of clathrate structures.

5.29 Make ball-and-spoke models of the nets in Figures 5.27a, b, 5.27d, e. For Figure 5.27b, make each node identical so that 8 branches emerge from the nodes at the corners of the unit cell.

5.30* Build a model of the ReO_3 structure using cardboard octahedra.

5.31* Build a ball model of the ideal perovskite structure using colored balls of the same size to represent the P and R types, and smaller balls to represent the Q type in the octahedral sites. **Build another model in which the P-type ball is smaller than the R type. Show that the P-type ball is too small to fill the large holes between the R-type balls. ***Using octahedra, build an approximate model of the distorted perovskite structure of $MgSiO_3$.

5.32** Build cardboard models of the close-packed arrays in Figure 5.28.

5.33** Pack together tetrahedra and truncated tetrahedra to represent the positions of the Cu atoms in $MgCu_2$ (or, alternatively, use just cardboard tetrahedra pierced by thin stiff rods). ***Read Pearson (1972) to discover other types of complex alloy structures.

ANSWERS

5.2 Primitive cubic, $a = 2.8$ Å; body-centered cubic, $a = 2.43$ Å; primitive hexagonal, $a = c = 2.8$ Å; hexagonal closest packing, $a = 2.8$ Å, $c = 4.57$ Å; cubic closest packing, $a = 3.96$ Å.

5.4 20.7 percent.

5.5 NaCl first, cubic directions; second, dodecahedral directions: CsCl first, octahedral directions; second, cubic directions.

5.7 There are many ways of placing Fe and Cu atoms in a regular array on the alternate nodes of the 4-connected net in blende. They have not been enumerated. There are many potential research problems of this type.

5.26 The other two structure types are found in natrolite and thomsonite.

6
Lattice and Point Group in Three Dimensions. Crystal Shape

Preview Although crystal structures can be studied without formal knowledge of lattices and point groups of symmetry elements, a systematic mathematical treatment of these concepts provides a rigorous basis for a deeper understanding. This chapter is restricted to separate consideration of an isolated lattice and an isolated group of symmetry elements passing through a single point, and Chapter 7 goes on to consider space groups of intersecting lattices and symmetry elements. The 14 isolated lattices are named after a French astronomer–physicist, A. Bravais (1811–1863). There is an infinite number of point groups of symmetry elements when the symmetry elements are not constrained by a lattice, but only 32 are required for description of the physical properties of crystals. Crystallographers found that the external shape of each crystal belongs to one of 32 *classes,* and that each class corresponds to a *crystallographic point group.* Each crystal *face* is parallel to a set of planes passing through atoms lying on the nodes of a lattice (cf. Figs. 1.14 and 1.15 for 2D with Section 4.8 for 3D). This results in control of the angular orientations of crystal faces by symmetry elements of order **1, 2, 3, 4,** and **6,** but not by other orders (as developed for 2D in Chapter 2). However, such a restriction does not apply to isolated bodies (e.g., molecules) and clusters (e.g., globular viruses); in particular, 5-fold symmetry is quite common in plants (though not in clover), and occurs in the regular icosahedron and regular dodecahedron. Most crystallographers use the International (Hermann–Mauguin) system of nomenclature for point groups and space groups, but most spectrographers use the Schoenflies system.

There is a complex literature on the derivation of the crystallographic point groups (Boisen and Gibbs, 1976), dating from the nineteenth century when emphasis was placed on polyhedra to the present day when group theory and equivalence relations are preferred. The rigorous treatment by Boisen and Gibbs is too lengthy to be given here, but the essential features are present. Thorough descriptions of point groups and space groups are given in *International Tables for X-Ray*

Crystallography (Henry et al., 1952) and in *The Mathematical Theory of Symmetry in Solids* (Bradley and Cracknell, 1972).

Symmetry elements are used in Section 6.1 on Bravais lattices before they are described rigorously in Section 6.2. It is expected that most readers will have no problem because of earlier exposure to symmetry elements. It is suggested that all readers return to Section 6.1 after reading Section 6.2.

6.1 Bravais Lattices

A *lattice* is an array of points, all of which have the same vectorial environment. The adjective *vectorial* is of key importance because it means that absolute direction as well as distance must be considered. The nodes of the 6-connected net in Figure 5.27a define a single Bravais lattice, but the nodes of the 4-connected net in Figure 5.27e do not (ignore the distinction between dot and circle). A 3D lattice can be specified by three nonplanar translation vectors, and these vectors correspond to three edges of a parallelepiped. Regular stacking of the parallelepiped will fill space, and the parallelepiped is described as a *unit cell*. When regularity of shape is considered, there are seven types of parallelepipeds (Fig. 3.14) that can be used as unit cells: namely, isometric, tetragonal, rhombohedral, hexagonal, orthorhombic, monoclinic, and triclinic. When stacked together in a regular array, the vertices of each type of parallelepiped produce lattice nodes related by a characteristic set of symmetry elements.

Consider an array of orthorhombic parallelepipeds (Fig. 6.1). Eight parallelepipeds meet at each lattice node, and any two lattice nodes are related by a vector $p\mathbf{a} + q\mathbf{b} + r\mathbf{c}$, where $\mathbf{a}$, $\mathbf{b}$, and $\mathbf{c}$ are the edge vectors of the unit cell and p, q, and r are any set of integers. Three diad rotation axes pass through each lattice node, and also through the midpoint of each cell edge and face. The diad axes are related by translations of $\frac{1}{2}(p\mathbf{a} + q\mathbf{b} + r\mathbf{c})$, where p, q, and r are any set of integers. Each face of a unit cell is part of a mirror plane of infinite extent. A lattice node that lies on a mirror plane automatically obeys the mirror symmetry. Because the lattice node is merely a point with no shape, the mirror operation leaves it in position. This orthorhombic lattice can be denoted $\mathbf{P2/m2/m2/m}$. The symbol $\mathbf{P}$ denotes a *primitive lattice* with only one lattice node for each unit cell, and the capitalization indicates three dimensions in contrast to small $\mathbf{p}$ for 2D. The first $\mathbf{2/m}$ (spoken as "two over em") denotes a diad rotation axis (2) perpendicular to (/) a mirror plane ($\mathbf{m}$). The diad axis is parallel to the first edge vector ($\mathbf{a}$), and the mirror plane is perpendicular to $\mathbf{a}$. Similarly, the second $\mathbf{2/m}$ refers to $\mathbf{b}$, and the third $\mathbf{2/m}$ to $\mathbf{c}$. Orthogonal geometry of the unit cell produces

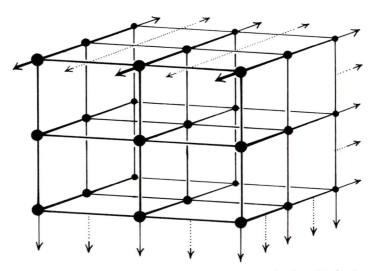

FIGURE 6.1 Eight unit cells of a primitive orthorhombic lattice drawn in clinographic projection. Some of the diad rotation axes are shown. Those in continuous line pass through lattice nodes, whereas those in dotted line pass half-way between the nodes.

the **2/m2/m2/m** symmetry, and vice versa. There is no symmetry control on lengths of the edge vectors **a**, **b**, and **c**, nor on the axial ratios a/b and c/b.

For brevity, it is sufficient to use the symbol **P222** for the lattice because the combined operations of three translations and three orthogonal rotation diads automatically generate the three mirror planes. Centers of symmetry are also generated automatically in the orthorhombic lattice, but will not be displayed in the lattice symbol.

Now here comes the key feature of ***Bravais lattices.*** Can both the ortho-rhombic symmetry and lattice congruency be maintained if further nodes are added? Try adding a point at the body center of a unit cell (Fig. 6.2*a*). For lattice congruency, a point must be added for each vector $p\mathbf{a} + q\mathbf{b} + r\mathbf{c}$, thereby occupying the body center of every cell. Does the body-centering point have the same vectorial environment as the point at the cell corner? Yes. The new lattice is denoted **I** from the German word *innenzentriertes,* and the population of the lattice nodes of the unit cell yields a ***multiplicity*** of 2. All symmetry elements are retained, and the lattice can be denoted **I2/m2/m2/m**. Figures 6.2*b* through 6.2*e* show that three new lattices can be obtained from centering either the **A**, **B**, or **C** faces of the unit cell, and a fourth one (**F**) by centering all the faces. As usual, crystallographers use an obvious symbolism: **A** refers to the face defined by the **b** and **c** edge vectors, and **F** refers to centering of all three faces. Is a (**B** + **C**) lattice possible? No, because the points would not then have an identical vectorial environment (Fig. 6.2*f*). Can a point be placed halfway along an edge, and still belong to a

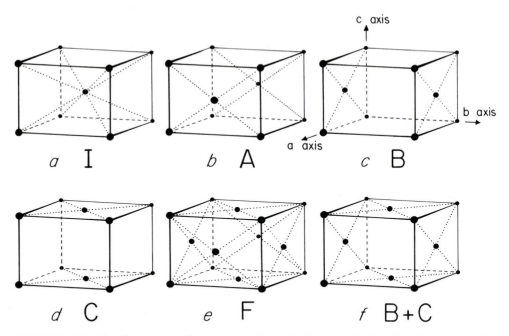

FIGURE 6.2   The five ways of centering the orthorhombic lattice plus the impossible combination $B + C$. Each diagram is a clinographic projection of one unit cell.

single lattice? No. To obtain a lattice, a point must then be added at face centers, and a new unit cell can then be chosen with a new edge vector one half of the starting value.

Crystallographers sometimes deliberately use an unnecessarily large cell when comparing families of superstructures and parent structures, but in normal work the *smallest unit cell is used that has the most regular geometry.* Thus, for three orthogonal diad axes, the most regular geometry is $\alpha = \beta = \gamma = 90°$ and $a \neq b \neq c$. For any nonprimitive orthorhombic lattice, it is possible to obtain a primitive lattice with a smaller unit cell, but the geometry is less regular. Thus for the orthorhombic **C** lattice, the corresponding primitive lattice would be monoclinic.

Although there are six orthorhombic lattices (**P, I, F, A, B,** and **C**), ask yourself whether the labels **A, B,** and **C** have any fundamental significance. Indeed, they do not, and from the viewpoint of lattice symmetry and geometry there is no difference between the orthorhombic **A, B,** and **C** lattices. So there are only four types of orthorhombic lattices. But once a crystallographer has assigned the symbols *a, b,* and *c* to the cell edges of an orthohombic crystal, then there is a distinction between the **A, B,** and **C** lattices. Crystallographers commonly prefer **C** to **A** and **B**, but all three symbols are used.

An obvious Bravais lattice is primitive cubic (Fig. 6.3). The unit cell is a cube, which becomes a square upon projection down the *c* axis. Passing through each lattice point are three tetrad axes, each perpendicular to a mirror plane, four triad axes, each lying at the intersection of three out of the six diagonal mirror planes, and six diad axes. To express the symmetry and geometry, the symbol **P4/m3m2** can be used.

What is the lattice type for Figure 5.2*b* and 5.27*b*? Certainly this body-centered cubic array could be considered as the intersection of two primitive

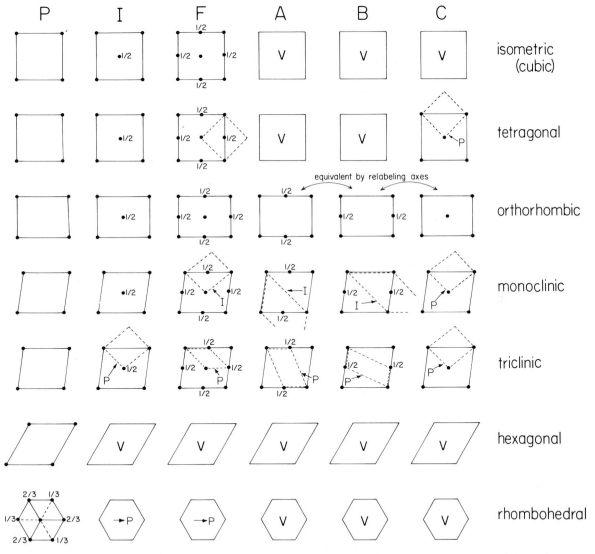

FIGURE 6.3 Plan of unit cells of Bravais lattices. Boxes marked with *v* do not yield Bravais lattices because of violation of lattice and symmetry requirements. Dashed lines show the smaller unit cell obtainable for the non-Bravais lattices.

cubic lattices displaced by $(\mathbf{a} + \mathbf{b} + \mathbf{c})/2$. But this does not utilize the identical vectorial environment of both lattices. Hence this is a new Bravais lattice (Fig. 6.3), which is denoted **I4/m3m2**. The CsCl structure (Fig. 5.7a) does not have a body-centered cubic lattice, because the corners and body centers are occupied by different atoms. It has two primitive lattices, occupied separately by Cs and Cl atoms. It is important to distinguish clearly between the simple mathematical concept of a lattice array of points, and the use of a lattice in crystal chemistry for placement of atoms. Even if two points are related by a lattice vector, they cannot belong to a single lattice if occupied by different atoms.

Cubic closest packing of spheres gives a face-centered cubic array of sphere centers (Fig. 5.3d), which are represented by a single face-centered cubic lattice **F4/m3m2** (Fig. 6.3). Each point has the same vectorial environment, and description by four primitive-cubic lattices would implicitly miss the congruency of the lattice positions. For both the **I** and **F** cubic lattices, it is possible to find a primitive oblique cell, but it is much easier to use the unit cell with the highest geometry. This is a cube for both the **I** and **F** lattices, and the former contains 2 lattice points and the latter contains 4.

Just to hammer home the basic principles, the reader should determine the Bravais lattice type for the Cu_3Au_1 structure (Fig. 5.6) and the NaCl structure (Fig. 5.8). Look at the environment of each type of atom. The gold atoms of Cu_3Au_1 fall on a primitive lattice, and the Cu atoms must be expressed by three separate primitive lattices—hence there are four intersecting primitive lattices. In NaCl, all Na atoms have the same vectorial environment, and so do the Cl atoms. This structure is based on two intersecting face-centered lattices.

The lattice geometry must obey the symmetry. The **P, I,** and **F** lattices can obey cubic symmetry, but can just one face be centered (e.g., the A face that is parallel to the b and c axes)? No, because the triad axes require centering of the B and C faces as well, and this gives the **F** lattice. Can lattice positions be added at fractions different from $\frac{1}{2}$ (e.g., $\frac{1}{3}$ and $\frac{1}{4}$)? No, because either the lattice condition is violated, or a smaller unit cell of **P, I,** or **F** type can be taken. So there are only three cubic lattices.

All but one of the other Bravais lattices can be obtained by degrading the symmetry of the isometric lattices. Remove a tetrad axis. This destroys the triad axes. Two tetrad axes are mutually inconsistent if the third tetrad axis is missing, and a single tetrad axis controls the tetragonal lattices. The cubic **P** lattice degrades to the tetragonal **P** lattice (Fig. 6.3), which looks like the cubic **P** lattice in projection down the tetrad axis c, but which has $c \neq a$ and $a = b$. The cubic **I** lattice degrades to the tetragonal **I** lattice. A C-centered tetragonal lattice is unnecessarily large and can be changed to a tetragonal **P** lattice by taking new axes with $\mathbf{a}' = (\mathbf{a} + \mathbf{b})/2$, $\mathbf{b}' = (\mathbf{a} - \mathbf{b})/2$, and $\mathbf{c}' = \mathbf{c}$ (unbracketed positions in Fig. 6.4a). Similarly, a tetragonal **F** lattice can be re-

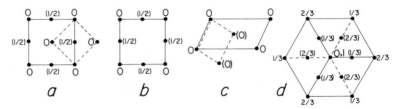

FIGURE 6.4 The illustrations of derivation of the Bravais lattices. (*a*) The addition of new points to a tetragonal lattice allows choice of small cell with dashed lines. (*b*) The centering of tetragonal *P* lattice on *A* and *B* faces does not give lattice points with the same vectorial environment. (*c*) *C* centering of monoclinic *P* lattice allows choice of a small unit cell such as the one illustrated with dashed lines. (*d*) The points with unbracketed heights represent a rhombohedral cell viewed down its triad axis. Centering of all the faces gives the points with bracketed heights, and these allow choice of a small unit cell whose vertices are the new points plus the top and bottom vertices of the original cell.

described as a tetragonal **I** lattice (bracketed and unbracketed positions). An A-centered lattice would violate tetragonal symmetry unless the *B* face were also centered. The vectorial environments of the lattice points would not be identical (Fig. 6.4*b*) until the *C* face was also centered. Both the **P** and **I** tetragonal lattices *when isolated* have symmetry **4/m2/m2/m**, or briefly **4/mmm**, where the first symbols **4/m** refer to the *c* axis, the second symbols **2/m** to both the *a* and *b* axes, and the third symbols **2/m** to the [110] directions at 45° to the *a* and *b* axes.

Monoclinic lattices (Fig. 6.3) are characterized by a 2-fold axis perpendicular to a mirror plane, and the resulting unit cell has three unequal axes and one oblique interaxial angle. Place the 2-fold axis along *c*, causing γ to be oblique. A **C** lattice turns into a **P** lattice merely by choosing new axes (Fig. 6.4*c*). An **I** lattice cannot be turned into a monoclinic **P** lattice. The reader can demonstrate that the **A, B,** and **F** lattices can be converted into an **I** lattice by choosing a new cell that retains two 90° angles. The symmetry of an *isolated* monoclinic lattice is **112/m** when the 2-fold axis is chosen along *c*. The two monad axes denote lack of symmetry along the *a* and *b* axes. If the 2-fold axis is placed along *b*, the symmetry is **12/m1**, and there are still only two irreducible lattices **P** and **I**. However, it is common to use an **A** or a **B** lattice rather than an **I** lattice when the 2-fold axis is along *c*, and an **A** or a **C** lattice when the 2-fold axis is along *b*.

Triclinic crystals (Fig. 6.3) can always be referred to a **P** lattice because the symmetry elements do not give any geometrical restrictions on the lattice geometry. An isolated triclinic lattice has a center of symmetry (denoted $\bar{1}$, see next section) at each lattice node, and halfway between each pair of nodes. Sometimes a non-**P** triclinic lattice is used (as in triclinic feldspars) in order to demonstrate a geometrical relationship with related structures (e.g., monoclinic feldspars), but all triclinic structures can be based on a **P$\bar{1}$** lattice. All Bravais lattices have a center of symmetry, but it is not listed when it is automatically produced by the other symmetry elements, such as 2 plus **m**.

Retention of one triad axis gives the rhombohedral lattice (Fig. 6.3) **R32/m** from the cubic lattice **P4/m3m2/m**. This is the only rhombohedral lattice because (i) new rhombohedral axes can be taken for **I** and **F** lattices, as illustrated by Figure 6.4*d* for the latter, and (ii) **A**, **B**, and **C** lattices violate the rhombohedral symmetry.

The hexagonal lattice (Fig. 6.3) cannot be derived from the isometric lattices. All attempts to center the primitive hexagonal lattice, **P6/m2/m2/m**, fail because the points violate the lattice condition of identical environment.

This completes the derivation of the 14 Bravais lattices. However, readers may encounter 29 types of lattices (Fig. 6.3), of which 15 can be converted to another Bravais lattice. Furthermore, they must be on the lookout for monoclinic lattices with oblique β instead of the oblique γ used in Figure 6.3.

Each Bravais lattice can be formally described by a ***translation group***. Any lattice translation can be expressed by a vector $\mathbf{t} = n_1\mathbf{a}_1 + n_2\mathbf{a}_2 + n_3\mathbf{a}_3 = n_i\mathbf{a}_i$ (tensor notation). The three base vectors $\mathbf{a}_i$ specify the edges of a unit cell, and the three numbers n_i are independent integers. The unit cell must be primitive, and is not the conventional unit cell in a centered Bravais lattice. Thus for a body-centered isometric lattice, the three base vectors could be **a**, **b**, and **(a + b + c)/2**. The infinite set of vector additions forms a translation group $T = \{n_k\mathbf{a}_k\}$, because: (i) the translation $(\mathbf{t}_k = \mathbf{t}_i + \mathbf{t}_j)$ belongs to the group, (ii) vector addition is associative: $(\mathbf{t}_i + \mathbf{t}_j) + \mathbf{t}_k = \mathbf{t}_i + (\mathbf{t}_j + \mathbf{t}_k)$, (iii) the zero vector provides the identity element, and (iv) each vector has an inverse vector. The multiplication table for a translation group would be triply infinite, and it is sufficient to use the above expression for T. A more convenient way to handle a centered Bravais lattice is to express the lattice translations by the product of $(n_i\mathbf{a}_i)$ and $\{\mathbf{S}\}$, where $\mathbf{a}_i$ are the primitive translations and **S** consists of the centering translation or translations. Thus for an orthorhombic **F** lattice the three $\mathbf{a}_i$ are the edge vectors of the orthogonal unit cell and **S** consists of the three vectors **(a + b)/2**, **(b + c)/2**, and **(c + a)/2**.

6.2   Symmetry Elements

Symmetry is important in the daily life of all human beings and, indeed, pervades science and art (Weyl, 1952; Shubnikov and Koptsik, 1974). The spokes of a bicycle wheel have rotation symmetry. An image is produced in a mirror. The North Pole is related to the South Pole by a center of symmetry. A left-handed spiral cannot be moved into coincidence with a right-handed spiral, as in the barren courtship of the columbine and the bindweed (Flanders and Swann, "Misalliance" in *At the Drop of a Hat,* Angel Record 35797, Original Cast Album). Two cubes can be oriented to have parallel edges and faces, but a left-handed and a right-handed snub cube cannot (Section 3.6 and Fig. 3.10*e*). All these facts are easy to demonstrate, but the abstract formal treatment of 3D symmetry elements is not trivial. Several technical terms will be used.

A one-to-one *mapping* occurs when every *point* (*P*) in space is related uniquely to an *image* (*P'*). Two points cannot be related to one image in a one-to-one mapping. The word "image" need not involve a mirror operation, as would be implied in common speech. A *symmetry operation* uses a special type of mapping in which the distance between any pair of points is preserved: that is, $P_1 - P_2 = P_1' - P_2'$. Thus a cube must be turned into another cube of identical size (but not necessarily the same orientation or position) by a symmetry operation, and cannot be turned into a triclinic parallelepiped. However, the two bodies related by a symmetry operation need not be *congruent*. A left-handed snub cube can be turned into either a left-handed or a right-handed snub cube by a symmetry operation. The distances are preserved in the snub cubes, but the left-handed one cannot be made identical to the right-handed one by the *mechanical* operations of rotation and translation. The conversion between them can be made only in the brain. A *proper mapping* retains the hand, whereas an *improper mapping* changes the hand. An *inversion operation* is involved in an improper mapping. Two bodies that are identical except for the change from a left hand to a right hand are *enantiomorphic*. A left-handed snub cube can be turned into a right-handed snub cube of the same size by an inversion, and this improper mapping is responsible for the enantiomorphic relationship. Two cubes of the same size can be related by a proper mapping, and there is not an enantiomorphic relationship.

To emphasize these important matters, consider the *orthorhombic tetrahedron* composed of four congruent scalene triangles (Fig. 6.5). Take three orthogonal axes *a, b,* and *c* in clinographic projection, and construct a rectangular parallelepiped so that the axes pass through the midpoints of the faces and the edges have three different lengths. Join points *A, B, C,* and *D* to form an orthorhombic tetrahedron. Each of the four faces is a scalene triangle. Construct an orthorhombic tetrahedron out of four scalene triangles with convenient angles of 80°, 65°, and 35°. Join points *E, F, G,* and *H* to form a second orthorhombic tetrahedron, and construct it from four more triangles. In Figure 6.4, the two orthorhombic tetrahedra are related by the **mirror plane m[010]**, where **m** stands for the mirror operation and

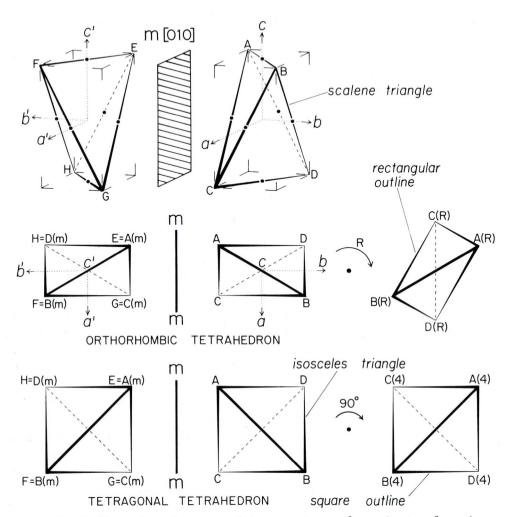

FIGURE 6.5 Illustration of proper and improper types of mapping transformations using enantiomorphic orthorhombic tetrahedra and a non-enantiomorphic tetragonal tetrahedron.

[010] is the axis perpendicular to the mirror plane. There is a one-to-one mapping: $A \to E$; $B \to F$; $C \to G$; $D \to H$. All distances are preserved. The mirror operation can be expressed by $E = A(\mathbf{m})$; $F = B(\mathbf{m})$; $G = C(\mathbf{m})$; $H = D(\mathbf{m})$. These operations are perhaps seen most clearly in c-axis projection where the rectangular outline is not distorted by the clinographic projection. It is impossible to turn *ABCD* into *EFGH* by any mechanical operation. Thus rotation of *ABCD* so that $A(\mathbf{R})$ and $B(\mathbf{R})$ match E and F leaves $C(\mathbf{R})$ and $D(\mathbf{R})$ out of position with respect to G and H. The mirror operation was improper because it involved an inversion from left hand to right hand. The orthorhombic tetrahedra *ABCD* and *EFGH* are enantiomorphs. How-

ever, *ABCD* and *A*(**R**)*B*(**R**)*C*(**R**)*D*(**R**) are related by a mechanical transformation that involves a proper mapping, and they are congruent and not enantiomorphic.

Now consider a ***tetragonal tetrahedron***. This could be drawn with the aid of a tetragonal parallelepiped whose upper and lower faces are squares, and can be constructed from four isosceles triangles. Use $72\frac{1}{2}°$, $72\frac{1}{2}°$, and $35°$ for comparison with the scalene triangle used earlier. The *c*-axis projections of *ABCD*, *EFGH*, and *A*(**R**)*B*(**R**)*C*(**R**)*D*(**R**) are now squares, and $E = A(\mathbf{m}) = A(\mathbf{R})$; $H = D(\mathbf{m}) = C(\mathbf{R})$, and so on. Thus, the tetragonal tetrahedron is double handed, and is nonenantiomorphic. The value of **R** is 90°, and is denoted by the symbol **4**.

The improper operation **m[010]** converted a right-handed set of axes *a, b,* and *c* into a left-handed set of axes *a', b',* and *c'* by the matrix transformation

$$\begin{vmatrix} 1 & 0 & 0 \\ 0 & \bar{1} & 0 \\ 0 & 0 & 1 \end{vmatrix}$$

It would be inconvenient to change the coordinate axes during symmetry transformations, and by agreed convention the axes are left in the starting position, and the matrix is used to transform positional coordinates with respect to a single set of axes.

To summarize so far: (i) the mapping transformation for a symmetry operation does not change the size of a body, but only its orientation; (ii) the mapping transformation can be proper or improper, respectively without or with a change of hand (or ***chirality***); and (iii) a body may be enantiomorphic or nonenantiomorphic.

The next step is to explore the geometrical features of the mapping transformation of a symmetry operation. So far, casual references have been made to the operations of rotation, reflection, and inversion, and it is quite easy to make empirical use of them. From the mathematical viewpoint, the key feature is that a *mapping transformation corresponds to a symmetry operation only if the starting point P is ultimately regenerated during successive application of the mapping transformation*. The ***order*** of the symmetry operation is given by the smallest number of mapping transformations required to regenerate the starting point *P*. Consider a 3-fold rotation (Fig. 6.6; first column, third down) at the point labeled with a solid triangle. The triangular motif is returned onto itself after three successive rotations of 120°. In particular, the filled circle labeled **1** moves anticlockwise first to 3^1, then to 3^2, and then to 3^3, which regenerates the starting position. Or in general: $P \rightarrow P(3^1) \rightarrow P(3^2) \rightarrow P(3^3) = P(1)$. *An N-fold rotation can be a mapping transformation for a symmetry operation only if N is an integer:* otherwise, *P* could not be regenerated. In the first column of Figure 6.6, five examples

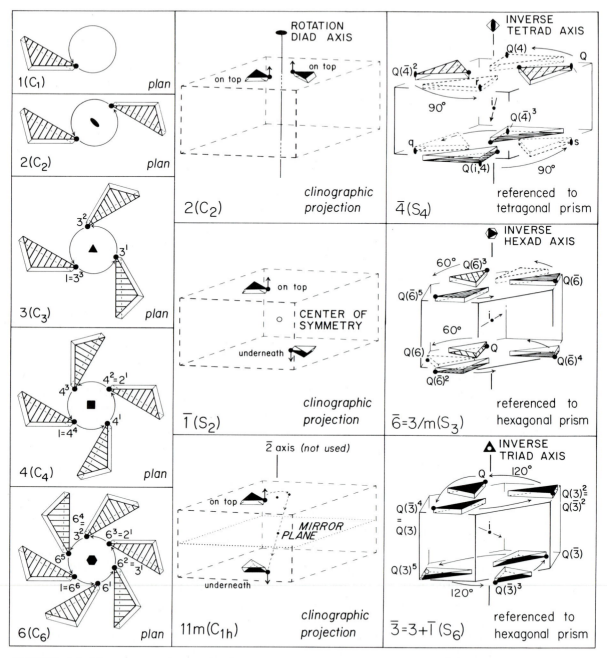

FIGURE 6.6 Mapping transformations for proper and improper rotations. For $\bar{6}$, four of the six intermediate positions are omitted to reduce overlap. Intermediate positions for $\bar{3}$ fall on the final positions: for example, intermediate position $Q(3)$ would fall on $Q(\bar{3})^4$.

are shown with N = 1, 2, 3, 4, and 6, but any other value of N is permissible except 0. The value of infinity corresponds to cylindrical symmetry. It is customary to consider rotations to be counterclockwise, and to use only positive values of N. The mapping transformations for a **rotation axis** can be simply expressed as $m \cdot 2\pi/N$, where m takes all integral values from 0 to (N – 1). In the *International* notation, N is used to denote the rotation operator and its point group of mapping transformations and equivalent positions. In the Schoenflies notation, C_N is used, where **C** is derived from Cyclic. An ellipse or appropriate polygon is used to show the position of a rotation axis.

Consider now the inversion operations. An ***inversion center of symmetry*** (Fig. 6.6, center) is denoted by a small open circle. For clarity, take a parallelepiped (dashed lines; clinographic projection), and place a center of symmetry at the body center. The one-to-one mapping transformation will take a triangular motif on top of the parallelepiped to a position below the parallelepiped. If the origin of coordinate axes is placed at the center of symmetry, a starting point P at (xyz) is mapped to an image P' at $(\bar{x}\bar{y}\bar{z})$. Thus, a black triangle is on the top of the upper motif and on the bottom of the lower motif. The open triangle of the upper motif rests on the upper face of the parallelepiped, and the open triangle of the lower motif touches the lower face. The mapping transformation of a diad rotation axis is illustrated at top center.

Can a center of symmetry be combined with a rotation axis? Consider the combination of a center of symmetry and a rotation diad (bottom center). The triangular motif is first rotated 180° as at top center, and then inverted through a center of symmetry at the body center of the parallelepiped to complete the double operation. The intermediate position is merely a temporary construction, and only the final one-to-one mapping is used. Actually, the double operation accomplishes the same result as a horizontal mirror plane halfway up the parallelepiped. Repetition of the double operation regenerates the starting motif, thereby closing the group of operations. Note that this closure occurs only if the inversion center lies on the rotation axis. In the *International* notation, the double operation of N-fold rotation plus inversion (**i**) is denoted $\bar{N}$ (spoken as "bar en"). Because $\bar{2}$ corresponds to the single operation of ***mirror reflection,*** the simpler symbol **m** is preferred. If the $\bar{2}$ axis is placed along the c coordinate axis, the point group is denoted **11m** to show that the mirror operation is perpendicular to the third coordinate axis. This is simpler than writing **m[001]**.

Each **N** axis of rotation can be combined with an inversion to give a $\bar{N}$ ***axis of rotoinversion.*** An inverse tetrad $\bar{4}$ (upper right) takes point Q to temporary position $Q(4)$ by a mapping transformation of 90° rotation. Inversion through **i** completes the double operation (**i**,4) to give $Q(\mathbf{i},4)$, or briefly $Q(\bar{4})$. A further rotation of 90° gives temporary position s, and inversion yields $Q(\bar{4})^2$. Similarly, 90° rotation produces r and inversion gives $Q(\bar{4})^3$. Further rotation gives q, and inversion regenerates Q, thereby closing the group of rotoinversion operations. Whereas simple rotation would have produced four triangular motifs on the top surface of the tetragonal prism, rotoinversion produces two on the top surface and two below the bottom surface. Because the top pair are related by a rotation diad, as also are the bottom pair, the *International* symbol for a $\bar{4}$ axis is an ellipse inscribed in a square. Note the convention that the second operation is listed first: for example, $(\mathbf{i},\mathbf{4})$ means that $\mathbf{i}$ follows $\mathbf{4}$.

Unfortunately, the Schoenflies notation is based on a different pair of operations, known as an ***improper rotation***. To obtain an improper rotation $\mathbf{S_n}$, rotate by $2\pi/n$ to obtain a temporary position, and then reflect across a plane perpendicular to the axis of rotation. Thus in Figure 6.6, $\mathbf{S_2} = \bar{\mathbf{1}}$, $\mathbf{S_4} = \bar{\mathbf{4}}$, $\mathbf{S_6} = \bar{\mathbf{3}}$, and $\mathbf{S_3} = \bar{\mathbf{6}}$. Of course, $\mathbf{S_1} = \bar{\mathbf{2}} = \mathbf{m}$. The symbol $\mathbf{S}$ is for Sphenoidisch, after the shape of a polyhedron (see later). It is obvious that care is needed to avoid confusion.

The diagrams for $\bar{6}$ and $\bar{3}$ are referred to a hexagonal prism. For $\bar{6}$, select any point on any triangular motif as Q. Rotate the chosen motif including Q by 60° anticlockwise and invert through a point on the inverse hexad axis to obtain the motif containing $Q(\bar{6})$. Rotate 60° anticlockwise and invert to obtain $Q(\bar{6})^2$, and so on to obtain $Q(\bar{6})^3$, $Q(\bar{6})^4$, $Q(\bar{6})^5$ and finally the starting position Q. For $\mathbf{S_3}$, the sequence of positions obtained by the improper rotation $(\mathbf{m},\mathbf{3})$ would be $Q(\mathbf{m},\mathbf{3}) = Q(\bar{6})^5$; $Q(\mathbf{m},\mathbf{3})^2 = Q(\bar{6})^4$; $Q(\mathbf{m},\mathbf{3})^3 = Q(\bar{6})^3$; $Q(\mathbf{m},\mathbf{3})^4 = Q(\bar{6})^2$; $Q(\mathbf{m},\mathbf{3})^5 = Q(\bar{6})$. For $\bar{3}$, again select any point on any triangular motif, and show that $(\mathbf{m},\mathbf{6}) = (\bar{3})^5$; $(\mathbf{m},\mathbf{6})^2 = (\bar{3})^4$; $(\mathbf{m},\mathbf{6})^3 = (\bar{3})^3$; $(\mathbf{m},\mathbf{6})^4 = (\bar{3})^2$; $(\mathbf{m},\mathbf{6})^5 = (\bar{3})$.

An inverse hexad axis corresponds to the combination of a rotation triad and a mirror plane perpendicular to it (i.e., $3/\mathbf{m}$). Thus $(\bar{6})^2 = 3$; $(\bar{6})^4 = 3^2$; $\bar{6} = (\mathbf{m},3^2)$; $(\bar{6})^3 = \mathbf{m}$; $(\bar{6})^5 = (\mathbf{m},3)$. Crystallographers prefer $\bar{6}$ to $3/\mathbf{m}$, and use a triangle inscribed inside a hexagon to show the 3-fold rotation.

An inverse triad axis corresponds to the combination of a rotation triad and a center of symmetry. Thus $3 = (\bar{3})^4$, and so on. Again crystallographers use an obvious symbol—a triangle enclosing a circle for the center of symmetry.

Before considering systematically all the ways in which symmetry elements can be combined together, it is convenient to list the matrix representations for the mapping transformations of the symmetry operations used in crystallography (Table 6.1). The coordinates $x'y'z'$ of the image P' are obtained from the coordinates xyz of the starting point by the method of matrix multiplication given in Section 4.9. Thus for a matrix operator

$$\begin{vmatrix} A & B & C \\ D & E & F \\ G & H & I \end{vmatrix}$$

Table 6.1 Matrix representations and mapping transformations

MONAD OPERATION (any coordinate axes)

1(E)
$$\begin{vmatrix} 1&0&0\\0&1&0\\0&0&1\end{vmatrix}$$
xyz

$\bar{1}(i)$
$$\begin{vmatrix} -1&0&0\\0&-1&0\\0&0&-1\end{vmatrix}$$
$\bar{x}\bar{y}\bar{z}$

DIAD OPERATION

[001] monoclinic (c), orthogonal, hexagonal

$2(C_2)$ $\begin{vmatrix} -1&0&0\\0&-1&0\\0&0&1\end{vmatrix}$ $\bar{x}\bar{y}z$
  $m(\sigma)$ $\begin{vmatrix} 1&0&0\\0&1&0\\0&0&-1\end{vmatrix}$ $xy\bar{z}$

[010] monoclinic (b), orthogonal

$2(C_2)$ $\begin{vmatrix} -1&0&0\\0&1&0\\0&0&-1\end{vmatrix}$ $\bar{x}y\bar{z}$
  $m(\sigma)$ $\begin{vmatrix} 1&0&0\\0&-1&0\\0&0&1\end{vmatrix}$ $x\bar{y}z$

[100] monoclinic (a), orthogonal

$2(C_2)$ $\begin{vmatrix} 1&0&0\\0&-1&0\\0&0&-1\end{vmatrix}$ $x\bar{y}\bar{z}$
  $m(\sigma)$ $\begin{vmatrix} -1&0&0\\0&1&0\\0&0&1\end{vmatrix}$ $\bar{x}yz$

[110] tetragonal, isometric

$2(C_2)$ $\begin{vmatrix} 0&1&0\\1&0&0\\0&0&-1\end{vmatrix}$ $yx\bar{z}$
  $m(\sigma)$ $\begin{vmatrix} 0&-1&0\\-1&0&0\\0&0&1\end{vmatrix}$ $\bar{y}\bar{x}z$

[100] hexagonal (120° prism)

$2(C_2)$ $\begin{vmatrix} 1&-1&0\\0&-1&0\\0&0&-1\end{vmatrix}$ $x-y,\bar{y},\bar{z}$
  $m(\sigma)$ $\begin{vmatrix} -1&1&0\\0&1&0\\0&0&1\end{vmatrix}$ $y-x,y,z$

[210] hexagonal (120° prism)

$2(C_2)$ $\begin{vmatrix} 1&0&0\\1&-1&0\\0&0&-1\end{vmatrix}$ $x,x-y,\bar{z}$
  $m(\sigma)$ $\begin{vmatrix} -1&0&0\\-1&1&0\\0&0&1\end{vmatrix}$ $\bar{x},y-x,z$

TRIAD OPERATION

[001] hexagonal (120° prism)

$3(C_3)$ $\begin{vmatrix} 0&-1&0\\1&-1&0\\0&0&1\end{vmatrix}$ $\bar{y},x-y,z$
  $\bar{3}(S_6^{\,5})$ $\begin{vmatrix} 0&1&0\\-1&1&0\\0&0&-1\end{vmatrix}$ $y,y-x,\bar{z}$

$3^2(C_3^{\,2})$ $\begin{vmatrix} -1&1&0\\-1&0&0\\0&0&1\end{vmatrix}$ $y-x,\bar{x},z$
  $\bar{3}^5(S_6)$ $\begin{vmatrix} 1&-1&0\\1&0&0\\0&0&-1\end{vmatrix}$ $x-y,x,\bar{z}$

[111] isometric, rhombohedral

$3(C_3)$ $\begin{vmatrix} 0&0&1\\1&0&0\\0&1&0\end{vmatrix}$ zxy
  $\bar{3}(S_6^{\,5})$ $\begin{vmatrix} 0&0&-1\\-1&0&0\\0&-1&0\end{vmatrix}$ $\bar{z}\bar{x}\bar{y}$

$3^2(C_3^{\,2})$ $\begin{vmatrix} 0&1&0\\0&0&1\\1&0&0\end{vmatrix}$ yzx
  $\bar{3}^5(S_6)$ $\begin{vmatrix} 0&-1&0\\0&0&-1\\-1&0&0\end{vmatrix}$ $\bar{y}\bar{z}\bar{x}$

[11$\bar{1}$] isometric

$3(C_3)$ $\begin{vmatrix} 0&1&0\\0&0&-1\\-1&0&0\end{vmatrix}$ $y\bar{z}\bar{x}$
  $\bar{3}(S_6^{\,5})$ $\begin{vmatrix} 0&-1&0\\0&0&1\\1&0&0\end{vmatrix}$ $\bar{y}zx$

$3^2(C_3^{\,2})$ $\begin{vmatrix} 0&0&-1\\1&0&0\\0&-1&0\end{vmatrix}$ $\bar{z}x\bar{y}$
  $\bar{3}^5(S_6)$ $\begin{vmatrix} 0&0&1\\-1&0&0\\0&1&0\end{vmatrix}$ $z\bar{x}y$

TETRAD OPERATION

[001] tetragonal, isometric

$4(C_4)$ $\begin{vmatrix} 0&-1&0\\1&0&0\\0&0&1\end{vmatrix}$ $\bar{y}xz$
  $\bar{4}(S_4^{\,3})$ $\begin{vmatrix} 0&1&0\\-1&0&0\\0&0&-1\end{vmatrix}$ $y\bar{x}\bar{z}$

$4^3(C_4^{\,3})$ $\begin{vmatrix} 0&1&0\\-1&0&0\\0&0&1\end{vmatrix}$ $y\bar{x}z$
  $\bar{4}^3(S_4)$ $\begin{vmatrix} 0&-1&0\\1&0&0\\0&0&-1\end{vmatrix}$ $\bar{y}x\bar{z}$

[010] isometric

$4(C_4)$ $\begin{vmatrix} 0&0&1\\0&1&0\\-1&0&0\end{vmatrix}$ $zy\bar{x}$
  $\bar{4}(S_4^{\,3})$ $\begin{vmatrix} 0&0&-1\\0&1&0\\1&0&0\end{vmatrix}$ $\bar{z}yx$

$4^3(C_4^{\,3})$ $\begin{vmatrix} 0&0&-1\\0&1&0\\1&0&0\end{vmatrix}$ $\bar{z}yx$
  $\bar{4}^3(S_4)$ $\begin{vmatrix} 0&0&1\\0&-1&0\\-1&0&0\end{vmatrix}$ $z\bar{y}\bar{x}$

[100] isometric

$4(C_4)$ $\begin{vmatrix} 1&0&0\\0&0&-1\\0&1&0\end{vmatrix}$ $x\bar{z}y$
  $\bar{4}(S_4^{\,3})$ $\begin{vmatrix} -1&0&0\\0&0&1\\0&-1&0\end{vmatrix}$ $\bar{x}z\bar{y}$

$4^3(C_4^{\,3})$ $\begin{vmatrix} 1&0&0\\0&0&1\\0&-1&0\end{vmatrix}$ $xz\bar{y}$
  $\bar{4}^3(S_4)$ $\begin{vmatrix} -1&0&0\\0&0&-1\\0&1&0\end{vmatrix}$ $\bar{x}\bar{z}y$

HEXAD OPERATION [001] hexagonal (120° prism)

$6(C_6)$ $\begin{vmatrix} 1&-1&0\\1&0&0\\0&0&1\end{vmatrix}$ $x-y,x,z$
  $\bar{6}(S_3^{\,5})$ $\begin{vmatrix} -1&1&0\\-1&0&0\\0&0&-1\end{vmatrix}$ $y-x,\bar{x},\bar{z}$

$6^5(C_6^{\,5})$ $\begin{vmatrix} 0&1&0\\-1&1&0\\0&0&1\end{vmatrix}$ $y,y-x,z$
  $\bar{6}^5(S_3)$ $\begin{vmatrix} 0&-1&0\\1&-1&0\\0&0&-1\end{vmatrix}$ $\bar{y},x-y,\bar{z}$

TRIAD OPERATION

[$\bar{1}11$] isometric

$3(C_3)$ $\begin{vmatrix} 0&-1&0\\0&0&1\\-1&0&0\end{vmatrix}$ $\bar{y}z\bar{x}$
  $\bar{3}(S_6^{\,5})$ $\begin{vmatrix} 0&1&0\\0&0&-1\\1&0&0\end{vmatrix}$ $y\bar{z}x$

$3^2(C_3^{\,2})$ $\begin{vmatrix} 0&0&-1\\-1&0&0\\0&1&0\end{vmatrix}$ $\bar{z}\bar{x}y$
  $\bar{3}^5(S_6)$ $\begin{vmatrix} 0&0&1\\1&0&0\\0&-1&0\end{vmatrix}$ $zx\bar{y}$

[$1\bar{1}1$] isometric

$3(C_3)$ $\begin{vmatrix} 0&-1&0\\0&0&-1\\1&0&0\end{vmatrix}$ $\bar{y}\bar{z}x$
  $\bar{3}(S_6^{\,5})$ $\begin{vmatrix} 0&1&0\\0&0&1\\-1&0&0\end{vmatrix}$ $yz\bar{x}$

$3^2(C_3^{\,2})$ $\begin{vmatrix} 0&0&1\\-1&0&0\\0&-1&0\end{vmatrix}$ $z\bar{x}\bar{y}$
  $\bar{3}^5(S_6)$ $\begin{vmatrix} 0&0&-1\\1&0&0\\0&1&0\end{vmatrix}$ $\bar{z}xy$

The International symbol is followed by the Schoenflies symbol in brackets. Underneath are the coordinates of the image P' produced from position P at xyz.

Note: $\bar{2}$ = m; $\bar{3}^2$ = 6^4 = $\bar{6}^4$ = 3^2; $\bar{3}^3$ = $\bar{1}$; $\bar{3}^4$ = 6^2 = $\bar{6}^2$ = 3; 6^3 = 2; $\bar{6}^3$ = m; $\bar{4}^2$ = 4^2 = 2.

$x' = Ax + By + Cz$; $y' = Dx + Ey + Fz$; $z' = Gx + Hy + Iz$, or in tensor notation $x_i' = s_{ij} x_j$, where **i** and **j** run independently from 1 to 3.

To use the matrix operators correctly, it must be realized that the reference coordinate axes are chosen in the special orientation resulting from the symmetry operator. This is true except for the monad operations 1 and $\bar{1}$ for which any coordinate axes can be chosen. For any tetrad operation, the axes must be orthogonal. For a single tetrad axis, it is conventional to place the tetrad axis along [001], that is, the c axis. Presence of three tetrad axes enforces isometric symmetry with $a = b = c$. A single hexad axis is placed along [001], and the other two axes are placed at 120° to each other (Fig. 3.14). In Table 6.1, a triad axis in hexagonal crystals is placed along [001], while in isometric crystals a triad axis can be placed along any one of the body diagonals of a cube for which coordinate axes are placed along edges. In monoclinic crystals, a diad axis is usually placed along [001] with $\gamma \neq 90°$ (Fig. 3.14). If placed along [100], α is not 90° and $\beta = \gamma = 90°$, and if along [010], β is not 90°.

The center of symmetry $\bar{1}$ (*International*) or i (Schoenflies) takes xyz through the origin at the center of symmetry to $\bar{x}\bar{y}\bar{x}$. A mirror operation **m** (International) or σ (Schoenflies) placed perpendicular to [001] requires that $\alpha = \beta = 90°$, but does not require that $\gamma = 90°$. The mirror operation leaves x and y unchanged, but transforms z to $\bar{z}$ when the mirror plane is placed at $z = 0$. When the mirror operation is perpendicular to [110] in tetragonal and isometric crystals, $xyz \rightarrow \bar{y}\bar{x}z$ (see Section 2.5 for a reminder of the meaning of this algebraic notation). For hexagonal axes, a mirror operation perpendicular to [100] gives $xyz \rightarrow y-x, y, z$, and when perpendicular to [210] gives $\bar{x}, y-x, z$. The reader should now be able to work out the remaining operations in Table 6.1.

6.3 Point Groups for a Singular Point

To enumerate all the point groups of symmetry elements and equivalent positions it is necessary to study how two or more symmetry elements can be compatible with each other. Each orthorhombic tetrahedron in Figure 6.5 has four faces related by three rotation diad axes, each of which passes through the body center and the midpoint of each of two opposing edges. Right-handed coordinate axes a, b, and c can be taken parallel to the rotation axes (upper right). The diad axis along c relates face ABC to ABD, and ACD to BCD. Similarly, the other two axes relate different pairs of faces. These mutually consistent relationships can occur only if the three diad axes are orthogonal. Furthermore, placement of two diad rotation axes at 90° automatically generates the third diad axis. This can be demonstrated by multiplication of matrices for 2[001] and 2[010] in Table 6.1:

$$\begin{vmatrix} -1 & 0 & 0 \\ 0 & -1 & 0 \\ 0 & 0 & 1 \end{vmatrix} \cdot \begin{vmatrix} -1 & 0 & 0 \\ 0 & 1 & 0 \\ 0 & 0 & -1 \end{vmatrix} \rightarrow \begin{vmatrix} 1 & 0 & 0 \\ 0 & -1 & 0 \\ 0 & 0 & -1 \end{vmatrix}$$

Remember that the row of the first matrix $\mathbf{D}_{ij}$ is multiplied by the column of the second matrix $\mathbf{D}_{ij}'$ to give the product matrix $\mathbf{P}_{ij}$ so that $\mathbf{P}_{ij} = \mathbf{D}_{i1} \mathbf{D}_{1j}' + \mathbf{D}_{i2} \mathbf{D}_{2j}' + \mathbf{D}_{i3} \mathbf{D}_{3j}'$.

The mapping transformations in Table 6.1 can also be used to show the relationship in a compact way:

$$2[001] \quad xyz \rightarrow \bar{x}\bar{y}z$$

$$2[010] \quad xyz \rightarrow \bar{x}y\bar{z}$$

To combine the operations, first change xyz to $\bar{x}\bar{y}z$ for $2[001]$. Then reverse the first symbol, retain the second symbol, and reverse the third symbol for the second mapping transformation for $2[010]$. Hence $xyz \rightarrow \bar{x}\bar{y}z \rightarrow x\bar{y}\bar{z}$, giving the operation for $2[100]$. This simple procedure avoids the tedium of matrix multiplication, and saves time because of the many zeros in the matrices.

6.3.1 DIHEDRAL POINT GROUPS

Consider now whether any rotation axis **N** (*International*) or $\mathbf{C_N}$ (Schoenflies) can be combined with a diad rotation axis 2 or C_2. This is possible for any **N** if the diad axis is placed perpendicular to the **N** axis. The easiest visual way to show this is with a stereographic projection (Fig. 6.7). To obtain the stereographic projection, join a starting position xyz to the origin of the coordinate axes 000 and project outward from the origin to the projection sphere as in Figure 4.4. Do the same for the successive images produced by mapping transformations. Because this is a book on crystal symmetry, Figure 6.7 shows only the stereograms for the crystallographic point groups (next section), but readers can develop any other point group of their choice by simple analogy. For $N = 2$, the point group **222** is produced, as demonstrated in the preceding paragraph. If direction D is obtained from point xyz, E would correspond to $x\bar{y}\bar{z}$, F to $\bar{x}y\bar{z}$, and G to $\bar{x}\bar{y}z$. For $N = 3$, the triad axis produces two more diad axes, and the three diad axes lie at 120° to each other. At first sight they appear to lie at 60°, but they are **unipolar** with a "positive" and a "negative" direction. The "positive" directions lie at 120°, and not at 60°. For $N = 4$, the tetrad axis combines with a diad rotation axis along (say) the a-axis to give another diad rotation axis at 90° along the b-axis. These axes are **bipolar**. Two more equivalent rotation diad axes are formed automatically at 45° to the first set of diad axes, but are not

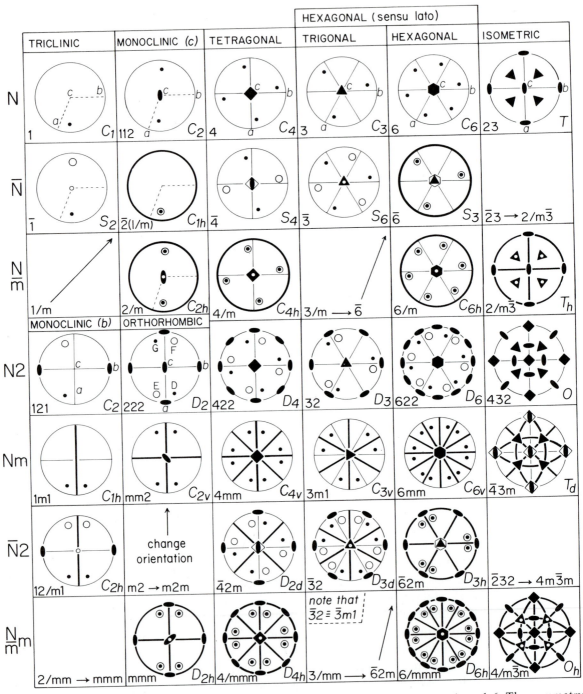

FIGURE 6.7 Point groups obtained from symmetry elements of order **1**, **2**, **3**, **4**, and **6**. The symmetry elements and equivalent directions are shown in stereograms. Seven boxes do not have a stereogram because the combination of symmetry elements corresponds to another preferred combination. The stereograms for the monoclinic point groups are shown in two orientations. The International and Schoenflies designations are given at the bottom left and right of each box.

equivalent to them, as shown by the spacing of the stereo points. In the *International* system, the group of symmetry elements could be listed as 4[001]2[100]2[110], but the directions are omitted for simplicity to give **422**. Actually **42** is sufficient because it gives $xyz \to \bar{y}xz \to \bar{y}\bar{x}\bar{z}$, and $\bar{y}x\bar{z}$ is the image of xyz for $2[1\bar{1}0]$. For $N = 6$, two sets of bipolar diad axes are formed. The long symbol would be 6[001]2[100]2[210], but **622** is sufficient, or even **62**. The Schoenflies symbol is $\mathbf{D_N}$ where $\mathbf{D}$ comes from Diëdergruppe, the German for dihedral group. $\mathbf{D_2}$ is unnecessary because it is described more simply as $\mathbf{C_2}$.

Must the diad rotation axes be perpendicular to the $\mathbf{N}$ rotation axis? No; axes can be combined at other angles in point groups **23, 432**, and **532** (see later). Can rotation axes with $N > 2$ be combined together? Only triad, tetrad, and pentad axes can be combined together, as in the Platonic solids: again it is easier to discuss this later.

6.3.2 POINT GROUPS WITH A ROTOINVERSION AXIS

The combination of rotation and inversion operations gives the improper operation of rotoinversion. Point group $\bar{1}$ provides the formal description of a center of symmetry, and the mapping transformation takes xyz to $\bar{x}\bar{y}\bar{z}$. The combination of a 2-fold rotation with an inversion center is the same as a mirror operation about a plane perpendicular to the diad operator: hence crystallographers replace $\bar{2}$ by 1/m. In the Schoenflies system, a horizontal mirror plane is denoted $\mathbf{C_{1h}}$. The stereogram for $\bar{4}$ has four points lying at the corners of a square, but two are above (dot) and two below (circle). This arrangement corresponds to the face normals of a ***tetragonal tetrahedron*** (Fig. 6.5). The stereogram for $\bar{3}$ has six points lying at the corners of a hexagon, but alternate ones are above and below the equatorial plane. For $\bar{6}$, the six points fall into three pairs that correspond to a triad axis perpendicular to a mirror plane, that is, 3/m. Except for $N = 1$, the Schoenflies system uses $\mathbf{S_N}$ where S denotes Sphenoidisch. [Note that $\bar{N} = \mathbf{S_N}$ only when N is a multiple of 4.]

6.3.3 POINT GROUPS WITH ROTATION AND MIRROR OPERATIONS

Crystallographers prefer to use mirror and rotoinversion operations in preference to an inversion center. Because $\bar{2} = 1/m$, an inversion center will now be ignored for enumeration of the remaining point groups. The $\mathbf{N/m}$ point groups (*International* notation) are denoted $\mathbf{C_{Nh}}$ in the Schoenflies notation, where $\mathbf{h}$ denotes a horizontal mirror plane perpendicular to a vertical $\mathbf{C_N}$ axis. All point groups N/m with N even have a center of symmetry shown by an open circle inside the filled polygon representing the rotation axis (Fig.

6.7). The horizontal mirror plane is shown by thickening the equator of the stereogram. All point groups N/m with N odd can be expressed more compactly as $\overline{2N}$: for example, $3/m \rightarrow \overline{6}$.

The images for $6/m$ are obtained as follows from Table 6.1:

$$
\left.
\begin{array}{lll}
xyz \rightarrow x-y,x,z & 6 \\
\rightarrow \bar{y},x-y,z & 6^2 = 3 \\
\rightarrow \bar{x},\bar{y},z & 6^3 = 2 \\
\rightarrow y-x,\bar{x},z & 6^4 = 3^2 \\
\rightarrow y,y-x,z & 6^5
\end{array}
\right\}
\begin{array}{l}
\text{Five successive anticlockwise} \\
\text{rotations of } 60° \text{ about } [001]
\end{array}
$$

$$
\left.
\begin{array}{l}
\text{6 positions with } \mathbf{m} \\
z \rightarrow \bar{z}
\end{array}
\right\}
\; xyz \rightarrow xy\bar{z} \text{ for } \mathbf{m}[001] \text{ at } z = 0
$$

The next type of point group is derived by placing a vertical rotation axis in a vertical mirror plane. In the *International* notation $\mathbf{Nm}$ is used, and in the Schoenflies notation $\mathbf{C_N^v}$ denotes a vertical mirror operation σ^v combined with a cyclic operator. In Figure 6.7, the mirror plane is placed perpendicular to [010] for convenience. Point group $\mathbf{m}[010]$ can be turned into $\bar{2} = 1/m$ merely by relabeling the coordinate axes. The combination of $\mathbf{m}[010]$ with $2[001]$ produces $\mathbf{m}[100]$ to give point group $\mathbf{mm2}$:

$$
\begin{array}{lll}
xyz \rightarrow x\bar{y}z & \mathbf{m}[010] & \text{horizontal arrows} \\
\downarrow \qquad \downarrow & 2[001] & \text{vertical arrows} \\
\bar{x}\bar{y}z \rightarrow \bar{x}yz & \mathbf{m}[100] & \text{diagonal relations } xyz \rightarrow \bar{x}yz;\ x\bar{y}z \rightarrow \bar{x}\bar{y}z
\end{array}
$$

When N is even, a second set of vertical mirror planes is produced halfway between the first set: thus $\mathbf{4mm}$ stands for $\mathbf{4}[001]\mathbf{m}[010]\mathbf{m}[110]$ and $\mathbf{6mm}$ stands for $\mathbf{6}[001]\mathbf{m}[010]\mathbf{m}[210]$. When N is odd, a second set of vertical mirror planes is not produced, as shown by the symbol $\mathbf{3m1}$, which stands for $\mathbf{3}[001]\mathbf{m}[0\bar{1}0]$. The symbol $\mathbf{31m}$ for $\mathbf{3}[001]\mathbf{m}[210]$ corresponds merely to choice of new horizontal axes at $30°$ to the axes for $\mathbf{3m1}$. Note that [010] can be replaced by [100] because of the trigonal symmetry.

Combination of both horizontal and vertical mirror operators with a vertical rotation operator provides the next type of point group. The same result is obtained by combining a vertical rotation operator with a horizontal diad rotor and a horizontal mirror plane, and the Schoenflies symbol is $\mathbf{D_{Nh}}$. For the *International* notation, $\mathbf{N}[001]\mathbf{m}[001]\mathbf{m}[100]$ is shortened to $\mathbf{N/mm}$. Interaction between the symmetry operators results in further symmetry operators. Thus $\mathbf{2/mm}$ becomes $\mathbf{2/m2/m2/m}$, or briefly $\mathbf{mmm}$, $\mathbf{4/mm}$ becomes $\mathbf{4/m2/m2/m}$ or $\mathbf{4/mmm}$, and $\mathbf{6/mm}$ becomes $\mathbf{6/m2/m2/m}$ or $\mathbf{6/mmm}$. The final $\mathbf{2/m}$ refers to a position halfway between the preceding

$2/m$. When N is odd, an inverse axis is used in the *International* notation: thus $3/mm \rightarrow \bar{6}2m$, where the 2 is along $[100]$ and the m is perpendicular to $[210]$.

6.3.4 POINT GROUPS WITH ROTOINVERSION AND ROTATION-DIAD OPERATIONS

A final type of point group is obtained by combining $\bar{N}$ with a perpendicular 2 axis. The combination of $\bar{1}$ with 2 gives $12/m1$, which was already obtained from 2 and m. The combination $\bar{2}2$ or m2 gives m2m, which is merely a different orientation for **mm2**. Diagonal mirror planes are generated from $\bar{4}2$ to give the *International* designation $\bar{4}2m$. The designation $\bar{4}m2$ merely refers to a reorientation with $m[100]2[110]$ instead of the $2[100]m[110]$ for $\bar{4}2m$. In the Schoenflies notation, $\mathbf{D_{2d}}$ is derived by addition of a diagonal mirror operation to $\mathbf{D_2}$. The combination of $\bar{6}[001]$ with $2[001]$ produces $m[210]$, while $\bar{3}[001]$ and $2[100]$ yield $m[100]$. Hence $\bar{3}2$ can be written $\bar{3}2/m$ or $\bar{3}m$. The point group $\bar{6}2m$ can be reoriented to $\bar{6}m2$. Because $\bar{6} = 3/m$, the Schoenflies designation for $\bar{6}2m$ is $\mathbf{D_{3h}}$.

Readers can reduce all combinations of vertical and horizontal operators to one of the point groups already described, except for the ones to be considered now.

6.3.5 POINT GROUPS WITH MULTIPLE TRIAD AXES

Historically, the concept of point group symmetry was developed from the study of polyhedra and crystal shapes, and the remaining point groups will be derived from the Platonic solids.

The *hexoctahedron* (Fig. 3.13*f*) was obtained by edge and face stellation of either the cube or regular octahedron, and can be converted back into these two Platonic solids by systematically flattening either 8 or 6 faces into a single face. It can be constructed from 48 congruent scalene triangles, and the stereogram for the face normals (Fig. 6.8) demonstrates point group $4/m\bar{3}2/m$ (*International*) or $\mathbf{O_h}$ for Octahedrisch and horizontal mirror plane (Schoenflies). Subsidiary stereograms show the operations of the symmetry operations, and representative transformation matrixes are given in Table 6.1. A rotation tetrad lies along each of the $[100]$, $[010]$, and $[001]$ directions of orthogonal coordinate axes, and a mirror plane is perpendicular to each axis. The $\bar{3}$ symbol represents four inverse triad axes along the body diagonals $[\pm 1, \pm 1, \pm 1]$ of the cube from which the hexoctahedron was obtained. The 2 symbol represents six rotation diad axes parallel to face diagonals of the cube, and the second **m** represents six mirror operators, each of

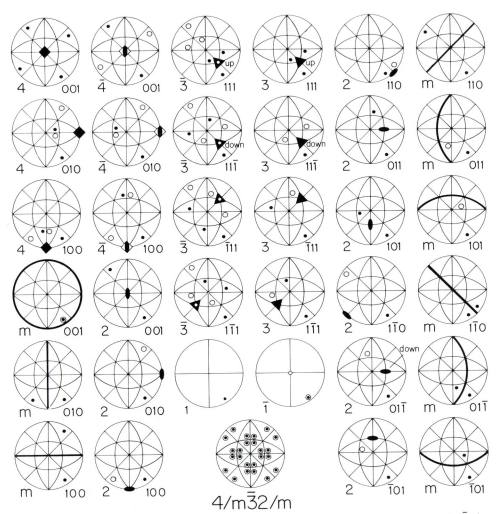

FIGURE 6.8 A stereogram of the equivalent general positions for point group **4/m$\bar{3}$2/m** (bottom center), and stereograms for the mapping transformations of each symmetry element. Note that the triad axis [11$\bar{1}$] and the diad axis [01$\bar{1}$] point downward. The equivalent general positions for the isometric point groups are given in Figure 6.9 (top left).

which is perpendicular to a diad axis and passes through two inverse triad axes. The presence of an inversion center is immediately obvious because $\bar{3} = 3 + \bar{1}$. All these elements fit together to obey the postulates of a group, as can be shown by tedious preparation of a multiplication table, or by inspection from the stereogram.

Actually it is not necessary to use the *full symbol* **4/m$\bar{3}$2/m**. The *short symbol* **m3m** is sufficient to derive the *equivalent positions* of the point group, and, indeed, it is sufficient to use only one rotation triad and four mirror operators as follows from the transformations in Table 6.1.

$$xyz \to zxy \to yzx$$ 3 and 3^2 about [111]: cyclic rotation

$$x y \bar{z};\ z x \bar{y};\ y z \bar{x}$$ m[001]: change sign of third symbol

$$\left.\begin{array}{l} x \bar{y} z;\ z \bar{x} y;\ y \bar{z} x \\ x \bar{y} \bar{z};\ z \bar{x} \bar{y};\ y \bar{z} \bar{x} \end{array}\right\}$$ m[010]: change sign of second symbol

$$\left.\begin{array}{l} \bar{x} y z;\ \bar{z} x y;\ \bar{y} z x \\ \bar{x} y \bar{z};\ \bar{z} x \bar{y};\ \bar{y} z \bar{x} \\ \bar{x} \bar{y} z;\ \bar{z} \bar{x} y;\ \bar{y} \bar{z} x \\ \bar{x} \bar{y} \bar{z};\ \bar{z} \bar{x} \bar{y};\ \bar{y} \bar{z} \bar{x} \end{array}\right\}$$ m[100]: change sign of first symbol

$$\left.\begin{array}{l} \bar{y} \bar{x} z;\ \bar{x} \bar{z} y;\ \bar{z} \bar{y} x \\ \bar{y} \bar{x} \bar{z};\ \bar{x} \bar{z} \bar{y};\ \bar{z} \bar{y} \bar{x} \\ y \bar{x} z;\ x \bar{z} y;\ z \bar{y} x \\ y \bar{x} \bar{z};\ x \bar{z} \bar{y};\ z \bar{y} \bar{x} \\ \bar{y} x z;\ \bar{x} z y;\ \bar{z} y x \\ \bar{y} x \bar{z};\ \bar{x} z \bar{y};\ \bar{z} y \bar{x} \\ y x z;\ x z y;\ z y x \\ y x \bar{z};\ x z \bar{y};\ z y \bar{x} \end{array}\right\}$$ m[110]: switch first two symbols and change sign of both

All other symmetry elements can be found in this set of 48 **general positions** for point group **m3m**. A tetrad operator 4[001] would give $xyz \to \bar{y}xz$, which was obtained from **m3m** by $xyz \to \bar{x}yz \to \bar{y}xz$. An inverse triad operator $\bar{3}^5$ [$\bar{1}$11] would give $xyz \to zx\bar{y}$, which was obtained from **m3m** by $xyz \to zxy \to zx\bar{y}$. The presence of $\bar{x}\bar{y}\bar{z}$ shows that **m3m** has a center of symmetry, and the combination of $3 + \bar{1}$ gives $\bar{3}$. It is customary to use $\bar{3}$ instead of 3 for the full symbol because $\bar{3}$ is a sixth-order symmetry element and 3 is only a third-order element. Although the presence of $y\bar{x}\bar{z}$ corresponds to $\bar{4}$[001], the operator 4 is used in the full symbol because it is simpler than $\bar{4}$ and both are fourth-order operators.

The 48 positions of **m3m** correspond to the maximum number of combinations of ±x, ±y, ±z for the coordinates of the *a, b,* and *c* axes. Thus, there are six ways of choosing the coordinate for the *a* axis. When *x, y,* or *z* is chosen, there are only four ways of choosing the coordinate for the *b* axis. After the second choice, there are only two choices for the *c* axis. Hence 6 × 4 × 2 = 48 combinations.

Four more point groups can be obtained that have four triad axes along the body diagonals of a cube (Fig. 6.7, right; Fig. 6.9, top left). Removal of the center of symmetry gives point group $\bar{4}$3m, which has also lost m[001], m[010], and m[100]: furthermore, the inverse triad axis is degraded to a rotation triad axis and the rotation tetrad axis to an inverse one. Removal of

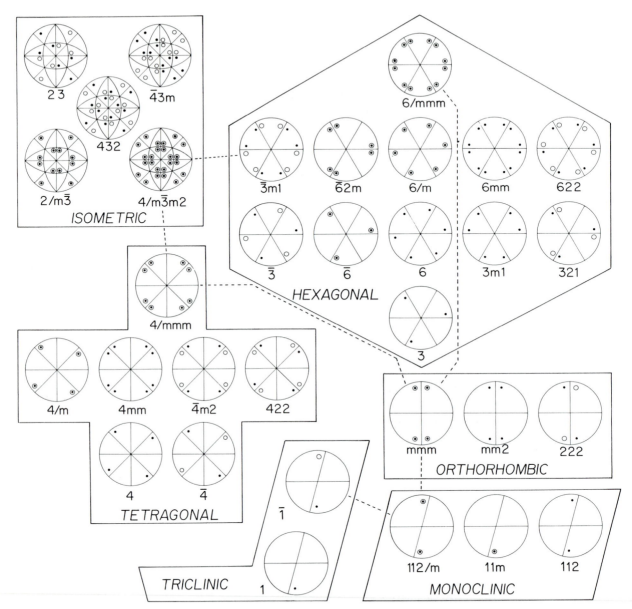

FIGURE 6.9 The hierarchy of crystallographic point groups arranged in the six crystal systems.

the six diagonal mirror planes yields point group $2/m\bar{3}$, in which the tetrad axes are degraded to diad rotation axes. Point group **432** is obtained by removal of both sets of mirror planes, and the final point group results from removal of the six diad axes. This last point group ends up with three diad axes because each tetrad axis in **432** is degraded, and its symbol **23** must not be confused with **32** for the point group with one rotation triad axis and three diad axes.

It must be emphasized that the feature common to $4/m\bar{3}2/m$ and these four derivatives is the presence of four triad rotation axes. These triad axes enforce isometric geometry in which the three coordinate axes *a, b,* and *c* are equivalent. Only three of these point groups contain tetrad axes.

The hierarchy of the five isometric point groups can be demonstrated with polyhedra. The stereogram for $4/m\bar{3}2/m$ corresponds to a hexoctahedron as already stated. That for $\bar{4}3m$ corresponds to a ***hextetrahedron*** (Fig. 3.13*j*), and the Schoenflies symbol $\mathbf{T_d}$ is derived from Tetraedrisch and the diagonal mirror planes. The stereogram for $2/m\bar{3}$ corresponds to a ***dyakis dodecahedron*** (Fig. 3.13*k*), for which the absence of a tetrad rotation axis is obvious from faces *p, q, r,* and *s:* this polyhedron is also called a ***diploid*** in recognition of the 2-fold symmetry. The ***pentagonal icositetrahedron*** (Fig. 3.13*h*), also called a ***gyroid,*** demonstrates point group **432.** Absence of all the mirror planes allows the faces to tilt or gyrate. Finally, the ***tetrahedral pentagonal dodecahedron*** (Fig. 3.13*l*), or ***tetartoid,*** is the representative of point group **23.** This rather ugly polyhedron has its axes parallel to those of the gyroid immediately above, and each diad axis emerges from the center of an edge instead of each tetrad axis emerging from a vertex between four faces of the gyroid.

Two more point groups can be obtained from the regular dodecahedron and icosahedron, and the Archimedean derivatives therefrom. The ***small and great rhombicosidodecahedra*** (Figs. 3.10*c,* 3.10*d*) represent a mutually consistent group of pentad, triad, and diad axes. Mirror planes pass through the axes and the point group can be designated **5m32.** The mirror planes are absent in the ***snub dodecahedron,*** which can exist in left- and right-handed varieties, and the point group is reduced to **532.** These two point groups are useful for clusters of globular viruses. Matrix representations involve tedious arithmetic.

6.4 Point Groups Associated with a Lattice Point; Crystallographic Point Groups; Crystal Class and System

Just as in 2D, there is a restriction on the number of point groups of symmetry elements that can be associated with a lattice point.

Consider a molecule PQ_5 with an atom P (circle) at the apex of a pentagonal pyramid, and five identical Q atoms at the base (Fig. 6.10). Place a rotation pentad axis vertically through P. The five Q atoms must lie at the corners of a regular pentagon. Now place an infinity of molecules so that the P atom of each molecule occupies each node of a lattice, and each molecule is congruent and parallel to each other molecule. A 3 × 3 × 3 array is shown in Figure 6.10, with *A–C* lying at the back and *G–I* at the front of the top 3 × 3 layer. Eight adjacent P atoms can be chosen for the corners of

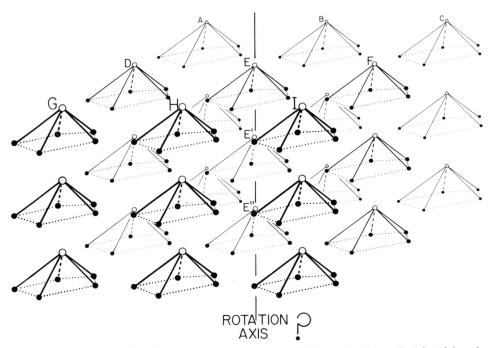

FIGURE 6.10   A 3 × 3 × 3 part of a lattice array of PQ_5 molecules with P (circle) and Q (dot) forming a regular pentagonal pyramid.

a unit cell that can reproduce the infinite array of molecules by integral translations. Can a rotation axis be placed through a P atom? From the viewpoint just of the symmetry of the PQ_5 molecule, the only possible rotation axis is a pentad. Such an axis must pass through a line of P atoms (e.g., E–E''), and each molecule must be oriented so that the axis passes through the centroid of the pentagonal base. However, the molecules are arranged in a lattice, and the symmetry must apply to the whole lattice array of molecules, not merely to one line of molecules. It is impossible to choose the positions of any 5 of the P atoms at A, B, F, I, H, and D so that they obey the requirements of both 5-fold rotation about E and lattice translation (e.g., of unit cell vectors **AD** and **AB**). If the angle DEH is chosen to be 72° (i.e., 360°/5), and DE is made equal to EH, the dimensions of a unit cell are fixed in the plane containing A–I. The position H can then be the image for a mapping transformation $D(5)$ about E, but the lattice translations do not yield the positions for an image $D(5^2)$ because the angle HEI is 108° instead of 72°. This conclusion is identical with that reached for a two-dimensional lattice. The 5-fold symmetry of the PQ_5 molecule has nothing to do with the lattice symmetry. The P atoms lie on an independent lattice, and each of the 5 Q atoms occupies its own lattice array of points. There are six independent lattices that are congruent and parallel but translated with respect to each other. The amount of the translations depends on the chemical forces between the P and Q atoms, as does the molecular symmetry. *When lattices intersect each other, it is not sufficient to look just at the nearest neighbors to find the point group of symmetry elements associated with a lattice point: rather, it is necessary to look at the entire array of intersecting lattices.* A molecule may have symmetry elements that are not present in the whole array of molecules. A molecule with a pentad rotation axis cannot transmit this symmetry to a lattice array of molecules. Just as in 2D, the only orders of symmetry elements for a lattice or array of intersecting lattices are **1, 2, 3, 4,** and **6,** and these orders impose their own characteristic geometry on the lattices, but not necessarily on atomic clusters and molecules [e.g., crystal and molecular structure of a pentagonal dodecahedrane (Paquette et al., 1981)].

In the next chapter, the space groups of symmetry elements of intersecting equivalent lattices will be considered systematically, and the above conclusions will be reinforced: in particular, it will be found that atoms and molecules can be placed either in special or general positions with respect to the symmetry elements of an array of equivalent lattices. There is no reason why atoms cannot be chemically bonded to form a polygon or polyhedron that is translated by integral lattice repeats, and rotated or inverted, or both, to yield further equivalent polygons or polyhedra, which are themselves translated by the integral lattice repeats. Thus in crystalline benzene, each benzene molecule has hexagonal symmetry in itself, but the molecules lie on intersecting monoclinic lattices.

It is now desirable to reexamine the symmetry of the Bravais lattices. Each point of an isometric lattice has point symmetry $4/m\bar{3}2/m$, irrespective of whether the lattice is primitive or centered. This latter condition is an obvious consequence of the equivalence of the points of the centered lattices. Because the point lies on all the symmetry elements, the 48-fold multiplicity of a general position xyz is reduced to the 1-fold multiplicity of 000. The points of the tetragonal, orthorhombic, monoclinic, and triclinic lattices respectively obey $4/m2/m2/m$, $2/m2/m2/m$, $2/m$ and $\bar{1}$ point symmetry. Even though the sides of a 120° prism do not obey $6/m2/m2/m$, the points of a hexagonal lattice do. A rhombohedral lattice obeys $\bar{3}2/m$. These seven point groups are the only ones that can be associated with a point in an *isolated* Bravais lattice.

But why are 32 point groups required to describe the shape of crystals? Should there not be only 7 if the atoms inside a crystal lie on a Bravais lattice? The answer to this apparent paradox is that atoms can lie on *intersecting* lattices, and this reduces the symmetry. Thus the C atoms of diamond (Fig. 5.5) lie on two intersecting **F** lattices, each of which would have point symmetry $4/m\bar{3}2/m$ if isolated, but which have point symmetry $\bar{4}3m$ when considered simultaneously. This point symmetry can be deduced easily from the tetrahedral coordination of each C atom: a tetrahedron is merely a limiting case of the hextetrahedron after removal of the stellation. In the next chapter, it will be shown how there are 230 space groups of intersecting equivalent lattices. Removal of the translations in and between the intersecting lattices does not affect the rotational elements of symmetry, and these rotational elements yield the *32 crystallographic point groups.* Twenty-five are subgroups of the 7 point groups for the isolated Bravais lattices, and are displayed by stereograms in Figure 6.7. These 32 point groups were derived by considering the ways in which proper and improper symmetry elements of order **1, 2, 3, 4,** and **6** could be combined together, and neglect of all other orders produces the same result as removal of the translational parts of the space groups of intersecting lattices.

Each of the 32 crystallographic point groups corresponds to a *class* of crystals, originally recognized from the morphology, and a complete list of *crystallographic polyhedra* is given in the next section. The 32 classes of crystals were grouped into *systems* on the basis of the order of the symmetry axes. The five point groups with four triad axes fall in the *isometric or cubic system,* and all crystals in the isometric system have atoms lying on either a single Bravais lattice with **4/m3̄2/m** point symmetry, or on one or more sets of intersecting Bravais lattices with an isometric unit cell. Seven point groups fall into the *tetragonal system,* which is characterized by a single tetrad axis, either of rotation or inverse-rotation type; the unit cell is a tetragonal prism. Three point groups fall into the *orthorhombic system,* which is characterized by three diad axes and no higher axis; the unit cell is a rectangular parallelepiped. Three point groups fall into the *monoclinic system,* which has a monoclinic parallelepiped as a unit cell; this is consistent with either one or two diad axes. In Figure 6.7, each monoclinic point group is shown twice because crystallographers may prefer to place the diad axes either along *c* or *b.* Two point groups with monad axes fall in the *triclinic system* with a triclinic parallelepiped as unit cell.

The classification of crystals with a single triad or hexad axis is a problem, and it is easiest to place all 12 classes into the *hexagonal system* (*sensu lato*). In the *International* notation, this broad system could be split up into a *trigonal system* of 5 point groups based on a triad axis and a *hexagonal system* of 7 point groups based on a hexad axis. However, this division does not fit with the Schoenflies notation for which S_3 and D_{3h} would fall in the hexagonal system. A *rhombohedron* can be used as a unit cell for crystals in classes **3** and **3̄**, and some crystals in classes **32, 3m,** and **3̄m.** For all other crystals, a hexagonal unit cell is needed. A primitive rhombohedral lattice can be redescribed as a triply primitive hexagonal lattice with a 120° prismatic unit cell. To avoid any confusion, the *hexagonal system* (*s.1*) will be used here, thus giving 6 crystal systems: however, many crystallographers may prefer to use 7 crystal systems, and there is no firm reason to argue against them.

The crystallographic point groups can be arranged in a hierarchy (Fig. 6.9). All can be derived from either or both of $4/m\bar{3}m2$ and $6/mmm$. In the isometric system, $2/m\bar{3}$, 432, and $\bar{4}3m$ are derived from $m3m$ by removing half the positions in the three possible ways, and 23 by removing another half to yield 12 equivalent positions. Half of the 24 positions for $6/mmm$ can be removed in six ways to give the upper row of point groups, and another half to give the lower row of six point groups. Removal of three positions from these six point groups yields point group **3**. Only $\bar{3}m1$ of the upper row of six can be derived from the isometric group $m3m$, and the other five must be derived from $6/mmm$. The tetragonal point groups can be derived from $4/mmm$, which is itself obtainable from $m3m$. The orthorhombic point group mmm is derivable from both $4/mmm$ and $6/mmm$, and serves as the parent for $mm2$ and 222. Finally, the hierarchy is completed by the monoclinic and triclinic groups. This hierarchy is unique once the two parents, $m3m$ and $6/mmm$, are specified. At the head of each system is the point group for the isolated Bravais lattice that yielded the characterization of each system. Each of these point groups gives rise to the ***holosymmetric class*** of crystals. Again there is the problem of the hexagonal point groups. Morphological crystallographers would use $\bar{3}m1$ as the holosymmetric class of the trigonal system (or trigonal division of the hexagonal system *sensu lato*). Each parent and daughter in the hierarchy can be called a ***supergroup*** and a ***subgroup***, respectively. A daughter may have more than one parent. It is easy to determine parents and daughters from Figure 6.9, and a complete list is given in *International Tables, Volume 1,* pages 36 and 37.

6.5   Crystallographic Polyhedra; Crystal Form

An alternative to a stereogram of equivalent directions for a point group is a polyhedron in which each face is congruent and in which each face is perpendicular to an equivalent direction. Thus the stereogram for $4/m\bar{3}2/m$ (Fig. 6.8) gives a hexoctahedron composed of 48 congruent scalene triangles. A complication arises because a face may be chosen in either a general or a special position with respect to the symmetry elements of the point group. Furthermore a distinction must be made between the polyhedra consistent with a general point group and those consistent with the restricted list of crystallographic point groups.

Each polyhedron consistent with a crystallographic point group may occur as part or all of the surface of a crystal whose space group can be reduced to the crystallographic point group. The growth of most crystals is hindered, and most faces are not equally developed even when the symmetry would allow. Fortunately the faces of some crystals are near planar with orientation close to the theoretical orientation, even though the area may be erratically developed. This is the basis of the **Law of Constancy of Angle** dating back to N. Steno (1669) and D. Guglielmini (1688). The original observation, that the angles between corresponding faces are constant for all crystals of the same crystalline species, is now known to result from the strong tendency for a crystal face to grow parallel to a lattice plane. Furthermore, this parallelism is responsible for the **Law of Rational Intercepts,** enunciated by R.J. Haüy (1801), which was associated with the early ideas of an internal subunit. Therefrom, Miller indices were assigned to relate the ratio of the intercepts made by faces on the crystal axes.

For a general point group, a chosen face may have any angle to the symmetry elements and the remaining faces must then be obtained as images of the mapping transformations. For a crystallographic point group, the initial face must have an orientation parallel to a lattice plane of the internal crystal structure, and this orientation will depend on the cell dimensions as well as the indices **hkl**.

The term *ideal form* is used for the assemblage of faces produced by the mapping transformations of a point group of symmetry elements. Thus, the regular hexoctahedron is an ideal form composed of 48 scalene triangles produced by the 48 mapping transformations of **m3m**. Inherent in this definition is congruency of the faces, but in real crystals the faces of a *form* will have unequal area. A round bracket (**hkl**) denotes a face, and addition of a curly bracket {(**hkl**)} denotes all the faces of a form: {**hkl**} is often used for brevity.

A *general form* has its faces in a nonspecial orientation to the symmetry elements, whereas a *special form* has its faces in a special orientation that leads to either a smaller number of faces than the general form, or to a different topology, or to both.

6.5.1 CRYSTAL FORMS

Table 6.2 lists the 48 crystal forms and their distribution among the 32 crystal classes. Until emphasis was placed on the internal structure of crystals, considerable time was spent in crystallography classes on the details of this distribution. Such emphasis is no longer desirable because it is easy though sometimes tedious to work out the forms for a particular crystal class. Only a few examples are given here, and readers are referred to Buerger (1956) and Phillips (1971) for details. Useful drawings of crystals are also given in Hurlbut and Klein (1977).

Figure 6.11 provides a systematic listing of all the forms for the non-

Table 6.2 Distribution of crystal forms among the crystal classes

CRYSTAL CLASS	GENERAL FORM	SPECIAL FORMS WITH MILLER INDICES
TRICLINIC SYSTEM		
1	A	none
$\bar{1}$	B	none
MONOCLINIC SYSTEM (SYMMETRY AXIS c)		
112	C	B hk0; A 001
11m	D	A hk0; B 001
112/m	E	B hk0; B 001
ORTHORHOMBIC SYSTEM		
222	F	E hk0 & 0k$\bar{l}$ & h0$\bar{l}$; B 100 & 010 & 001
mm2	G	E hk0; D 0k$\bar{l}$ & h0$\bar{l}$; B 100 & 010; A 001
mmm	H	E hk0 & 0k$\bar{l}$ & h0$\bar{l}$; B 100 & 010 & 001
TETRAGONAL SYSTEM		
4	I	I hh$\bar{l}$ & h0$\bar{l}$; J hk0 & 110 & 100; A 001
$\bar{4}$	K	K hh$\bar{l}$ & h0$\bar{l}$; J hk0 & 110 & 100; B 001
4/m	L	L hh$\bar{l}$ & h0$\bar{l}$; J hk0 & 110 & 100; B 001
422	M	L hh$\bar{l}$ & h0$\bar{l}$; N hk0; J 110 & 100; B 001
4mm	O	I hh$\bar{l}$ & h0$\bar{l}$; N hk0; J 110 & 100; A 001
$\bar{4}$2m	P	K hh$\bar{l}$; L h0$\bar{l}$; N hk0; J 110 & 100; B 001
4/mmm	Q	L hh$\bar{l}$ & h0$\bar{l}$; N hk0; J 110 & 100 B 001
HEXAGONAL SYSTEM		
3	R	R h0$\bar{l}$ & hh$\bar{l}$; S hk0 & 100 & 110; A 001
$\bar{3}$	T	T h0$\bar{l}$ & hh$\bar{l}$; U hk0 & 100 & 110; B 001
32	V	T h0$\bar{l}$; W hh$\bar{l}$; X hk0; U 100; S 110; B 001
3m1	Y	R h0$\bar{l}$; Z hh$\bar{l}$; X hk0; U 100; S 110; A 001
$\bar{3}$2/m	α	T h0$\bar{l}$; β hh$\bar{l}$; γ hk0; U 100 & 110; B 001
6	Z	Z h0$\bar{l}$ & hh$\bar{l}$; U hk0 & 100 & 110; A 001
$\bar{6}$	W	W h0$\bar{l}$ & hh$\bar{l}$; S hk0 & 100 & 110; B 001
6/m	β	β h0$\bar{l}$ & hh$\bar{l}$; U hk0 & 100 & 110; B 001
622	δ	β h0$\bar{l}$ & hh$\bar{l}$; γ hk0; U 100 & 110; B 001
6mm	ε	Z h0$\bar{l}$ & hh$\bar{l}$; γ hk0; U 100 & 110; A 001
$\bar{6}$m2	ζ	W h0$\bar{l}$; β hh$\bar{l}$; X hk0; U 100; S 110; B 001
6/mmm	η	β h0$\bar{l}$ & hh$\bar{l}$; γ hk0; U 100 & 110; B 001
ISOMETRIC SYSTEM		
23	θ	ι hh$\bar{l}$(h>$\bar{l}$); κ hh$\bar{l}$(h<$\bar{l}$); λ hk0; μ 111; ν 110; ξ 100
432	π	ρ " ; σ " ; τ " ; ϕ " ; ν " ; ξ "
2/m$\bar{3}$	χ	ρ " ; σ " ; λ " ; ϕ " ; ν " ; ξ "
$\bar{4}$3m	ψ	ι " ; κ " ; μ " ; ν " ; ξ "
4/m$\bar{3}$m	ω	ρ " ; σ " ; τ " ϕ " ; ν " ; ξ "

KEY

A	pedion
B	pinacoid
C	sphenoid
D	dome
E	prism (monoclinic and orthorhombic)
F	orthorhombic tetrahedron
G	orthorhombic pyramid
H	orthorhombic bipyramid
I	tetragonal pyramid
J	tetragonal prism
K	tetragonal tetrahedron
L	tetragonal bipyramid
M	tetragonal trapezohedron
N	ditetragonal prism
O	ditetragonal pyramid
P	tetragonal scalenohedron
Q	ditetragonal bipyramid
R	trigonal pyramid
S	trigonal prism
T	rhombohedron
U	hexagonal prism
V	trigonal trapezohedron
W	trigonal bipyramid
X	ditrigonal prism
Y	ditrigonal pyramid
Z	hexagonal pyramid
α	hexagonal scalenohedron
β	hexagonal bipyramid
γ	dihexagonal prism
δ	hexagonal trapezohedron
ε	dihexagonal pyramid
ζ	ditrigonal bipyramid
η	dihexagonal bipyramid
θ	tetrahedral pentagonal dodecahedron
ι	trapezoidal dodecahedron
κ	tristetrahedron
λ	pentagonal dodecahedron
μ	tetrahedron
ν	rhombic dodecahedron
ξ	cube
π	pentagonal icositetrahedron
ρ	trisoctahedron
σ	trapezoidal icositetrahedron
τ	tetrahexahedron
ϕ	octahedron
χ	dyakis dodecahedron
ψ	hextetrahedron
ω	hexoctahedron

isometric systems. In the first column are seven polygons that provide a key to the name. The rhombus leads to forms in the orthorhombic system, and the assignment of the next six polygons is obvious. Note carefully the use of the prefix *di-,* and do not confuse it with *bi-.* In some books, *di-* is used for both horizontal and vertical orientations, but this usage should be abandoned. In the second column is a systematic listing of **pyramids** whose prefix is obvious. Thus a pyramid with a di-hexagonal cross section is a dihexagonal pyramid. This is the general form for class **6mm** whose stereogram is given in row **Nm** of Figure 6.7. Each face of a pyramid should go to infinity, and it is necessary to terminate a pyramid with some other form in an actual crystal. A *closed form* can enclose space by itself, whereas an *open form* cannot. The pyramids in Figure 6.11 are terminated by a *pedion,* which is a single plane. The pedion is a general form in class **1**, and at least four pedions are needed to enclose space. In **6mm**, the pedion is a special form, and it can only lie perpendicular to the hexad axis in order to satisfy the **6mm** point symmetry; thus in Figure 6.7, the six dots in the stereogram would be merged into the center. The dihexagonal pyramid is denoted {**(hkl)**} to show it is a general form with any integral but nonequal values of **h, k,** and **l.** The pedion is described as {**(001)**} to show it is a special form: indeed, it consists only of one face—either (**001**) if on top, or (**00$\bar{1}$**) if on the bottom as in Figure 6.11.

A *bipyramid* is obtained from a pyramid by adding a horizontal plane of symmetry. It is a closed form. A *dihexagonal bipyramid* is the general form for **6/mmm,** and the 24 faces are represented by the dots and circles in the stereogram at the bottom of Figure 6.7.

The names for the **prisms** are obvious. Just as for pyramids, the faces should go to infinity, and the ones in Figure 6.11 are terminated by a *pinacoid,* which is a pair of parallel faces. A pinacoid is the general form for class $\bar{1}$ (Table 6.2), and like a prism is an open form. Each prism, but one, is a special form, and is represented by points in the equator of a stereogram. The exception with a rhombic cross section can be a special form in the three classes of the orthorhombic system, but is a general form for class **2/m** of the monoclinic system. Once the prefixes orthorhombic and monoclinic have been assigned there is no ambiguity.

There is considerable confusion in the nomenclature of some of the remaining polyhedra, but the historical usages are not worth pursuing here. The word *sphenoid* is derived from the Greek word for *wedge,* and is used here for the pair of planes related by a rotation diad axis (Fig. 6.11, middle right). A pair of planes is also produced by a mirror plane (shaded), and this is known to crystallographers as a *dome,* even though architects use the word for a roof with a circular cross section; a triangular vault is the architectural equivalent of a dome. Obviously there is no way of distinguishing a single sphenoid from a crystallographic dome, but two or more sphenoids on a crystal can be distinguished from two or more domes because the type of symmetry is then revealed. Four scalene triangles can be assembled into either a left-handed or a right-handed *orthorhombic tetrahedron,* as shown separately in Figure 6.5. This has also been called the orthorhombic sphenoid or orthorhombic bisphenoid. The term *bisphenoid* is etymologically better than sphenoid for the orthorhombic tetrahedron, but it is best to ignore both terms because the word *tetrahedron* is unambiguous and gives a reminder that the orthorhombic tetrahedron is a distortion of the Platonic

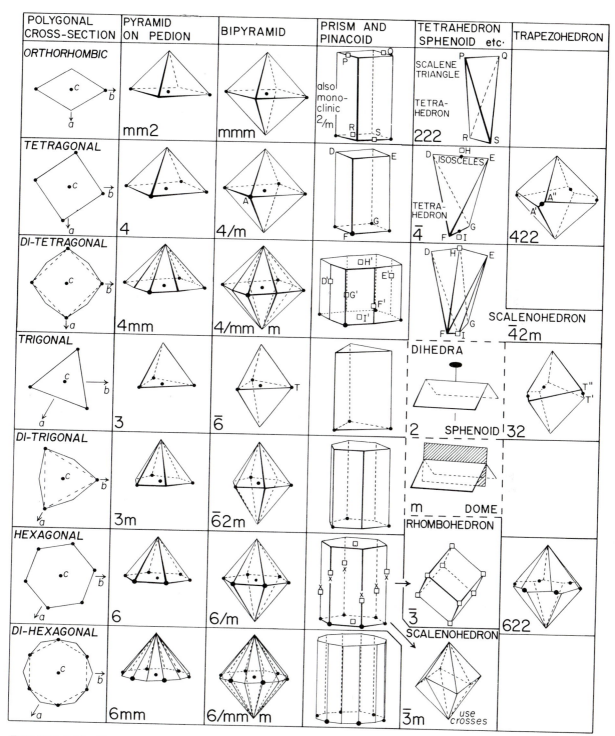

FIGURE 6.11 The crystal forms for the nonisometric systems. For convenience, two dihedra are inserted in the fifth column. The dots show symmetry-related positions.

tetrahedron. The orthorhombic tetrahedron can be obtained from an orthorhombic prism by joining positions *P–S* (Fig. 6.11). The tetragonal tetrahedron can be derived in a similar way from a tetragonal prism, and also by joining vertices instead of midpoints of edges. Isosceles triangles result therefrom, and splitting of each of the four isosceles triangles into two scalene triangles yields the ***tetragonal scalenohedron.*** This polyhedron, which is the general form for $\overline{4}2m$, can also be derived from the ditetragonal prism as shown by *D'–I'*. The ***rhombohedron*** can be obtained from a cube by homogeneous compression or extension along a body diagonal, or from a hexagonal prism by using the squares that are alternately one third and two thirds up the vertical edges. Joining the crosses that are alternately at fractional heights of $\tfrac{1}{3} - z$ and $\tfrac{2}{3} + z$ yields the ***hexagonal scalenohedron,*** which is composed of 12 scalene triangles, and is the general form of class $\overline{3}m$. This form can be reduced to the rhombohedron by turning each pair of scalene triangles into a rhombus.

Three ***trapezohedra,*** each composed of congruent quadrilaterals with four nonparallel edges, complete the nonisometric polyhedra. The name may be confusing because the term *trapezium* is sometimes used for a quadrilateral with two parallel edges. To obtain the ***tetragonal trapezohedron,*** convert vertex *A* of a tetragonal bipyramid into vertices *A'* and *A''*, and repeat for the three other vertices related by the tetrad axis. A similar conversion yields the ***trigonal trapezohedron*** from the trigonal bipyramid $(T \rightarrow T'T'')$, and the ***hexagonal trapezohedron*** from the hexagonal bipyramid.

6.6 Combinations of Crystallographic Polyhedra; Stereogram and Clinographic Projection

Most crystals are bounded by a combination of forms whose relative areal development varies from one crystal to another of the same crystalline species. Drawings of thousands of unidealized crystals are given in the *Atlas der Kristallformen* by V. Goldschmidt (1913–1923). Some hundreds of idealized drawings are given in Buerger (1956) and Phillips (1971) to represent the 32 crystallographic point groups. Only a few examples are given here.

Crystal drawing is fascinating but time-consuming and exacting. To obtain satisfactory results for complex crystals it is necessary to maintain an accuracy of $<0.2°$ and <0.2 mm for individual lines—otherwise errors can accumulate so that lines do not close at a vertex. A drafting machine is desirable, but parallel rulers are adequate. A sharp pencil of #3 grade and high-quality drawing paper should be used. The simplest drawing uses a clinographic projection with the axes of Figure 3.4. I choose units of 10 cm on the projections of the *b* and *c* axes and 3.5 cm for the projection of the *a*-axis. The projection of the *c*-axis lies at 108.4° to *a* and 92.1° to *b*. It must be emphasized that these projections are for orthogonal axes, and that new projections must be calculated for nonorthogonal axes, as explained later. A crystal edge is parallel to a lattice direction [**UVW**], and its orientation is obtained by joining the origin to the clinographic projection of the point U*a,* V*b,* W*c.* The **UVW** triplet is calculated from the **hkl** triplets of the two faces responsible for the edge. Edges must be added in the correct sequence, and systematic use of the point group symmetry is necessary. Arbitrary choices must be made about the relative length of edges unrelated by symmetry. A stereographic projection is invaluable. The easiest way to learn is to copy the following examples.

6.6.1 HABIT

This is a technical term used to describe the overall shape of a crystal. A prismatic *habit* results when prism faces have greater area than other faces. Each of the names of a form can be used as an adjective. In addition, the words *tablet, lath, and needle* are used to describe crystals with respective lengths: $l_1 \sim l_2 > l_3$; $l_1 > l_2 > l_3$; and $l_1 \sim l_2 < l_3$. Corresponding adjectives are tabular, lathy, and acicular. An *equant* crystal has $l_1 \sim l_2 \sim l_3$. A *euhedral* crystal is bounded by planar faces, in contrast to an *anhedral* one. A symmetrical habit corresponds to equal area of each face of a form.

6.6.2 POINT GROUP mmm

The three diad operators enforce orthogonal symmetry, but not equality of the axial repeats. Gem quality crystals of the peridot variety of *olivine* illustrate the forms of this point group (Fig. 6.12). The three forms {(**100**)}, {(**010**)}, and {(**001**)} are pinacoids (Table 6.2, *B*); {(**110**)}, {(**101**)}, and {(**021**)} are orthorhombic prisms (*E*); and {(**111**)} is an orthorhombic bipyramid (*H*). In the orthorhombic system {(**111**)} is a general form even though **h**, **k**, and **l** are equal: this is not true in the isometric system. The position of each face normal can be readily plotted on a stereogram using the cell dimensions $a = 4.76$ Å, $b = 10.20$ Å, $c = 5.98$ Å. Axes are taken in the conventional orientation with c at center and b at right. The three pinacoids yield poles in the orthogonal directions with (**100**) along +a, ($\bar{1}$**00**) along –a, and so on for b and c. The tangent formulas (Section 4.9) give (**100**) $\wedge$ (**110**) $= \tan^{-1}(a/b) = 25.02°$; (**100**) $\wedge$ (**101**) $= \tan^{-1}(a/c) = 38.52°$; (**010**) $\wedge$ (**021**) $= \tan^{-1}(b/2c) = 40.46°$. The (**111**) pole lies at the intersection of the great circles between (**110**) and (**001**), and between (**010**) and (**101**); the other seven poles of the orthorhombic bipyramid {(**111**)} are located by symmetry. In Figure 6.12, only the poles with positive l (i.e., dots) are indexed. For the circles, use negative l.

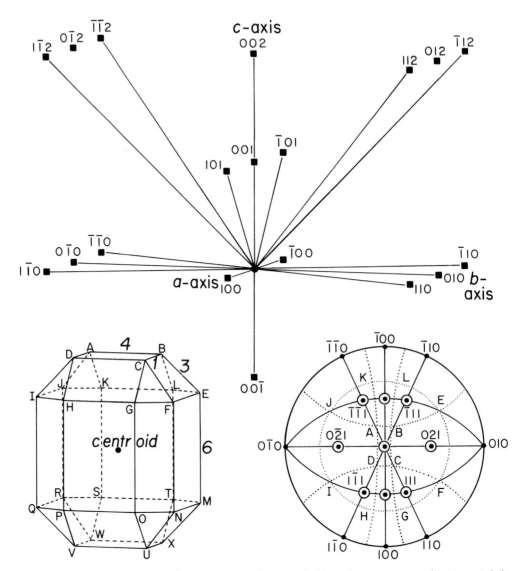

FIGURE 6.12 A clinographic projection (bottom left) and a stereogram (bottom right) of an olivine crystal with point symmetry **mmm**. The construction of edges from positions **Ua**, **Vb**, and **Wc** is shown at the top. Whereas a triplet **UVW** in the upper diagram should be enclosed in square brackets to show that it corresponds to a set of lattice rows and a zone axis, a triplet **hkl** of the lower-right diagram should be enclosed in round brackets to denote a normal to a set of lattice planes and a crystal face.

The crystal drawing was constructed as follows: To accommodate the inequality of the orthorhombic axes, b was retained at 10 cm in the original drawing, and a and c were reduced by the axial ratios to 1.63 and 5.86 cm, respectively. The linkage of the faces was chosen from the stereogram. Each dotted line corresponds to an edge, and each area to a face. The crystal can now be drawn using the dotted lines as guides. At the top of the crystal is face *ABCD* representing (**001**), whose face normal is represented by the dot at the center of the stereogram. Edge *CD* joins this face to *DCGH*, which represents (**101**). Face *CFG* is triangular, and so on. However, the stereogram does not control the relative areas of the faces, and these must be decided upon during the drawing. Edge *DC* is between the (**101**) and (**001**) faces, and its [**UVW**] triplet is **010** from the Weiss zone law (Section 4.8): hence *DC* is parallel to the clinographic projection of the *b*-axis. Actually all faces (**h0l**) must belong to the [**010**] zone, and their edges must be parallel to *b*, as shown by *AB, CD, HG, PO,* and *VU*. Similarly, *BC, AD,* and all edges between (**0kl**) faces are parallel to *a*. To start the drawing, *AB* and *CB* are drawn parallel to *b* and *a* with arbitrary lengths of 4 and 1 cm, respectively. Edge *BE* is between (**021**) and ($\bar{1}$**11**), which give [**1**$\bar{1}$**2**]. The orientation is obtained by joining the coordinate point $1a, \bar{1}b, 2c$ to the origin in the upper diagram. Edge *BE* is drawn with length 3 cm. Edge *CF* between (**021**) and (**111**) is parallel to [$\bar{1}\bar{1}$**2**], and edge *EF* is parallel to [**100**], that is, the *a*-axis. The vertex *F* is constructed from the intersection of *CF* and *EF*. Similarly, *FG* is parallel to [**1**$\bar{1}$**0**] and *CG* to [$\bar{1}$**01**], whence *G* is obtained. The face *CGHD* is a parallelogram, thereby giving *H*. The vertex *I* is constructed from *HI* [$\bar{1}\bar{1}$**0**] and *ID* [$\bar{1}$**12**]. The edge *IJ* is parallel to *CB*, and *AJ* is [**112**]. The vertex *K* comes from *JK* [$\bar{1}$**10**] and *AK* [**101**]. To complete the upper half of the crystal drawing, construct *L* from *BL* [**101**] and *EL* [$\bar{1}\bar{1}$**0**], and check that *ABLK* is a parallelogram within the accuracy of drawing. The remainder of the crystal drawing is obtained by symmetry. All the edges of the zone [**001**] are parallel, and are obtained quickly from *EM*, which is given the arbitrary length of 6 cm. Each point in the circuit *EFGHIJKL* should be diametrically opposite to a point on the circuit *QRSTMNOP* because there is a center of symmetry in point group **mmm**. This center of symmetry lies at the centroid. Before proceeding, check that all points in *E–L* and *Q–P* obey the center of symmetry. If so, construct *UVWX* from *ABCD,* or draw edges parallel to directions in the upper diagram.

It is obvious that the crystal drawing can be modified merely by changing the arbitrary lengths of 4, 3, 1, and 6 cm. Such changes will not affect the topology if the above procedure is followed. However, enlargement of face *CFG* at the expense of neighboring faces would create new edges as the triangle turned into a hexagon. Such new edges would correspond to splitting of the intersections of dotted lines in the stereogram.

6.6.3 POINT GROUP 2/m

This holosymmetric class of the monoclinic system is represented by the *sanidine* variety of K-feldspar. The three observed habits in Figure 6.13 differ in the relative development of the pinacoids {(010)}, {(001)}, {($\bar{1}$01)}, and {($\bar{2}$01)} and the monoclinic prism {(110)}. The historical choice for feldspar of b as the symmetry axis (Fig. 4.18c) requires that B, **hk0** and **001** in Table 6.2 must be changed to B,**h0l** and **010**. The construction of the stereogram follows the procedure given in Figure 4.18c. It is necessary to calculate (**100**) ∧ (**20$\bar{1}$**) = 36.00°. To construct the axial cross: (i) retain the b and c axes in the orientation of Figure 3.4, but shorten the c length by $c/b = 0.551$; this moves point s to [**001**]; (ii) join the coordinate point $(a \sin\beta, 0, -c \cos\beta)$ to the origin to obtain the direction of the monoclinic a-axis, and shorten the length by a/b. For sanidine, move point p at 3.5 cm from the origin to q at 3.5 × 0.899 = 3.15 cm, and then move q by 10 × 0.438 = 4.38 cm along the −c axis to point r. The resulting length along

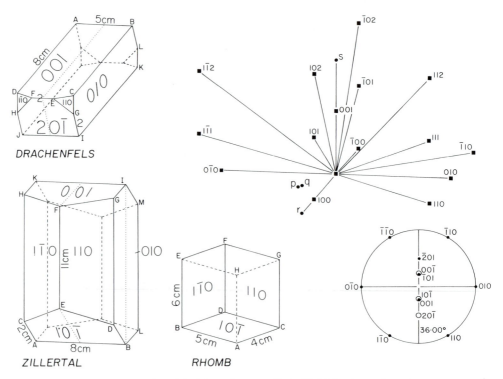

FIGURE 6.13 Three idealized habits of monoclinic K-feldspar, and a stereogram and axial cross used for construction. See Smith (1974, Figs. 17-17 to 17-29) for further drawings.

the projection of the monoclinic *a* axis is shortened from 4.54 cm by the factor 0.657 to give 2.98 cm for [**100**]. An unlimited number of sanidine crystals can now be constructed by (i) using the zone law for each pair of adjacent edges, and (ii) changing the edge lengths.

Sanidine crystals with the Zillertal habit are elongated along the *c*-axis. To construct a representative, follow the sequence *AL*. The edge *AB* between (10$\bar{1}$) and (00$\bar{1}$) is parallel to [**010**] and of arbitrary length 8 cm. Edges *AC* and *BD* are parallel to [**101**] and 2 cm long. Vertex *E* is fixed by drawing *CE* and *DE* parallel, respectively, to [**111**] and [**1$\bar{1}$1**]. If the drawing is correct, the dotted line from *E* to the midpoint of *AB* should be parallel to *AC* and *BD*. Edge *EF* is parallel to [**001**] and of arbitrary length 11 cm. Edges *CH* and *DG* are also parallel to [**001**], and edges *FG* and *FH* to [$\bar{1}$**10**] and [**110**], respectively. Vertex *I* is obtained by drawing *GI* parallel to [**100**] and the dotted line *BI* parallel to [**001**]. The rest follows from the 2/m symmetry.

Crystals with the Drachenfels habit are elongated along the *a*-axis. Edge *AB* (5 cm) is parallel to [**010**], and *AD* (8 cm) to [**100**]. To locate *E* and *F*, draw a line from *C* parallel to [$\bar{1}$**10**] and one from *D* parallel to [**110**]. Draw the dotted line through the middle of the (**001**) face, and draw *FE* parallel to [**010**] so that the dotted line bisects *FE*. A length of 2 cm is used. Point *G* is constructed from *CG* parallel to [**001**] and *EG* parallel to [**1$\bar{1}$2**]. Edge *GI* is parallel to [**102**] and is of arbitrary length 2 cm. The remainder of the construction proceeds in a similar way.

Feldspar crystals from some alkali basalts show only the six faces comprising the prism {(**110**)} and the pinacoid {(**101**)}. These six faces combine to give rhomb-shaped crystals, and the host rocks are called rhomb porphyries.

6.6.4   POINT GROUP $\bar{1}$

In old crystallographic literature, the term **anorthic** is used instead of triclinic. The calcium-feldspar, **anorthite**, occurs in complex crystals composed of many pinacoids (Fig. 6.14). **Axinite** (Hurlbut and Klein, 1977, p. 367) occurs in wedge-shaped crystals dominated by three pinacoids whose edges are beveled by other pinacoids.

To construct an axial cross for triclinic crystals: (i) draw the clinographic projection *a'*, *b'*, and *c'* of equal orthogonal axes (Fig. 3.4); (ii) retain *c'*; (iii) join the coordinate points ($a'\sin\beta,0,c'\cos\beta$) and ($-a'\sin\alpha\cos\gamma^\star$, $b'\sin\alpha\sin\gamma^\star$, $c'\cos\alpha$) to the origin to obtain *a''* and *b''*, respectively; (iii) retain *b''*, and multiply *a''* and *c'* by *a/b* and *c/b*, respectively. This procedure can be deduced from the angular relations in Figure 4.14; note that care is needed to keep track of positive and negative directions.

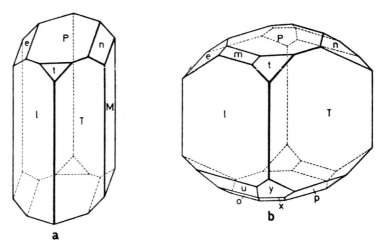

FIGURE 6.14 Two anorthite crystals composed of different combinations of pinacoids P {(001)}, M {(010)}, l {(110)}, T {(1$\bar{1}$0)}, e {(021)}, n {(0$\bar{2}$1)}, t {(201)}, x {($\bar{1}$01)}, y {($\bar{2}$01)}, m {(111)}, p {($\bar{1}$11)}, o {($\bar{1}\bar{1}$1)} and u {($\bar{2}\bar{2}$1)}. From Smith (1974, Fig. 17–30). Reprinted by permission.

6.6.5   POINT GROUPS 32 AND 62

Quartz occurs as a high-temperature variety with point group **62** and a low-temperature variety **32**. The only structural difference is the Si—O—Si angle (Fig. 7.17), and for construction of stereograms and crystal drawings the cell dimensions of low quartz ($a = 4.91$ Å, $c = 5.41$ Å) will also be used for high quartz. Both types of quartz are enantiomorphic, but most crystals of quartz have equal or near-equal development of enantiomorphic forms, thereby giving the impression that the point group is nonenantiomorphic.

The first column of Figure 6.15 shows accurate stereograms of the forms {(10.0)}, {(10.1)}, {(2$\bar{1}$.1)}, and {(51.1)} for high quartz. Most actual crystals of quartz from acidic volcanic rocks are bounded mainly or entirely by the bipyramid {(10.1)}. To construct it, draw a clinographic projection of equal orthogonal axes. Multiply the length for the projection of c by $c/a = 1.102$, and of a by $\sqrt{3}/2$. This gives points B–G about center A. Draw IH and KJ symmetrically about F and G so that IH and KJ are parallel and equal to AB. The lines IJ and KH represent the hexagonal a and u axes at 120° to the b-axis. Join the clinographic projections of the cell edges to get the bipyramid. Some crystals also show the hexagonal prism {(10.0)}, which can be constructed merely by drawing the six parallel lines BB', and so on.

Upon cooling below ~573°C, the SiO$_4$ tetrahedra rotate so that the rotation hexad is degraded to a rotation triad. The hexagonal prism is re-

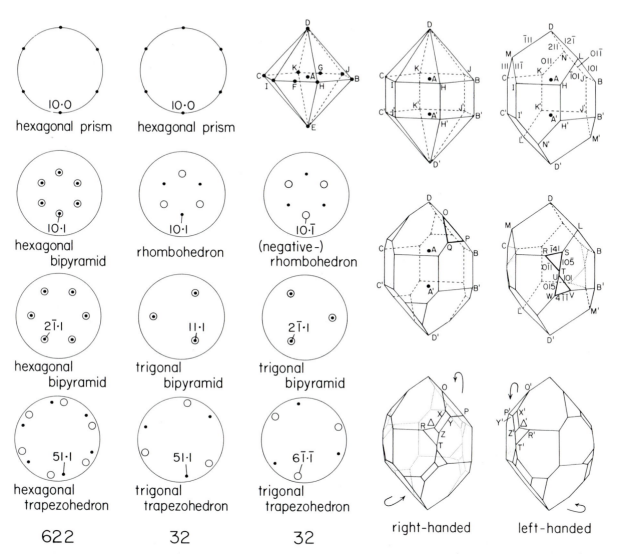

FIGURE 6.15 Stereograms of some forms in point groups 622 and 322; indices should be in round brackets. Drawings of possible quartz crystals; indices should be in square brackets.

tained as a form, but the 12 faces of the hexagonal bipyramid break up into the 6 faces of the rhombohedron {(10.1)} and the 6 faces of the rhombohedron {(10.$\bar{1}$)}. The latter can be regarded as being in a negative orientation with respect to the positive orientation of {(10.1)}. Because the reduction of symmetry is not accompanied by relative change of area of the faces, the two sets of rhombohedral faces look exactly like a set of bipyramidal faces, and the morphological symmetry still conforms to point group **622**. This illustrates a severe problem for determination of point group symmetry from the morphology. *The true symmetry can be hidden if two or more forms combine together to simulate a form representative of higher symmetry.* Thus equal development of three orthorhombic pinacoids {(100)}, {(010)}, {(001)} would correspond to a cube whose symmetry can correspond to any one of the five isometric point groups. Returning to quartz, the hexagonal bipyramid {(2$\bar{1}$.1)} could become degraded to the two trigonal bipyramids {(11.1)} and {(2$\bar{1}$.1)} during the high–low inversion, and the hexagonal trapezohedron {(51.1)} to the two trigonal trapezohedra {(51.1)} and {(6$\bar{1}$.1)}.

Below 573° C quartz crystals would grow with point group symmetry 32, but only a few crystals have clear-cut morphological evidence of the lower symmetry. Representative examples of such crystals are shown at the bottom of the last two columns of Figure 6.15. One is right handed and the other is left handed. The construction of these complex crystals is understood most easily by adding forms one by one. The crystal at the top right is a combination of the hexagonal prism {(10.0)} and the rhombohedron {(10.1)}. To construct the drawing, calculate the zone symbols [UV.W] for each pair of adjacent faces and obtain the direction by joining the coordinate point UV.W to the origin of the clinographic axes. Take the vertices from the drawing immediately to the left, and eliminate the unwanted faces. Line *DL* is between faces (10.1) and ($\bar{1}$1.1), and the direction is given by [12$\bar{1}$]. Similarly *BL* is [101], and a cross check on point *L* is obtained from *HL* [$\bar{1}$01]. Point *M* is derived from *MD* [$\bar{1}$111], *CM* [111], and *MI* [11$\bar{1}$], and *N* from *ND* [211], *KN* [011], and *NJ* [01$\bar{1}$]. Points *L'*, *M'*, and *N'* follow by symmetry. Small faces of the (negative) rhombohedron {(10.$\bar{1}$)} are now added. Point *O* is taken arbitrarily halfway between *D* and *L*. Point *Q* is halfway between *H* and *L*, and *OQ* is parallel to *DH*. The triangular face *OPQ* is completed by taking *P* halfway between *L* and *B*. Addition of five more triangular faces completes the negative rhombohedron. Care is needed to follow the symmetry operations between the different clinographic projections of the six triangles. Expansion of *OPQ* until *O*, *P*, and *Q* reach *D*, *B*, and *H* would recreate the hexagonal bipyramid. The adjacent drawing shows the construction of faces of the trigonal trapezohedron {(51.1)} by vertex truncation. Take point *R* an arbitrary one third from *H* to *I*, and find *S* on *HL* from direction [$\bar{1}$41], and *T* on *HH'* from [011]. A cross check is obtained from *ST* = [10$\bar{5}$]. Similarly *V* is one third along *H'B'*, and the edge directions are *VU* [101], *VW* [4$\bar{1}\bar{1}$], and *WU* [015]. The remaining four triangles are obtained in a similar way. It is very difficult to make an accurate drawing of faces strongly distorted at the boundary, and the truncation of *C'* yields a triangle foreshortened almost to a line.

Finally, the bottom drawing of the fourth column shows the right-handed combination of the hexagonal prism {(10.0)}, a large positive rhombohedron $\{(10.1)\}$ and a small negative one $\{(10.\bar{1})\}$, a small trigonal trapezohedron $\{(51.1)\}$ and a small trigonal bipyramid $\{(11.1)\}$. The parallelogram $XYZ\Delta$ for (11.1) truncates the trapezium $OPYX$ for $(10.\bar{1})$ and $R\Delta ZT$ for (51.1). Finally, the bottom drawing of the fifth column shows the left-handed enantiomorph in which the face $X'Y'Z'\Delta'$ $(1\bar{1}.1)$ lies between $R'\Delta'Z'T'$ $(6\bar{1}.1)$ and $O'P'Y'X'$ $(1\bar{1}.1)$. The original choice of left and right hands was arbitrary. Remember that the small faces correspond to an anticlockwise rotation (arrows) for the right-handed variety. Many crystals of quartz consist of an intimate assemblage of right- and left-handed domains, and this leads to equal development of positive and negative forms.

6.6.6 POINT GROUP 2/m3̄

Pyrite commonly occurs either as a cube with striated faces, or as a ***pentagonal dodecahedron*** (Fig. 6.16). The latter form results from $\{(hk0)\}$, and occurs in either a "positive" or a "negative" orientation. This form is so typical of pyrite that it is often called the ***pyritohedron***. The term pentag-

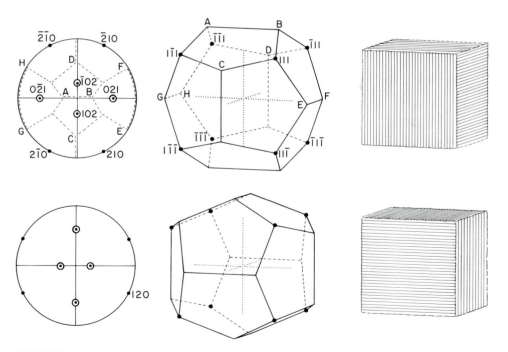

FIGURE 6.16 Stereograms and clinographic projections of crystals of pyrite. Top left and center: pentagonal dodecahedron $\{(210)\}$. Bottom left and center: pentagonal dodecahedron $\{(120)\}$. Top and bottom right: cubes with striations corresponding to $\{(210)\}$ and $\{(120)\}$, respectively.

onal dodecahedron is preferred here, but it must not be confused with the regular pentagonal dodecahedron, which is a Platonic solid. Several pyritohedra can occur in pyrite, and {(210)} and its "negative" {(120)} are more abundant than the others. A pentagonal dodecahedron is easy to draw. First construct a stereogram and index the faces. Only one angle needs calculation: (100) ∧ (210) = $\tan^{-1}(\frac{1}{2})$ = 26.57°. The dashed lines show the topological relations of the faces of {(210)} when equally developed. Eight of the vertices correspond to the triad axes [±1,±1,±1]. Construct an axial cross for clinographic projection (dotted lines), and locate the coordinate points (±1,±1,±1) to obtain 8 out of the 20 vertices. Calculate the **UVW** indices for each pair of adjacent faces {e.g., (102) and (210) give [$\bar{2}$41] }, and use the coordinate point (not shown) to obtain the orientation of the edge. Draw appropriate lines from the above 8 vertices to locate the remaining 12 vertices. A check on the accuracy of the construction is obtained because each of the last six edges (e.g., *AB, EF*) should be parallel to an arm of the axial cross. This general procedure of constructing edges from the coordinates [±1,±1,±1] is applicable to many of the crystallographic polyhedra of the isometric system, including the cube. In Figure 6.16, the stereogram and drawing for {(120)} are shown below those for {(210)}. Note that the polyhedra are related by a 90° rotation about any one of the diad axes. The two drawings at the right illustrate striated cubes. Each striation corresponds to the faces of a pentagonal dodecahedron sticking out of the cube face. The presence of the striations rules out a rotation tetrad axis, and the parallelism of the striations to the cube edges is required by the mirror planes (**100**), (**010**), and (**001**).

A few words about crystal drawing are desirable to complete this section. The procedure of (i) drawing a stereogram, (ii) deciding which faces meet to form edges, (iii) calculating **UVW** triplets and plotting the coordinate point *U***a**, *V***b**, *W***c** on an axial cross, and (iv) building up a drawing edge by edge, is probably the best for budding crystallographers. Most old hands prefer to use an orthographic projection and the angle–point construction described in Chapter 9 of Phillips (1971), to which readers are referred. This permits easy change of perspective if the first drawing looks awkward because of foreshortening of faces at the perimeter. It is not difficult to program an electronic computer and graphic terminal for crystal drawing, and I hope that some readers will do so or use an existing program (Dowty, 1980). Architects can rotate drawings of buildings using existing programs, and crystallographers would profit from a similar facility. Stereo plots would also be useful. Even when these outputs become readily available, I still recommend that all readers draw at least one crystal in order to reinforce their understanding of crystal symmetry and geometry.

EXERCISES   (*More Difficult)

6.1   After reading the section on Bravais lattices, close the book and develop the 14 Bravais lattices by testing all combinations of **P, I, F, A, B**, and **C** lattices with all seven types of unit cells.

6.2   What Bravais lattices are involved in hexagonal closest packing? cubic closest packing? diamond structure? Cu_1Au_1 structure? CsCl structure? NaCl structure? blende structure? CaF_2 structure? wurtzite structure? spinel structure? corundum structure? *cuprite structure?

6.3*   Show that a rhombohedral lattice remains primitive if additional points are added at the body center of each unit cell. [Remember that adjacent cells must be centered as well as the one that you will draw.] What will be the shape of the new primitive unit cell if the original cell had interaxial angles of 60° before body centering?

6.4   Construct orthorhombic and tetragonal tetrahedra as described in Section 6.2, and demonstrate that the former occurs as an enantiomorphic pair.

6.5   Which Archimedean polyhedra are enantiomorphic?

6.6   Cut out scalene triangles from cardboard, and demonstrate the mapping transformations for rotation symmetry (Fig. 6.6, col. 1). Draw the transformations for **5 (C_5)**.

6.7   Take a box (e.g., a cereal carton) and stick scalene triangles onto it to demonstrate the operations in Figure 6.6, columns 2 and 3. For demonstration to a class, a box made from plastic sheets and metal rods is convenient.

6.8   Determine the point symmetry of (i) orthorhombic tetrahedron, (ii) tetragonal tetrahedron, (iii) regular tetrahedron, (iv) regular octahedron, (v) rhombohedron, (vi) tetragonal bipyramid, and (vii) cube. Which of the above do not have a center of symmetry? Which are enantiomorphic?

6.9   Use the matrix operations in Table 6.1 to find the coordinates of the images produced by the mapping transformations of a rotation tetrad axis on coordinate point (xyz). Repeat for an inverse tetrad axis.

6.10   Draw a sketch stereogram for each of the dihedral crystallographic point groups and check with row 4 of Figure 6.7.

6.11   Determine the images for **422** either by inspection of the stereogram or by use of the transformations in Table 6.1. *Repeat for **322**.

6.12   Draw a sketch stereogram for each of the crystallographic point groups with a single rotoinversion axis and check with row 2 of Figure 6.7.

6.13   Determine the images for **4/m** and **4m**. Draw the corresponding stereograms, and compare with those in Figure 6.7.

6.14   Draw a sketch stereogram for each of the crystallographic point groups of type **N/mm** and compare with the bottom row of Figure 6.7. Draw a stereogram for $\bar{6}$2m and show that it corresponds to that for 3/mm. Determine the images for $\bar{6}$2m, and check that they correspond to those for 3/mm.

6.15   Draw a sketch stereogram for $\bar{4}$m2 and show that it is merely rotated 45° to that for $\bar{4}$2m in Figure 6.7.

6.16   Draw a sketch stereogram for the five isometric point groups found in crystals. Which have tetrad axes? Which are acentric?

6.17   Show how the arrangement of atoms on a lattice places a restriction on the number of the crystallographic point groups. What is the relation between a class and a point group. How are classes organized into systems? Pay particular attention to the hexagonal system.

6.18   Draw stereograms showing the hierarchy of isometric point groups and check with Figure 6.9. Repeat for the hexagonal point groups. Which point groups cannot be derived from **m3m**?

6.19   Describe the distinction between general and special forms. Examine the list of general forms in Table 6.2 and look at the drawings in Figure 6.11 and Chapter 3. A set of wooden models is useful but expensive. A set of cardboard models can be constructed cheaply, but is less durable. Particular care is needed to understand the operations of the symmetry elements in the trapezohedra, scalenohedra, tetrahedra, and the complex isometric forms. Draw a stereogram for each general form, and compare with the appropriate one in Figure 6.7. Which general forms are open? Which lack a center of symmetry? Which polyhedra can be built from scalene triangles?

6.20   Draw a symmetrical idealized crystal of peridot as in Figure 6.12, but change the relative length of the free edges.

6.21   Draw a symmetrical idealized crystal with the Zillertal habit of sanidine (Fig. 6.13, bottom left). *Add small faces of {(**100**)} and {($\bar{2}$**01**)}, and check against Figure 17–19 (Smith, 1974).

6.22**   Draw an anorthite crystal like one of those in Figure 6.14. Use $a = 8.17$ Å, $b = 12.87$ Å, $c = 7.08$ Å, $\alpha = 93.12°$, $\beta = 115.91°$, $\gamma = 91.26°$. Modify the axial cross, calculate the **UVW** triplets, find the coordinate point, and so on.

6.23   Construct an axial cross for quartz, and invent a symmetrical crystal of quartz with the hexagonal prism developed more strongly than the bipyramid. Index all the poles of the stereograms, and check the answers as far as possible against Figure 4.18*d*. *Make an accurate drawing of a crystal of low quartz, but with the prism faces larger than for Figure 6.15 (bottom right).

6.24   Draw a pyritohedron {(**210**)} and modify it with small octahedral faces. The following exercises are useful but not crucial.

6.25   Sulfur (**mmm**: a = 10.47 Å, b = 12.87 Å, c = 24.49 Å) has a pyramidal habit (Hurlbut and Klein, 1977, Fig. 7.9). Draw a crystal with large {(**111**)}, small {(**001**)}, and {(**011**)}.

6.26   Wulfenite (**4/m**: a = 5.42 Å, c = 12.10 Å) occurs in square tablets with large {(**001**)} and small {(**101**)}. Draw such a crystal and compare with Figure 9.33a of Hurlbut and Klein (1977).

6.27   Rutile (**4/mmm**: a = 4.59 Å, c = 2.96 Å) occurs in crystals dominated by the {(**110**)} prism and capped by the {(**101**)} bipyramid. Draw a crystal and compare with Figure 8.11 of Hurlbut and Klein (1977), which also shows small faces of {(**100**)} and {(**111**)}.

6.28   Scapolite (**4/m**: a = 12.2 Å, c = 7.6 Å) occurs in stout prisms composed of {(**010**)} and {(**110**)} capped by the {(**011**)} bipyramid. Draw a crystal and compare it with Figure 10.95 of Hurlbut and Klein (1977), which also shows small faces of another bipyramid.

6.29   Apophyllite (**4/mmm**: a = 9.02 Å, c = 15.8 Å) occurs as prisms dominated by {(**110**)} capped by the bipyramid {(**101**)} and the pinacoid {(**001**)}. Compare your drawing with Hurlbut and Klein (1977, Fig. 10.67).

6.30   Chalcopyrite, $CuFeS_2$, has a crystal structure in which Cu and Fe atoms occupy alternate positions for Zn atoms in the sphalerite polymorph of ZnS. Show that the orthorhombic tetrahedron {(**112**)} of chalcopyrite (**$\bar{4}$2m**: a = 5.25 Å, c = 10.32 Å) has almost the same shape as the Platonic tetrahedron of sphalerite (**$\bar{4}$3m**). Sphalerite occurs as crystals with prominent {(**111**)} and small {(**1$\bar{1}$1**)} and {(**100**)}. Draw a crystal and compare with Hurlbut and Klein (1977, Fig. 7.19). What information could be obtained about the point-group symmetry from a crystal with accidentally equal development of the positive and negative tetrahedra?

6.31   Sphene was named from its wedge-shaped crystals. Using a = 6.56 Å, b = 8.72 Å, c = 7.44 Å, β = 119.72°, **12/m1**, draw a crystal with approximately equal development of the pinacoid {(**001**)} and the monoclinic prisms {(**110**)} and {(**111**)}. Compare with Figure 10.20 of Hurlbut and Klein (1977).

6.32   Hornblende (**12/m1**: a = 9.87 Å, b = 18.01 Å, c = 5.33 Å, β = 105.73°) occurs as stout monoclinic prisms {(**110**)} terminated by the prism {(**011**)} and modified by the pinacoid {(**010**)}. Compare your drawing with Figure 10.51 of Hurlbut and Klein (1977).

6.33   Spinel (**m3m**) usually occurs in simply octahedral crystals, but rare octahedra have the edges beveled by faces of the rhombic dodecahedra

{(**110**)}. Make an accurate drawing. Note that the drawing in Figure 8.14*c* of Hurlbut and Klein (1977) has been made in a nonstandard orientation.

6.34 Cuprite (**m3m**) can occur as crystals showing cube faces truncated by small octahedral and dodecahedral faces. Compare your drawing with Figure 8.6 of Hurlbut and Klein (1977).

6.35 Apatite (**6/m**: $a = 9.39$ Å, $c = 6.89$ Å) commonly occurs as stout prisms {(**10.0**)} terminated by a bipyramid {(**10.1**)} and a pinacoid {(**00.1**)}. Compare your drawing with Figure 9.35*b* of Hurlbut and Klein (1977).

6.36 Chabazite (**3̄2/m**: $a = 13.78$ Å, $c = 14.97$ Å) occurs as the rhombohedron {(**10.1**)}. Make a drawing of the rhombohedron. Calculate (**10.1**) $\wedge$ (**1̄0.1**) and show that the rhombohedron is pseudocubic. What axial ratio for chabazite would turn the rhombohedron into a cube?

6.37 Calcite (**3̄2/m**: $a = 4.99$ Å, $c = 17.06$ Å) shows three important habits: prismatic, rhombohedral, and scalenohedral. Draw crystals with the following combinations of forms, and check against Figure 9.3 of Hurlbut and Klein (1977): (i) prominent prism {(**10.0**)} and small rhombohedron {(**01.2**)}, and (ii) prominent scalenohedron {(**21.1**)} and small rhombohedron {(**10.1**)}.

6.38 In addition to the coordination polyhedra listed in Table 5.1, a disphenoidal octahedron (Fig. 3.16*d*) also occurs. Construct two Archimedean trigonal prisms from cardboard. There are two ways to attach them across a square face. Reject the way that gives a 120° hexagonal prism. The other way gives the disphenoidal octahedron. What are its face and vertex symbols? What is the point symmetry of the centroid? What crystallographic polyhedra are present?

ANSWERS

6.2 hcp—two intersecting primitive hexagonal lattices. ccp—one cubic **F** lattice. Diamond—two cubic **F** lattices. Cu_1Au_1—two tetragonal **P** lattices. CsCl—two cubic **P** lattices. NaCl—two cubic **F** lattices. Blende—two cubic **F** lattices. CaF_2—three cubic **F** lattices. Wurtzite—four hexagonal **P** lattices. Spinel—all atoms in cubic **F** lattices. Corundum—all atoms in rhombohedral **R** lattices. Cuprite—the copper atoms taken individually are in a cubic **F** lattice and the oxygen atoms in a cubic **I** lattice, but when the lattices intersect, each set of individual lattice points must be segregated into either 4 or 2 sets of points, respectively, on primitive cubic lattices. Thus the atom at the origin is enclosed by four copper atoms at the corners of a tetrahedron in a negative orientation whereas the atom at the body center lies in a positively oriented tetrahedron.

6.3 Cube. This question can be solved by either tedious trigonometry or easy geometry.

6.5 Snub cube and snub dodecahedron.

6.8 (i) **222** (**D$_2$**), (ii) **$\bar{4}$2m** (**D$_{2d}$**), (iii) **$\bar{4}$3m** (**T$_d$**), (iv) **4/m$\bar{3}$m**, (v) **$\bar{3}$2** (**D$_{3d}$**); **$\bar{3}$m** is also correct, (vi) **4/mmm** (**D$_{4h}$**), (vii) **4/m$\bar{3}$m**. The following are acentric: (i), (ii), (iii). Only (i) is enantiomorphic.

6.9 $xyz \rightarrow \bar{y}xz \rightarrow \bar{x}\bar{y}z \rightarrow y\bar{x}z \rightarrow xyz$: $xyz \rightarrow y\bar{x}\bar{z} \rightarrow \bar{x}\bar{y}z \rightarrow \bar{y}x\bar{z} \rightarrow xyz$.

6.11 **422** $xyz \rightarrow \bar{y}xz \rightarrow \bar{x}\bar{y}z \rightarrow y\bar{x}z$ (tetrad); $x\bar{y}\bar{z} \rightarrow \bar{y}\bar{x}\bar{z} \rightarrow \bar{x}y\bar{z} \rightarrow yx\bar{z}$ (diad at [**100**] acting on preceding positions). **322** $xyz \rightarrow \bar{y}, x-y, z \rightarrow y-x, \bar{x}, z$ (triad operations); $x-y, \bar{y}, \bar{z} \rightarrow yx\bar{z} \rightarrow \bar{x}, y-x, \bar{z}$ (diad at [**100**] acting on preceding positions).

6.13 **4/m**: $xyz \rightarrow \bar{y}xz \rightarrow \bar{x}\bar{y}z \rightarrow y\bar{x}z$ (tetrad); $xyz \rightarrow \bar{y}xz\bar{z} \rightarrow \bar{x}\bar{y}\bar{z} \rightarrow y\bar{x}\bar{z}$ (mirror [**001**]. **4m**: $xyz \rightarrow \bar{y}xz \rightarrow \bar{x}\bar{y}z \rightarrow y\bar{x}z$ (tetrad); $\bar{x}yz \rightarrow yxz \rightarrow x\bar{y}z \rightarrow \bar{y}\bar{x}z$ (mirror [**100**].

6.14 **$\bar{6}$2m**: $xyz \rightarrow y-x, \bar{x}, \bar{z} \rightarrow \bar{y}, x-y, z \rightarrow xy\bar{z} \rightarrow y-x, \bar{x}, z \rightarrow \bar{y}, x-y, \bar{z}$ (inverse hexad); $\bar{x}, y-x, z \rightarrow x-y, \bar{y}, \bar{z} \rightarrow y, x, z \rightarrow \bar{x}, y-x, \bar{z} \rightarrow x-y, \bar{y}, z \rightarrow yx\bar{z}$ (m [**120**]).

6.16 **432, $\bar{4}$3m, m3m; 23, 432, $\bar{4}$3m.**

6.18 **6/mmm, $\bar{6}$2m, 6/m, 6mm, 622, $\bar{6}$, 6.**

6.19 Open: *A, B, C, D,* all prisms and pyramids. Acentric: *A, C, D, F, G, I, K, M, O, P, R, S, V–Z, γ, δ, ε, ζ, θ, ι, κ, μ, π, ψ.* Scalene: *F, O, P, Q, Y, α, ε, ζ, η, ψ, ω.*

6.24 Figure 2.165 of Hurlbut and Klein (1977) shows an octahedron modified by small pyritohedron faces.

6.25 Your drawing should look like Figure 7.9 (right) of Hurlbut and Klein (1977) after removal of the *s* faces.

6.30 The angle (112) ∧ (11$\bar{2}$) for chalcopyrite (108.67°) is close to 109.47° for **111** ∧ **1$\bar{1}$1** for the regular tetrahedron. Equal development of the positive and negative tetrahedra corresponds to the regular octahedron. This occurs as a special form in **432, 2/m$\bar{3}$,** and **4/m$\bar{3}$m**, and a misleading conclusion might be drawn. Etching the pseudo-octahedral crystal of sphalerite would probably reveal the true symmetry.

6.36 85.22°. $c/a = \sqrt{3/2} = 1.225$.

6.38 [$3^4 4^4$]($3^4 4^4$). **$\bar{4}$2m**. Tetragonal tetrahedron and prism.

7
Band, Rod, Layer and 3D Space Groups. Anti- and Color Symmetry

Preview   So far the repeated motif in symmetry and translation groups has been a point that remains a point after a symmetry operation or a translation. Some atoms have an uncompensated electron or nuclear spin, and it is necessary to consider not only the position of a point upon which an atom is placed but also the orientation of the spin. Consequently, it is necessary to distinguish between symmetry and translation groups for "neutral" points whose only property is position and those for "cambiant" points which are associated with some nonpositional property. The simplest such property is a change from black to white or positive to negative. Hence the literature contains the terms **black–white symmetry** and **antisymmetry.** For symmetry operations of order greater than 2, three or more aspects may be needed to describe the cambiant property, and the term **color symmetry** is used. A **classical group** involves only neutral points. The lattices, point groups, and space groups in Chapters 2 and 6 are classical groups. This chapter begins with the 31 two-sided band groups, of which 7 relate neutral motifs and 24 involve black–white ones. It proceeds to the classical rod groups composed of neutral motifs, and briefly discusses the extension to cambiant motifs. These band and rod groups are useful for description of polymers and linear subunits of crystal structures. Two-sided layer groups are needed for planar subunits of crystal structures and for interfaces between domains and twins; 16 relate neutral motifs and 64 relate black–white ones. The major emphasis is placed on the 230 classical space groups in 3D, which are the key to the crystal structures of "neutral" atoms. Whereas these classical space groups were enunciated independently by E.S. Fedorov (Russian crystallographer, 1885–1890), A.M. Schoenflies (German mathematician, 1891) and W. Barlow (English businessman, 1894) almost a century ago, and have been used for 70 years to describe crystal structures determined from X-ray diffraction, the two-sided layer groups were not enumerated until 1929 by the German mathematicians E. Alexander, K. Herrmann, and E. Weber. Furthermore, the idea of antisymmetry discovered by H. Heesch in

1929 lay dormant until Shubnikov rediscovered it in 1945. All types of symmetry, including color symmetry (N.V. Belov, 1956) are now incorporated into a generalized scheme (Shubnikov and Koptsik, 1974), and the pilot issue for a new *International Tables for X-Ray Crystallography* refers to nonclassical symmetry. The prefixes *Fedorov* and *Shubnikov* are sometimes applied to classical and to classical plus black–white space groups.

The present treatment uses simple mathematical procedures that are correct and easy to understand. Some readers may wish to proceed to the matrix representations of space groups in Bradley and Cracknell (1972)—see also Boisen and Gibbs (1978). Sphere packings and point complexes associated with rod and layer groups are developed by Fischer and Koch (1978) and Koch and Fischer (1978a,b).

7.1 Band Groups

A two-sided **band** is an infinite periodic figure related to a singular plane and a singular axis lying therein. A railway line with its ties is an example. Fig-two faces in some way (e.g., black versus white) to obtain a **cambiant motif**, and make the two faces identical (e.g., add a dot) to obtain a **neutral motif**. two faces in some way (e.g., black versus white) to obtain a cambiant motif, and make the two faces identical (e.g., add a dot) to obtain a neutral motif. In Figure 7.1, the singular plane is that of the drawing, and the singular axis *b* is east-west. Take two more coordinate axes *a* toward the reader and *c* out of the paper. The axes need not be orthogonal unless constrained by symmetry elements. Band group **p1** is generated merely by regular translation (**p**) of a black/white triangle along the unique axis *b*. A rotation diad axis can be placed in the singular plane and perpendicular to the singular axis to give **p211**. Alternate black/white triangles must be turned over. The triangles need not touch, but were deliberately drawn so that the contact point illustrates a special position on the symmetry element. Band **p112** requires 180° rotation between adjacent triangles, but retains the same side. For **p121**, there is only one axis in contrast to the infinite number for **p211** and **p112**: note that the spacing of diad axes for the latter two bands is half that for translated triangles. Two rotation diads automatically yield a third one to give **p222** (axes not shown). Band **p1̄** is generated by a center of symmetry, which must be placed in both the singular plane and singular axis. There is only one plane of symmetry in **pm11** in contrast to an infinite number in **p1m1**. Band **p11m** cannot be demonstrated with scalene triangles whose faces differ. It is necessary to use triangles with identical faces (marked by a dot). Mirror operations can be combined to yield four band groups: **pmm2** is constructed from triangles with dissimilar faces, whereas **p2mm**, **pm2m**, and

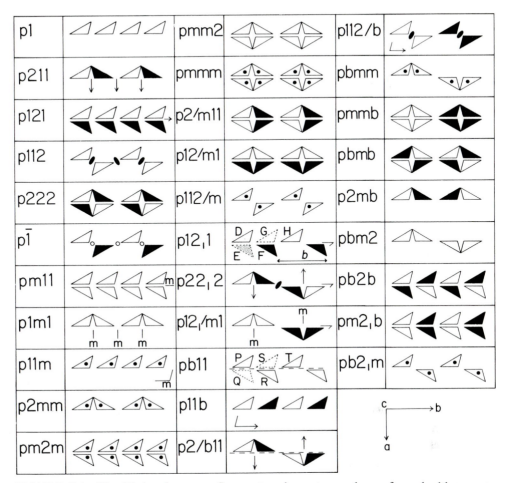

FIGURE 7.1 The 31 band groups. Symmetry elements are shown for only 11 groups. The *a*-axis need not be orthogonal to the *b*-axis for **p1, p112, p1̄, p11m, p112/m, p11b,** and **p112/b.**

p2/m2/m2/m require triangles with identical faces. The generation of the three band groups with **2/m** should be clear: note that a center of symmetry is automatically produced by **2/m**.

The next band group **p12₁1** is obviously symmetrical, but a new concept is needed. A ***screw diad axis* 2₁** involves the double operation of 180° rotation about the axis and translation of one half the repeat distance along the axis. There is an infinite number of screw axes **N_m**, where **N** and **m** are integers, as described next for rod groups. Only the screw diad is needed for band groups, and it can be placed only along the singular axis *b* of the band. Triangle *D* is rotated 180° to *E*, which involves turning it over, and the double operation is completed by translation of *b*/2 to *F*. The double operation is closed by rotation to *G* and translation to *H*. Addition of two rotation diads or one mirror plane gives **p22₁2** and **p12₁/m1**. A half-arrow is used for a screw diad axis.

The next band group **pb11** utilizes the concept of glide symmetry introduced in Chapter 2. A *glide plane* **b11** involves the double operation of mirror reflection perpendicular to the first axis (hence the position of the symbol **b**) and translation of half the repeat distance down the *b* axis. In a band group, the only possible translation is along *b*', but in a space group the translation can be along other lattice directions. For **pb11**, triangle *P* is reflected to *Q* and translated to *R*. Reflection to *S* and translation to *T* generates the lattice repeat *b* and closes the operations. Band group **p11b** is generated by reflection in the singular plane and translation by *b*/2. Band group **p1b1** does not exist, because translation cannot occur normal to the mirror plane. A glide plane lying perpendicular to the paper with its glide operation in the plane of the paper is conventionally shown by a row of dashes. A horizontal mirror plane is shown with a corner sign as for **p11m**, and a horizontal glide plane with a horizontal glide is shown by a corner sign with the appropriate arrow as for **p11b**.

Band groups **p2/b11** and **p112/b** have a center of symmetry on the singular axis midway between each pair of adjacent diad axes. The 8 groups with both mirror and glide planes complete the 31 two-sided band groups.

7.2   Rod Groups; Screw Axes

A *rod* is an infinite periodic figure related to a singular axis lying therein. Many examples occur in everyday life and in science: tube, chain, spiral staircase, plant stem, helix (including the double helix of DNA), and church tower. Of course, these examples should be of infinite length to strictly satisfy the mathematical concept of an infinite periodic figure.

Figure 7.2 shows near-clinographic projections of points related by the five possible symmetry axes of order **4**. In the rod group **p4**, the rotation tetrad successively rotates starting point *D* anticlockwise by 360°/4 to *E*, *F*, *G*, and back to *D*. The lattice operation **p** translates *D–G* to *D'–G'* by the repeat *c*. In **p4₁**, *D* undergoes the double operation of anticlockwise rotation 360/4° and translation of *c*/4 to *H, I, J*, and on to *D'* to close the operations with the periodic translation *c*. These points lie on an anticlockwise helix. In **p4₃**, the double operation of anticlockwise rotation 360°/4 and the translation 3*c*/4 produces an enantiomorphic sequence of points that lie on a clockwise helix. Rod group **p4₂** is obtained from the double operation of rotation by 360°/4 and translation by 2*c*/4. This gives pairs of points at the same height but rotated by 90° with respect to each other. Each pair obeys a diad axis, but the two pairs generate a tetrad axis. The fifth rod group **p4̄** uses the double operation of 360°/4 and inversion through a point

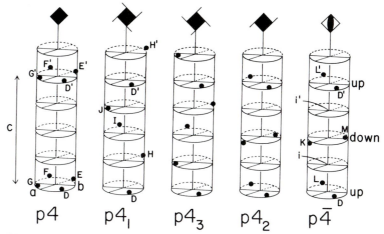

FIGURE 7.2 The five possible symmetry axes of order **4** in rod groups. The inversion points i and i' lie on the rod axis.

lying on the axis. In Figure 7.2, the point i (or i') was chosen, and the successive double operations produce $D \to K \to L \to M \to D$. The inversion operation causes K and M to lie as much below the reference planes as D and L lie above. This sequence of double operations is exactly the same as for point group $\bar{4}$, and it is necessary to use the periodic translation c to obtain rod group p$\bar{4}$. For the three screw axes, the spikes on the square symbol denote the operations: two spikes at 180° for **4**$_2$; four spikes at 90° pointing anticlockwise for **4**$_1$; and four spikes at 90° pointing clockwise for **4**$_3$ (90° clockwise = 270° anticlockwise).

It is obvious that there is an unlimited number of symmetry axes and classical rod groups obtained therefrom. The operations will close for any axis $\mathbf{N_m}$ or $\mathbf{\bar{N}}$ where both $\mathbf{N}$ and $\mathbf{m}$ are integers, and $\mathbf{m}$ runs from 1 to $\mathbf{N}$. After $\mathbf{N}$ operations, the double operation of 360/$\mathbf{N}$° rotation and $\mathbf{m}c/\mathbf{N}$ translation will close at the lattice repeat of $\mathbf{m}c$. All axes with $\mathbf{m}$ running from 1 to ($\mathbf{N}$ – 1) are **screw axes**, and those with $\mathbf{m}$ and ($\mathbf{N}$ – $\mathbf{m}$) are enantiomorphic pairs. An axis with $\mathbf{m}$ = $\mathbf{N}$ will generate the lattice repeat of a rod group as well as the rotation operations given by $\mathbf{m}$ = 0. However, it is easiest not to use $\mathbf{m}$ = $\mathbf{N}$, and the combination of a rotation axis (denoted merely as $\mathbf{N}$ without a subscript) and a translation $\mathbf{p}$ is preferred.

If a rod group is embedded in a 3D space group, $\mathbf{N}$ can assume only the values of **1, 2, 3, 4,** and **6** because of interaction with the other two lattice translations. Hence a subset of *crystallographic rod groups* is used by crystallographers. In everyday life, an architect could design a spiral staircase belonging to a noncrystallographic rod group. Isolated polymers, including ones important in biology, need not belong to a crystallographic rod group.

Figure 7.3 shows formal descriptions of screw axes. The clinographic projection of the operations of **4**$_1$ (upper right) is replaced by a symbolic projection down the axis. Point *p* is rotated anticlockwise and translated by *c*/4 to *q* and so on. In projection (middle right), the points *p–s* lie at 90° on a circle. If *p* is given the algebraic coordinate +*z* for a general position, *q* lies at height ¼ + *z*. Following the conventions of the *International Tables for X-Ray Crystallography,* this is formalized by the second diagram in the third row. Coordinate axes are taken in the usual orientation with *a* toward the reader, *b* to the right and *c* perpendicular to the paper. A starting posi-

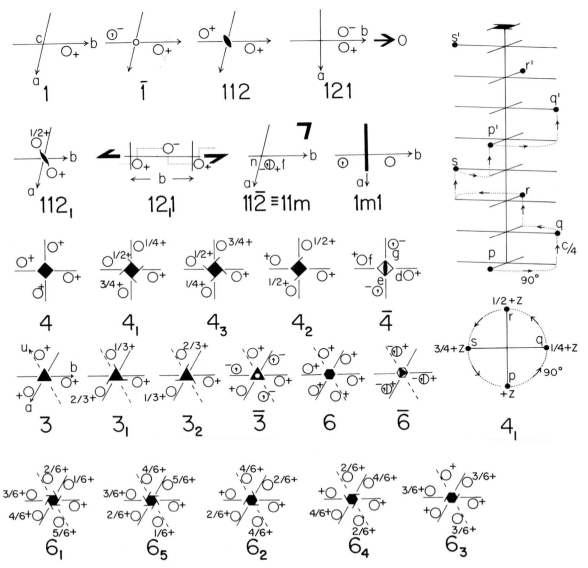

FIGURE 7.3 Summary of operations of symmetry axes. At the right is a clinographic drawing and plan of the operation of a **4₁** axis.

tion is shown by a circle marked with a plus. The center of the circle lies at algebraic coordinates x and y, and the + stands for $+z$. The screw operations generate points marked by circles at $\bar{y},x,\frac{1}{4} + z$; $\bar{x},\bar{y},\frac{1}{2} + z$; $y,\bar{x},\frac{3}{4} + z$; and back to xyz. Recall that the spikes on the square guide your eye around the helix. For the inversion axis $\bar{4}$, point d (xyz) is rotated 90° anticlockwise to $\bar{y}xz$ and inverted through 000 to point e ($y\bar{x}\bar{z}$), and so on to f ($\bar{x}\bar{y}z$), g ($\bar{y}x\bar{z}$), and back to d. The inversion point must lie on the inverse axis, and its height must be specified. By convention, the height is assumed to be zero if not given. To show that an inversion has taken place from d to e, a comma is placed in the circle at e: this is removed to show the inversion from e to f, and is replaced again at g.

The symbolic descriptions for the axes of order **3** and **6** should be self-explanatory. Axes 3_1 and 3_2 are enantiomorphic, as are 6_1 and 6_5, and 6_2 and 6_4. Axis 3_1 is a subgroup of 6_4, not 6_2. Commas are shown for half of the positions in $\bar{3}$ and $\bar{6}$. For the latter, superposition of points requires that each circle be split into half. At the top of Figure 7.3 are shown the operations for axes of order **1** and **2**. A vertical 2_1 axis is shown by two spikes added to the symbol for a rotation diad. A horizontal mirror plane generates two superimposed points l and n at xyz and $xy\bar{z}$, represented by the split circle with a comma in one half.

The mapping transformations for screw axes can be generated easily from Table 6.1. Because 4_1 [001] = 4[001] + 00¼, $xyz \rightarrow \bar{y},x,\frac{1}{4} + z$. Similarly, 6_5 [001] = 6[001] + 00⅚, and $xyz \rightarrow x-y,x, \frac{5}{6} + z$.

Table 7.1 lists the 75 classical groups for crystallographic rods. Because they all occur embedded in space groups, they are not drawn out here. For order one, only **p1** and **p1̄** are possible. For order 2, the 20 possibilities follow directly from the band groups after (i) changing the unique axis from b in Figure 7.1 to c in Table 7.1, and (ii) removing the singular plane. Thus **p121** (band) → **p112** (rod). Band groups **p211** and **p112** lose their distinction and become rod group **p211** in which the diad axis is arbitrarily placed in the first position. Similarly, **p1m1** → **p11m**, **pb11** → **pc11**, **p12₁/m1** → **p112₁/m**, **pbmb** → **pccm**, and so on. The coordinate axes for rod groups 13 through 22 must be orthogonal, but can be monoclinic for 3 through 12. The trigonal rod groups are enumerated by first using one axis (42–45), then adding a perpendicular rotation axis (46–48), and finally either a mirror or a c glide plane (49–52). Similarly the tetragonal and hexagonal rod groups can be enumerated. Some combinations of symmetry elements are incompatible, and some lead to a rod group of higher symmetry.

 Rod groups allow an easy introduction to color symmetry and antisymmetry. Consider the addition of color to the classical rod groups of order 4. One color corresponds merely to classical symmetry in which the symmetry operation involves only change of position (e.g., black dots in Fig. 7.2). To

Table 7.1 The 75 groups for crystallographic rods[a]

No.	Symbol	No.	Symbol	No.	Symbol	No.	Symbol
1*	**p1**	23	**p4**	42	**p3**	53	**p6**
2*	**p1̄**	24	**p4₁** ⎫	43	**p3₁** ⎫	54	**p6₁** ⎫
		25	**p4₃** ⎭	44	**p3₂** ⎭	55	**p6₅** ⎭
3*	**p112**	26	**p4₂**	45	**p3̄**	56	**p6₂** ⎫
4*	**p211**	27	**p4̄**	46	**p32**	57	**p6₄** ⎭
5*	**p112₁**	28	**p422**	47	**p3₁2** ⎫	58	**p6₃**
6*	**p11m**	29	**p4₁22** ⎫	48	**p3₂2** ⎭	59	**p6̄**
7*	**pm11**	30	**p4₃22** ⎭	49	**p3c**	60	**p622**
8*	**pc11**	31	**p4₂22**	50	**p3m**	61	**p6₁22** ⎫
9*	**p112/m**	32	**p4̄2m**	51	**p3̄c**	62	**p6₅22** ⎭
10*	**p2/m11**	33	**p4̄2c**	52	**p3̄m**	63	**p6₂22** ⎫
11*	**p112₁/m**	34	**p4/m**			64	**p6₄22** ⎭
12*	**p2/c11**	35	**p4₂/m**			65	**p6₃22**
		36	**p4mm**			66	**p6̄2m**
13*	**p222**	37	**p4cc**			67	**p6̄2c**
14*	**p222₁**	38	**p4₂mc**			68	**p6/m**
15*	**pmm2**	39	**p4/mcc**			69	**p6₃/m**
16*	**p2mm**	40	**p4₂/mmc**			70	**p6cc**
17*	**pcc2**	41	**p4/mmm**			71	**p6₃mc**
18*	**pmc2₁**					72	**p6mm**
19*	**p2cm**					73	**p6/mcc**
20*	**pccm**					74	**p6₃/mmc**
21*	**pcmm**					75	**p6/mmm**
22*	**pmmm**						

[a] For nos. 1–22, the fourth symbol refers to the rod axis. For nos. 23–75, the second symbol or cluster (e.g., $6_3/m$) refers to the rod axis. A space group with an asterisk can be matched with either one or two space groups for bands. Bracketed rod groups are enantiomorphic.

illustrate two colors for **p4[II]**, change E, E', G, and G' to another color (e.g., white). The symmetry operation now comprises anticlockwise rotation of 90° and a black–white color change. An international nomenclature for color symmetry has not yet been published, and the suffix **II** is used to show the two-color interchange. To illustrate *antitranslation* use different symbols for points D–G and D'–G' (e.g., unprimed dot for D–G and primed dot for D'–G'). The repeat distance is doubled for the new group **p4′**. Rod group **p4[IV]** uses four colors (e.g., red, yellow, green, blue) that must change in a cyclic fashion: for example, D (red), E (90° rotation plus a cyclic color change to yellow), F (90° rotation plus a cyclic color change to green),

G (90° rotation plus a cyclic color change to blue), and closure to *D* (90° rotation plus a cyclic color change to red). Three colors do not allow closure, and **p4^{III}** cannot occur.

7.3 Two-Sided Plane (or Layer) Groups

The 17 2D space groups (Chapter 2) were derived by considering only one side of a plane surface. As such, they are useful to crystallographers because they represent projections of the 230 3D space groups onto a plane. They can also be used to describe the symmetry of neutral motifs (i.e., with sides of the same color) in a two-sided plane. Another 63 two-sided plane groups are obtained using cambiant motifs to give a total of *80 two-sided plane groups.* These symmetry groups (Holser, 1958a) are important for description of layers of atoms, either in films or as components of crystals (e.g., silicate layers in clay minerals). They are also important for description of planar mistakes in crystals, such as occur at the boundaries of twin crystals and domain intergrowths. Figure 7.4 shows a unit cell of each two-sided plane group using a scalene triangle as a motif (Weber, 1929). They are arranged according to the one-sided parents, thereby permitting easy enumeration. The 46 diagrams with both black and white triangles also represent the operations of antisymmetry on *one side* of a plane. Table 7.2 shows a classification based on the shape of the 2D lattice (hence the system) and the point group (i.e., space group minus all translation operations). For a history of plane groups and a listing of subgroups, see Holser (1958b). Coordinate axes are taken in the orientations shown in Figure 7.4. Four symbols or clusters are used to designate each group. The first is **p** or **c** for the two possible types of lattice. The remaining three refer to the *a, b,* and *c* axes for the oblique and rectangular systems, and to *c, a,* and either [**110**] or [**210**] for the square and hexagonal systems.

In Figure 7.4, the one-sided parent **p1** yields two-sided **p1** whose two-sided scalene motifs form a single oblique lattice. Removal of the difference between the sides of the scalene motif generates **p11m**. Not so obvious is two-sided **p11a,** whose scalene motifs show alternate sides. The *a* repeat is doubled over **p1**, and the double symmetry operation of an **a glide plane** involves exchange of the sides of the scalene motif by mirror reflection in (**001**) followed by glide translation of *a*/2. A second double operation closes the symmetry group with the cell translation *a.* From the viewpoint of symmetry, two-sided plane group **p11b** is merely the same as **p11a** after relabeling the oblique axes.

One-sided plane group **p2** leads to **p112** (all motifs up), **p112/m** (neutral motif), **p$\bar{1}$** and **p112/a** (alternating motifs). Again, **p112/b** would be merely a relabeling of **p112/a**. One-sided **pm** yields seven two-sided rectangular

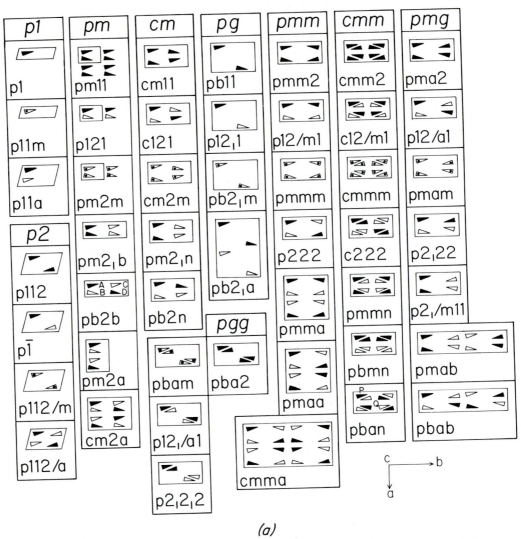

(a)

FIGURE 7.4 The 80 two-sided plane groups arranged according to their one-sided parent (italics). Motifs with identical sides are shown with dots. Note that the area of the unit cell is doubled for **p11a, pm2₁b, pb2b, pm2a, pb2₁a, pmma, pmaa, pmab, pbab, p4/n, p4/nbm, p4/nmm,** and is quadrupled for **cm2a** and **cmma.** Although four cells from the parent are shown for three square two-sided plane groups, a new unit cell of double area (dots) is chosen. Of course, the dots do not represent a glide plane.

groups. Three of them (**pm11, p121, pm2m**) retain the same unit cell, but **pm2₁b** and **pb2b** require doubling of the *b* repeat, **pm2a** requires doubling of the *a* repeat, and **cm2a** doubling of both repeats. Lattice centering is needed for the last group in order to maintain the rectangular geometry in accord with the conception of a Bravais lattice. In **pb2b,** the first glide plane

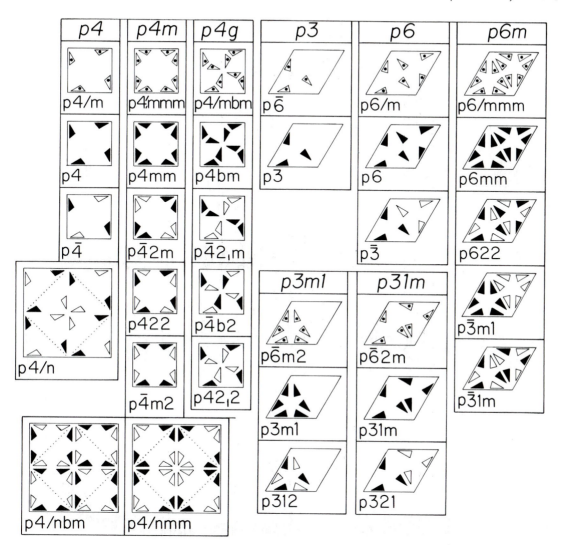

(b)

b11 involves mirror reflection in a plane perpendicular to the *a* axis (hence the first position for the symbol **b**) and translation of *b*/2. Motif *A*(filled) is reflected across the plane (**100**) at *x* = 0.5 to the same position as motif *B*, but filled instead of open, and is translated to final position *D*(filled). Glide plane **11b** turns *A*(filled) to *A*(unfilled) and translates to final position *C*(unfilled), which is in turn related to *B*(unfilled) by **b11**. A diad rotation axis **121** relates *A*(filled) to *B*(unfilled), and *C*(unfilled) to *D*(filled). The operations of all glide planes are shown systematically in Figure 7.20. If time is short, it is recommended that readers proceed now to the next section, and return to this section when a two-sided plane group is needed for a particular problem.

Table 7.2 The 80 two-sided plane groups arranged by system and point group

Oblique System		Square System		Rectangular System		Hexagonal System	
1	p1	4	p4	121	p121; p12$_1$1; c121	3	p3
$\bar{1}$	p$\bar{1}$	$\bar{4}$	p$\bar{4}$	1m1	p1m1*; p1b1*; c1m1*[a]	$\bar{3}$	p$\bar{3}$
112	p112	4/m	p4/m; p4/n	12/m1	p12/m1; p12$_1$/m1*; c12/m1; p12/a1; p12$_1$/a1	32	p312, p321
11m	p11m; p11a	422	p422; p42$_1$2	222	p222; p2$_1$22; p2$_1$2$_1$2; c222	3m	p3m1; p31m
112/m	p112/m; p112/a	4mm	p4mm; p4bm	mm2	pmm2; pma2; pba2; cmm2	$\bar{3}$m	p$\bar{3}$1m; p$\bar{3}$m1
		$\bar{4}$2m	p$\bar{4}$2m; p$\bar{4}$2$_1$m p$\bar{4}$m2; p$\bar{4}$b2	m2m	pm2m; pm2$_1$b; pb2$_1$m; pm2a; pm2$_1$n; pb2b; pb2$_1$a; pb2n; cm2m; cm2a	6	p6
		4/mmm	p4/mmm; p4/nbm p4/mbm; p4/nmm	mmm	pmmm; pmma; pmam; pmmn; pbam; pmaa; pbmn; pmab; pbab; pban; cmmm; cmma	$\bar{6}$	p$\bar{6}$
						6/m	p6/m
						622	p622
						6mm	p6mm
						$\bar{6}$m2	p$\bar{6}$m2; p$\bar{6}$2m
						6/mmm	p6/mmm

[a] *Means that first and second axes are interchanged in Figure 7.4.

244

One-sided plane group **cm** yields 5 two-sided groups, all with the same size of unit cell. Doubling of a cell repeat is not allowed, because the motifs would not have the same environment when distance and relative orientation of neighbors are considered. Recall from Chapter 2 that absolute orientation is required only for an isolated Bravais lattice, and that only relative orientation is needed for intersecting lattices of a space group. For example, two points are equivalent in space group **pm** if the surrounding positions and more distant neighbors are related by the mirror symmetry. One-sided **pg** yields 4 two-sided plane groups, of which **pb2₁a** has a doubled a axis. One-sided **pmm** yields 7 two-sided groups, for which a is doubled in **pmma** and **pmaa**, and b is also doubled in **cmma**. Doubling does not occur in the 7 derivatives of one-sided **cmm**, and the systematic enumeration is visually obvious. For one-sided **pmg**, 2 of the 7 derivatives have doubled b, and for one-sided **pgg**, no doubling occurs in the 4 derivatives.

Returning to **cmm**, 2 derivatives **pbmn** and **pban** contain a **11n** diagonal glide plane involving a mirror operation in the plane and a translation of $(a + b)/2$. Thus motif P undergoes glide reflection to Q.

In the square system, one-sided **p4** yields two-sided **p4**, **p4̄**, **p4/m** (identical sides) and **p4/n** (45° cell of double area). One-sided **p4m** yields 7 two-sided groups, and **p4g** yields 5.

In the hexagonal system, one-sided **p3** yields two-sided **p3** for triangles with different faces, and **p3/m** (= **p6̄**) for triangles with identical faces. One-sided **p6** yields two-sided **p6**, **p3̄**, and **p6/m** for isofacial triangles. One-sided **p3m1** gives two-sided **p3m1**, **p312**, and **p6̄m2** (identical faces). Similarly, **p31m** gives **p31m**, **p321**, and **p6̄2m** (identical faces). Finally, **p6m** yields **p6mm**, **p3̄m1**, **p3̄1m**, **p622**, and **p6/mmm** (identical faces).

This completes the two-sided plane groups, and sets the stage for development of the 230 classical 3D space groups. These will be discussed in clusters of increasing complexity.

7.4 Simple 3D Space Groups with a Primitive Lattice and a Single Rotation Axis

In 2D, the combination of a primitive lattice (**p**) with a 2-fold rotor (**2**) gave two intersecting lattices in space group **p2** (Fig. 2.6). In 3D, the combination of a primitive lattice with a 2-fold rotor gives the space group **P2** (Fig. 7.5). Because there are three axes for the unit cell in 3D, it is necessary to specify the location of the diad axis. For convenience, take a standard orientation of the unit cell with the projection parallel to c, and b pointing to the right (Fig. 4.1). The symbol **P112** shows that the diad axis is placed along c, which requires a monoclinic cell with $a \neq b \neq c$, $\alpha = \beta = 90°$, $\gamma \neq 90°$. To place the

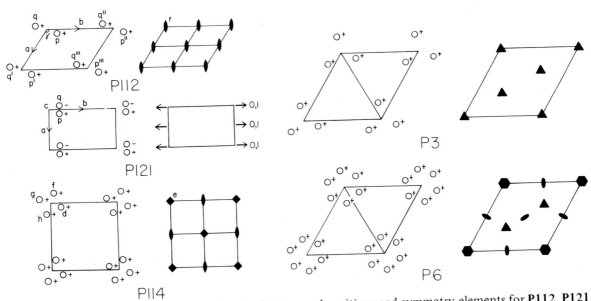

FIGURE 7.5 Space group diagrams of equivalent general positions and symmetry elements for **P112, P121, P4, P3,** and **P6.** The right-hand diagram for **P112** was quartered to locate the central diad axis. Even though the diad axes lie at the corners of the quarter-cells, the true unit cell is given by the repeats of the general positions. For **P3** and **P6** the 120° parallelogram can be divided into two equilateral triangles, but these cannot be used as unit cells.

diad axis along *b*, use **P121** for which the monoclinic unit cell has $\alpha = \gamma = 90°$, $\beta \neq 90°$. From the viewpoint of symmetry, the choice is irrelevant. For each space group, it is necessary to determine the shape of the appropriate projection of the unit cell, and draw two unit cells, one for the general positions and one for the symmetry elements. For **P112**, the unit cell is represented by an oblique parallelogram, and for **P121** by a rectangle.

It is customary to place symmetry elements at corners or faces of unit cells whenever possible. Consider first **P112**, and place a rotation diad axis along a vertical edge of a cell. It is represented in projection by a lens-shaped or elliptical symbol. A general position *p* at (x,y,z) is represented by a circle centered on (x,y) and a plus sign to represent height +z. The lattice vectors **a** and **b** generate positions *p'*, *p"*, and *p'"*. Although positions are often shown just outside the unit cell to aid visualization, they are not necessary. Operation of the diad axis *r* gives position *q* at $(\bar{x},\bar{y},z)$ represented by a circle with a plus sign, and the lattice repeats give *q'*, *q"*, and *q'"*. Diad axes must be placed at all the halfway positions in the right-hand diagram. Just as in 2D, it is often convenient to divide a unit cell into quadrants to aid in placement of positions and symmetry elements. These two diagrams represent two 3D primitive lattices related by a 2-fold rotor. The points *p* and *q* can move so long as they obey the 2-fold rotor. In a crystal structure with space group **P2**, the atoms will occur in pairs (x_1,y_1,z_1) and $(\bar{x}_1,\bar{y}_1,z_1)$; (x_2,y_2,z_2) and

$(\bar{x}_2,\bar{y}_2,z_2)$; and so on, where (x_1,y_1,z_1), $(x_2,y_2,z_2),\ldots$ must be determined experimentally by diffraction of X-rays, neutrons, or electrons. The general position has point symmetry **1** and multiplicity 2. There is a special position of point symmetry **2** and multiplicity 1 at $(0,0,z)$, and three others at $(\frac{1}{2},0,z)$, $(0,\frac{1}{2},z)$ and $(\frac{1}{2},\frac{1}{2},z)$.

For the orientation **P121**, the diad axis is horizontal and is represented by arrow symbols. The general position p at (x,y,z) is rotated to position q at $(\bar{x},y,\bar{z})$, and there are special positions at $(0,y,0)$, $(\frac{1}{2},y,0)$, $(0,y,\frac{1}{2})$, and $(\frac{1}{2},y,\frac{1}{2})$.

The only possible types of rotation axes that are consistent with a lattice are **1**, **2**, **3**, **4**, and **6**. Space group **P114** is usually listed with the short symbol **P4**, because all crystallographers place the 4-fold rotor along c. The 4-fold symmetry enforces $a = b$ and $\alpha = \beta = \gamma = 90°$. In projection down c (Fig. 7.5), a general position (x,y,z) is represented by circle d centered on (x,y) labeled with a plus sign to represent height $+z$. The tetrad axis at e produces positions $(\bar{y},x,z)$ at f, $(\bar{x},\bar{y},z)$ at g, and $(y,\bar{x},z)$ at h. Further tetrad and diad axes are produced just as in space group **p4** (Fig. 2.6). Special positions occur on the rotation axes.

Space groups **P3** and **P6** can be generated by analogy with Figure 2.9 for **p3** and **p6**; just place a plus sign against each general position.

7.5 Space Group P222 and the Crystal Structure of AIPS₄

Can rotation axes (ignoring monads) be placed along more than one direction? Yes. The simplest space group is **P222**, in which the three diad axes must be orthogonal and in an orthorhombic unit cell with $a \neq b \neq c$. In the usual c-axis projection (Fig. 7.6), general position p at (x,y,z) is rotated to q

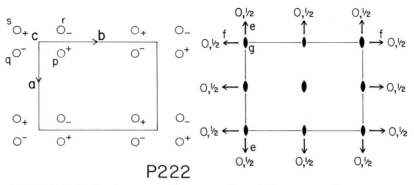

P222

FIGURE 7.6 Space group diagrams of equivalent general positions and symmetry elements for **P222**. The axes are labeled a, b, and c in the usual way. For this space group any assignment leads to the same pattern of general positions, but in a real crystal structure, the observed values of x_1, y_1, and z_1 must be assigned to the correct axes.

at $(x,\bar{y},\bar{z})$ by rotation axis *ee*. Rotation axis *ff* produces position *r* at $(\bar{x},y,\bar{z})$ from *p* and position *s* at $(\bar{x},\bar{y},z)$ from *q*. The resulting four positions automatically generate rotation axis *g* along the *c* axis. To complete the diagrams, add symmetry elements at all halfway positions, and use the lattice repeats. A general position has 4-fold multiplicity; there are special positions of 2-fold multiplicity on all the diad axes, and special positions of 1-fold multiplicity and point symmetry **222** at the intersections of the diad axes.

The crystal structure of aluminum thiophosphate, $AlPS_4$, illustrates space group **P222** rather nicely. The cell dimensions are $a=5.61$ Å, $b=5.67$ Å, $c=9.05$ Å, and the atomic coordinates are:

> Al   000   ½0½
>
> P   0½0   00½
>
> S(1) xyz $\bar{x}\bar{y}z$ $x\bar{y}\bar{z}$ $\bar{x}y\bar{z}$ with x_1 0.200 y_1 0.260 z_1 0.125
>
> S(2) xyz $\bar{x}\bar{y}z$ $x\bar{y}\bar{z}$ $\bar{x}y\bar{z}$ with x_2 0.740 y_2 0.800 z_2 0.630

Let us accept this description even though it does not fit the conventional morphological choice of axial lengths with $c < a < b$. Furthermore, let us view the unit cell down the *b* axis just to give practice in projecting a unit cell down an axis that is not *c* (Fig. 7.7). Three different symbols are used to represent the three types of atoms, and the crosses for S are either hori-

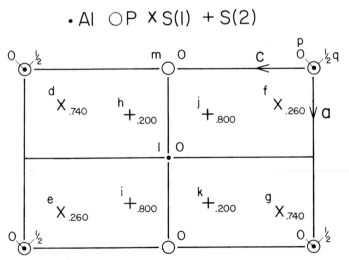

FIGURE 7.7 Unit cell of the crystal structure of $AlPS_4$. Note the unusual orientation produced by projection down the *b*-axis. To obtain the tetrahedron of S atoms around the Al atom at *l*, use *h* and *k* at height .200, and *i* and *j* at height – .200.

zontal or diagonal to distinguish the two types of positions $x_1 y_1 z_1$ and $x_2 y_2 z_2$. All atoms obey the symmetry of **P222**. Thus S atoms at d, e, f, and g [denoted S(1)] occupy one set of general positions, while those at h, i, j, and k [denoted S(2)] occupy a second set. The Al atom at (000) is in a special position at the intersection of three diad axes, and the Al atom at $(\frac{1}{2}0\frac{1}{2})$, labeled l, is in a different special position at the intersection of three other diad axes. The P atom at $(0\frac{1}{2}0)$, labeled m, also lies at the intersection of three diad axes, as does the P atom at $(00\frac{1}{2})$. Note that (i) a single chemical species can occupy more than one set of symmetry-equivalent positions in a crystal structure, and (ii) special positions with one kind of point symmetry can be occupied by more than one type of chemical species, though each specific subset of special positions is occupied by only one chemical species. The second statement must be modified if two types of chemical species occur randomly in just one *crystallographic* position.

The atomic coordinates in a crystal structure are determined by the chemical bonding between the atoms. The Al atom at l lies at the center of a tetrahedron of four S(2) atoms at h, i, j, and k. The Al–S(2) distance is 2.12 Å, and the tetrahedron is in a negative orientation. The Al atom at the origin is surrounded by four S(1) atoms at 2.17 Å in a tetrahedron with positive orientation. Thus these two Al atoms have similar but slightly different chemical bonding and must be labeled Al(1) and Al(2). Similarly, there are two P atoms, one surrounded by four S(1) at 2.09 Å, and the other by four S(2) at 2.19 Å. Undoubtedly, these distances suffer from experimental error, but even if P(1)–S(1) = P(2)–S(2), the P(1) and P(2) atoms would still be crystallographically distinct in space group **P222**.

7.6 Simple Space Groups with a Nonprimitive Lattice and a Single Rotation Axis

Space group **I4** (Fig. 7.8) is constructed as follows: Draw two squares. Place circle p with a plus sign to represent a general position (xyz) and insert three equivalent circles in adjacent unit cells at p', p'', p'''. Body centering requires addition of the circle marked $\frac{1}{2}+$ at position q to represent general position $(\frac{1}{2}+x, \frac{1}{2}+y, \frac{1}{2}+z)$. Operation of the 4-fold rotor (r) produces additional positions s, t, and u from p, and the body-centering vector produces v, w, and d as well as rotor e.

At first sight, the space group diagrams are completed by adding rotors e, f, g, and h, and by filling in positions in the left-hand diagram. But there are simple relationships that involve a screw axis. If points p at (xyz) and w at $(\frac{1}{2}-x, \frac{1}{2}-y, \frac{1}{2}+z)$ had the same height, a rotation diad axis would be placed at i. Similarly, points p, v, t', and d' would be related by a rotation tetrad axis at j if they had the same height. For both sets of positions, the

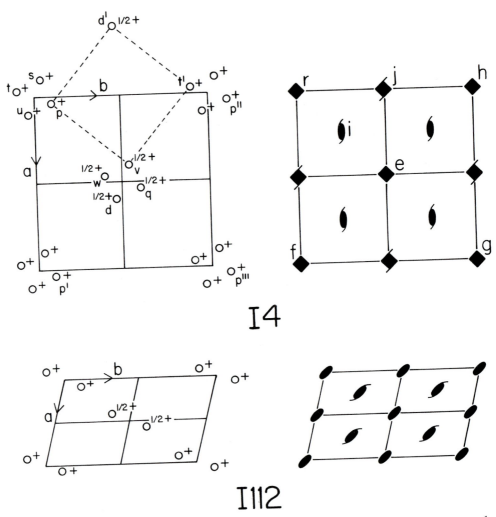

FIGURE 7.8 Space group diagrams of equivalent general positions and symmetry elements for **I4** and **I112**. The dashed lines show the tilted square produced by the operations of the screw tetrad at *j*.

heights differ by simple fractions, and the positions *p* and *w* are related by a screw diad, denoted 2_1, and shown symbolically by a lens symbol with two extensions, and the positions *p, v, t',* and *d'* by a screw tetrad, denoted 4_2, and shown by a square with extensions at opposite corners (Fig. 7.3).

Space group **I112** also develops a 2_1 axis (Fig. 7.8), and, indeed, the combination of any centered lattice with any rotation axis produces a screw axis.

7.7 Space Groups with Rotation, Screw and Inverse Axes

Table 7.3 contains a systematic list of all the space groups obtained by combining a Bravais lattice with one or more axes. For the triclinic lattice, the only possible space groups are **P1** and **P$\bar{1}$**, the latter with a center of symmetry, and the former with no rotation symmetry except the formal monad. For the monoclinic primitive lattice, a rotation diad along c gives **P112**, and a screw diad gives **P112$_1$**. The axes can be relabeled to give **P121** (commonly used) or **P211** (rarely used). The combination of a monoclinic **I** lattice with a rotation diad automatically produces a screw diad (Fig. 7.8) so that **I112** is identical with **I112$_1$**. Relabeling would give **I121**, and so on.

For the primitive orthorhombic lattice, three intersecting rotation diads give **P222** (Fig. 7.6). Relabeling the axes will not change the space group symbol, but will change the atomic coordinates $x_1 y_1 z_1$ in a crystal structure. Thus if new a is old b, new x_1 is old y_1. The only possible combinations of rotation and screw diads are **P222$_1$** and **P2$_1$2$_1$2** with relabeled variants, and the list is completed by **P2$_1$2$_1$2$_1$**, whose symbol is invariant to relabeling of axes. When drawing space group diagrams, it is necessary to select the relative position for the first two axes in order that the third one is produced automatically. In order to obtain **P222$_1$** (Fig. 7.9), insert a diad axis ee at $y = z = 0$ and then add a diad axis ff at $x = 0$, $z = \frac{1}{4}$ or $\frac{3}{4}$. The first axis relates position p to q, and the second axis relates position q to r so that

Table 7.3 Systematic list of space groups with a rotation, screw, or inverse axis[a]

Triclinic	**P1, P$\bar{1}$**
Monoclinic	**P112** (= **P121**), **P112$_1$** (= **P12$_1$1**), **I112** ($\equiv$ **I112$_1$** = **I121** = **I12$_1$1**)
Orthorhombic	**P222, P222$_1$** (= **P2$_1$22** = **P22$_1$2**), **P2$_1$2$_1$2** (= **P22$_1$2$_1$** = **P2$_1$22$_1$**), **P2$_1$2$_1$2$_1$**, **C222** ($\equiv$ **C2$_1$2$_1$2** = **A222** = **B222** = **A22$_1$2$_1$** = **B2$_1$22$_1$**), **C2$_1$2$_1$2$_1$** ($\equiv$ **C222$_1$** = **A2$_1$2$_1$2$_1$** = **B2$_1$2$_1$2$_1$** = **A2$_1$22** = **B22$_1$2**), **F222**, first type of **I222** ($\equiv$ **I2$_1$2$_1$2$_1$**) with intersecting rotation diads, second type of **I222** ($\equiv$ **I2$_1$2$_1$2$_1$**) with nonintersecting rotation diads
Tetragonal	**P4, P4$_1$, P4$_2$, P4$_3$, P$\bar{4}$, I4** ($\equiv$ **I4$_2$**), **I4$_1$** ($\equiv$ **I4$_3$**), **I$\bar{4}$, P422, P4$_1$22, P4$_3$22, P4$_2$22, P42$_1$2, P4$_1$2$_1$2, P4$_3$2$_1$2, P4$_2$2$_1$2, I422, I4$_1$22**
Hexagonal	(120° prism for unit cell) **P3, P3$_1$, P3$_2$, P$\bar{3}$, P6, P6$_1$, P6$_2$, P6$_3$, P6$_4$, P6$_5$, P$\bar{6}$, P312, P321, P3$_1$12, P3$_1$21, P3$_2$12, P3$_2$21, P622, P6$_1$22, P6$_5$22, P6$_2$22, P6$_4$22, P6$_3$22.** (rhombohedral unit cell) **R3, R$\bar{3}$, R32**
Isometric	**P23, I23, F23, P2$_1$3, I2$_1$3, P432, P4$_1$32, P4$_3$32, P4$_2$32, I432, I4$_1$32, F432, F4$_1$32**

[a] This list of space groups does not utilize mirror planes, even though **m $\equiv$ $\bar{2}$**.

p and r are related by the screw axis g at $x = y = 0$. If the diad axis ff had been placed at $x = 0, z = 0$, the space group $P222$ would have been produced.

Space group $P2_1\, 2_1\, 2$ must have intersecting screw diads in order to give a rotation diad. Insert the starting position p at (xyz) and put the rotation diad e at $x = y = 0$ to produce position q at $(\bar{x}\bar{y}z)$. The screw axes (symbolized by half-arrows) must be placed at ff and gg so that positions r and s are produced. Thus axis gg relates p to s and q' to r, and axis ff relates p to r and s to q'. No matter what sequence is used, all the operations fit together. There is only one answer, and every element in a space group must be related to every other element in a multiplication table, as in Chapter 2.

Is the absolute position of symmetry elements important? Place a screw diad at $y = z = 0$. The next screw diad must be placed at height $z = 0$, but x is free: choose $x = z = 0$. The rotation diad then appears at $x = y = \frac{1}{4}$, and everything fits together when all the operations are completed (Fig. 7.9, $P2_1\, 2_1\, 2$, different origin). At first sight, the general positions look quite different for the two choices of origin for $P2_1\, 2_1\, 2$, and indeed the absolute coordinates are different.

$P2_1\, 2_1\, 2$ (first choice) $xyz;\ \bar{x}\bar{y}z;\ \frac{1}{2} + x, \frac{1}{2} - y, \bar{z};\ \frac{1}{2} - x, \frac{1}{2} + y, \bar{z}$

$P2_1\, 2_1\, 2$ (second choice) $xyz;\ \frac{1}{2} - x, \frac{1}{2} - y, z;\ \frac{1}{2} + x, \bar{y}, \bar{z};\ \bar{x}, \frac{1}{2} + y, \bar{z}$

But remember that x, y, and z are algebraic quantities that can vary independently from 0 to 1, *so long as the same chosen values are used for all four positions.* For the first choice of axes, select values of x_1, y_1, and z_1. For the second choice of axes, first move the origin to $x = \frac{1}{4}$, $y = \frac{1}{4}$, $z = 0$ to give positions $x_2 - \frac{1}{4}, y_2 - \frac{1}{4}, z_2;\ \frac{1}{4} - x_2, \frac{1}{4} - y_2, z_2;\ \frac{1}{4} + x_2, -(y_2 + \frac{1}{4}), -z_2;$ $-(x_2 + \frac{1}{4}), \frac{1}{4} + y_2, \bar{z}_2$. Then choose $x_3 = x_2 - \frac{1}{4}$, $y_3 = y_2 - \frac{1}{4}$, $z_3 = z_2$. The new coordinates $x_3 y_3 z_3;\ \bar{x}_3 \bar{y}_3 \bar{y}_3;\ \frac{1}{2} + x_3, \frac{1}{2} - y_3, \bar{z}_3;\ \frac{1}{2} - x_3, \frac{1}{2} + y_3, \bar{z}_3$ are the same as those for the first choice, and x_3 is equal to x_1, $y_3 = y_1$, and $z_3 = z_1$.

Space group $P2_1\, 2_1\, 2_1$ has nonintersecting screw diads. Relative displacements of one fourth of a cell edge result in positions whose coordinates differ by the half-unit needed for a screw diad.

For the C-centered orthorhombic lattice, there are only two space groups, apart from the trivial variations produced by relabeling the axes to give either A- or B-centered lattices. Space group $C222$ also has 2_1 axes along the a and b axes and a second rotation diad axis along c (at $r, s, t,$ and u in Fig. 7.9) because of the interaction between the C-centering vector and the rotation diad axes. Similarly, space group $C222_1$ has screw diad axes along a, b, and c. Whereas space group $C222$ is developed automatically by inserting rotation diad axes at $y = z = 0$ and $x = z = 0$, $C222_1$ requires nonintersecting rotation diads at $y = z = 0$ and $x = 0, z = \frac{1}{4}$.

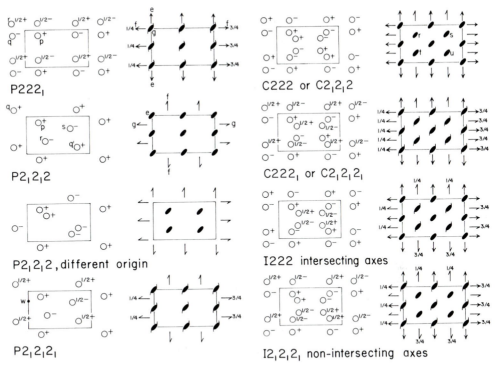

FIGURE 7.9 Diagrams of equivalent general positions and symmetry elements for orthorhombic space groups with rotation and screw diads.

An I-centered orthorhombic lattice and three intersecting rotation diads yield space group **I222**, which also has intersecting 2_1 axes in all three directions. To the despair of unsuspecting students, there is another space group **I222** with nonintersecting rotation diads and screw diads. These are the only two space groups with exactly the same complement of symmetry elements. In *International Tables,* the second space group is listed as $I2_1 2_1 2_1$ to distinguish it from the first space group, but this is merely a convenient distinction. All other space groups can be developed uniquely from the *International* (Hermann–Mauguin) symbol, which is used here. The Schoenflies symbol has been abandoned by most crystallographers because of arbitrary assignment of many sequence numbers. The only combination of 2-fold axes with an orthorhombic F lattice gives space group **F222**. Because it is tedious to draw, it is omitted from Figure 7.9.

Any combination of axes and relative positions with the orthorhombic lattices yields a set of general positions that either fits one of the space groups in Table 7.3, or yields a space group of higher symmetry. If the reader tries to develop diagrams for space group $P222_1$ by placing the first two axes at height $z = 0$, the resulting space group will be **P222** if the reader stops right there. But if the 2_1 axis is then added, the space group will end up with eight general positions and will involve planes of symmetry or a centered lattice. Orthorhombic space groups with three rotation axes (no matter whether pure or inverse) must have the third axis produced automatically by operation of the first two axes.

Tetragonal space groups (Fig. 7.10) can be developed with just a single tetrad axis, or with addition of a diad axis placed along *a* (and automatically on *b*). Both primitive and body-centered lattices can occur, and all possible combinations are listed in Table 7.3. Space group $P4_1$ is straightforward, and screw diad axes appear automatically at $x = \frac{1}{2}$, $y = 0$, and equivalent positions. Space group $P4_3$ is not shown because it is the same as $P4_1$ except for switching heights $\frac{1}{4}$ and $\frac{3}{4}$ and changing spikes on the square symbols. Rotation diads appear automatically for $P4_2$ and for $P\bar{4}$. For the latter space group, the inversion point on the tetrad was placed at height $z = 0$. The combination of body centering with an inverse tetrad axis in $I\bar{4}$ results in a further set of inverse tetrad axes whose inversion point is at height $\frac{1}{4}$ or $\frac{3}{4}$. Thus axis *d* relates points *e* to *h*. Screw diads at $x = \frac{1}{4}$, $y = \frac{1}{4}$, and so on are also produced automatically. In space group $I4_1$ positions *f, g,* and *h* are derived from operation of the 4_1 axis at *d* on the position *xyz* at *e*, and positions *i, j, k,* and *l* from the combination of the operations of the I lattice and the

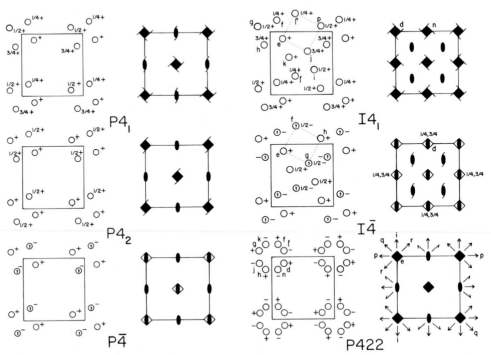

FIGURE 7.10 Diagrams of equivalent general positions and symmetry elements for tetragonal space groups with rotation and screw diads.

inverse tetrad. A 4_3 axis is automatically produced at n from the positions $e, l', p,$ and j.

To obtain space group **P422**, start with general position xyz at d and operate with the tetrad axis at e to produce positions $f, g,$ and h. Insert rotation diad ii at $y = z = 0$, and operate on d to give position j, on f to give k, g to l, and h to n. Rotation diad pp along the y axis is produced automatically by the combination of the tetrad axis e and the diad axis ii. Further rotation diads qq and rr are produced automatically in the 45° horizontal positions, and it is to these axes that the second 2 is referring in the space group symbol. Table 7.3 lists nine more space groups with a tetrad axis and diad axes along the horizontal directions. Development is fairly straightforward but too time-consuming to be worth pursuing here. Note that **P4₁22** and **P4₃22** are enantiomorphic, as are **P4₁2₁2** and **P4₃2₁2**. Space group **I422** also has 4_2 axes, and **I4₁22** also has 4_3 axes. To construct **P4₁2₁2** or **P4₃2₁2**, it is necessary to place the 4_1 or 4_3 axis along c and the 2_1 axes along a and b so that they do not intersect. All other space groups are produced automatically by placing the first two axes so that they intersect. Readers may find that they choose a different origin for the unit cell than the one in *International Tables*.

For the hexagonal Bravais lattice, some selected space groups are illustrated (Fig. 7.11). Space group **P3** is a subgroup of **P6**, because the latter contains all the general positions and symmetry elements of the former. Space group **P3₂** is the same as its enantiomorph **P3₁** except for interchange of heights $\frac{1}{3} + z$ and $\frac{2}{3} + z$, and reversal of spikes on the triangular symbol. Similarly, **P6₂** is enantiomorphous with **P6₄**. In the hexagonal groups, a triad occurs at the centroid of each triangle, and a diad at the midpoint of each edge. Space group **P3₁** is a subgroup of **P6₁** and **P6₄**, and, similarly, **P6₂** and **P6₅** yield subgroup **P3₂**. Space group **P3̄** has an inverse triad at each corner of the unit cell, a rotation triad at the centroid of each triangle, and a center at the midpoint of each edge. Removal of all centers of symmetry, including the one in the $\bar{3}$ axis, converts **P3̄** into subgroup **P3**. Space group **P6̄** has inverse hexads at each corner and centroid of the triangles. It is equivalent to **P3/m**, and can be converted into subgroup **P3** by removal of the horizontal plane of symmetry. A 6_3 axis can be regarded as the combination of 3 and 2_1 axes, and removal of the operations of both the separate screw diad and the one inherent in the 6_3 axis converts **P6₃** into **P3**. Twelve space groups (Table 7.3) can be enumerated with horizontal diad axes perpendicular to a vertical triad or hexad axis.

It is obvious that there is a hierarchy of space groups, and, indeed, it is possible to develop all 230 space groups by degrading certain isometric and hexagonal space groups successively until **P1** is reached.

In the isometric system, axes can be placed in cubic, octahedral, or dodecahedral directions. Thus the cube and octahedron have three rotation

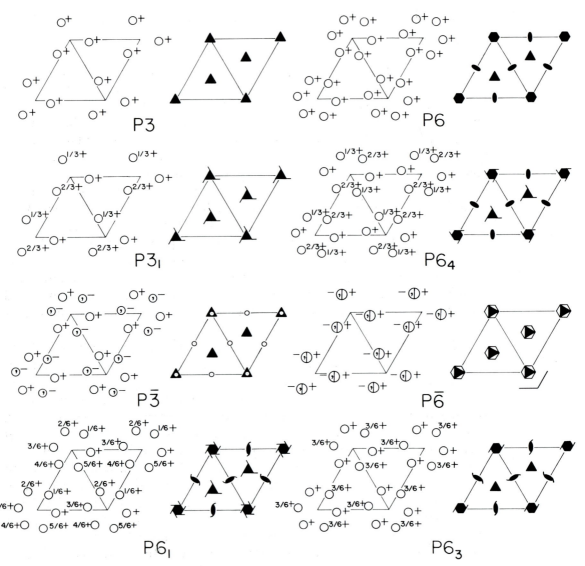

FIGURE 7.11 Diagrams of equivalent general positions and symmetry elements for hexagonal space groups with a **P** lattice (120° prism as unit cell) and a rotation axis parallel to the *c*-axis.

tetrads, four rotation triads and six rotation diads, whereas the tetrahedron has three diads and four triads. These symmetry elements pass through a point at the center of a regular polyhedron and can be regarded as a subgroup of space groups for which the lattice translations occur as well. Because there are three kinds of isometric Bravais lattices, the following space groups can be expected, and indeed do occur (Table 7.3):

23 → **P23, I23, F23** (rotation axes only)

P2$_1$3, I2$_1$3, F2$_1$3 (≡ **F23**) (conversion of rotation diad into screw diad)

432 → **P432, I432, F432** (rotation axes only)

P4$_1$32, P4$_3$32, P4$_2$32 (conversion of rotation tetrad to screw tetrad)

I4$_1$32 (≡ **I4$_3$32**), **I4$_2$32** (≡ **I432**) (conversion of rotation tetrad to screw tetrad)

F4$_1$32 (≡ **F4$_2$32** ≡ **F4$_3$32**) (conversion of rotation tetrad to screw tetrad)

Just as for the tetragonal and orthorhombic space groups, the combination of an **I** lattice with a **4$_1$** axis produces a **4$_3$** axis, and so on for the other identical space groups listed above. Conversion of a rotation triad into a screw triad is not possible, because four screw triads are not mutually compatible. Conversion to an inverse triad is possible, but gives space groups with symmetry planes to be considered in a later section. Note that **P23** is an isometric space group and **P32** is a hexagonal space group.

It is difficult to depict isometric space groups, and, indeed, there are no diagrams for these space groups in the *International Tables for X-ray Crystallography*. Some readers may wish to examine the diagrams in *Elementary Crystallography* by M.J. Buerger (1956). Another approach is purely abstract, and can be applied to all space groups. A rotation diad at $y = z = 0$ (i.e., parallel to the *a* axis) converts a general position at xyz into $x\bar{y}\bar{z}$ (c.f. Table 6.1), and one at $y = z = \frac{1}{4}$ produces the position at $x, \frac{1}{2} - y, \frac{1}{2} - z$, and so on for all the symmetry elements.

Space group **P23** will have a 12-fold general position. Position xyz is converted to $x\bar{y}\bar{z}$ by a diad along *a*, and these two positions are converted to $\bar{x}y\bar{z}$ and $\bar{x}\bar{y}z$ by the diad along *b*. Positions xyz and $\bar{x}\bar{y}z$ automatically produce the diad along *c*, and similarly for $x\bar{y}\bar{z}$ and $\bar{x}y\bar{z}$. A triad axis lying symmetrically between the *a*, *b*, and *c* axes produces a cyclic rotation xyz → zxy → yzx, and similarly for the other three positions.

$$x\bar{y}\bar{z} \rightarrow \bar{z}x\bar{y} \rightarrow \bar{y}\bar{z}x$$

$$\bar{x}y\bar{z} \rightarrow \bar{z}\bar{x}y \rightarrow y\bar{z}\bar{x}$$

$$\bar{x}\bar{y}z \rightarrow z\bar{x}\bar{y} \rightarrow \bar{y}z\bar{x}$$

The resulting 12 positions and symmetry elements are mutually consistent and allow closure of the group **P23**.

These operations can be envisaged with the aid of a tetrahedron (Fig.

7.12). Recall that a tetrahedron is obtained by drawing half the face diagonals of a cube (Fig. 3.5). Take a tetrahedron in either the positive or negative orientation, and place three dots on any face such that they obey the symmetry of the triad axis normal to the face. Then use the diad axes to locate corresponding dots on the other three faces. The resulting 12 dots fall in a general position for **P23** with the original cube as its unit cell. Points *d, e,* and *f* demonstrate the cyclic rotation $xyz \rightarrow zxy \rightarrow yzx$.

For space group **I23**, it is merely necessary to obtain another 12 positions by adding $\frac{1}{2}, \frac{1}{2}, \frac{1}{2}$ to the 12 positions for **P23**, and for **F23** three sets of 12 additional positions are obtained by adding $0, \frac{1}{2}, \frac{1}{2}$; $\frac{1}{2}, 0, \frac{1}{2}$; $0, \frac{1}{2}, \frac{1}{2}$ to the positions for **P23**. Space group **P2₁3** will close only if the screw diads do not intersect, just as for the orthorhombic space group **P2₁2₁2₁**, which is a subgroup of the isometric **P2₁3**. The extra positions for **I2₁3** are obtained from those of **P2₁3** by adding $\frac{1}{2}, \frac{1}{2}, \frac{1}{2}$ to those of **P2₁3**, and so on for all the isometric space groups based just on the combination of a Bravais lattice with symmetry axes.

 Returning to hexagonal space groups, there are three space groups with a rhombohedral unit cell (Table 7.3). Space group **R3** can be obtained by removing the diad axes and two out of the three triad axes from **P23**. The resulting unit cell is no longer constrained to be a cube, and turns into a

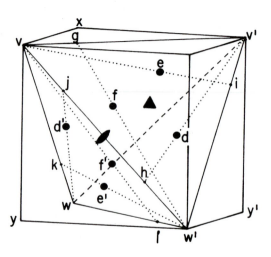

FIGURE 7.12 The illustration of operation of symmetry elements on a general position in space group **P23**. A tetrahedron is drawn inside a cube, and the vertices are labeled as in Figure 3.5. Position *i* is taken arbitrarily one-quarter of the way from *v′* to *w′*, and position *e* is one-third of the way from *i* to *v*. Positions *d* and *f* were constructed in a similar manner, and are related to *e* by the triad axis in the center of the face *vv′w′*. Positions *d′*, *e′*, and *f′* are related to *d*, *e*, and *f* by the diad axis midway between *v* and *w′*. Corresponding positions on the back two faces are not shown. The symmetry operations can be seen best on a model. A general position in a space group can move off the tetrahedral surface, but all points must move the same distance perpendicular from the nearest tetrahedral surface.

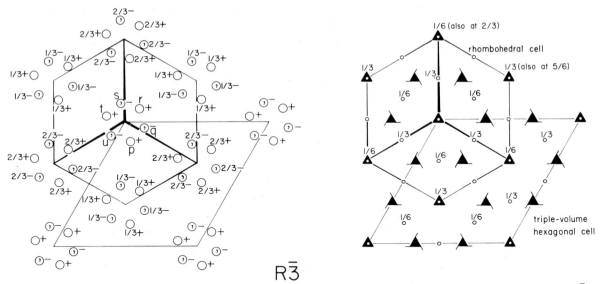

$R\bar{3}$

FIGURE 7.13 The diagrams of equivalent general positions and symmetry elements for space group $R\bar{3}$. The rhombohedral unit cell appears like a hexagon because it is viewed down the inverse triad axis, and the hexagonal cell projects onto the usual 120° rhombus. Positions p to u fall in one rhombohedral cell, and all other positions belong to other unit cells.

rhombohedron whose six faces are identical rhombuses. The general position degrades from 12-fold to 3-fold, and the coordinates of the general position are the cyclic triplet: xyz, zxy, yzx. Space group $R\bar{3}$ is produced from $R3$ merely by adding a center of symmetry at the origin of the cell, just as $P\bar{3}$ was produced from $P3$, and the extra positions are $\bar{x}\bar{y}\bar{z}$, $\bar{z}\bar{x}\bar{y}$, and $\bar{y}\bar{z}\bar{x}$. It is customary to project the rhombohedral unit cell down its triad axis to give a hexagonal outline (Fig. 7.13), and distance calculations are usually made from the hexagonal unit cell, which has three times the volume. This projection is confusing at first because the hexagonal outline of a single rhombohedral cell actually encompasses additional positions from pieces of other rhombohedral cells that project into the same hexagon. In Figure 7.13, only the six positions labeled p through u belong to one rhombohedral cell, and the others with extra height of ⅓ or ⅔ belong to adjacent cells. In addition to the $\bar{3}$ axis, there are 3_1 and 3_2 axes and centers of symmetry that are not inherent in the $\bar{3}$ axis.

7.8 Crystal Structures of Zn(OH)₂, Ice III, and High and Low Quartz

In *Crystal Structures*, Volume 1 by Wyckoff (1963), the crystal structure of Zn(OH)₂ is described as orthorhombic with space group $P2_12_12_1$ and $a = 8.53$ Å, $b = 5.16$ Å, $c = 4.92$ Å. The Zn atoms lie in a general position with $x = 0.125$, $y = 0.100$, $z = 0.175$, and the oxygen atoms occur in two general positions: O(1) 0.025,0.430,0.085; O(2) 0.325,0.125,0.370. Wyckoff gives the following algebraic coordinates for the general position of $P2_12_12_1$:

$$xyz; \tfrac{1}{2} - x, \bar{y}, \tfrac{1}{2} + z; \tfrac{1}{2} + x, \tfrac{1}{2} - y, \bar{z}; \bar{x}, \tfrac{1}{2} + y, \tfrac{1}{2} - z$$

These do not check with the coordinates in Figure 7.9, which would be:

$$xyz; \bar{x}, \bar{y}, \tfrac{1}{2} + z; \tfrac{1}{2} + x, \tfrac{1}{2} - y, \bar{z}; \tfrac{1}{2} - x, \tfrac{1}{2} + y, \tfrac{1}{2} - z$$

This difference results merely from Wyckoff choosing an origin (Fig. 7.9, point w) displaced by $x = \tfrac{1}{4}$. Retaining Wyckoff's origin, three unit cells of the crystal structure are projected down the c axis in Figure 7.14. At the left, Zn, O(1), and O(2) atoms are plotted with different symbols marked with the fractional height z. The three nonintersecting diad axes are shown by lensoid and half-arrow symbols. Those parallel to b occur at $z = \tfrac{1}{4}$ and $\tfrac{3}{4}$, whereas those parallel to a are unlabeled because they occur at $z = 0$ and $\tfrac{1}{2}$.

By inspection, each Zn atom is found to have four oxygen atoms as first neighbors, and each oxygen atom has two Zn atoms as first neighbors. A line

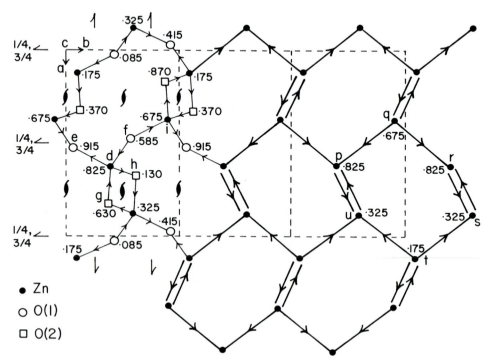

FIGURE 7.14 Crystal structure of Zn(OH)$_2$. In and around the unit cell at the left, lines connect Zn and O atoms with the arrow showing the upward direction. For the two unit cells at the right, lines connect Zn atoms that are bonded to the same oxygen atom.

is drawn between each pair of neighbors, and an arrow shows the direction of increasing height. Each Zn atom has two arrows pointing inward and two pointing outward, and the four oxygen atoms must lie at the corners of a tetrahedron. For the Zn atom d (0.625,0.400,0.825) and the O atoms e (0.525,0.070,0.915), f (0.475,0.570,0.585), g (0.825,0.375,0.630), and h (0.675,0.625,0.130), the calculated distances are: $de = 1.95_5$, $df = 1.95_0$, $dg = 1.96_1$, $dh = 1.94_5$, $ef = 3.07_8$, $eg = 3.31_5$, $eh = 3.31_0$, $fg = 3.15_8$, $fh = 3.19_1$, $gh = 3.05_8$ Å. Although the Zn–O(1) and Zn–O(2) distances are essentially equal within the experimental error, this is the result of chemical bonding, and has nothing to do with space-group symmetry. Although Zn–O distances are near equal, the O–O distances are unequal. Each O atom is bonded to a proton to give a hydroxyl ion OH⁻, but the hydrogen position is not discussed here.

The Zn and O atoms generate a framework of corner-shared tetrahedra. Thus the oxygen at corner f is shared between the two tetrahedra centered on Zn atoms at d and i. To simplify the interpretation, oxygen atoms are removed in the two unit cells at the right, and each oxygen-sharing pair of Zn atoms is linked by a heavy line. The Zn atoms p–u form a 6-ring. Double lines join p and u because there are superimposed linkages that form a zigzag along the c axis; thus an upward zig joins u at height 0.325 to p at 0.825, and an upward zag joins p at 0.825 to u at 1.325 and so on. There are also zigzag chains along b formed by p, q, r, and lattice repeats and by u, t, s, and lattice repeats. In projection down c, the 3D net of linked tetrahedral nodes forms a simple hexagonal net, but distorted with respect to the regular net in Figure 2.1c. Topologically, the net is identical to that of cristobalite (Fig. 5.22), but this is difficult to visualize because of geometrical distortion and difference in perspective.

Ice III (Wyckoff, 1963) is tetragonal, pseudocubic, with $a \cong c = 6.80$Å. The unit cell contains 12 H_2O molecules, and the oxygen atoms occur in a 4-fold special position and an 8-fold general position of **P4₁2₁2** (or its enantiomorph **P4₃2₁2**):

O(1) $u,u,0; \bar{u},\bar{u},\frac{1}{2}; \frac{1}{2} - u,\frac{1}{2} + u,\frac{1}{4}; \frac{1}{2} + u,\frac{1}{2} - u,\frac{3}{4}$ with $u = 0.392$

O(2) $x,y,z; \bar{x},\bar{y},\frac{1}{2} + z; \frac{1}{2} - y,\frac{1}{2} + x,\frac{1}{4} + z; \frac{1}{2} + y, \frac{1}{2} - x,\frac{3}{4} + z; y,x,\bar{z};$

   $\bar{y},\bar{x},\frac{1}{2} - z; \frac{1}{2} - x,\frac{1}{2} + y,\frac{1}{4} - z; \frac{1}{2} + x,\frac{1}{2} - y,\frac{3}{4} - z$ with $x = 0.095$,

   $y = 0.30, z = 0.29$

The structure is difficult to visualize (Fig. 7.15) even though the components are quite simple. The eight O(2) atoms form two spirals per cell, one for each **4₁** axis (e.g., n at height $z = 0.21$, p at 0.46, q at 0.71, i at 0.96, n at 1.21,

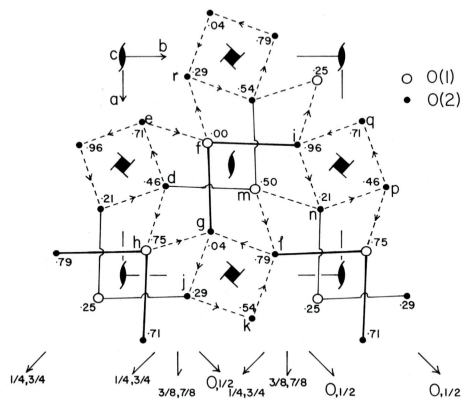

FIGURE 7.15 Crystal structure of ice III. Continuous lines join oxygen atoms at nearly the same height, and dashed lines connect aoms whose height differs by about $z = \frac{1}{4}$. Some representative axes in horizontal directions are shown at the bottom.

etc.). The O(2)–O(2) distance of 2.72 Å corresponds to chemical bonding. The four O(1) atoms also form a spiral around each 4_1 axis, but their paired relation to each vertical screw diad (e.g., f and m) is seen more easily. Each O(1) atom (e.g., f) is bonded to four O(2) atoms, with two (g and i) at 2.79 Å, and two (e and r) at 2.89 Å. When all the bonds are considered, each oxygen has four neighbors, and the bonds generate an infinite net. Although the net is obtained by simple cross linking of the tetragonal spirals, the linkage is complicated. The oxygens at e, d, h, g, and f generate a 5-ring when g is chosen at height 1.04 and f at 1.00. An 8-ring is formed by f (1.00), i (0.96), n (1.21), m (1.50), l (1.79), k (1.54), j (1.29), g (1.04), and back to f. A few of the horizontal symmetry elements are shown. In the space group symbol $\mathbf{P4_1 2_1 2}$, the $\mathbf{2_1}$ refers to the screw diads parallel to [100] and [010] and the $\mathbf{2}$ to the rotation diads parallel to [110] and [1$\bar{1}$0]. There are also 2_1 axes parallel to [110] and [1$\bar{1}$0]. Silicon atoms in the keatite variety of SiO_2 have the same topologic relationship as the oxygen atoms of ice III.

Quartz is another polymorph of silica. Low quartz inverts near 846 K (573°C) to high quartz. The latter occurs in two enantiomorphic varieties; left handed with space group **P6₄22** and right handed with **P6₂22**. High quartz has a framework of corner-shared SiO_4 tetrahedra that twists during inversion to low quartz, which has space group **P3₁22** (left handed) and **P3₂22** (right handed). The choice of left handed and right handed was based on crystal morphology and optical rotation before the crystal structure was known.

Wyckoff (1963, p. 313) gives the following data for high quartz:

$a = 5.01$ Å, $c = 5.47$ Å at ~600°C; coordinates for **P6₂22**

Si ½,½,⅓; ½,0,0; 0,½,⅔

O $u,\bar{u},\tfrac{5}{6}$; $\bar{u},u,\tfrac{5}{6}$; $u,2u,\tfrac{1}{2}$; $\bar{u},2\bar{u},\tfrac{1}{2}$; $2u,u,\tfrac{1}{6}$; $2\bar{u},\bar{u},\tfrac{1}{6}$; $u = 0.197$

In the c-axis projection (Fig. 7.16), the Si atoms form a 6-fold anticlockwise helix at heights ⅙, ⅖, ⅘, ⅚, ⅞, and ¹⁰⁄₆ about each **6₂** axis, and a 3-fold

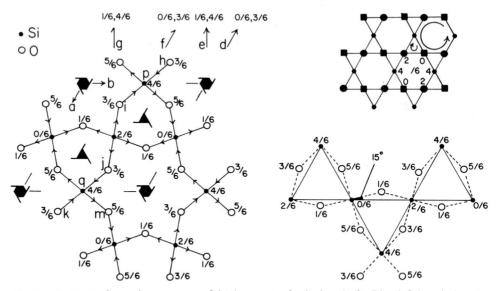

FIGURE 7.16 Crystal structure of high quartz (right-handed). The left-hand drawing shows both Si and O atoms together with lines joining first neighbors. Four representative axes in horizontal directions are shown at the top. The upper-right drawing shows how the Si atoms project onto a kagomé net, and the three symbols represent different heights. The lower-right diagram shows how idealized positions of oxygen atoms (circle) are obtained by drawing dashed lines at 15° to the continuous lines of the kagomé net of the silicon atoms (all depicted with a dot).

clockwise helix at heights $\%$, $\%$, and $\%$ about each 3_2 axis. The oxygen atoms also lie in helices but at heights $\frac{1}{6}$, $\frac{3}{6}$, and $\frac{5}{6}$. Each silicon atom lies at the center of a tetrahedron of four oxygen atoms, and a vertical rotation diad axis passes through each silicon atom. There are many horizontal axes, and the four basic sets are represented at the top. Axis *d* is parallel to *a* and is represented by the first **2** in the space group symbol. The second **2** represents axis *e*, which lies at 30° to axis *d*. Two sets of screw diads also occur; axis *f* relates oxygen atoms at *h*, *i*, *j*, and *k*, while axis *g* relates oxygen atoms at *i* and *m*, and silicon atoms at *p* and *q*. To understand the operation of axis *g*, remember that the translation for a screw diad axis is half the repeat distance in the direction of the axis. Axis *g* crosses two unit cells to obtain a repeat, and position *i* must be translated by $a \cdot \sin 60°$ to give position *m*, and similarly to obtain *q* from *p*.

As in many simple crystal structures, all the atoms in high quartz occupy special positions. Each Si atom lies at the intersection of three diad axes, giving point symmetry **222**, and each O atom has point symmetry **2** because it lies on a single horizontal diad axis. The structure can be related to a 2D tessellation or net (Fig. 7.16, upper right). Take a kagomé net or tessellation from Figure 2.1*h*, and place the heights $\%$, $\%$, $\%$, and so on around a 6-ring of nodes in the net. These nodes then develop into an anticlockwise hexagonal helix with symmetry 6_2. A clockwise triangular helix with symmetry 3_2 is developed automatically. The nodes are occupied by Si atoms. To obtain the oxygen atoms (Fig. 7.16, lower right), draw dashed lines at 15° to produce the circles. In the upper diagram, the three heights of the Si atoms were distinguished by dot, circle, and square, but in the lower diagram, all the Si atoms were marked with a dot. Four oxygens surround each dot, and by choosing the heights of $\frac{1}{6}$, $\frac{3}{6}$, and $\frac{5}{6}$, a tetrahedral configuration is obtained. Each oxygen is linked to two silicon atoms, and high quartz can be regarded as an infinite molecule that consists of a 4-connected net of corner-sharing SiO_4 tetrahedra.

Low quartz, $a = 4.913$ Å, $c = 5.405$ Å at 298 K (25°C), has the following atomic coordinates in $\mathbf{P3_2 21}$:

Si $\bar{u}, \bar{u}, \frac{1}{3}; u, 0, 0; 0, u, \frac{2}{3}$ with $u = 0.470$

O $x, y, z; y - x, \bar{x}, \frac{1}{3} + z; \bar{y}, x - y, \frac{2}{3} + z; x - y, \bar{y}, \bar{z}; y, x, \frac{2}{3} - z; \bar{x}, y - x, \frac{1}{3} - z$

with $x = 0.415$, $y = 0.266$ and $z = 0.119$

The crystal structure (Fig. 7.17, left) has each Si atom almost, but not quite, in the same position as in high quartz. The oxygen atoms have rotated horizontally (curved arrow) and moved vertically so that the 6_2 axis has become degraded to a 3_2 axis. Whereas the six atoms related by the 6_2 axis projected onto the corners of a hexagon (Fig. 7.16, left), they project in low quartz onto the corners of two equilateral triangles, *def* and *ghi.* Horizontal rotation diad axes relate the two groups of positions. Because an oxygen atom can twist either clockwise or anticlockwise (so long as it moves symmetrically with the other oxygen atoms), there are two possible positions for the atoms in low quartz after undergoing inversion from high quartz. These two positions (Fig. 7.17, left and right) are related by the 2-fold operation lost when the 6_2 axis was degraded to a 3_2 axis. During the structural inversion from high to low quartz, nuclei of low quartz can grow at random in the high quartz. Two nuclei with clockwise rotation will merge, but two nuclei with opposite sense of rotation cannot merge and will meet at an out-of-phase boundary. The final product of inversion from high to low quartz is a twin intergrowth, in which twin domains of low quartz (clockwise) and low quartz (anticlockwise) are intimately arranged. Both components of the inversion twin of low quartz are right handed if the original high quartz was right handed $P6_2 22$, and left handed if the high quartz was left handed. The name Dauphiné is given to the inversion twin of quartz, which is distinguished from the Brazil twin that relates left-handed to right-handed domains of quartz, irrespective of whether it is in the high or low structural state. At a twin boundary there must be a misfit, and the boundary region of a Dauphiné twin must approximate statistically to the

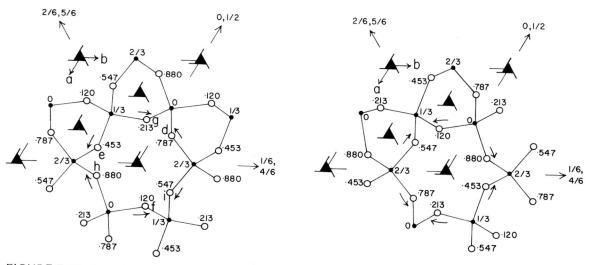

FIGURE 7.17 Crystal structure of low quartz (right-handed). Reduction of symmetry of the 6_2 vertical axis of high quartz to the 3_2 axis of low quartz allows the oxygen atoms to rotate horizontally (curved arrow) and to move vertically from the positions in high quartz (Figure 7.16). The two possible directions of rotation are shown in the left-hand and right-hand drawings. The resulting structures are related by the 2-fold axis lost when the 6_2 became degraded to a 3_2 axis. Representative diad axes are shown in the horizontal direction.

symmetry of high quartz as the continuous framework of linked SiO_4 tetrahedra flexes to accommodate the strain.

7.9   Simple Space Groups with a Primitive Lattice and One or More Mirror Planes

The space groups in Table 7.4A provide a nice introduction to the operation of mirror symmetry in space groups. In the monoclinic system, space group **P11m** has a mirror plane perpendicular to the c axis. The unit cell must have $\alpha = \beta = 90°$, γ not equal to 90° except by chance, and $a \neq b \neq c$. A general position at xyz is reflected to $xy\bar{z}$ by a mirror plane at $z = 0$ and $\frac{1}{2}$. A split circle is needed to accommodate the two superimposed positions, and a comma is useful but not really necessary to show the inversion operation (Fig. 7.18). The mirror plane is shown by the angle sign. Placement of the mirror plane perpendicular to the b axis is specified by **P1m1**, and the mirror plane is shown by a heavy line. Intersection of a rotation diad with a perpendicular mirror plane gives a center of symmetry in **P112/m**, and replacement of the rotation diad with a screw diad in **P112₁/m** causes relative displacement of the center of symmetry and the mirror plane by $z = \frac{1}{4}$.

The orthorhombic space groups **Pmm2** and **P2/m2/m2m** have 4-fold and 8-fold general positions, respectively. In **Pmm2**, the rotation diad appears automatically.

Table 7.4   Systematic list of space groups with one or more mirror planes

A   Space Groups with No Glide Planes

Monoclinic	**P11m** (= P1m1), **P112/m** (= P12/m1), **P112₁/m** (= P12₁/m1)
Orthorhombic	**Pmm2** (= P2mm = Pm2m), **P2/m2/m2/m**
Tetragonal	**P4/m, P4₂/m, P4̄m2**
Hexagonal	**P6/m** (note P6̄ ≡ P3/m), **P6₃/m**
Isometric	**P2/m3̄**

B   Space Groups with Automatic Glide Planes

Monoclinic	**I11m** (= I11n), **I112/m** (≡ I112₁/m = I112/n = I112₁/n)
Orthorhombic	**Cmm2** (≡ Cba2), **Amm2** (≡ Anc2₁), **Imm2** (≡ Inn2₁), **Fmm2** (≡ Fbc2₁ ≡ Fca2₁ ≡ Fnn2), **Cmmm** (≡ Cban), **Immm** (≡ Innn), **Fmmm** (≡ Fbca ≡ Fcab ≡ Fnnn)
Tetragonal	**I4/m** (≡ I4/n), **P4mm, I4mm** (≡ I4nc), **P4̄2m, P4̄2₁m, I4̄m2** (≡ I4̄n2₁), **I4̄2m** (≡ I4̄2₁c), **P4/mmm, I4/mmm** (≡ I4/nnc)
Hexagonal	**P3m1, P31m** (≡ P31a), **R3m** (≡ R3₁m ≡ R3₂m), **P3̄1m, P3̄m1,** **R3̄m, P6mm, P6/mmm**
Isometric	**I2/m3̄** (≡ I2/n3̄), **F2/m3̄** (≡ F2/a3̄), **P4̄3m, I4̄3m, F4̄3m, P4/m3̄2/m,** **I4/m3̄2/m, F4/m3̄2/m**

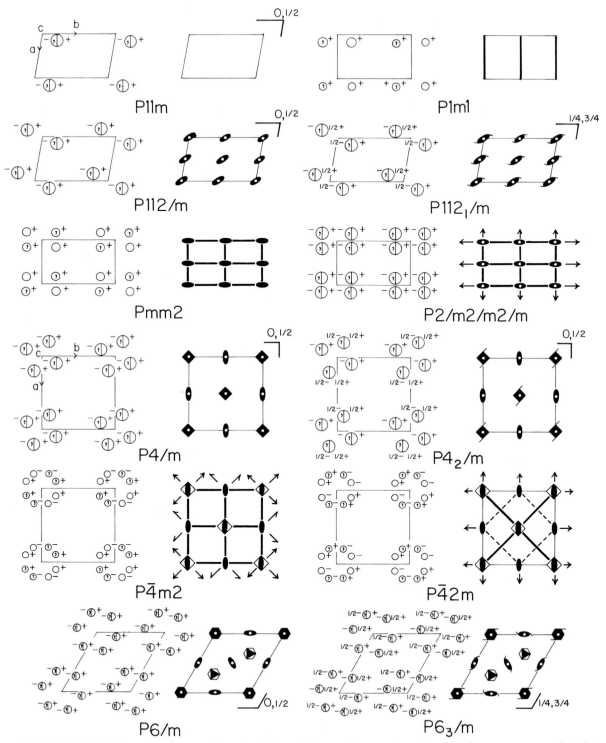

FIGURE 7.18 Diagrams of general equivalent positions and symmetry elements for space groups listed in Table 7.4A. Space group **P112₁/m** is drawn with the mirror plane at $z = \frac{1}{4}$, $\frac{3}{4}$ so that the center of symmetry is at height zero. Space group **P̄42m** from Table 7.4B is also shown; dashed lines represent glide planes described in section 7.10.

The tetragonal space groups **P4/m** and **P4$_2$/m** have a center of symmetry at each intersection of a tetrad axis with a mirror plane. Space group **P$\bar{4}$m2** has a mirror plane perpendicular to the a and b axes and a rotation axis at the two 45° positions. Horizontal screw diads are also produced. Space group **P$\bar{4}$2m** is not a 45° rotation of **P$\bar{4}$m2**. The dashed lines represent a glide plane (next section).

The hexagonal space groups **P6/m** and **P6$_3$/m** contain an inverse hexad axis. Because the origin of a space group is usually placed on a center of symmetry if one is available, the mirror plane in **P6$_3$/m** lies at $z = \frac{1}{4}, \frac{3}{4}$.

The isometric space group **P2/m$\bar{3}$** can be handled quite easily by algebraic manipulation:

xyz, zxy, yzx Cyclic rotation for a rotation triad symmetrical to $+a$, $+b$, and $+c$ axes

$\left. \begin{array}{l} \bar{x}yz, \bar{z}xy, \bar{y}zx \\ x\bar{y}z, z\bar{x}y, y\bar{z}x \\ \bar{x}\bar{y}z, \bar{z}\bar{x}y, \bar{y}\bar{z}x \end{array} \right\}$
Mirror plane perpendicular to a axis

Mirror plane perpendicular to b axis, operating on previous six positions

$\left. \begin{array}{l} xy\bar{z}, zx\bar{y}, yz\bar{x} \\ \bar{x}y\bar{z}, \bar{z}x\bar{y}, \bar{y}z\bar{x} \\ x\bar{y}\bar{z}, z\bar{x}\bar{y}, y\bar{z}\bar{x} \\ \bar{x}\bar{y}\bar{z}, \bar{z}\bar{x}\bar{y}, \bar{y}\bar{z}\bar{x} \end{array} \right\}$
Mirror plane perpendicular to c axis, operating on previous twelve positions

The diad axes are generated automatically: for example, a diad axis parallel to the a axis relates two positions such as xyz and $x\bar{y}\bar{z}$. Similarly, a center of symmetry relates two positions such as xyz and $\bar{x}\bar{y}\bar{z}$. Because a rotation triad and a center of symmetry combine together to give the same positions as an inverse triad, the space group of general positions consists of the 24 positions generated merely from the rotation triad and three mirror planes.

Crystallographers often use shortened symbols, and, indeed, **Pm$\bar{3}$** or even **Pm3** is sufficient for generation of the positions for **P2/m3**. Similarly, **P2/m2/m2/m** can be shortened to **Pmmm**.

7.10   Space Groups with Automatic Glide Planes: Systematic List of Glide Planes

Just as screw axes were generated automatically by the combined operations of rotation axes and lattice centering, so do glide planes develop from the combined operations of mirror planes and lattice centering.

Consider space group **Cmm2** (Fig. 7.19). The starting position d is operated on as follows:

xyz by **m11** (i.e., mirror plane perpendicular to *a* at *x* = 0) → *x̄yz* (position *e*)

by **1m1** (i.e., mirror plane perpendicular to *b* at *y* = 0) → *xȳz* (position *f*)

by **m11** and **1m1** → *x̄ȳz* (position *g*) ≡ operation of **112** at *x* = *y* = 0

by **C** → ½ + *x*, ½ + *y*,*z* (position *h*)

by **C** and **m11** → ½ − *x*, ½ + *y*,*z* (position *j*) ≡ operation of **b11** (i.e., *b* glide plane at *x* = ¼ that first reflects and then translates *b*/2)

by **C** and **1m1** → ½ + *x*, ½ − *y*,*z* (position *k*) ≡ operation of **1a1** (i.e., *a* glide plane at *y* = ¼ that first reflects and then translates *a*/2)

by **C**, **m11**, and **1m1** → ½ − *x*, ½ − *y*,*z* (position *l*) ≡ operation of a second **112** at *x* = *y* = ¼

The three operations **C**, **m11**, and **1m1** are sufficient to generate the eight general equivalent positions, and the rotation diad at *x* = *y* = 0 (and equiv-

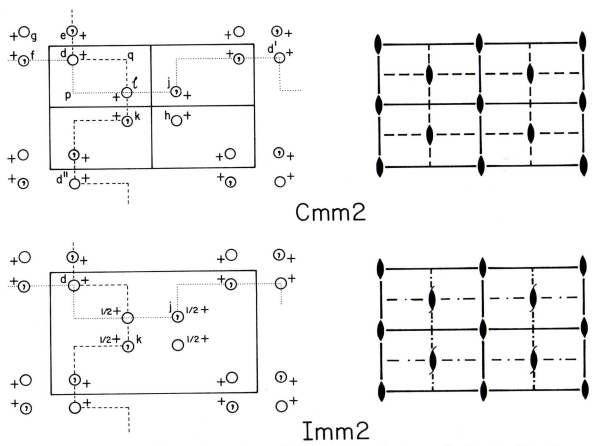

FIGURE 7.19 Diagrams of general equivalent positions and symmetry elements for **Cmm2** and **Imm2**.

alent positions at $x = y = \frac{1}{2}$; $x = 0$, $y = \frac{1}{2}$; $x = \frac{1}{2}$, $y = 0$) is generated automatically from the combination of the **m11** and **1m1** operations. The combination of C and **m11** can be expressed more neatly by the operation **b11**, which stands for a **b** glide plane perpendicular to the *a* axis. This glide plane lies at $x = \frac{1}{4}$, and position *d* is first reflected to imaginary position *p* at $(\frac{1}{2}-x,y,z)$ and then translated by $y = \frac{1}{2}$ to final position *j* at $(\frac{1}{2}-x, \frac{1}{2}+y,z)$. Continued operation gives *d'*. Similarly, the glide plane at $y = \frac{1}{4}$ produces imaginary position *q* at $(x,\frac{1}{2}-y,z)$ by reflection of *d* and then translates *q* to *k*. A further double operation takes *k* to *d"* thereby closing the operation with lattice repeat *a*.

A similar treatment for **Imm2** leads to the concept of a ***diagonal glide plane***. Thus position *j* is related to position *d* at (xyz) by an initial reflection in the plane $x = \frac{1}{4}$ to imaginary position $(\frac{1}{2} - x,y,z)$, and a subsequent translation of $(\mathbf{b} + \mathbf{c})/2$ to the position $(\frac{1}{2} - x,\frac{1}{2} + y,\frac{1}{2} + z)$. Position *k* is similarly related to position *d* by initial reflection in the plane $y = \frac{1}{4}$, and subsequent translation of $(\mathbf{a} + \mathbf{c})/2$.

Glide planes do not need to be generated by the combined operations of other symmetry elements, and can stand alone in some space groups. Figure 7.20 shows all possible glide planes in space groups. The general positions and symmetry elements are placed in the same drawing to make it easier to see the operations. A general position *xyz* is inserted in a projection of a unit cell with the usual orientation of *b* to the right and *a* toward the reader. A glide plane **b11** must lie perpendicular to the *a* axis because the *b* is in the first position. The translation operation is $b/2$. Position *p* is reflected in vertical glide plane *qq* and translated $b/2$ to position *r*. Further reflection and translation to *p'* closes the operations. Another glide plane *ss* is automatically produced by the operations depicted by the row of circles that link *p"*, *r*, and *p'''*. The **c11** glide plane is also perpendicular to *a*, but the translation operation is $c/2$. A series of dashed lines is used to show a vertical glide plane with a horizontal translation, and a row of dots is used when a vertical glide plane has a vertical translation. Glide planes **1a1** and **1c1** (not shown) would be perpendicular to the *b* axis, and glide planes **11a** and **11b** are perpendicular to the *c* axis with an arrow on a corner sign to show the direction of translation. The operations of a **11a** glide plane at $z = 0$ and $\frac{1}{2}$ and at $z = \frac{1}{4}$ and $\frac{3}{4}$ are depicted, as well as the operations of a **11b** glide plane at $z = 0$ and $\frac{1}{2}$. Glide must occur parallel to the plane of a glide plane and the symbols **a11**, **1b1**, and **11c** are not allowed.

Closure of reflection-glide operations can also occur for diagonal glide: thus, **n11** stands for a diagonal glide plane perpendicular to the *a* axis with a translation of $(\mathbf{b} + \mathbf{c})/2$. A vertical glide plane is symbolized by alternating dots and dashes. When the plane is horizontal, as for **11n** at heights $0,\frac{1}{2}$ or $\frac{1}{4},\frac{3}{4}$, the glide operation is denoted by the diagonal arrow. If the glide

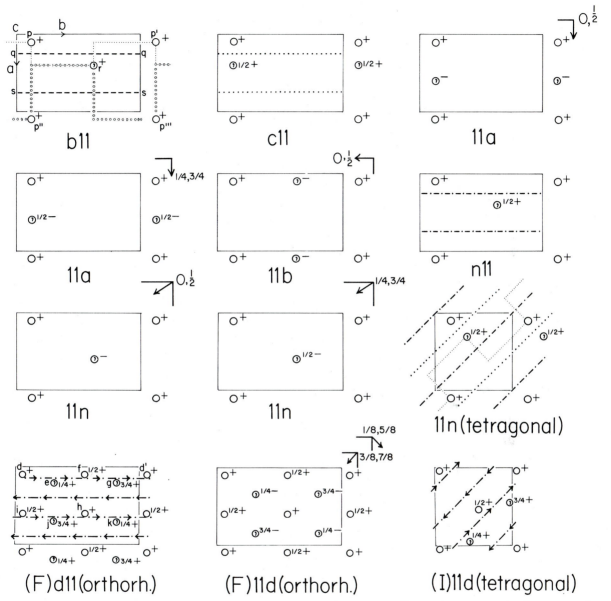

FIGURE 7.20 Operation of glide planes. The projection of the unit cell must be rectangular for a vertical glide plane, but need not be for a horizontal glide plane unless other symmetry elements (e.g., in an orthorhombic or tetragonal group) enforce orthogonal geometry.

plane lies in (110) or (1$\bar{1}$0), as in tetragonal and isometric space groups, the translation is $(a + b + c)/2$ because the repeat distance in the horizontal direction is $(a + b)$ instead of b for a glide plane perpendicular to the a axis. The diagonal glide planes in 11n (tetragonal) alternate with 11c glide planes.

Finally, in a few space groups with an I or F lattice, a *diamond glide plane* occurs in which the operation is exactly the same as for the ordinary *diagonal glide plane* except that the translation for a diamond glide plane perpendicular to the a axis is $(b + c)/4$ and for one at 45° to a and b axes of tetragonal and isometric space groups is $(a + b + c)/4$. Whereas two reflection-translation operations are needed for closure of a diagonal glide plane, four operations are needed for the diamond glide plane. A vertical diagonal glide plane is shown by a row of alternate dots and dashes, to which an arrow is added for the diamond glide plane to show increasing height.

In the diagram for d11, position d at xyz is reflected and translated in the diamond glide plane at $x = \frac{1}{8}$ to give successively e ($\frac{1}{4} - x, \frac{1}{4} + y, \frac{1}{4} + z$), f ($x, \frac{1}{2} + y, \frac{1}{2} + z$), g ($\frac{1}{4} - x, \frac{3}{4} + y, \frac{3}{4} + z$), and d'. The glide plane at $x = \frac{5}{8}$ is produced automatically. Position f is related to d by A centering. Addition of positions i and h adds B and C centering to give an F lattice. The positions i, e, h, and g then generate a diamond glide plane at $x = \frac{3}{8}$ whose arrows are reversed with respect to those for the initial glide plane. Positions j and k are needed to relate i and h by the initial glide plane. Diamond glide planes always occur in quadruplets like this in space groups. Horizontal diamond glide planes are shown in the diagram for 11d (orthorhombic), and again the F-centering operation is used. A diamond glide plane can occur in five body-centered tetragonal groups (I4$_1$md, I4$_1$cd, I$\bar{4}$2d, I4$_1$/amd, I4$_1$/acd) and the isometric groups I$\bar{4}$3d, Fd3m, Fd3c, and Ia3d. In Figure 7.20, the bottom-right drawing shows only one set of diamond glide planes running SW–NE, but the tetrad axis enforces the existence of diamond glide planes in the SE–NW direction for the tetragonal groups, and the tetrad and triad axes do the same for the dodecahedral directions of the isometric groups.

Table 7.4B lists all the space groups with one or more mirror planes whose interaction with a centered lattice produces one or more glide planes automatically.

In space group I112/m (Fig. 7.21), the combination of the mirror plane 11m and the body-centering vector produces a diagonal glide plane 11n. A set of screw diads 112$_1$ and two sets of centers of symmetry are also produced.

Space group Fmm2 must have automatic glide planes, and indeed perpendicular to the a axis there is an n glide plane at $x = 0$ and $x = \frac{1}{2}$, which is the same location as the mirror plane. Also there are superimposed b and c glide planes at $x = \frac{1}{4}$ and $\frac{3}{4}$. In the middle diagram it is necessary to place the symbols next to each other even though the planes superimpose exactly. Similarly, there are **m** and **n** planes at $y = 0$ and $\frac{1}{2}$, and **a** and **b** planes at $y = \frac{1}{4}$ and $\frac{3}{4}$. Two sets each of rotation and screw diad axes are produced automatically along the c axis (right-hand diagram).

Space group **Cmmm** has an automatic **b** glide plane perpendicular to the a axis, an automatic **a** glide plane perpendicular to b, and an **n** glide plane perpendicular to c. Space group **Immm** is illustrated by the structure of Rb_2O_2 (next section). Space group **Fmmm** must include all the symmetry elements of **Fmm2** as a subgroup, and in addition has automatic **a**, **b**, and **n** glide planes perpendicular to the c axis. It is not illustrated, but the diagram of general positions can be obtained from that of **Fmm2** by applying the operation of **11m**. Replace each circle at height + by two half-circles with + on one side and − and comma on the other; a height of $\frac{1}{2}+$ is changed to $\frac{1}{2}-$ by the mirror operation.

Space group **I4/m** develops an automatic diagonal glide plane perpendicular to c, and extra axes appear parallel to c. Space group **I4mm** can be generated from **P4mm** by adding eight general positions from the body-centering operation. New symmetry elements result therefrom, and the 2 axes are upgraded to 4_2 axes. The general positions of **I4mm** could be generated from the symbol **I4nc** because of the automatic presence of these two sets of glide planes in **I4mm**. Both **I4/m** and **I4mm** must be subgroups of **I4/mmm**, and, indeed, the reader can find all the glide planes of **I4/m** and **I4mm** in the symmetry diagram for **I4/mmm** (Fig. 7.22).

At first sight, the diagrams for **I4/mmm** seem complex, but the general positions can be developed quite simply as follows:

xyz; $\bar{y}xz$; $\bar{x}\bar{y}z$; $y\bar{x}z$	Anticlockwise rotation of tetrad axis
$\bar{x}yz$; yxz; $x\bar{y}z$; $\bar{y}\bar{x}z$	**m11** operation at $x = 0$ on preceding 4 positions
$xy\bar{z}$; $\bar{y}x\bar{z}$; $\bar{x}\bar{y}\bar{z}$; $y\bar{x}\bar{z}$	**11m** operation at $z = 0$ on preceding 8 positions
$\bar{x}y\bar{z}$; $yx\bar{z}$; $x\bar{y}\bar{z}$; $\bar{y}\bar{x}\bar{z}$	
$\frac{1}{2} + x, \frac{1}{2} + y, \frac{1}{2} + z$; . . .	Operation of **I** body-centering vector on preceding 16 positions

From these coordinates for the 32-fold general position, all the other symmetry elements can be developed, as shown by the following examples:

$$(\textbf{m11}) \times (\textbf{11m}) \rightarrow (\textbf{1m1}), \text{ or } xyz \rightarrow x\bar{y}z$$

$$\textbf{I} \times (\textbf{m11}) \rightarrow (\textbf{n11}), \text{ or } xyz \rightarrow \tfrac{1}{2} - x, \tfrac{1}{2} + y, \tfrac{1}{2} + z$$

$$\textbf{I} \times (180° \text{ rotation in tetrad axis}) \rightarrow 2_1 \text{ axis parallel to } c \text{ axis,}$$
$$\text{or } xyz \rightarrow \tfrac{1}{2} - x, \tfrac{1}{2} - y, \tfrac{1}{2} + z$$

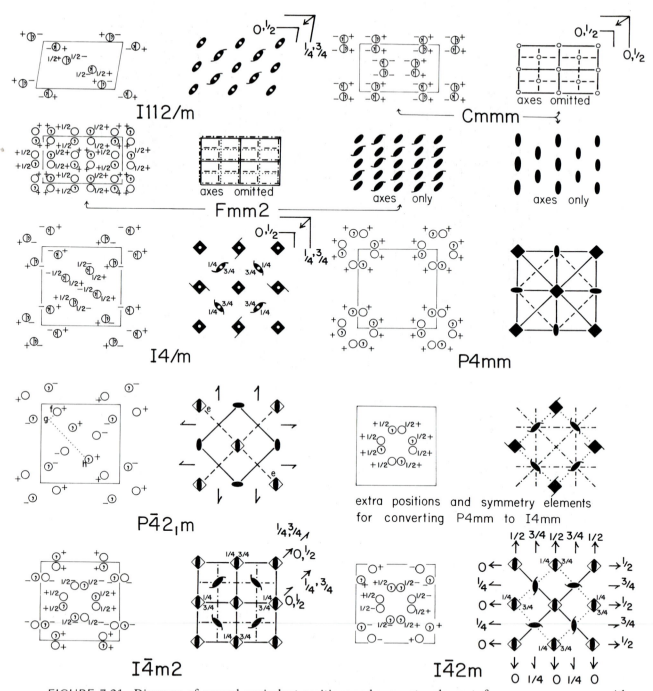

FIGURE 7.21 Diagrams of general equivalent positions and symmetry elements for some space groups with automatic glide planes. In **I112/m**, the centers of symmetry (small open circles enclosed by lensoid symbols for diad axes) are at height $z = 0$ and $\frac{1}{2}$. For **Fmm2**, the symmetry elements are separated into two diagrams to reduce confusion. Nevertheless the middle diagram is confused by overlap of pairs of planes. In **Cmmm**, the horizontal mirror and diagonal glide planes lie at the same heights. In **I4/m**, note the alternation of **4** and 4_2 axes. There are two sets of centers of symmetry, one set lying on tetrad axes at $z = 0$ and $\frac{1}{2}$ and the other lying on screw diad axes at $z = \frac{1}{4}$ and $\frac{3}{4}$. Space group **P4mm** has mirror planes both perpendicular to

Obviously it is rather tedious developing the complete list of relations between the symmetry elements, but here is a complete list of the types of symmetry elements:

Related to the *c* axis—**4, 4$_2$, 2$_1$, m, n**

Related to the *a* (or *b*) axis—**2, 2$_1$, m, n**

Related to a horizontal axis at 45° to *a* and *b*—**2, 2$_1$, m, c**

Centers of symmetry—intersection of **4/mmm** at 000, ½½½, and at 00½, ½½0

—intersection of **mmm** at 0½0 and related positions

—intersection of **2/m** at ¼¼¼ and related positions

A formal mathematical treatment of the above can be developed elegantly with matrices, but the above visual treatment is sufficient here.

Other tetragonal space groups can be developed from **I4/mmm** by systematic removal of symmetry elements. Thus **I4/mmm** goes to **I4/m** by removal of the mirror planes perpendicular to *a* and *b*, and automatic removal of the mirror planes in the 45° positions. Space group **P4/mmm** is produced by simple removal of the set of 16 positions at the center of the unit, and consequent removal of the **n** and **c** glide planes as well as some axes and centers of symmetry. Not so obvious is the conversion of **I4/mmm** to **I$\bar{4}$m2**; the **$\bar{4}$** axis is not a subelement of **4** but is a subelement of **4/m**. Let it suffice that all the details of all the tetragonal space groups listed in Table 7.4 can be interpreted in terms of appropriate selection of features from **I4/mmm**. Space group **P$\bar{4}$2m** is depicted in Figure 7.18 and **P$\bar{4}$2$_1$m, I$\bar{4}$m2**, and **I$\bar{4}$2m** in Figure 7.21.

Similar hierarchies of space groups occur for the isometric system, and, indeed, **I4/mmm** can be developed by removal of two tetrad axes, four triad axes, and various symmetry planes and diad axes from **I4/m$\bar{3}$2/m**. In Table 7.4, **F4/m$\bar{3}$2/m** can be degraded to **F$\bar{4}$3m** and **F2/m$\bar{3}$**, while **I4/m$\bar{3}$2/m** can be degraded to **I$\bar{4}$3m** and **I2/m$\bar{3}$**. Alternatively, the F and I lattices can be degraded to a primitive lattice giving **P4/m$\bar{3}$2/m, P$\bar{4}$3m**, and **P2/m$\bar{3}$**. General positions for all these space groups can be generated easily by algebraic manipulation, but enumeration of all the symmetry elements is tedious. Section 8.1 will show that simple crystal structures such as the NaCl struc-

the *a*- and *b*-axes (first **m**) and at 45° to them (second **m**). Space group **I4mm** can be constructed by adding the positions and symmetry elements shown in the diagrams below those for **P4mm**. For space group **P$\bar{4}$2$_1$m**, a glide-reflection operation of the automatic glide plane (*ee*) at 45° to the *c*- and *b*-axes is shown by dotted line *fgh*. For **I$\bar{4}$m2**, only a few representative axes are shown in the horizontal directions. There are two kinds of inverse tetrads, one with inversion points at *z* = 0 and ½ and the other with inversion points at *z* = ¼ and ¾. For **I$\bar{4}$2m**, general positions are shown only inside one unit cell. Again, there are two kinds of inverse tetrad axes.

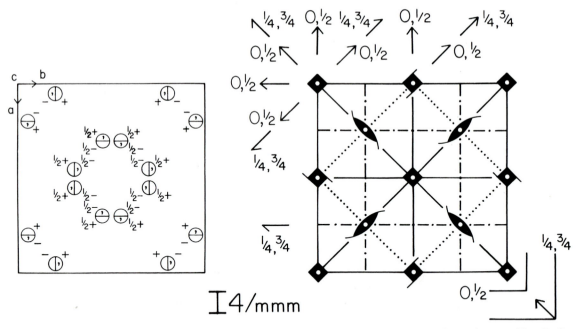

I4/mmm

FIGURE 7.22 Diagrams of general equivalent positions and symmetry elements for **I4/mmm**. Merely for elegance, the half-circles in the left-hand drawing have been arranged to obey the symmetry, but the halving can be carried out at any orientation, since positions related by the horizontal mirror plane superimpose in projection. Only a few representative axes are shown in the horizontal direction.

ture can be handled quite easily without going into all the gory details. Space group **F4/m$\bar{3}$2m** has a 192-fold general position, but the coordinates can be developed rapidly as follows :

xyz, zxy, yzx	Cyclic rotation for triad axis gives 3 positions
$\bar{y}xz, \bar{x}zy, \bar{z}yx$ $\bar{x}\bar{y}z, \bar{z}\bar{x}y, \bar{y}\bar{z}x$ $y\bar{x}z, x\bar{z}y, z\bar{y}x$	90° anticlockwise rotation for **114** axis 180° anticlockwise rotation for **114** axis 270° anticlockwise rotation for **114** axis
$\bar{x}yz, \bar{z}xy, \bar{y}zx$ $\dots . \bar{z}\bar{y}x$	Operation of **m11** plane adds 12 more positions by reversing the sign of the first symbol
$\bar{x}y\bar{z}, \bar{z}x\bar{y}, \bar{y}z\bar{x}$ $\dots . \bar{z}\bar{y}\bar{x}$	Operation of **11m** plane adds 24 more positions by reversing the sign of the third symbol
($\frac{1}{2} + x, \frac{1}{2} + y, z$), ($x, \frac{1}{2} + y, \frac{1}{2} + z$), ($\frac{1}{2} + x, y, \frac{1}{2} + z$), . . ., and so on	Addition of ($\frac{1}{2}, \frac{1}{2}, 0$), ($0, \frac{1}{2}, \frac{1}{2}$), and ($\frac{1}{2}, 0, \frac{1}{2}$) for the **F** centering gives 144 more positions

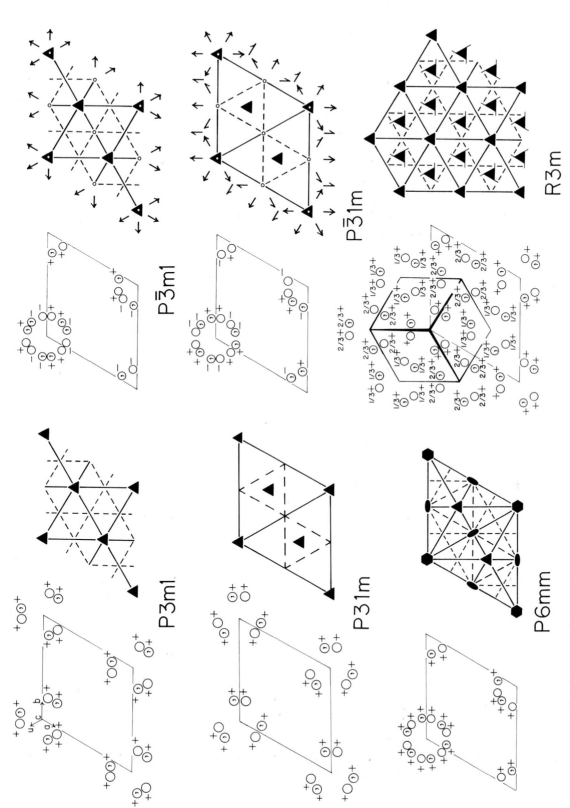

FIGURE 7.23 The diagrams of general equivalent positions and symmetry elements for some hexagonal space groups with automatic glide planes.

In the hexagonal space groups (Fig. 7.23), **P3m1** and **P31m** differ in the orientation of the vertical mirror plane and the automatic glide plane. Space group **P3̄m1** can be constructed merely by addition of a center of symmetry to **P3m1** in order to convert 3 to 3̄, and similarly for **P3̄1m** and **P31m**. Space group **P6mm** has automatic glide planes, and space group **P6/mmm** (not shown) can be obtained from **P6mm** by adding a mirror operation perpendicular to the hexad. Rotation triads are converted to inverse triads because of the automatic development of centers of symmetry. Space group **R3m** has 3_1 and 3_2 axes alternating in the triangles formed by the rotation triad axes, and automatic glide planes. Addition of a center of symmetry converts **R3m** to **R3̄m** (not shown) with automatic development of horizontal diad axes, both of rotation and screw type. **R3̄** (Fig. 7.13) is also a subgroup of **R3̄m**.

7.11   Crystal Structures of Rb_2O_2, $NaNO_2$, As, and Se

Rubidium peroxide, Rb_2O_2, is orthorhombic with $a = 5.983$ Å, $b = 7.075$ Å, $c = 4.201$ Å (Wyckoff, 1963, p. 169). In space group **Immm**, the atoms occur in special positions:

Rb $\pm(0,u,0; \frac{1}{2},u,\frac{1}{2})$ with $u = \frac{1}{4}$ [note the plus and minus]

O $\pm(v,0,0; v + \frac{1}{2},\frac{1}{2},\frac{1}{2})$ with $v = 0.375$

The c-axis projection (Fig. 7.24) shows that the O atoms occur in pairs separated by 1.49_6 Å. Each Rb atom has six oxygen neighbors, two each at 2.85 Å from two O_2 doublets, and one each at 3.07 Å from an end of two

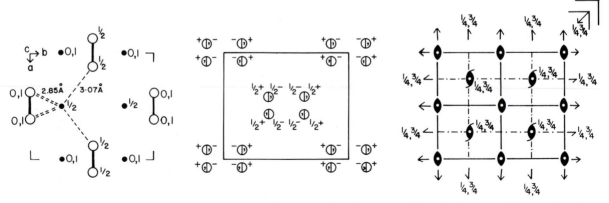

FIGURE 7.24 Projection of crystal structure of Rb_2O_2 (Rb, dot; O, circle), and diagrams of general equivalent positions and symmetry elements for **Immm**.

doublets. Because of the strong bonding in the peroxide ion O_2^-, the O–O distance is much less than twice the ionic radius of the O^{2-} ion (2.7 Å: Table 1.1). This results in a very distorted coordination polyhedron of oxygen atoms around each Rb atom, in spite of the nearly equal Rb–O distances.

Like many simple structures, the Rb$_2$O$_2$ structure has atoms in special positions. Here, there are two accidental features that could confuse an unskilled crystallographer. The following is a list of general and special positions in **Immm**. Add $\frac{1}{2}, \frac{1}{2}, \frac{1}{2}$ to all positions to obtain the positions developed by the body-centering translation.

Multiplicity	Wyckoff Notation	Point Symmetry	Coordinates of Equivalent Positions
16	o	1	$xyz, \bar{x}yz, x\bar{y}z, \bar{x}\bar{y}z, xy\bar{z}, \bar{x}y\bar{z}, x\bar{y}\bar{z}, \bar{x}\bar{y}\bar{z}$
8	n	11m	$xy0, \bar{x}y0, x\bar{y}0, \bar{x}\bar{y}0$
8	m	1m1	$x0z, \bar{x}0z, x0\bar{z}, \bar{x}0\bar{z}$
8	l	m11	$0yz, 0\bar{y}z, 0y\bar{z}, 0\bar{y}\bar{z}$
8	k	$\bar{1}$	$\frac{1}{4}\frac{1}{4}\frac{1}{4}, \frac{3}{4}\frac{3}{4}\frac{1}{4}, \frac{1}{4}\frac{3}{4}\frac{1}{4}, \frac{3}{4}\frac{1}{4}\frac{1}{4}$
4	j and i	mm2	$\frac{1}{2}0z, \frac{1}{2}0\bar{z}$ and $00z, 00\bar{z}$
4	h and g	m2m	$0y\frac{1}{2}, 0\bar{y}\frac{1}{2}$ and $0y0, 0\bar{y}0$
4	f and e	2mm	$x\frac{1}{2}0, \bar{x}\frac{1}{2}0$ and $x00, \bar{x}00$
2	$d; c; b;$ and a	mmm	$\frac{1}{2}0\frac{1}{2}; \frac{1}{2}\frac{1}{2}0; 0\frac{1}{2}\frac{1}{2};$ and 000

The Rb atoms occur in Wyckoff position g, and there is no symmetry restriction on the value of u. From the packing viewpoint u need not be exactly $\frac{1}{4}$ because a value of (say) 0.249 would not change the topology of the chemical bonding. The O atoms occur in position e and the value of v is not constrained at $\frac{3}{8}$ by the symmetry. In spite of these fractional coordinates for the O and Rb atoms, it is not possible to find a symmetry group higher than **Immm**. Note that reduction of the v coordinate and increase of the u coordinate would allow approach of the two types of Rb–O distances: perhaps the structure should be examined again to check whether u and v really do lie exactly at fractional values.

Sodium nitrite, NaNO$_2$, has an orthorhombic unit cell with a = 5.384 Å, b = 5.563 Å, c = 3.569 Å (Wyckoff, 1964, p. 302). The atoms occur in space group **Im2m** at

$$\text{Na}\quad 0,0.5856,0; \tfrac{1}{2},0.0856,\tfrac{1}{2} \qquad \text{N}\quad 0,0.1202,0; \tfrac{1}{2},0.6202,\tfrac{1}{2}$$

$$\text{O}\quad 0.1944,0^{\#},0; 0.8056,0^{\dagger},0; 0.6944,\tfrac{1}{2}^{\dagger},\tfrac{1}{2}; 0.3056,\tfrac{1}{2}^{\dagger},\tfrac{1}{2}$$

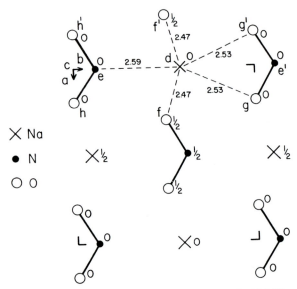

DIAGRAM 7.25 Crystal structure of $NaNO_2$, which illustrates space group **Im2m**. Mirror planes occur at $x = 0$ and $\frac{1}{2}$, and $z = 0$ and $\frac{1}{2}$. Diagonal glide planes occur at $x = \frac{1}{4}$ and $\frac{3}{4}$, and at $z = \frac{1}{4}$ and $\frac{3}{4}$. Rotation diad axes occur at $x = z = 0$, and equivalent positions, and screw diad axes occur at $x = z = \frac{1}{4}$, and equivalent positions.

(See later for $^{\#}$ and †.) From the c-axis projection, Figure 7.25, NO_2 molecules with $N-O = 1.24_2$ Å and $O-N-O = 114.9°$ can be recognized. Each Na atom lies at 2.47_4 Å to four oxygens (labeled f and f') at $\pm z = \frac{1}{2}$, and at the following distances to atoms at $z = 0$: N at e, 2.58_9; O at g and g', 2.53_2; N at e', 2.97; O at h and h', 3.42 Å. Taking only distances below 2.6 Å, each Na atom would be coordinated to six oxygens and one nitrogen atom. The shorter distances between oxygen atoms of adjacent molecules are typified by $ff' = 3.29$ Å and $fg = f'g' = 3.36$ Å. Thus, the $NaNO_2$ structure can be interpreted in terms of (i) covalent bonding between N and O, which produces a lone pair of electrons and the elbow-shaped NO_2 molecules, (ii) the impossibility of packing Na atoms and NO_2 molecules to produce a simple coordination, and (iii) a compromise so that Na—O distances are near 2.5 Å and O—O distances between adjacent molecules are near 3.3 Å.

The following is the full list of general and special positions for **Im2m**. Add $\frac{1}{2}, \frac{1}{2}, \frac{1}{2}$ to obtain the positions related by the body-centering vector.

Multiplicity	Wyckoff Notation	Point Symmetry	Coordinates of Equivalent Positions
8	e	m11	$xyz, \bar{x}yz, xy\bar{z}, \bar{x}y\bar{z}$
4	d	11m	$xy0, \bar{x}y0$
4	c	m11	$0yz, 0y\bar{z}$
2	b and a	m2m	$0y0$ and $\frac{1}{2}y0$

The Na and N atoms are in special positions of Wyckoff type b, and the O atoms are in type-d positions with y arbitrarily chosen as zero. In a non-centrosymmetric space group like **Im2m**, there is no obvious place for the origin of the unit cell. Any of the y coordinates can be chosen arbitrarily as zero, and for $NaNO_2$, the y coordinate of an oxygen atom was chosen. Once this choice is made (# in the above list), all other y coordinates for the same set of equivalent positions (†) are fixed by the symmetry operations. The y coordinates for Na and N must then be referred to the same origin.

 Arsenic is a semimetal whose crystal structure can be interpreted in terms of strong covalent bonding to three neighbors and much weaker bonding to additional neighbors. There are two atoms at uuu and $\bar{u}\bar{u}\bar{u}$, where u is 0.226, in a rhombohedral unit cell with $a = 4.131$ Å, $\alpha = 54.17°$ (Wyckoff, 1963, p. 32). The corresponding hexagonal unit cell has $a = 3.760$ Å, $c = 10.548$ Å, and in projection down c, the 120° rhombus contains 6 As atoms from three rhombohedral unit cells (Fig. 7.26). Add $\frac{1}{3}$ and $\frac{2}{3}$ to 0.226 and 0.774 to obtain the other z coordinates.

In cubic closest packing, each atom has 12 neighbors. When viewed down a triad axis of the cubic unit cell, the hexagonal closest-packed layers have the $\dot{A}\dot{B}\dot{C}$ sequence and a hexagonal unit cell (Fig. 5.3c) can be chosen. For covalent bonding to 3 neighbors, it is necessary to attract 3 of the 12 neighbors and repel the other 9. This can be achieved by bringing together adjacent layers as follows:

$$\ldots \; A_1 \; B_1 \; C_1 \; A_2 \; B_2 \; C_2 \; A_1 \; \ldots$$
$$\rightarrow \quad \leftarrow \quad \rightarrow \quad \leftarrow \quad \rightarrow \quad \leftarrow \quad \rightarrow$$

Layer B_1 approaches A_1 and moves away from C_1; layer A_2 approaches C_1 and moves away from B_2, and so on. The cell repeat is no longer **ABC** but is now $A_1 B_1 C_1 A_2 B_2 C_2$. An atom in an A_1 layer has 3 close neighbors from a B_1 layer, 6 more distant neighbors from its own A_1 layer, and 3 neighbors from a C_2 layer at an even greater distance.

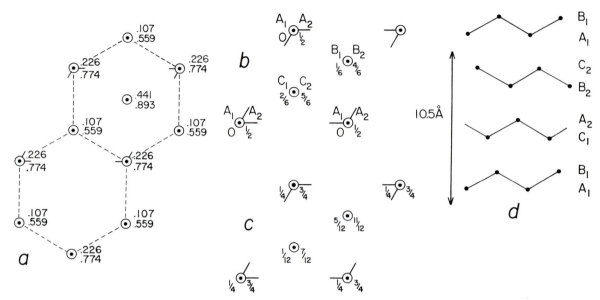

FIGURE 7.26   The crystal structure of arsenic metal referred to a hexagonal unit cell (*a*). Cubic closest-packing is depicted by two sets of positions of type *A, B,* and *C* to give a double unit cell with hexagonal geometry. The proper unit cell, of course, is face-centered cubic. Translation of the coordinates for the 6-layer hexagonal cell by $z = \frac{1}{4}$ gives the coordinates in diagram (*c*). These are displaced by $z = 0.024$ from those in the arsenic structure (*a*). A vertical section (*d*) shows schematically the double layers A_1B_1, C_1A_2, and B_2C_2.

This arrangement is exactly that of arsenic. Consider the 120° prism for two unit cells of cubic closest packing and refer the heights to a new repeat $c' = 2c$ (Fig. 7.26*b*). An A_1 layer would occur at height 0; B_1 at $\frac{1}{6}$; C_1 at $\frac{2}{6}$; A_2 at $\frac{3}{6}$; B_2 at $\frac{4}{6}$; and C_2 at $\frac{5}{6}$. To obtain the arsenic structure, add $z = \frac{1}{4}$ to all positions (Fig. 7.26*c*) and then subtract 0.024 from the heights of $\frac{1}{4}$, $\frac{7}{12}$, and $\frac{11}{12}$, and add 0.024 to the heights of $\frac{1}{12}$, $\frac{5}{12}$, and $\frac{3}{4}$. A vertical section through the arsenic structure (Fig. 7.26*d*) shows the three double layers that span the body diagonal of the rhombohedral unit cell.

The space group is not easy to determine because the As atoms occur in a 2-fold special position along the body diagonal of the rhombohedral unit cell, and no general position is occupied. A special position $\pm(uuu)$ occurs only in those rhombohedral space groups with a center of symmetry. This rules out a rotation triad **3** and enforces an inverse triad $\bar{3}$. Mirror planes can intersect along the body diagonal when they lie at 120° to each other, and the space group is therefore **R$\bar{3}$m**. To obtain general positions for this space group, add a center of symmetry to space group **R3m** (Fig. 7.23).

Selenium, another semimetal, also has a structure that is based on a distortion of cubic closest packing. In order to obtain covalent bonding to two neighbors, the selenium structure has space group **P3$_1$21** (or its enantio-

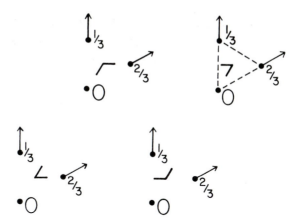

FIGURE 7.27 The crystal structure of selenium metal referred to a hexagonal unit cell. The dashed lines show the projection of a helix, and the arrows show the movements required to obtain a cubic closest-packed array.

morph **P3₂21**) in which the links between nearest neighbors form infinite helices, each with the symmetry of a 3_1 (or a 3_2) axis (Wyckoff, 1963, p. 38). The atomic positions in Figure 7.27 are for **P3₂21**; $u00, \bar{u}\bar{u}\frac{1}{3}, 0u\frac{2}{3}$, with $u = 0.217$ in a cell with $a = 4.36$ Å, $c = 4.95$ Å. The projection of one of the helices is shown by dashed lines. Each Se atom has two near neighbors from a helix at 2.32 Å. Arrows show the movements required to obtain a cubic closest-packed array. Atoms at height 0 were left in position, and those at height $\frac{1}{3}$ and $\frac{2}{3}$ were moved.

Both the arsenic and selenium structures have the simplest possible distortions of the ccp array, and illustrate the essential economy of the stacking array of a crystal structure. Usually it is possible to interpret a crystal structure in terms of some simple way (often the simplest way) of packing together atoms, ions, or molecules such that they satisfy requirements of chemical bonding.

7.12 Systematic List of Space Groups and Point Groups;
Axial Transformation; General and Special Positions

Removal of all translation operations from a space group of symmetry elements produces a *crystallographic point group of symmetry elements,* as described in Chapter 6. The lattice becomes irrelevant; each set of glide planes is converted to a mirror plane; and each set of screw axes becomes a rotation axis. Mirror planes, rotation and inverse axes, and centers of symmetry survive unscathed except that each infinite set of symmetry elements in a space group is degraded to one element of symmetry passing through a singular point. Thus:

$P112_1 \rightarrow 112$

Ccc2 and Ama2 $\rightarrow$ mm2

Pnnn, Ccca, and Fddd $\rightarrow$ mmm

$P4_2$ and $I4_1 \rightarrow 4$

$P\bar{4}$ and $I\bar{4} \rightarrow \bar{4}$

$P4_2$nm and $I4_1$md $\rightarrow$ 4mm

$P\bar{4}2c$ and $I\bar{4}2d \rightarrow \bar{4}2m$

$P\bar{4}c2$ and $I\bar{4}m2 \rightarrow \bar{4}m2 \rightarrow \bar{4}2m$
  upon 45° rotation about the $\bar{4}$ axis

$P4_2$/nbc and $I4_1$/acd $\rightarrow$ 4/mmm

P321 and $P3_1$21 $\rightarrow$ 321

P3m1 $\rightarrow$ 3m1

P31m $\rightarrow$ 31m $\rightarrow$ 3m1 upon 30°
  rotation about the 3 axis

$P\bar{6}m2 \rightarrow \bar{6}m2$

$P\bar{6}2m \rightarrow \bar{6}2m \rightarrow \bar{6}m2$
  upon 30° rotation about the 6 axis

P23 and $I2_1$3 $\rightarrow$ 23

$F2/d\bar{3}$ and $I2_1/a\bar{3} \rightarrow 2/m\bar{3}$

Fm3m and Ia3d $\rightarrow$ m3m, which
  is the short symbol for $4/m\bar{3}2/m$

The crystallographic point group is useful because it controls the symmetry of all macroscopic physical properties, such as electrical conductivity and elasticity. It should not be confused with the symmetry of an individual position or set of positions in a space group, though the symmetry operations for a position with point symmetry (say) **4mm** are exactly the same as for the physical properties of a crystal with crystallographic point group **4mm**. Note that the point symmetry of the special position with the lowest multiplicity in a space group must be the same as the crystallographic point group that controls the physical properties. Also note that crystallographers usually omit the word "crystallographic" before point group.

Table 7.5 contains a systematic list of the 230 space groups arranged in the 32 point groups and the 6 crystal systems. Each space group is given the sequence number (italicized) used in *International Tables for X-Ray Crystallography.*

The triclinic system (with a monad axis) has only two space groups, one in point group **1**, and one in point group $\bar{1}$ (i.e., a center of symmetry).

The monoclinic system has three point groups characterized by either a rotation diad, or an inverse diad (i.e., a mirror plane), or by a rotation diad perpendicular to a mirror plane. Following *International Tables for X-Ray Crystallography,* the diad axis is placed in the third position (i.e., along the *c* axis), and **B** centering of the lattice is used instead of the **I** centering used in Tables 7.3 and 7.4. Many crystallographers have placed the diad axis along the *b* axis, and others have used **A** centering instead of **B** centering. All nontrivial axial transformations are shown in Figure 7.28 for space group **B112/b**. The first diagram shows the B centering by a point in the middle of a *B* face, the diad along the *c* axis by the arrowed line with ellipse, and the **b** glide plane perpendicular to the *c* axis by the *H* with an arrow. Retention of the *c* axis, and interchange of *a* and *b* (second diagram) requires relabeling of the space group from **B112/b** to **A112/a**. It is necessary to change the relative directions of the axes in order to retain a right-handed system, but this is irrelevant for the space group symbol. Placement of the

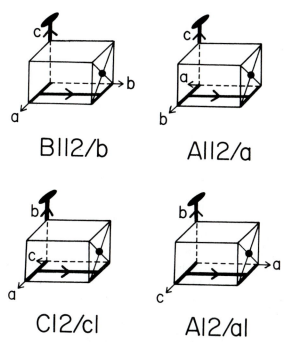

FIGURE 7.28 Diagrams illustrating relabeling of axes and space group symbol for **B112/b**.

diad axis along the **b** axis gives two possible symbols **C12/c1** and **A12/a1** (third and fourth diagrams).

The orthorhombic system (with three orthogonal diad axes) has three point groups. In **222** and **mm2**, the third symbol is produced automatically when the first two are specified. In **mmm**, all three mirror planes operate independently. Axial transformations of the space group symbol can be determined graphically as in Figure 7.28, but an algebraic method is quicker.

Table 7.5 Systematic list of space groups arranged in point groups and listed in the sequence of *International Tables for X-Ray Crystallography*

Triclinic

1	*1* P1
$\bar{1}$	*2* P$\bar{1}$

Monoclinic

112	*3* P112	*4* P112$_1$	*5* B112			
11m	*6* P11m	*7* P11b	*8* B11m	*9* B11b		
112/m	*10* P112/m	*11* P112$_1$/m	*12* B112/m	*13* P112/b	*14* P112$_1$/b	*15* B112/b

Orthorhombic

222	*16* P222	*17* P222$_1$	*18* P2$_1$2$_1$2	*19* P2$_1$2$_1$2$_1$	*20* C222$_1$	*21* C222
	22 F222	*23* I222	*24* I2$_1$2$_1$2$_1$			
mm2	*25* Pmm2	*26* Pmc2$_1$	*27* Pcc2	*28* Pma2	*29* Pca2$_1$	*30* Pnc2
	31 Pmn2$_1$	*32* Pba2	*33* Pna2$_1$	*34* Pnn2	*35* Cmm2	*36* Cmc2$_1$
	37 Ccc2	*38* Amm2	*39* Abm2	*40* Ama2	*41* Aba2	*42* Fmm2
	43 Fdd2	*44* Imm2	*45* Iba2	*46* Ima2		
mmm	*47* Pmmm	*48* Pnnn	*49* Pccm	*50* Pban	*51* Pmma	*52* Pnna
	53 Pmna	*54* Pcca	*55* Pbam	*56* Pccn	*57* Pbcm	*58* Pnnm
	59 Pmmn	*60* Pbcn	*61* Pbca	*62* Pnma	*63* Cmcm	*64* Cmca
	65 Cmmm	*66* Cccm	*67* Cmma	*68* Ccca	*69* Fmmm	*70* Fddd
	71 Immm	*72* Ibam	*73* Ibca	*74* Imma		

Tetragonal

4	*75* P4	*76* P4$_1$	*77* P4$_2$	*78* P4$_3$	*79* I4	*80* I4$_1$
$\bar{4}$	*81* P$\bar{4}$	*82* I$\bar{4}$				
4/m	*83* P4/m	*84* P4$_2$/m	*85* P4/n	*86* P4$_2$/n	*87* I4/m	*88* I4$_1$/a
422	*89* P422	*90* P42$_1$2	*91* P4$_1$22	*92* P4$_1$2$_1$2	*93* P4$_2$22	*94* P4$_2$2$_1$2
	95 P4$_3$22	*96* P4$_3$2$_1$2	*97* I422	*98* I4$_1$22		
4mm	*99* P4mm	*100* P4bm	*101* P4$_2$cm	*102* P4$_2$nm	*103* P4cc	*104* P4nc
	105 P4$_2$mc	*106* P4$_2$bc	*107* I4mm	*108* I4cm	*109* I4$_1$md	*110* I4$_1$cd
$\bar{4}$2m	*111* P$\bar{4}$2m	*112* P$\bar{4}$2c	*113* P$\bar{4}$2$_1$m	*114* P$\bar{4}$2$_1$c	*115* P$\bar{4}$m2	*116* P$\bar{4}$c2
	117 P$\bar{4}$b2	*118* P$\bar{4}$n2	*119* I$\bar{4}$m2	*120* I$\bar{4}$c2	*121* I$\bar{4}$2m	*122* I$\bar{4}$2d
4/mmm	*123* P4/mmm	*124* P4/mcc	*125* P4/nbm	*126* P4/nnc	*127* P4/mbm	*128* P4/mnc
	129 P4/nmm	*130* P4/ncc	*131* P4$_2$/mmc	*132* P4$_2$/mcm	*133* P4$_2$/nbc	*134* P4$_2$/nnm
	135 P4$_2$/mbc	*136* P4$_2$/mnm	*137* P4$_2$/nmc	*138* P4$_2$/ncm	*139* I4/mmm	*140* I4/mcm
	141 I4$_1$/amd	*142* I4$_1$/acd				

Table 7.5 (*Continued*)

Hexagonal

3	*143* P3	*144* P3$_1$	*145* P3$_2$	*146* R3	
$\bar{3}$	*147* P$\bar{3}$	*148* R$\bar{3}$			
32	*149* P312	*150* P321	*151* P3$_1$12	*152* P3$_1$21	*153* P3$_2$12
	155 R32				
3m	*156* P3m1	*157* P31m	*158* P3c1	*159* P31c	*160* R3m
$\bar{3}$m	*162* P$\bar{3}$1m	*163* P$\bar{3}$1c	*164* P$\bar{3}$m1	*165* P$\bar{3}$c1	*166* R$\bar{3}$m
6	*168* P6	*169* P6$_1$	*170* P6$_5$	*171* P6$_2$	*172* P6$_4$
$\bar{6}$	*174* P$\bar{6}$				
6/m	*175* P6/m	*176* P6$_3$/m			
622	*177* P622	*178* P6$_1$22	*179* P6$_5$22	*180* P6$_2$22	*181* P6$_4$22
6mm	*183* P6mm	*184* P6cc	*185* P6$_3$cm	*186* P6$_3$mc	
$\bar{6}$m2	*187* P$\bar{6}$m2	*188* P$\bar{6}$c2	*189* P$\bar{6}$2m	*190* P$\bar{6}$2c	
6/mmm	*191* P6/mmm	*192* P6/mcc	*193* P6$_3$/mcm	*194* P6$_3$/mmc	

Additional sixth-column entries (Hexagonal): *154* P3$_2$21; *161* R3c; *167* R$\bar{3}$c; *173* P6$_3$; *182* P6$_3$22.

Isometric

23	*195* P23	*196* F23	*197* I23	*198* P2$_1$3	*199* I2$_1$3
2/m$\bar{3}$	*200* P2/m$\bar{3}$	*201* P2/n$\bar{3}$	*202* F2/m$\bar{3}$	*203* F2/d$\bar{3}$	*204* I2/m$\bar{3}$
	206 I2$_1$/a$\bar{3}$				
432	*207* P432	*208* P4$_2$32	*209* F432	*210* F4$_1$32	*211* I432
	213 P4$_1$32	*214* I4$_1$32			
$\bar{4}$3m	*215* P$\bar{4}$3m	*216* F$\bar{4}$3m	*217* I$\bar{4}$3m	*218* P$\bar{4}$3n	*219* F$\bar{4}$3c
4/m$\bar{3}$2/m	*221* Pm3m	*222* Pn3n	*223* Pm3n	*224* Pn3m	*225* Fm3m
	227 Fd3m	*228* Fd3c	*229* Im3m	*230* Ia3d	

Additional sixth-column entries (Isometric): *205* P2$_1$/a$\bar{3}$; *212* P4$_3$32; *220* I$\bar{4}$3d; *226* Fm3c.

There are six possible ways of assigning axes to an orthorhombic unit cell because the first axis can be assigned in three ways, the second one in two ways, and the third one in only one way giving $3 \times 2 \times 1 = 6$. Suppose the space group symbol is **Pmna** for choice 1 with axes a_1, b_1, c_1. The other five choices are found as follows:

		$a_1b_1c_1$	$c_1a_1b_1$	$b_1c_1a_1$	$a_1\bar{c}_1b_1$	$b_1a_1\bar{c}_1$	$\bar{c}_1b_1a_1$
			$a_2b_2c_2$	$a_3b_3c_3$	$a_4b_4c_4$	$a_5b_5c_5$	$a_6b_6c_6$
Step 1		**Pmna**	**P.mn**	**Pn.m**	**Pm.n**	**Pnm.**	**P.nm**
Step 2			**Pbmn**	**Pncm**	**Pman**	**Pnmb**	**Pcnm**

Transform $a_1b_1c_1$ to $c_1a_1b_1$ and $b_1c_1a_1$ by cyclic rotation, and then to $a_1c_1b_1$ by switching b_1 and c_1. Change c_1 to $\bar{c}_1$ in order to retain right-handed axes, and then obtain $b_1a_1\bar{c}_1$ and $\bar{c}_1b_1a_1$ by cyclic rotation. Write $a_2b_2c_2 \ldots a_6b_6c_6$ for the five choices of new axes. Thus for the second choice $c_1 \to a_2$, $a_1 \to b_2$ and $b_1 \to c_2$. In step 1, list all the symbols that do not depend on the axial labels, that is, **P**, **F**, **I**, **m**, **n**, **d**, and symmetry axes. **Pmna** goes to **P.mn** because the **m** is carried with the a_1 axis to the second position b_2, and **n** with the b_1 axis to the third position c_2; and similarly for the other four axial changes. In step 2, transform any glide planes and variable lattice symbols. Thus **a** in the third position must go to **b** in the first position because the glide component $a/2$ is now in the second position following $a_1 \rightarrow b_2$. Four more examples follow:

Step 1	Pmc2₁	P2₁m.	P.2₁m	Pm2₁.	P.m2₁	P2₁.m
Step 2		P2₁ma	Pb2₁m	Pm2₁b	Pcm2₁	P2₁am

Step 1	Ima2	I2m.	I.2m	Im2.	I.m2	I2.m
Step 2		I2mb	Ic2m	Im2a	Ibm2	I2cm

Step 1	Cmma	..mm	.m.m	.m.m	.mm.	..mm
Step 2		Abmm	Bmcm	Bmam	Cmmb	Acmm

Step 1	Cmca	..m.	...m	.m..	..m.	...m
Step 2		A.ma	Bb.m	Bm.b	Ccm.	A.am
Step 3		Abma	Bbcm	Bmab	Ccmb	Acam

Some space groups do not change their symbol upon axial transformation (e.g., **Immm**), but the cell dimensions and atomic coordinations must change. Although the three planes in **Immm** apparently stay the same, strictly speaking they have undergone axial transformation so that $m_1 \rightarrow m_2$, $m_2 \rightarrow m_3$, $m_3 \rightarrow m_1$; and so on.

Tetragonal space groups belong to seven point groups. Because the tetrad axis is always placed along the c axis, there is no axial transformation for a tetragonal space group unless a **C** cell is chosen instead of a **P** cell or an **F** cell instead of an **I** cell for some special reason. However, space groups numbered 111 to 122 give either point group $\bar{4}2m$ or $\bar{4}m2$ if the point group is referred to the same axes as the space group: thus **P$\bar{4}$2m** $\rightarrow \bar{4}2m$ and **P$\bar{4}$c2** $\rightarrow \bar{4}m2$. These two point groups differ merely in the choice of the a and b axes, and a rotation of 45° about the c axis converts one into the other. Morphological crystallographers had no knowledge of the space group before the advent of X-ray diffraction, and preferred $\bar{4}2m$ in which diad axes lie along the a and b axes. It is simplest to refer the morphology of a crystal with a space group like **P$\bar{4}$c2** to point group $\bar{4}2m$.

Hexagonal space groups belong to 12 point groups. Many crystallographers place the 6 point groups with a triad axis into the trigonal system and those with a hexad axis into the hexagonal system. Seven space groups have a rhombohedral unit cell, which can be referred to a 120° prism with three times the volume of the rhombohedron. The other 18 space groups with a triad axis are referred to a 120° prism, just like the space groups with a hexad axis. Morphological crystallographers had no knowledge of the unit cell before X-ray diffraction allowed measurement, and were able to use either rhombohedral or hexagonal axes to describe the morphology of crystals with a triad axis in the point group. Point groups $\bar{6}m2$ and $\bar{6}2m$ are related in the same way as $\bar{4}m2$ and $\bar{4}2m$, except that the rotation angle is 30° instead of 45°.

Isometric space groups reduce to five point groups. Diad axes along *a, b,* and *c* are represented by a **2** before the triad symbol, and those in dodecahedral directions are represented by a **2** after the triad symbol. The symbols **/m** before the triad symbol refer to mirror planes perpendicular to *a, b,* or *c,* and the **m** after the triad symbol refers to mirror planes parallel to faces of the rhombic dodecahedron. If the point group has a center of symmetry, the triad axis should be designated $\bar{3}$ instead of **3**.

Each space group has a set of general positions with $x \neq y \neq z$ whose point symmetry is **1** and whose multiplicity can be calculated simply from the point-group diagrams in Chapter 6. Thus from Figure 6.7, point group **222** has four equivalent directions. Hence, space group **P222** has $1 \times 4 = 4$ general equivalent positions; space groups **C222** and **I222** have $2 \times 4 = 8$ positions; and space group **F222** has $4 \times 4 = 16$. When drawing out the diagrams for a space group, the reader should profit by first drawing the stereogram for the point group. This will reveal how many general positions there should be in the space group after multiplying by the number of points in a unit cell of the Bravais lattice. Furthermore, it will tell the reader what are the relationships between the symmetry elements. Thus point group **222** has only two independent operations because the third one is automatic; furthermore, there is no center of symmetry. If the Bravais unit cell contains two points, a second set of symmetry elements is needed for each set in the Hermann–Mauguin symbol in order to complete the operations relating the general positions. It will be necessary to determine what is the nature of the second set because it depends on the interaction of the first type of symmetry element and the lattice operation. Thus, **Cmca** will turn out to have the additional operations of **bnb**. In Table 7.5, all space groups with a centered lattice can be described with a different set of symbols, just as **Cmca** can be described as **Cbnb** or **Cmnb** or any other of the eight possible combinations. Arbitrary but sensible rules were used by the editors of *International Tables for X-Ray Crystallography* in choosing the symbols in Table 7.5.

Most space groups have one or more sets of special positions obtained by moving a general position to a symmetry element or the intersection of two or more symmetry elements, thereby causing reduction of the multiplicity. Glide planes and screw axes are not regarded as potential sites for special positions, because the multiplicity is not reduced. Only mirror planes, rotation and inverse axes, and centers of symmetry are possible sites for special positions. If there is a center of symmetry in the point group, there must be one or more in the space group. It is easy to overlook the center of symmetry in $P112_1/b$, because the other symmetry elements do not provide special positions. Atoms do not have commas attached to them. Experimental data provide only the atomic positions, and a crystallographer when given just a table of atomic positions for a crystal structure must look for inversion operations directly from the atomic positions: thus if all positions occur in pairs with coordinates $x_iy_iz_i$, $\bar{x}_iy_iz_i$, where i can be 1, 2, ..., there is a mirror plane at $x = 0$. The point symmetry at a center of symmetry is $2/m$ and not $\bar{1}$ if the center of symmetry lies at the intersection of a rotation diad and a mirror plane, but is $\bar{1}$ for the centers of symmetry in $P112_1/m$.

7.13   Hierarchies of Space Groups with Decreasing Symmetry

The space-group symmetry is reduced when the Cu_1Au_1 and Cu_3Au_1 superstructures are developed from a disordered alloy with random occupancy of sites on a face-centered cubic lattice (Fig. 5.6). Alternate occupancy of the tetrahedral nodes of the diamond net to give the blende structure of ZnS also results in reduction of the space-group symmetry. Brief references have already been made to the derivative relationship between space groups such as $P6$ and $P3$, and to hierarchies of space groups. It is impossible to find room for a complete description of the hierarchies, but readers should be able to work out their own examples after studying the following procedure, kindly developed by T.J. McLarnan.

The ideal structure of feldspar (Section 8.6.7) is described with monoclinic space group $C12/m1$. Ordering of Al and Si atoms on the nodes of the tetrahedral framework, together with angular distortions, leads to a hierarchy of structure types with lower symmetry. To develop all possible space groups of lower symmetry, consider first the group table of the symmetry elements in $C12/m1$.

In Figure 7.29, the two diagrams for equivalent general positions and equivalent symmetry elements are developed in the usual way. A third diagram replaces the equivalent general positions with the symmetry operations relating each position to the starting position p at x,y,z. Positions p', p'', p'''

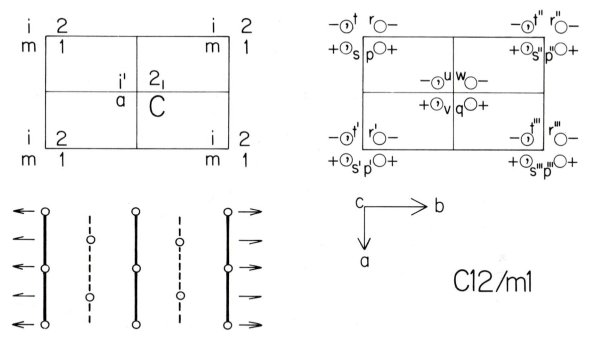

FIGURE 7.29 The general equivalent positions and symmetry operations in space group **C12/m1**.

are related to p by the translations of a monoclinic primitive lattice, and do not need to be considered in the group table. Position q is related to position p by the operator **C**. Positions s and r are related to p by the operators **m** and **2**, respectively. The center of symmetry that relates p to t is denoted by **i**. A second center of symmetry relates p to u and is denoted **i'**. The space group operations are completed by the **a** glide plane and the 2_1 screw axis, which generate positions v and w, respectively.

To generate the general positions of **C12/m1**, only the three operators **C**, **2**, and **m** were used. All the other operators were obtained by looking at the relationships between the positions. Consequently, all the elements of the space group can be generated from **C**, **2**, and **m**: **i = 2m**, 2_1 = **2C**, **a = mC**, and **i' = iC = 2mC**. Furthermore, the three generators satisfy the relations

$$22 = 1 \quad mm = 1 \quad CC = 1$$

$$2C = C2 \quad mC = Cm \quad 2m = m2$$

Readers are reminded that **2m** means that the operation **m** is followed (i.e., "multiplied") by the operation **2** to yield the operation **i**, and so on. The multiplication **22** can be written 2^2.

The group table for **C12/ml** can be written entirely in terms of C, 2, and **m**:

	1	2	m	2m	C	2C	mC	2mC
1	1	2	m	2m	C	2C	mC	2mC
2	2	1	2m	m	2C	C	2mC	mC
m	m	2m	1	2	mC	2mC	C	2C
i = 2m	2m	m	2	1	2mC	mC	2C	C
C	C	2C	mC	2mC	1	2	m	2m
2_1 = 2C	2C	C	2mC	mC	2	1	2m	m
a = mC	mC	2mC	C	2C	m	2m	1	2
i' = iC = 2mC	2mC	mC	2C	C	2m	m	2	1

The group is commutative (Abelian) because the elements are permutable with each other. It is not cyclic, because the elements do not form a single cycle. Space group **P6** is cyclic because the six elements are based on a single cycle of 60° rotations. Because the three generators **2**, **m**, and **C** behave in the same way in the table, the group can be written $Z_2 \times Z_2 \times Z_2$.

Systematic removal of rows and columns from the group table yields the hierarchy of space groups in Figure 7.30. First remove the operation **C** from the group table. Only the 4 × 4 block of rows and columns remains with operators **1**, **2**, **m**, and **i = 2m**. The resulting space group is **P12/m1**. Removal of operation **m** leaves the operations **1**, **2**, **C**, and 2_1 = **2C** in space group **C121**. Removal of operation **2** leaves **1**, **m**, **C**, and **a = mC** in **C1m1**. The next space group is obtained by retaining **2** and **mC = a** to give **P12/a1**. This space group also contains **i' = 2mC**. Similarly, **P12₁/m1** is obtained by retaining **m** and **2C**, and **C1̄** by retaining **C** and **2m**. Finally retention of **2m = i**, **2C = 2₁**, and **mC = a** gives space group **P12₁/a**. These seven space groups form the first level of subgroups. There are also seven space groups in the next level. Retention of **2** gives **P121**; **m** gives **P1m1**; **C** gives **C1**; **2m** gives **P1̄**; **2C** gives **P12₁1**; **mC** gives **P1a1**; and **2mC** gives **P1̄**. Finally, all the space groups degrade into **P1**.

What is the meaning of the two space groups **P1̄**? The first one retains the center of symmetry of type **i** obtained by the interaction of operators **2** and **m**, whereas the second one (denoted **P1̄'** in Fig. 7.30) retains the center of symmetry **i'** obtained by the interaction of **i** and **C** (or of **2**, **m**, and **C**). Thus, these two space groups are different when the relative positions of the symmetry elements with respect to **C12/m1** are important, but are identical when just the formal symmetry is considered. Domain intergrowths are con-

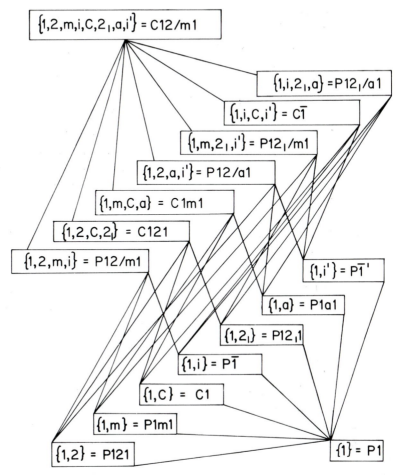

FIGURE 7.30 The hierarchy of cell-equivalent subgroups derived from **C12/m1**.

trolled by the relative positions of the symmetry elements, and $P\bar{1}$ is different from $P\bar{1}'$ in determining the nature of the domain boundaries. Space group $C\bar{1}$ can be converted into $P\bar{1}$ by choice of new axes to give a cell of half the volume, and similarly **C1** can be converted into **P1**. Microcline and low albite have space group $C\bar{1}$ because of ordering of the Al atom onto the $T_1(0)$ node of the tetrahedral framework (Section 8.6.7).

The subgroups in Figure 7.30 are known as the ***cell-equivalent*** or ***Zellengleich subgroups*** of **C12/m1**, because they retain the same unit cell (*gleich* is German for "equal"). Four subgroups **P12/m1**, **P12/a1**, **P12₁/m1**, and **P12₁/a1**) have the same point group **12/m1** as the parent group **C12/m1**, and can be described as ***class-equivalent*** or ***Klassengleich subgroups***. The name *Klass* is the German for "class," and morphological crystallographers assigned crystals to classes, each of which is related to a point group of symmetry (Chapter 6).

The symmetry of a space group may be degraded by taking a larger cell repeat. Thus the c repeat of the celsian ($BaAl_2 Si_2 O_8$) and anorthite ($CaAl_2 Si_2 O_8$) varieties of feldspar is twice that of the sanidine ($KAlSi_3 O_8$) and albite ($NaAlSi_3 O_8$) varieties because of difference in ordering patterns of Al and Si atoms on the nodes of the tetrahedral framework. Celsian has the space group $I12/c1$, which is a class-equivalent subgroup of $C12/m1$ with new $c = 2 \times$ old c. For clarity the subgroup can be denoted $I12/c1(2c)$, where the bracketed $2c$ denotes the doubled axis. Note the use of 2 and c, first to show symmetry operations, and then as 2 and c to show the doubled repeat.

How many subgroups can be obtained by doubling the c repeat of $C12/m1$? Figure 7.31 shows that there are four class-equivalent subgroups with centered lattices. Retention of different positions allows new lattices that are of C, I, or P type, but not of type A, B, or F. The point group $12/m1$ is retained by keeping either or both of the 2 and 2_1 axes from $C12/m1$, and either or both of the $1m1$ and $1a1$ planes of symmetry. For the glide planes, a complication arises because the doubling of the c repeat results in the possibility of new symmetry planes. Thus the mirror operation $(xyz) \rightarrow (x\bar{y}z)$ for the small cell can become either the mirror operation $(xyz) \rightarrow (x\bar{y}z)$ or the $1c1$ glide operation $(xyz) \rightarrow (x,\bar{y},\frac{1}{2} + z)$ of the doubled cell. Similarly, the glide operation $1a1$ that related (xyz) and $(\frac{1}{2} + x, \frac{1}{2} - y, z)$ in the small cell can become either the glide operation $1a1$ in the double cell for $(xyz) \rightarrow (\frac{1}{2} + x, \frac{1}{2} - y, z)$ or the glide operation $1n1$ for $(xyz) \rightarrow (\frac{1}{2} + x, \frac{1}{2} - y, \frac{1}{2} + z)$. Hence the class-equivalent subgroups for double c are represented formally as

$$\{C,I,P\}\{2,2_1\}/\{m,a,c,n\}$$

In spite of this formidable formula, there are only two class-equivalent subgroups with a C lattice and two with an I lattice. Figure 7.31 shows that space group $C12/m1$ contains the 2_1 and a operations, while $C12/c1$ contains the 2_1 and n operations. Hence

$$C12/m1 \equiv C12_1/m1 \equiv C12/a1 \equiv C12_1/a1$$
$$C12/c1 \equiv C12_1/c1 \equiv C12/n1 \equiv C12_1/n1$$
$$I12/m1 \equiv I12_1/m1 \equiv I12/n1 \equiv I12_1/n1$$
$$I12/c1 \equiv I12_1/c1 \equiv I12/a1 \equiv I12_1/a1$$

For each of these class-equivalent subgroups with a centered lattice, there are class-equivalent subgroups with a primitive lattice and cell-equivalent subgroups with the same cell but lower point-group symmetry. Those for

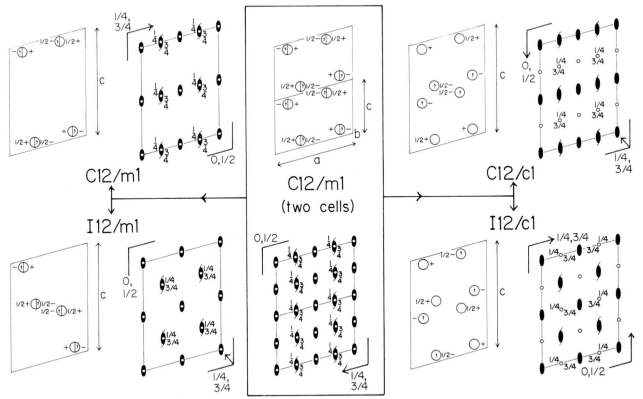

FIGURE 7.31 The derivation of the four class-equivalent subgroups with a centered lattice from **C12/m1** after doubling of the c repeat. The middle pair of diagrams show two unit cells of general equivalent positions and symmetry elements for **C12/m1** projected down the b-axis. The four pairs of diagrams at the left or right show the subgroups.

C12/m1 were listed in Figure 7.30, and those for **C12/c1**, **I12/m1**, and **I12/c1** can be determined in a similar way.

Crystallographers usually refer space groups to an origin that lies at a center of symmetry if one is available. Hence the origin of the diagrams in Figure 7.31 for **C12/c1** and **I12/c1** would be shifted. Furthermore, a shift of origin by $c/4$ of the double cell produces isomorphous copies of the four **C** and **I** class-equivalent subgroups, and these copies are themselves subgroups of **C12/m1**.

Observed space groups in feldspars (Smith, 1974) are:

C12/m1 ($c = 7$ Å)	Sanidine
C$\bar{1}$	($c = 7$ Å)	High and low albite; microcline
I12/c1	($c = 14$ Å)	Celsian
I$\bar{1}$	($c = 14$ Å)	Body-centered Ca-rich plagioclase
P$\bar{1}$	($c = 14$ Å)	Primitive anorthite

7.14   Further Remarks on Classical Space Groups

 Each classical space group can be described with either a ***short symbol*** or a ***long symbol***. Thus space group *64* (Table 7.5) can be described by either **Cmca** or **C2/m2/c2$_1$/a**. The operators in the short symbol are sufficient to generate all the symmetry elements and equivalent positions, as demonstrated for **C2/m** in the preceding section. Those in **Cmca** can be written {**P**}{**C,m,c,a**} where {**P**} is the infinite translation group of edge vectors and {**C,m,c,a**} is the symmetry group.

Each space group can be derived by integration of the operations of a Bravais lattice and a point group. The glide and screw operators result from the combination of a translation operator from a Bravais lattice and a symmetry operator from a point group. In principle, enumeration of the 230 space groups consists merely of systematically combining these operators in consistent ways. In practice, this becomes more difficult as the number of operations increases. It is easy to demonstrate that the union of translation group {**P**} and point symmetry group {**6**} generates **P6**, **P6$_1$**, **P6$_2$**, **P6$_3$**, **P6$_4$**, and **P6$_5$**. The union of {**P**} and {**$\bar{6}$**} is even easier because $\bar{6}$ cannot be combined with a translation operator; hence **P$\bar{6}$** is the only product. A ***symmorphic*** space group is obtained directly from the operations of the translation and point-symmetry groups (e.g., **P6** and **P$\bar{6}$**), and a ***nonsymmorphic*** group is generated by combining operations from translation and point-symmetry groups (e.g., **P6$_1$**).

Orthorhombic point group 222 can combine with the four Bravais lattices to yield four symmorphic space groups **P222**, **I222**, **F222**, and **C222** (≡**A222**≡**B222** by relabeling axes). The rotation operator can combine with a translation operator of half an edge vector to give the screw operator **2$_1$**. When relabeling of axes is considered, the only nonsymmorphic space groups involving **P** and **2$_1$** are **P222$_1$**, **P2$_1$2$_1$2**, and **P2$_1$2$_1$2$_1$**. Because a centering vector interacts with **2** to generate **2$_1$**, the only other nonsymmorphic space groups in point-symmetry group 222 are **C222$_1$** and **I2$_1$2$_1$2$_1$**.

Enumeration for **mmm** is tedious but straightforward. Point-symmetry group **mmm** yields four symmorphic space groups **Pmmm**, **Immm**, **Fmmm**, and **Cmmm**. Replacement of one mirror plane by a simple glide plane yields **Pmma**, **Imma**, **Cmma**, and **Cmcm**. Note that for the **P** and **I** lattices all other combinations are topologically identical and involve merely relabeling of axes: for example, **Pmma** → **Pbmm** → **Pmcm** → **Pmmb** → **Pcmm** → **Pmam**. Because the **C** lattice places a label on one direction, **Cmma** and **Cmcm** (≡**Ccmm**) are distinct. For an **F** lattice, only the diamond glide operation is consistent with the lattice translations, and **Fddd** results. The reader might like to use the following sequence as a guide to completing the enumeration:

Pmmm	3 mirror planes
Pnnn	3 diagonal glide planes
Pcca, Pbca	3 axial glide planes
Pmna, Pnma	1 of each
Pmmn, Pmma	2 mirror planes
Pnnm, Pnna	2 diagonal glide planes
Pccm, Pccn, Pbcm, Pbcn, Pbam, Pban	2 axial glide planes
Cccm, Cmca; Ibam	1 mirror plane
Ccca; Ibca	no mirror plane

Space-symmetry groups can also be enumerated from the systematic union of rod and plane groups. The union of **P1** (rod) with the 80 two-sided plane groups immediately yields 80 space groups. Plane group **p4** combines with rod groups **p4, p4$_1$, p4$_2$,** and **p4$_3$** to generate tetragonal space groups **P4, P4$_1$, P4$_2$,** and **P4$_3$,** while rod and plane groups **p4̄** yield **P4̄.** The isometric space groups result from the union of appropriate rod and space groups of order **4.** In some combinations the 4-fold operators become degraded to 2-fold operators.

7.15 Nonclassical Lattices, Point Groups, and Space Groups

Because antisymmetry and color symmetry are encountered rarely in crystallographic research, readers are referred to Shubnikov and Koptsik (1974) and Shubnikov and Belov (1964) for details.

There are *22 antisymmetrical lattices* with antisymmetrical translations (Fig. 7.32). A body-centered isometric Bravais lattice is transformed into an antilattice by conversion of the body-centering vector. This antilattice provides a formal description of the CsCl structure in which the antisymmetrical operation is the interchange of Cs and Cl. Similarly, the second isometric antilattice provides a formal description of the halite structure.

There are *58 antisymmetrical crystallographic point groups* with antisymmetrical operations consistent with lattice translations in a space group (Table 7.6), and there is an infinite number of noncrystallographic ones. An antisymmetrical operation is denoted by italicizing the corresponding neutral operator, and an extra dimension is used in a matrix operator. To distinguish between a mirror plane **m** (Table 6.1) and an antimirror plane **m'** perpendicular to [**100**] use the following:

$$\mathbf{m} \begin{vmatrix} -1 & 0 & 0 & 0 \\ 0 & 1 & 0 & 0 \\ 0 & 0 & 1 & 0 \\ 0 & 0 & 0 & 1 \end{vmatrix} \qquad \mathbf{m'} \begin{vmatrix} -1 & 0 & 0 & 0 \\ 0 & 1 & 0 & 0 \\ 0 & 0 & 1 & 0 \\ 0 & 0 & 0 & -1 \end{vmatrix}$$

or briefly:

$$\mathbf{m} \quad xyz(1) \rightarrow \bar{x}yz(1) \qquad\qquad \mathbf{m'} \quad xyz(1) \rightarrow \bar{x}yz(1')$$

where (1) and (1') represent the operations of identity and anti-identity.

There are *1321 antisymmetrical space groups* comprised of (i) 230 neutral (Fedorov) groups extended by the anti-identity group $1' = \{1,1'\}$, (ii)

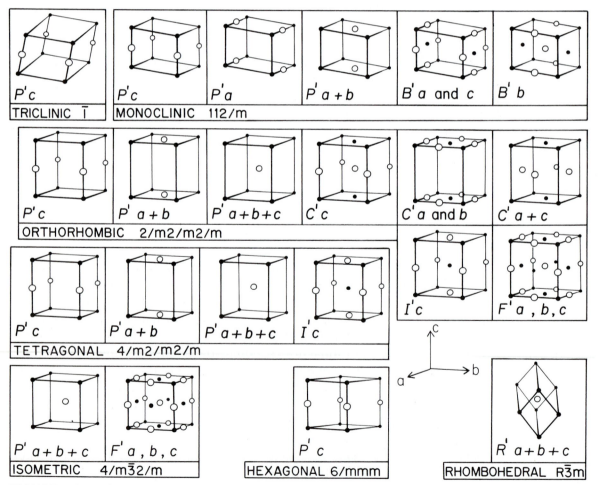

FIGURE 7.32   The 22 antisymmetrical lattices arranged according to the geometry of the unit cell and the point symmetry. The lattice type and the antisymmetry vector are shown.

Table 7.6 The 58 antisymmetrical crystallographic point groups[a]

Triclinic	Tetragonal	Hexagonal (*sensu lato*)	
$\bar{1}$	4	$\bar{3}$*	6/mmm*
	$\bar{4}$	3m	6/mmm
Monoclinic	4/m	32	6/mmm*
	4/m*	$\bar{3}$2/m	6/mmm*
m	4/m*	$\bar{3}$2/m*	6/mmm
2	4mm	$\bar{3}$2/m*	
2/m*	4mm	6	Isometric
2/m*	$\bar{4}$2m	$\bar{6}$	
2/m	$\bar{4}$2m	6/m*	2/m$\bar{3}$*
	$\bar{4}$2m	$\bar{6}$/m*	$\bar{4}$3m
Orthorhombic	422	$\bar{6}$/m	432
	42$\bar{2}$	$\bar{6}$m2	4/m$\bar{3}$2/m*
mm2	4/mmm	$\bar{6}$m2	4/m$\bar{3}$2/m
mm2	4/mmm*	$\bar{6}$m2	4/m$\bar{3}$2/m*
222	4/mmm*	6mm	
mmm*	4/mmm	6mm	
mmm	4/mmm*	622	
mmm*		622	

[a] Italic operators are antisymmetrical. An asterisk denotes a group with an anticenter of symmetry $\bar{1}$.

674 groups lacking antitranslation and anti-identity operators but containing antisymmetry operators, and (iii) 517 groups lacking antisymmetry operators but containing antitranslation operators.

The magnetic dipole of an atom with uncompensated electron spins may be oriented either parallel or antiparallel to a magnetic field vector when constrained by interaction with the other atoms. If the dipole vector obeys neutral symmetry operations, a classical (Fedorov) space group is applicable, but if an antisymmetrical operator is obeyed, an antisymmetric space group is needed. Collectively, the classical and antisymmetrical groups can be called **magnetic groups**. There are 36 magnetic lattices, 90 magnetic point groups, and 1651 magnetic space groups. In the original literature, the term Shubnikov is applied collectively to the classical and antisymmetrical groups, but it might be easier to restrict it to the antisymmetrical groups. A systematic list of the Shubnikov space groups is given in Shubnikov and Belov (1964, pp. 198–210).

There are 15 polychromatic (Belov) plane groups related to crystallographic groups (Shubnikov and Belov, 1964, pp. 228–237) and 2942 space groups based on 3, 4, and 6 colors (Shubnikov and Koptsik, 1974, p. 302).

EXERCISES   (*More Difficult)

7.1   Cut scalene triangles out of cardboard, and mark the sides in some manner. Use the triangles to demonstrate the symmetry elements in the band groups of Figure 7.1. Locate the symmetry elements on the diagrams. Look for band groups in art and architecture.

7.2   Place dots on rods (e.g., wooden dowel rods) to represent the rod groups in Figure 7.2. After studying Figure 7.3, close the book and draw the operations of the crystallographic axes of symmetry. *After studying Table 7.1, enumerate the 75 space groups for crystallographic rods in some systematic manner.

7.3   Use scalene triangles cut out of cardboard to help you study the two-sided plane groups in Figure 7.4. Locate the symmetry elements on the diagram. *Close the book, and systematically enumerate the 80 two-sided plane groups.

7.4   After studying Figure 7.5, close the book and draw c-axis projections of the space groups of equivalent general positions and symmetry elements for **P112**, **P121**, **P114**, **P3**, and **P6**. Make sure that the unit cell has the correct shape. Choose a position that is asymmetric with respect to axial directions. Systematically develop equivalent positions when one symmetry element is chosen, and then look for equivalent symmetry elements. Always look halfway along the cell when twofold and fourfold axes are used, and in the center of equilateral triangles when triad and hexad axes are used.

7.5   Draw the two space group diagrams for **P222** and check with Figure 7.6.

7.6   After studying Figure 7.7, close the book and generate the b-axis projection of the crystal structure of $AlPS_4$ from the atomic coordinates in the text.

7.7*   Alternatively, draw the a-axis or the c-axis projection of the crystal structure of $AlPS_4$, and identify the coordination polyhedra of S atoms about the Al and P atoms.

7.8   Draw the two space group diagrams for **I112** and **I4**, and check with Figure 7.8. For **I4** be careful to choose a starting position with $y \neq x$.

7.9   To check your understanding of space groups with three intersecting diad axes, study Figure 7.9 and then generate the space groups one by one until you have complete confidence. You should not feel confident until you generate a pair of space group diagrams with no errors at all.

7.10   Draw diagrams for space groups in Figure 7.10 until you gain confidence. Remember that centered lattices must have extra symmetry elements, and that a comma signals an inversion operation.

7.11   Draw diagrams for space groups in Figure 7.11. Look for symmetry elements at the centers of equilateral triangles, and at midpoints of cell edges.

7.12* Draw a b-axis projection of the crystal structure of $Zn(OH)_2$, and construct branches of the 4-coordinated net that links the Zn atoms. Use the origin of Figure 7.14. **Identify the Zn–O linkages.

7.13* The keatite polymorph of SiO_2 has its Si atoms in positions similar to those of the oxygen atoms of ice III. Draw the c-axis projection using $a = 7.46$ Å, $c = 8.60$ Å, and $u = 0.41$, $x = 0.12$, $y = 0.33$, $z = 0.25$. You will find that each Si atom lies approximately 3 Å from four adjacent Si atoms. Oxygen atoms lie approximately halfway between adjacent Si atoms to give a tetrahedral framework, but their positions have not been measured with high accuracy.

7.14* Draw the c-axis projection for the variety of high quartz with space group $P6_4 22$.

7.15* Boron phosphate BPO_4 has crystal structures similar to those of SiO_2 except for alternation of B and P on the tetrahedral nodes. The BPO_4 analog of high cristobalite (Fig. 5.22) has the following structural parameters:

tetragonal $a = 4.33$ Å, $c = 6.64$ Å; space group $I\bar{4}$;

atomic coordinates B $0,\frac{1}{2},\frac{1}{4}$; $\frac{1}{2},0,\frac{3}{4}$

P $0,0,0; \frac{1}{2},\frac{1}{2},\frac{1}{2}$

O $u,v,w; \bar{v},u,\bar{w}; \bar{u},\bar{v},w; v,\bar{u},\bar{w}$

$\frac{1}{2} + u, \frac{1}{2} + v, \frac{1}{2} + w; \frac{1}{2} - v, \frac{1}{2} + u, \frac{1}{2} - w;$

$\frac{1}{2} - u, \frac{1}{2} - v, \frac{1}{2} + w; \frac{1}{2} + v, \frac{1}{2} - u, \frac{1}{2} - w$

$u = 0.14, v = 0.26, w = 0.13$

Draw the c-axis projection and identify the tetrahedral framework. Show that the atomic positions obey the operations of $I\bar{4}$.

7.16** The rhombohedral variety of elemental sulfur has a rhombohedron as unit cell with $a=6.46$ Å and $\alpha=115.3°$. However, it is easier to refer the crystal structure with space group $R\bar{3}$ to a hexagonal cell with triple volume (cf. Fig. 7.13). The resulting atomic coordinates in the hexagonal cell with $a = 10.82$ Å, $c = 4.28$ Å are:

$\pm(uvw; \bar{v},u - v,w; v - u,\bar{u},w); u = 0.145, v = 0.188, w = 0.105$

Draw a 120° rhombus, and plot the six atomic positions about each of the four corners of the rhombus. Show that the positions generate a crown-shaped S_6 molecule. To complete the structure add two more sets of six positions displaced by $x = \frac{1}{3}$, $y = \frac{2}{3}$, $z = \frac{2}{3}$ and $x = \frac{2}{3}$, $y = \frac{1}{3}$, $z = \frac{1}{3}$. Show that the positions obey the rhombohedral lattice and the $\bar{3}$ axis. The shortest distance between S atoms of different molecules is 3.5 Å. Assuming that this gives twice the atomic radius, draw spheres (actually circles) of radius 1.75 Å about all six atoms of a molecule. Show that the strong bonding in the S_6 molecule causes strong compression of the spheres representing the S atoms.

7.17** Potassium dithionate, $K_2 S_2 O_6$, is hexagonal with a=9.78 Å, c=6.29 Å. The space group is **P321**. Draw the c-axis projection from the coordinates in Wyckoff (1965, p. 292), and check that they obey the symmetry elements. Show that there are two crystallographically distinct types of $S_2 O_6$ molecules and determine the symmetry of the molecules.

7.18 Draw diagrams for space groups in Figure 7.18 until you gain confidence.

7.19 After studying Figure 7.20, close the book and reconstruct the diagrams of the first three rows. *Reconstruct the diagrams of the bottom row.

7.20 After studying Figure 7.21, close the book and reconstruct the diagrams for **P4mm, Cmmm, I4/m,** and **I112/m.** *Reconstruct the diagrams for the remaining space groups.

7.21* After studying Figure 7.23, close the book and reconstruct the diagrams.

7.22 Draw the b-axis projection of the $Rb_2 O_2$ structure, and identify the operations of the symmetry elements. Calculate the O–O distances. Show that oxygen spheres with radius corresponding to 0.74 Å (Table 1.1) will almost touch in a molecule, but not between molecules.

7.23 Determine all the possible space group symbols produced by axial transformations of the following space groups: **Pmm2, Pmn2₁, Pba2, Pnc2, Amm2, Pmmm, Pccm, Pnma.**

7.24* Show that space group **Cmc2₁** is the same as **Cbn2₁**, and that **Cccm** is the same as **Cnnn.**

7.25 Determine point group symbols for space groups in Tables 7.3 and 7.4, and check with Table 6.3.

7.26* After reading Section 7.13, close the book and construct the group table for **C12/m1** and the hierarchy of space groups in Figure 7.30. What space groups are cell-equivalent subgroups of **Pmmm**?

ANSWERS

7.7 See Figure 7.33. Some illustrative tetrahedra are shown. Heights are in hundredths.

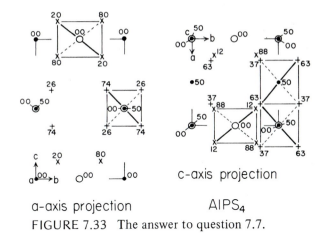

c-axis projection

AlPS₄

a-axis projection

FIGURE 7.33 The answer to question 7.7.

7.12 See Figure 7.34. Lines with two arrows represent superimposed zigzag branches.

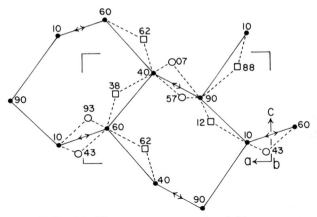

FIGURE 7.34 The answer to question 7.12.

7.13 Your figure should be similar to Figure 7.15. All heights are in multiples of 0.25.

7.14 This is the mirror image of the left-hand diagram in Figure 7.16. The easiest way is to merely exchange the heights ³⁄₆ and ⁵⁄₆, and ⁰⁄₆ and ²⁄₆, while leaving heights ¹⁄₆ and ⁴⁄₆ in place. This changes the chirality of the screw axes.

7.15 See Figure 7.35. Dashed lines show B–O and P–O bonds.

7.16 See Figure 7.36. The upper S atoms of each crown are shown by filled circles and the lower ones by open circles. An S₆ molecule with atomic radius of 1.75 Å is shown at the lower right.

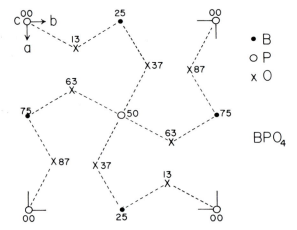

FIGURE 7.35   The answer to question 7.15.

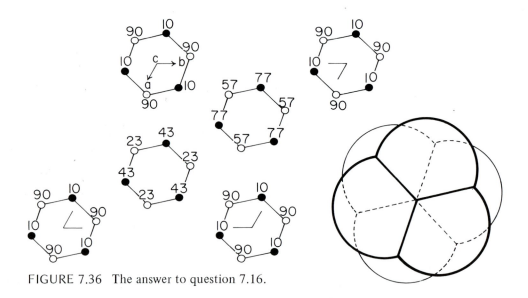

FIGURE 7.36   The answer to question 7.16.

7.17 See Wyckoff (1965, p. 392) for drawings. Two S_2O_6 molecules have the S atoms lying on a rotation triad axis, and the O atoms related in threes by the axis: hence the molecular symmetry is **3**. The third S_2O_6 molecule comprised of S(1) and O(1) atoms has molecular symmetry **32** because the atoms are related by both the triad and diad axes. Note that the coordinates of the S atoms in Wyckoff's Figure IXC,29a are slightly different from those in his Table IXC,22.

7.22 See Figure 7.37. The symmetry elements are in the same positions as on Figure 7.24 (right). The O—O distances are $ps = 1.50$ Å, $pq = 4.20$ Å, and

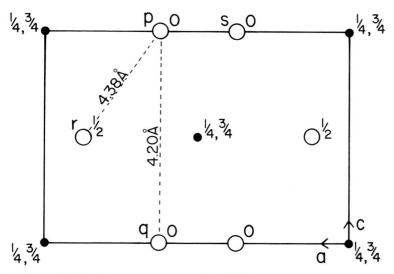

FIGURE 7.37   The answer to question 7.22.

pr = 4.38 Å. Thus spheres of radius 0.74 Å will almost touch in a molecule, but will be far apart between molecules.

7.23   **Pmm2** → **P2mm, Pm2m; Pmn2$_1$** → **Pnm2$_1$, P2$_1$mn, P2$_1$nm, Pm2$_1$n, Pn2$_1$m; Pba2** → **P2cb, Pc2a; Pnc2** → **P2na, Pb2n, Pn2b, Pcn2, P2an; Amm2** → **Am2m, B2mm, Bmm2, Cm2m, C2mm; Pmm,** no change but in a real crystal structure the mirror planes relate different atoms and do change positions; **Pccm** → **Pmaa, Pbmb; Pnma** → **Pbnm, Pmcn, Pnam, Pmnb, Pcmn.**

7.26   **P2mm, Pm2m, Pmm2, P222, P2/m11, P12/m1, P112/m, Pm11, P1m1, P11m, P211, P121, P112, P$\bar{1}$, P1.** Space groups **P2mm, Pm2m,** and **Pmm2** are nonidentical with respect to a specific set of axes, but are formally identical just from the topological viewpoint. Similarly, the monoclinic space groups are related in threes.

8

Atomic Packing and Symmetry I. Simple and Tetrahedral Structures

Preview The concepts of atomic packing (Chapter 5) and space-group symmetry (Chapter 7) are now considered simultaneously using important crystal structures as examples. Structures with close-packing or tetrahedral linkages are discussed in this chapter, and ones with octahedral linkages in Chapter 9.

8.1 Simple Structures with Isometric Symmetry

The space groups of simple cubic packing, body-centered cubic packing, and face-centered cubic packing (i.e., cubic closest packing) are readily derived.

Simple cubic packing (Figs. 5.2a, 8.1) has a primitive lattice, and the single sphere center at (0,0,0) must lie at a special position in an isometric space group. It is easy to find that the sphere center lies at the intersection of the following symmetry elements:

Rotation tetrad axes along [100], [010], [001], or briefly {[100]}

Rotation triad axes along {[111]}

Rotation diad axes along {[110]}

Mirror planes in {(100)}

Mirror planes in {(110)}

Perusal of the isometric space groups in Table 7.5 leads unequivocally to **Pm3m**, whose long symbol is **P4/m$\bar{3}$2/m**. Note that the rotation triad is sufficient in the short symbol, but must be upgraded to an inverse triad in the long symbol because of the presence of a center of symmetry.

The following information provides a sufficient description of simple cubic packing:

cell edge (= sphere diameter)

space group: **m3m**

coordinates of equivalent positions: 0,0,0

Because there is only one equivalent position, the ***multiplicity*** is 1. In the *International Tables for X-Ray Crystallography,* the position (0,0,0) is listed as (*a*) in Wyckoff notation, and the point symmetry is **m3m**. When the multiplicity of an equivalent position is 1, all positions fall on a single Bravais lattice, and the point symmetry of the equivalent position is given immediately by the symmetry elements in the space group symbol.

For body-centered cubic packing, a similar procedure leads immediately to space group **Im3m**, whose long symbol is **I4/m$\bar{3}$2/m**. The formal description is given by the cell edge (= $2/\sqrt{3}$ × sphere diameter) and the following information:

multiplicity: 2

Wyckoff notation: (*a*)

coordinates of equivalent positions: 0,0,0; ½,½,½

point symmetry: **m3m**

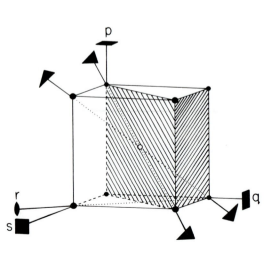

FIGURE 8.1 A clinographic projection of a primitive unit cell of simple cubic packing with one sphere-center at each corner. Three rotation tetrad axes (*p, q,* and *s*) are shown; readers can find others displaced by one-half the cell repeat. Two out of the four rotation triad axes are shown along body diagonal directions. Just one (labeled *r*) out of six orientations of rotation diad axes is shown along a face diagonal. The right-hand face of the cube is hatched to show the mirror plane perpendicular to axis *q*, and a diagonal plane perpendicular to axis *r* is also shown by hatching. The open circle at the body-center represents a center of symmetry. Other centers of symmetry can be found displaced by one-half the cell repeat, including ones lying at the sphere-centers.

For face-centered cubic packing, the space group is **Fm3m** (or **F4/m$\bar{3}$2/m**) and the formal description is:

cell edge = $2/\sqrt{2}$ × sphere diameter

multiplicity: 4

Wyckoff notation: (*a*)

coordinates of equivalent positions: 0,0,0; 0,½,½; ½,0,½; ½,½,0

point symmetry: **m3m**

The *CsCl (or β-brass) structure* (Fig. 5.7*a*) is based on two primitive lattices, and the space group is **Pm3m** with the following positions:

Cs   multiplicity 1   coordinates   0,0,0
Cl   multiplicity 1   coordinates   ½,½,½

Both positions have point symmetry **m3m**, and the second position is given the Wyckoff notation (*b*).

The *fluorite (CaF$_2$)* and *Li$_2$O structures* (Fig. 5.12) are based on three intersecting cubic lattices, and can be described as follows, using CaF$_2$ as the example (Fig. 8.2):

Space group **Fm3m**

Ca   multiplicity 4   coordinates   000; 0½½; ½0½; ½½0
F   multiplicity 8   coordinates   ¼¼¼; ¼¾¾; ¾¼¾; ¾¾¼;
                                ¼¼¾; ¼¾¼; ¾¼¼; ¾¾¾

Rotation tetrad axes pass through the Ca positions, and inverse tetrads through the F positions. Mirror planes of both types pass through Ca to give point symmetry **4/m$\bar{3}$2/m**, which is abbreviated to **m3m**. Only the second type of mirror planes passes through the F positions [Wyckoff notation (*c*)] that have point symmetry **$\bar{4}$3m**. There are many glide planes and screw axes.

The *diamond* structure is obtained by removing half the F sites in the CaF$_2$ structure, and replacing the remaining F and Ca by C atoms (Fig. 5.12*c*). The resulting atomic coordinates define two face-centered cubic lattices (Fig. 8.2*c*): 0,0,0; 0,½,½; ½,0,½; ½,½,0; and ¼,¼,¼; ¼,¾,¾; ¾,¾,¼; ¾,¼,¾. The space group is **Fd3m** (long symbol **F4$_1$/d$\bar{3}$2/m**). Figure 8.2*d* shows some representative symmetry elements, whose operations are explained in the legend.

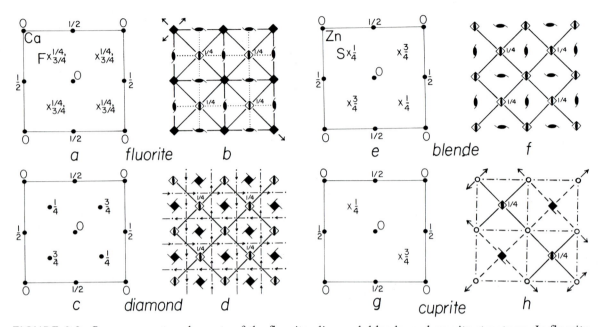

FIGURE 8.2 Some symmetry elements of the fluorite, diamond, blende, and cuprite structures. In fluorite, each Ca atom (dot) lies at the intersection of three rotation tetrad axes (of which only the vertical axis is shown), three mirror planes perpendicular to these axes (of which only two are shown), four triad axes (not shown), and six rotation diad axes (of which only two are shown). Each F atom lies at the intersection of inverse tetrad axes, triad axes, and mirror planes in the diagonal orientation. Screw diad axes and vertical glide planes are also shown. In diamond, inverse tetrad axes alternate with 4_1 and 4_3 screw axes. Mirror planes lie in diagonal directions, and diamond glide planes are parallel to the faces of the unit cell. The symmetry elements of blende are obtained from diamond by (i) inheriting the inverse tetrad axes and mirror planes, and (ii) reducing the 4_1 and 4_3 screw axes to screw diad axes. In cuprite, each copper atom (dot) lies on a center of symmetry and each oxygen atom (cross) lies at the intersection of three inverse tetrad axes. Note that the triad axes are omitted from all diagrams, and that the centers of symmetry are not shown in the diagrams for fluorite and diamond. There are no centers of symmetry in blende. The heights of the inversion points in the $\bar{4}$ axes are either ¼ (and ¾) or 0,½ (not shown).

The space group of the ***blende*** structure (Fig. 5.12e) is a cell-equivalent subgroup of the space group of the diamond structure. The space group $F\bar{4}3m$ is derived (Fig. 8.2f) from $F4_1/d\bar{3}2/m$ by removal of the diamond glide plane and conversion of the 4_1 and 4_3 axes to screw diad axes. Furthermore, the inverse triad axis has become degraded to a rotation triad together with loss of the center of symmetry. Inverse tetrad axes occur in the diamond structure, but are not listed in the space group symbol because the 4_1 axis is given priority. Ordering of Cu and Fe atoms on the Zn positions gives the ***chalcopyrite*** structure, $CuFeS_2$, which has tetragonal symmetry and a doubled c repeat. Its space group, $I\bar{4}2d$, is a subgroup of $F\bar{4}3m$ and $Fd3m$.

Although the ***cuprite*** structure (Fig. 5.12e) has Cu atoms on a face-centered lattice and oxygen atoms on a body-centered lattice, the space group is based on a primitive lattice. Figures 8.2*g* and 8.2*h* demonstrate the operations in **P4$_2$/n$\bar{3}$2/m**. Because there is a center of symmetry the triad axis is of the inverse type. Both $\bar{4}$ and **4$_2$** axes occur perpendicular to the *n* glide planes. Mirror planes and simple glide planes alternate with each other in face-diagonal directions, and are perpendicular to rotation diad axes.

Readers may now like to look back over this section and note how simple changes in choice of atomic positions produce complex changes in the space group symmetry.

8.2 Simple Structures with Hexagonal Symmetry

Simple hexagonal packing (Fig. 5.2*c*) has the sphere center (Fig. 8.3*a*) at the special position (0,0,0) with point symmetry **6/mmm** in space group **P6/m2/m2/m**. Symmetry elements are shown in Figure 8.3*b*. The space group contains $\bar{6}$ and **2$_1$** axes and glide planes in addition to the axes listed in the Hermann–Mauguin symbol.

The *AlB$_2$ structure* (Fig. 5.10*b*) was described initially in terms of a hexagonal net of B atoms and a triangular net of Al atoms. Both types of atoms are in special positions in space group **P6/mmm** (Fig. 8.3*c*):

Al	multiplicity 1	point symmetry **6/mmm**	coordinates 0,0,0
B	multiplicity 2	point symmetry $\bar{6}$m2	coordinates $\frac{1}{3},\frac{2}{3},\frac{1}{2}$; $\frac{2}{3},\frac{1}{3},\frac{1}{2}$

If the hexagonal layer of B atoms were distorted by moving B atoms alternately up and down to $\frac{1}{3},\frac{2}{3},u$; $\frac{2}{3},\frac{1}{3},\bar{u}$, where $u \neq \frac{1}{2}$, the space group would become **P$\bar{3}$m1** (Fig. 7.23).

The *WC structure* (Fig. 5.10) consists of atoms on two triangular nets (Fig. 8.3*d*) with special positions in **P$\bar{6}$m2** (Fig. 8.3*e*):

W	multiplicity 1	point symmetry $\bar{6}$m2	coordinates 0,0,0
C	multiplicity 1	point symmetry $\bar{6}$m2	coordinates $\frac{1}{3},\frac{2}{3},\frac{1}{2}$

Readers should recall that $\bar{6}$ is equivalent to 3/m.

For hexagonal closest packing of equal spheres (Fig. 5.3*b*), the 2-fold position of point symmetry $\bar{6}$m2 in space group **P6$_3$/mmc** is occupied. Because space groups are usually referred to an origin at a center of symmetry, the atomic coordinates for hcp are $\frac{1}{3},\frac{2}{3},\frac{1}{4}$; $\frac{2}{3},\frac{1}{3},\frac{3}{4}$ in Figure 8.3*f* rather than 0,0,0 and $\frac{1}{3},\frac{2}{3},\frac{1}{2}$ in Figure 5.3.

The coordinates of the WC structure are exactly the same as for the hcp structure, except for a translation of ($\frac{1}{3},\frac{2}{3},\frac{1}{4}$), and the WC structure can be

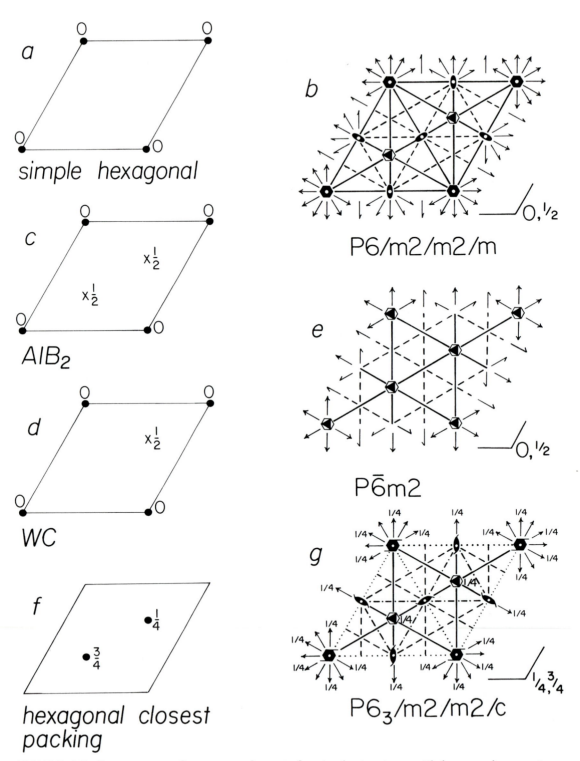

a
simple hexagonal

c
x½
x½
AlB₂

d
x½
WC

f
•¼
•¾
hexagonal closest
packing

b
0,½
P6/m2/m2/m

e
0,½
P6̄m2

g
P6₃/m2/m2/c

FIGURE 8.3 Space groups of symmetry elements for simple structures with hexagonal symmetry. For **P6/mmm** and **P6̄m2**, the inversion point for the $\bar{6}$ axes is at height 0, whereas in **P6₃/mmc** the in-

regarded as a derivative structure from the parent hcp structure. Therefore, the space group of the WC structure should be a subgroup of the space group of the hcp structure, and indeed all the symmetry elements of **P6̄m2** can be found in **P6$_3$/mmc** after translation by ($\frac{1}{3},\frac{2}{3},\frac{1}{4}$). Note that $\bar{6}$ is a subelement of **6$_3$/m**.

The **niccolite** structure type (Fig. 5.10) is based on the following types of special positions in **P6$_3$mc** (Fig. 8.4c):

Multiplicity	Wyckoff type	Point symmetry	Coordinates
2	a	3m	$0,0,u$; $0,0,\frac{1}{2}+u$
2	b	3m	$\frac{1}{3},\frac{2}{3},v$; $\frac{2}{3},\frac{1}{3},\frac{1}{2}+v$

Because there is no center of symmetry, it is necessary to define an origin by setting the z coordinate of one position to zero. After choosing $u = 0$ for Ni in 2(a), $v = \frac{1}{4}$ is needed for As in 2(b).

Carbon atoms in four sites of **graphite** (**P6$_3$/mmc**) give infinite hexagonal layers (Figs. 1.1b, 8.4a) that are tightly bonded at 120° angles by hybridization of sp^2 electron orbitals. Van der Waals bonding between the layers is very weak. The relative length of C–C linkages in (1.42 Å) and between (3.35 Å) the layers is not controlled just by the space group symmetry, and depends also on the measured cell dimensions of $a = 2.46$ Å, $c = 6.70$ Å.

The **wurtzite** structure (Fig. 5.11) uses two sets of special positions of type 2(b) from **P6$_3$mc**, one with $v = 0$ and the other with $v = \frac{3}{8}$. This is not immediately obvious because of a different choice of origin between Figures 5.11 and 8.4a. Make a relative translation of $x = \frac{1}{3}$ and $y = \frac{2}{3}$.

The **molybdenite** structure (Fig. 8.5) is a nice example of another weakly bonded layer structure, and has the same space group **P6$_3$/mmc** as hexagonal closest packing of equal spheres. Let us envisage how the atomic packing and chemical bonding were deduced by crystallographic research. X-ray diffraction studies yielded the cell dimensions, $a = 3.16$ Å, $c = 12.30$ Å, and space group **P6$_3$/mmc**. Chemical analysis yielded the formula MoS_2, and from the density and atomic weights it was deduced that there are 2 Mo and 4 S atoms

---

version point is at height $\frac{1}{4}$. In **P6/mmm**, glide planes with horizontal translation (dashed lines) lie midway between the two sets of vertical mirror planes. In **P6$_3$/mmc**, diagonal glide planes (dash-dot lines) lie midway between the glide planes with vertical translations (dotted lines). Centers of symmetry (open circle inside symbols for hexad and diad axes) occur in **P6/mmm** and **P6$_3$/mmc**, but not in **P6̄m2**. Both sets of horizontal diad axes lie at height $0,\frac{1}{2}$ in **P6/mmm**, but one set lies at height $\frac{1}{4},\frac{3}{4}$ in **P6$_3$/mmc**. Screw diad axes were deliberately omitted in **P6$_3$/mmc** to reduce congestion, but they can be located by analogy with **P6/mmm**.

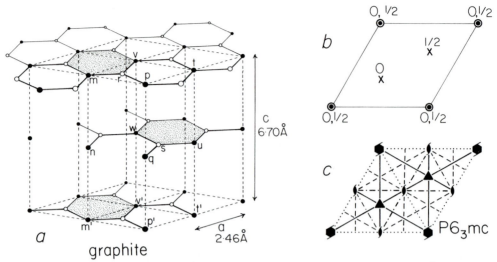

FIGURE 8.4 The crystal structure of graphite, and symmetry elements of space group **P6$_3$mc**. (*a*) A perspective drawing showing three hexagonal layers of carbon atoms joined by the sp^2 hybridization bonds (continuous line). Four unit cells are outlined by dashed lines. Weak van der Waals bonds (vertical dashed lines) connect atoms (*m, p, t, v*) at (0,0,0) to atoms (*n, q, u, w*) at (0,0,½). These C atoms are in Wyckoff position 2(*a*). Carbon atoms marked with open circles are in Wyckoff position 2(*b*). The stippled hexagons are related by a **c** glide plane passing through *v, p, p′*, and *v′*. (*b*) A plan of unit cell projected down the c-axis. The dot and circle correspond to the dots in diagram (*a*), and the crosses correspond to the circles. (*c*) Symmetry elements of **P6$_3$mc**. Planes with horizontal glide lie halfway between mirror planes, and planes with diagonal glide lie halfway between *c* glide planes.

per cell. From measurements of the X-ray diffraction intensities, the atoms were located as follows:

	Wyckoff Site	Coordinates
Mo	(*2c*)	$\pm(\frac{1}{3}, \frac{2}{3}, \frac{1}{4})$
S	(*4f*)	$\pm(\frac{1}{3}, \frac{2}{3}, u; \frac{2}{3}, \frac{1}{3}, \frac{1}{2} + u)$, $u = 0.63$

The c-axis projection is confusing because the atoms overlap in threes (diagram 8.5*a*), but the *a*-axis projection is clearly resolved. In the *a*-axis projection (diagram 8.5*c*), the unit cell is a rectangle outlined by the repeats down the *c* and *b*⋆ axes, where *b*⋆ is the axis at 90° to both the *a* and *c* axes (diagram 8.5*a*). Sulfur atoms at heights $^{13}/_{100}$ and $^{37}/_{100}$ of *c* form a double layer in simple hexagonal packing (diagram 8.5*c*), and the molybdenum atoms at height $^{25}/_{100}$ lie halfway between. In diagram 8.5*b*, the S atoms are approximated by circles in contact in the horizontal plane. Each molybdenum atom occupies the center of a trigonal prism of sulfur atoms (dashed lines), and each vertical edge is shared by three prisms. The prisms do not

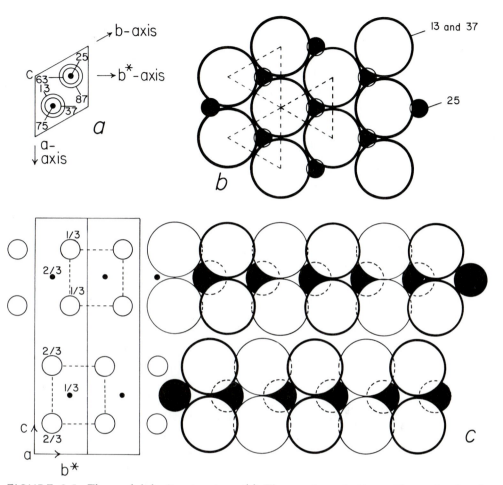

FIGURE 8.5 The molybdenite structure. (*a*) The *c*-axis projection with *a*-axis placed up-down on the paper to allow direct comparison with the *a*-axis projection below. The *z*-coordinate in hundredths. Mo is represented by the filled circle; S by the open circles. (*b*) The plan of the S-Mo-S layer. Sulfur atoms are superimposed at 13 and 37 hundredths, and are represented by touching circles. Their centers form trigonal prisms (dashed lines) about the Mo atoms, which are represented by filled circles smaller than required for close packing of spheres. (*c*) The *a*-axis projection showing atomic positions in and around two unit cells (rectangles) at left, and a packing model at right. Two trigonal prisms are outlined by dashes. S atoms (open circles) at the right touch each other in the S-Mo-S layers, and the Mo atoms are represented by filled circles large enough to give close packing with the S atoms.

share faces. To show the double layers in cross section, the atomic positions are shown as small circles and dots at the left of diagram 8.5c, for which two unit cells are outlined by rectangles. The x coordinates are shown for atoms in one of the unit cells. At the right of diagram 8.5c, the atoms are shown full size. The sulfur atoms were drawn as circles just large enough to touch along the vertical edges of the trigonal prisms, and the Mo atoms were drawn just large enough to touch the sulfur atoms. The circles for the S atoms are not large enough to touch between the double layers. This arises because the bonding between Mo and S atoms is much stronger than between S atoms, thereby ruling out a simple model of packing of spheres. The Mo–S distance is 2.35 Å, and the S–S distance between the nearest atoms in a trigonal prism is 2.98 Å, which is much less than the S–S distance of 3.66 Å between layers. For an ionic model, the radius for the S^{2-} ion is usually about 1.8 Å. Subtraction from 2.35 Å would give ~0.55 Å for the radius of the molybdenum ion. The resulting radius ratio of ~0.3 is less than the value of 0.528 for sphere filling of a trigonal prism (Table 5.1). It is obvious that concepts based on sphere packing provide only an approximate guide to understanding the crystal structure of molybdenite.

Taken alone, the Mo atoms occupy the sphere centers of hexagonal closest packing (Fig. 8.3f), and the S atoms are related by vertical translations of $\pm(u - \frac{1}{4}) = \pm 0.38$. Consequently the MoS_2 structure can be described symbolically by using the A, B, a, and b symbolism already used in Chapter 5. See Wells (1975, Fig. 4.10) for the relation to other structures with trigonal prismatic coordination.

Again, readers may like to look back over this section and marvel at the complex changes in space-group symmetry produced by simple changes in atomic positions. Furthermore, they should note that some knowledge of chemical bonding is needed in order to relate measured atomic coordinates to inferred atomic packing.

8.3   The Halite Structure and Its Derivatives—CaC_2, Pyrite, Cobaltite, Calcite, and Dolomite: Marcasite and Aragonite Polymorphs

The Na atoms of the *halite* structure (Fig. 5.8) occupy the same positions as face-centered cubic packing (Section 8.1). The Cl atoms lie on another isometric F lattice and their positions are given Wyckoff notation (b) in Fm3m.

The halite structure can be explained by close packing of Na^+ and Cl^- spherical ions (Fig. 1.1a). The crystal structure of *calcium carbide, CaC_2*, corresponds to the packing of spherical Ca atoms and dumbbell C_2 molecules. Figure 1.9a shows a possible way of packing equal numbers of circles and doublets in a 2D sheet, and, indeed, this is the key to the CaC_2 structure. The clinographic projection of a unit cell (Fig. 8.6*a*) corresponds to the following crystallographic data in Wyckoff (1963, p. 350): tetragonal, space group **I4/mmm**, $a = 3.87$ Å, $c = 6.37$ Å;

Ca Wyckoff position (*2a*) Coordinates $0,0,0; \frac{1}{2},\frac{1}{2},\frac{1}{2}$
C   (*4c*) $0,0,u; \quad 0,0,\bar{u}; \quad \frac{1}{2},\frac{1}{2},u; \quad \frac{1}{2},\frac{1}{2},\bar{u};$
                                                    $u = 0.406$

Two C atoms and one Ca atom overlap in threes in the *c*-axis projection (Fig. 8.6*b*). Although the body-centered cell (continuous lines) yields the

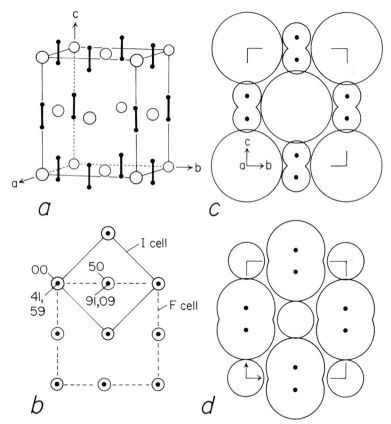

FIGURE 8.6 The tetragonal CaC$_2$ structure. (*a*) Clinographic projection of **F**-centered unit cell showing Ca atoms (large circles) at corners and face centers, and C$_2$ doublets at edge and body centers. (*b*) The *c*-axis projection showing **I**- and **F**- centered unit cells. Atomic heights to nearest hundredth. (*c*) The *a*-axis projection of **F** cell showing one layer of Ca atoms and C$_2$ doublets. (*d*) The alternative to diagram (*c*) in which each small circle represents a Ca^{2+} ion of radius 1 Å and the C$_2$ doublets are drawn from part-circles with radius 1.55 Å.

simplest description of the crystal structure, the face-centered cell (dashed lines) with doubled volume is needed for direct comparison with the NaCl structure. Then the Ca atoms (large circles in Fig. 8.6*a*) correspond to the Na atoms of the halite structure, and the C_2 doublets correspond to the Cl atoms. Because the doublets are parallel to each other, the CaC_2 structure cannot have isometric symmetry, and the axis of the doublets becomes the *c* axis of the tetragonal unit cell. The space group **I4/m3m** of the halite structure becomes degraded to **I4/mmm** (Fig. 7.22). The Ca atoms lie in a special position with point symmetry **4/mmm**. Absence of the horizontal mirror plane results in point symmetry **4mm** for the C atoms.

Interpretation of the atomic packing depends on chemical models. X-ray diffraction techniques yielded the cell dimensions and atomic coordinates of the CaC_2 structure, but did not yield direct information on the chemical bonding. Figure 8.6*c* shows an interpretation based on the assumption that the Ca and C atoms can be represented by spheres with atomic radii of 1.97 and 0.77 Å, respectively (Table 1.1). Because each pair of C atoms is separated only by 1.2 Å, the C atoms are truncated. The slight misfit between the Ca atoms and C_2 doublets requires some distortion from spherical shape. For an ionic model with a radius of 1.0 Å for the Ca^{2+} ion, a radius near 1.55 Å would be needed for the supposed carbon ion (Fig. 8.6*d*).

At high temperature, the C_2 doublets are not parallel, and appear to "rotate." The averaged position has a spherical distribution of electron density, and the crystal structure is of halite type with statistical isometric symmetry.

 The crystal structure of ***pyrite, FeS$_2$***, is also based on the halite structure. Iron atoms occupy the same positions as the Na atoms of halite and the Ca atoms of the CaC_2 structure. Each S_2 doublet lies parallel to one of the four triad axes of the isometric unit cell (*a* = 5.41 Å), and the relative orientations obey the symmetry of space group **P2$_1$/a$\bar{3}$** (short symbol **Pa3**). The atomic coordinates (Wyckoff, 1963, p. 346) are:

Fe $0,0,0; 0,\frac{1}{2},\frac{1}{2}; \frac{1}{2},0,\frac{1}{2}; \frac{1}{2},\frac{1}{2},0$

S $\pm(u,u,u; \frac{1}{2}+u,\frac{1}{2}-u,\bar{u}; \bar{u},\frac{1}{2}+u,\frac{1}{2}-u; \frac{1}{2}-u,\bar{u},\frac{1}{2}+u)$, $u = 0.386$

A projection of the unit cell (Fig. 8.7*a*) shows a zigzag pattern of doublets. Some readers may be able to visualize the structure in 3D just from the 2D projection, but most will prefer a stereo view from Figure 8.8. Each Fe atom lies at the center of an octahedron of six S atoms (Fig. 8.7*b*), and each S atom is bonded to one S atom at 2.14 Å and three Fe atoms at 2.26 Å.

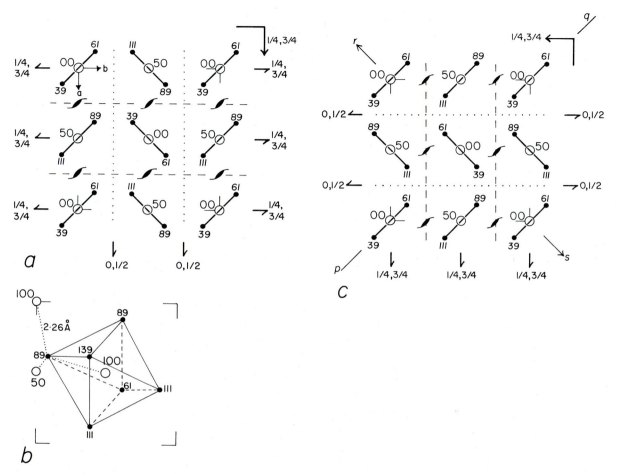

FIGURE 8.7   The crystal structure of pyrite. (*a*) The *c*-axis projection of Fe atoms (large circles) and S_2 doublets (filled small circles connected by bars) related by symmetry elements of **Pa3**. Height to nearest hundredth. (*b*) An octahedron of S atoms about an iron atom at $\frac{1}{2}, \frac{1}{2}, 0$. The S atom at 0.39, 0.11, 0.89 is bonded to three Fe atoms at 2.26 Å (dotted lines). (*c*) The *c*-axis projection of atomic positions and symmetry elements for **Pb3**.

The octahedra share vertices, but not edges or faces, to give an infinite framework of linked octahedra.

Space group $\mathbf{P2_1/a\bar{3}}$ requires careful study. In Figure 8.7*a,* there is a **b** glide plane (dashed lines) perpendicular to the *a* axis, a **c** glide plane (dotted lines) perpendicular to the *b* axis, and an **a** glide plane at height $\frac{1}{4}, \frac{3}{4}$ perpendicular to the *c* axis. There are screw diad axes parallel to all three cube-edge directions, and inverse triad axes (not shown) parallel to all four body-diagonal directions. Centers of symmetry (not shown) lie at all corners and centers of faces and edges. In an orthorhombic space group, the correspon-

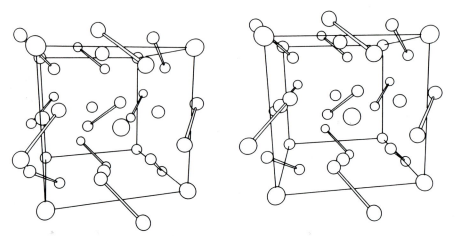

FIGURE 8.8   A stereo-view of pyrite structure.

ding symmetry planes and diad axes would be expressed by the symbols $2_1/b\ 2_1/c\ 2_1/a$. In the cubic space group, the three screw axes and the three glide planes are equivalent because of the operations of the triad axis, whereas in the orthorhombic space group $P2_1/b\ 2_1/c\ 2_1/a$ they are not equivalent. The symbol for the cubic space group $P2_1/a\bar{3}$ ignores the $2_1/b\ 2_1/c$ because these elements are generated automatically. Actually, all the symmetry elements and the multiplication table for the symmetry group can be derived just from the short symbol **Pa3**.

Is there a space group $P2_1/b\bar{3}$? Yes, as shown by the symmetry operations and atomic coordinates for pyrite in Figure 8.7c. However, this is not an independent space group, because the positions in Figure 8.7c can be converted into those of Figure 8.7a by 180° rotation about the line *rs*. Reflection in the plane *pq* will also accomplish the conversion. Although $P2_1/b\bar{3}$ is not an independent space group, it is useful in explaining the twinning of pyrite.

In *calcite,* the low-pressure variety of $CaCO_3$, Ca atoms and CO_3 groups replace the Na and Cl atoms of halite. Each CO_3 group is an equilateral triangle (Fig. 1.1e), and in calcite all CO_3 groups are parallel with consequent reduction of the symmetry to rhombohedral. Figure 8.9a is a perspective drawing of a unit cell of the NaCl structure with one of the triad axes vertical, and Figure 8.9b is a corresponding view of the atomic positions of calcite in which the Ca and C atoms, respectively, match the Na and Cl atoms. Although all CO_3 groups are parallel and lie in horizontal layers, the CO_3 groups of adjacent layers point alternately to the left and to the right. Thus the CO_3 groups in layers 11 and 7 point to the right, whereas those in layer 9 point to the left. This change of direction allows each Ca atom to be bonded to six oxygen atoms, one each from three CO_3 groups in an upper layer and three CO_3 groups in a lower layer. To see this, first

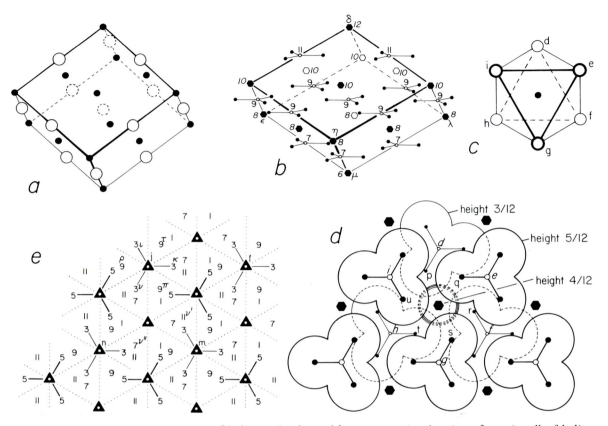

FIGURE 8.9   The crystal structures of halite and calcite. (*a*) A perspective drawing of a unit cell of halite with a triad axis vertical. An Na atom (filled circle) is placed at each corner and face-center, and a Cl atom (open circle) at each edge-center and the body-center. (*b*) A corresponding drawing of calcite with Ca atoms (hexagon) and CO_3 groups (C open circle; O filled circle). (*c*) An octahedron of Cl atoms around an Na atom, viewed down a triad axis. (*d*) A projection down the triad axis of layers of CO_3 groups at heights $3/12$ (thin line) and $5/12$ (thick line). The centers of calcium atoms at height $4/12$ are shown by a hexagon, but the size is shown for only one of them (triple circle). (*e*) A projection down the triad axis of the oxygen atoms of calcite. Inverse triad axes and vertical glide planes are shown, but other symmetry elements are omitted.

examine Figure 8.9*c*, which shows an octahedron of Cl atoms in halite arranged about an Na atom, and viewed down a triad axis. In calcite (Fig. 8.9*d*), each of the three upper Cl atoms (*e*, *g*, and *i*) is replaced by a CO_3 triangle pointing to the left while each of the three lower Cl atoms (*d*, *f*, and *h*) is replaced by a CO_3 triangle pointing to the right. Because a CO_3 group is nonspherical, the octahedron *defghi* in diagram 8.9*c* is replaced by a trigonal antiprism *defghi* in diagram 8.9*d*. The cubic unit cell of NaCl is correspondingly changed to the rhombohedron of calcite, which has *a* = 6.41 Å and $\alpha = 101.9°$. The Ca atom at the center of the trigonal antiprism of six C atoms is bonded to six oxygen atoms that form the octahedron *pqrstu* (diagram 8.9*d*) with Ca–O = 2.4 Å. This octahedral coordination is the reason for the particular arrangement of CO_3 groups and Ca atoms in calcite. In diagram 8.9*d,* the oxygen atoms are represented by spheres (actually circles) corresponding to an ionic radius of 1.35 Å and the calcium atoms by circles of radius 1.0 Å. The oxygen atoms are truncated because of the shortness of the C–O distance. Oxygen spheres of adjacent CO_3 groups do not touch.

Although the atomic packing in diagram 8.9*d* is simple and elegant, careful attention is required for understanding of the space–group symmetry. The rhombohedron in diagram 8.9*b* is not the unit cell of calcite. For convenience, number the atoms and molecules from 12 to 6 in order of decreasing height. The Ca atoms lie at heights 12, 10, 8, and 6 and the CO_3 groups at heights 11, 9, and 7. Groups at height 11 point to the right, those at 9 to the left, and those at 7 to the right. Continuing the alternation, groups at height 5 must point to the left. Because the groups at height 11 point to the right, the cell repeat cannot be merely 6 units. Continuing the sequence:

$$11 \quad 9 \quad 7 \quad 5 \quad 3 \quad 1 \quad -1 \quad -3$$

$$R \quad L \quad R \quad L \quad R \quad L \quad R \quad L$$

the layer at –1 matches the layer at 11, thereby giving a repeat of 12 layers. This, of course, is the reason why the layers in diagram 8.9*b* were numbered up to 12. To obtain the edges of the rhombohedral unit cell, draw the lines δε, δη, and δλ. The Ca atom labeled μ lies at the body center of the unit cell.

The space-group symmetry is seen best from the projection of the oxygen atoms down the body diagonal of the unit cell (diagram 8.9*e*). Each oxygen atom is represented by its height in twelfths. The C–O bonds at heights $^3/_{12}$ and $^5/_{12}$ in diagram 8.9*d* are carried over into diagram 8.9*e*, but lines are omitted for C–O bonds at heights $^1/_{12}$, $^7/_{12}$, $^9/_{12}$, and $^{11}/_{12}$. Readers might wish to insert these lines using colored pencils. From this simplified diagram, it is now easy to locate symmetry elements. Each set of three oxygen atoms (e.g., ι, κ, and ν) obeys a rotation triad axis. However, there are three other oxygen atoms (e.g., π, ρ, and τ) that require the higher symmetry of an inverse triad axis $\bar{3}$. This axis, of course, is inherited from the space group **F4/m3̄m** of halite. Pairs of atoms like ι and τ, or ι and ρ, or ρ and ν, satisfy the three orientations of a vertical glide plane shown by dotted lines. This **c** glide plane was not listed in **F4/m3̄m** but is present in that space group. Diagonal glide planes (not shown) occur midway between the **c** glide planes, but need not be specified in the space-group symbol because they are generated automatically from the other crystallographic elements. There are rotation and screw diad axes in horizontal directions and centers of symmetry (not shown). The resulting space-group symbol of calcite is **R$\bar{3}$2/c**, or briefly **R$\bar{3}$c**.

It is easiest to work with an hexagonal unit cell such as *jlmn* in diagram 8.9*e*. This cell has $a = 4.99$ Å, $c = 17.1$ Å, and the atomic coordinates are:

Ca　(6*b*)　　0,0,0; 0,0,½; rh

C　　(6*a*)　　0,0,¼; 0,0,¾; rh

O　　(18*e*)　$u,0,¼; \bar{u},0,¾; 0,u,¼; 0,\bar{u},¾; \bar{u},\bar{u},¼; u,u,¾$; rh; $u = 0.26$

The abbreviation rh means that there are two more sets of positions that are obtained merely by adding ⅓,⅔,⅔ and ⅔,⅓,⅓. Thus position ν in diagram 8.9*e* at $x = 0.26$, $y = 0$, $z = ¼$ (i.e., ³⁄₁₂) is repeated at $x = ⅓ + 0.26$, $y = ⅔$, $z = ⅔ + ¼$ (position ν') and at $x = ⅔ + 0.26$, $y = ⅓$, $z = ⅓ + ¼$ (position ν'').

Dolomite, CaMg(CO₃)₂, has essentially the same structure as calcite except for regular alternation of Ca and Mg. Because the ionic radius of Mg^{2+} is smaller (0.72 Å) than for Ca^{2+}, the cell dimensions shrink to $a = 4.81$ Å, $c = 16.0$ Å. The space group loses the c glide plane and becomes just **R$\bar{3}$**. The CO_3 groups move slightly up or down to accommodate to the two sizes of cations, and the coordinates are:

Ca　(3*a*)　　0,0,0, rh

Mg　(3*b*)　　0,0,½; rh

C　　(6*c*)　　±0,0,u; rh; $u \cong ¼$

O　　(18*f*)　$\pm(x,y,z; \bar{y},x - y,z; y - x,\bar{x},z)$; rh; $x = 0.237$; $y = 0.965$; $z = 0.244$

Thus the dolomite superstructure retains the rhombohedral cell of the parent calcite structure (except for minor shrinkage) but adopts a lower symmetry.

The *cobaltite (CoAsS)* and *ullmanite (NiSbS)* superstructures of pyrite have regular alternation of S with As or Sb which eliminates the glide planes. In the resulting space group **P2₁3**, all atoms are in a 4-fold position with co-ordinates u,u,u; $½ + u, ½ - u, \bar{u}$; $\bar{u}, ½ + u, ½ - u$; $½ - u, \bar{u}, ½ + u$, where $u(Ni) = 0.976$, $u(Sb) = 0.625$, and $u(S) = 0.39$.

Marcasite (FeS₂) has the same chemical composition as pyrite, and its structure likewise consists of S_2 doublets linked by Fe atoms. Because the topological relationships are different, marcasite and pyrite are polymorphs. The reader should now do Exercise 4.4 if this has not already been done. Examination of Figure 4.20*d* should reveal a glide plane of type **n11** at $x = ¼, ¾$ that relates atom q to atom r. A second glide plane **1n1** lies at $y = ¼, ¾$, and a mirror plane **11m** lies at $z = 0, ½$. The lattice is primitive, and the space group is **Pnnm**. Each Fe atom lies in an octahedron of S atoms, but the shape is different from that of the octahedron in pyrite. *Arsenopyrite,*

FeAsS, is a superstructure of the marcasite parent structure. Removal of the
S_2 doublets from pyrite leaves a face-centered cubic array like that of the
taenite variety of iron metal, except for the greater interatomic distance.
Removal of the S_2 doublets from marcasite leaves a body-centered array of
Fe irons, reminiscent of the kamacite variety of iron metal; however, the
iron atoms of marcasite lie on the nodes of an orthorhombic lattice instead
of the cubic lattice of kamacite.

The *aragonite* variety of $CaCO_3$ is also orthorhombic, but its space group
is **Pmcn**. Whereas the calcite structure is related to the halite structure, the
aragonite structure is related to the niccolite structure. The following data
are taken from Wyckoff (1964, p. 364) after changing the labeling of the
axes:

$$a=4.96 \text{ Å}, b=7.97 \text{ Å}, c=5.74 \text{ Å}$$

Position (4c) $\pm(\frac{1}{4},y,z; \frac{1}{4}, \frac{1}{2} - y, \frac{1}{2} + z)$

Position (8d) $\pm(x,y,z; \frac{1}{2} - x,y,z; x, \frac{1}{2} - y, \frac{1}{2} + z; \frac{1}{2} - x, \frac{1}{2} - y, \frac{1}{2} + z)$

Atom	Position	x	y	z
Ca	(4c)	1/4	0.417	0.750
C	(4c)	1/4	0.750	0.917
O(1)	(4c)	1/4	0.917	0.917
O(2)	(8d)	0.48	0.67	0.917

Each CO_3 group is bisected by a mirror plane (Figs. 8.10a, 8.10b). All
groups are parallel, but alternately point to left and right. Those at heights
92/100 and 108/100 are approximately coplanar. To demonstrate the packing, circles are drawn about the oxygen atoms with radius of 1.35 Å appropriate for ionic bonding. The oxygen atoms fit tightly along the a axis, but
there are gaps along the b axis. Readers may like to draw circles around the
oxygen atoms at heights 42/100 and 58/100, which form a second near-
planar layer. Each Ca atom has nine oxygen neighbors, and the coordination
polyhedron (diagram 8.10c) has point symmetry **m11**. There is no symmetry
control over the z coordinate of Ca and the y coordinate of C, and the values
of 0.75 are controlled by the chemical bonding, and are only fortuitously
equal to ¾.

The structural relation to niccolite is demonstrated by comparing diagram 8.10d for part of the aragonite structure with diagram 8.10e for a unit
cell of niccolite. Change of the height of the C atoms from 42/100 and
92/100 to 50/100 and 100/100, plus a small change in the axial ratio a/b
from $4.96/7.97 = 0.622$ to $1/\sqrt{3} \cong 0.577$ would equalize the diagram. In-

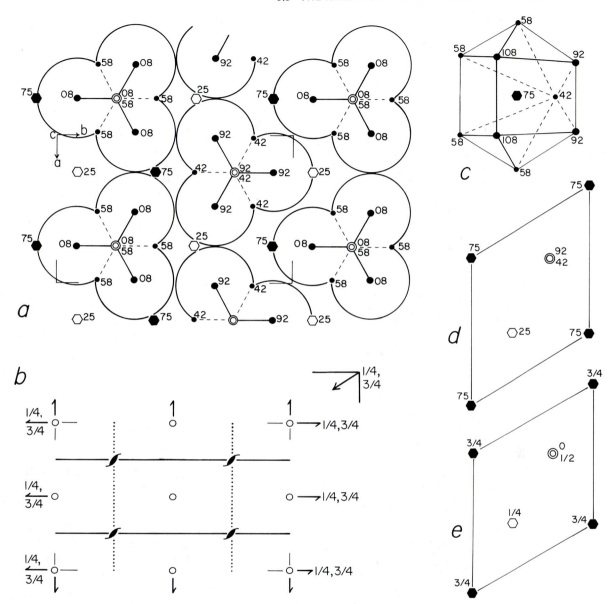

FIGURE 8.10 The crystal structure of aragonite. (*a*) The *c*-axis projection showing positions of calcium (hexagon), carbon (open circle), and oxygen (filled circle) atoms. Height in hundredths. The circles show the packing of one layer of oxygen anions (radius 1.35 Å). (*b*) The symmetry elements of **Pmcn**. (*c*) The coordination polyhedron around Ca atom at 0.25, 0.417, 0.75. (*d* and *e*) A comparison of positions of Ca and C atoms with those of As and Ni atoms of niccolite (cf. Figure 5.10*d*).

deed the pseudohexagonal nature of the aragonite structure is responsible for the mimetic twinning.

Why does $CaCO_3$ occur in two structural modifications? Aragonite is thermodynamically stable only at high pressure, and calcite only at low pressure (actually aragonite is precipitated metastably by certain creatures). Aragonite is denser (2.95 g/cm^3) than calcite (2.71), and this arises from the closer packing of the CO_3 groups and the higher coordination number of the Ca atom (aragonite **9**; calcite **6**). The space group of calcite will not allow the CO_3 molecules to pack together more closely, and the lower symmetry of **Pmcn** allows the tighter packing of aragonite. Unfortunately, this explanation assumes knowledge of the structure types. A more fundamental explanation would involve a priori invention of all possible ways of packing together Ca atoms and CO_3 molecules and demonstration that the calcite and aragonite structures are the stable ones. An even more fundamental explanation would merely assume the chemical composition $CaCO_3$. At the current stage of thinking about crystal structures it is possible only to work with simple geometrical and topological ideas, and the next stage will present a great challenge.

The reader may now wish to look back over this section and note how the simple highly symmetric halite structure is the parent of less-symmetric structures (including more complex ones not described here). In turn, the calcite structure is the parent for the dolomite superstructure.

8.4 Structural Principles and Classifications

Now that some structures have been analyzed from the viewpoint of both geometry and symmetry, it is time to reflect on structural principles and classifications. It is expected that readers will be familiar with the principles of chemical bonding, as exemplified by Pauling's rules for ionic bonding.

At the deepest philosophical level, a crystal can be viewed as a beautiful object in the tradition of the Greek thinkers. A mystical feeling emanates from the writings of the Swiss–German crystallographers typified by P. Niggli and F. Laves (e.g., 1930), and, indeed, there can be great psychic satisfaction in the study of crystals. Even the most complicated crystal structure fits together neatly, and discovery of its structural principles provides deep satisfaction. Unfortunately, however, it seems doubtful that there is a single grand principle, and it is necessary to develop a variety of classifications.

The extensive literature of structural types is reviewed by Hellner (1979), and the most thorough modern classifications of crystal structures are by Wells (1975) and Pearson (1972). Many structures can be classified on the basis of sheets (e.g., Lima-de-Faria, 1978; Figueiredo and Lima-de-Faria, 1978).

Laves (1955) presented three general principles that are interrelated. The ***principle of economical use of space*** is particularly exemplified by the closest packing of spheres. Ionic bonding, and to a lesser degree van der Waals bonding, results in attractive forces that pull the structural units together. This principle is less applicable to covalent crystals because the restricted interatomic angles are not necessarily compatible with economical use of space. Thus, a benzene ring contains an empty hole, and the tetrahedral bonding of diamond is *topologically* inefficient with respect to closest packing of equal spheres. However, some covalent structures can be dense (e.g., diamond) because of the shortness of the bond distances. The ***principle of highest symmetry*** is difficult to evaluate. As time passes, crystallographers are able to solve crystal structures of increasing complexity, and these tend to have low symmetry. Perhaps it might be demonstrated that a crystal structure has the highest symmetry compatible with economical use of space and the specific requirements of chemical bonding between nearest neighbors. The ***principle of the greatest number of connections*** is also difficult to evaluate. Certainly the coordination number is always high in alloys and many oxides as a result of the principle of economy, but it is not immediately obvious how the principle of connections is applied to the frameworks of some zeolites. In spite of the above sceptical comments, there can be no doubt that these three principles are built into the brain of many crystallographers, and that they help in gaining a "sixth sense" about the properties of crystals. The science of complex phenomena transforms imperceptibly into art, and the distinction between science and magic can be hazy.

Following this rather philosophical introduction, some specific concepts are now discussed. First, what properties determine whether crystal structures are "equal," "similar," or "different" (Laves in Hellner, 1979, p. 82)?

Two crystals "have the same atomic arrangement" and are ***isotypic*** (or ***isostructural***) if their crystal structures are congruent with respect to space group and occupancy of equivalent positions. To understand this definition it is desirable to describe what is meant by a lattice complex. In morphological crystallography, the ***form*** consists of the assemblage of faces necessitated by repetition of a chosen face when the symmetry operations of the point group are applied. In a crystal structure, a ***lattice complex*** is the assemblage of points necessitated by repetition of a chosen point when the operations of the space group of symmetry elements are applied. A lattice complex is invariant when the x, y, and z coordinates are fixed by symmetry, as for the lattice complex at $(0,0,0)$ in primitive cubic packing. An invariant lattice complex is analogous to a crystallographic polyhedron with fixed interfacial angles, such as a cube. A lattice complex with unfixed x, y, or z is analogous to a crystallographic polyhedron with unfixed interfacial angles. Consider now how invariant and variant lattice complexes are related to the concept of ***isotypism*** in classification of crystal structures. Occupancy of the same invariant lattice complexes by chemically similar species (e.g., for NaCl and MgO with the halite structure type) clearly fits the definition of isotypism. If one or more of the lattice complexes has unfixed coordinates, there may be problems in deciding whether the structures are isotypic. Because the bonding to nearest neighbors is the most important chemical property of a crystal structure, the topological nature of the nearest-neighbor bonds is the main criterion for making a decision. Unfortunately, it is impossible to draw a firm line: thus, in feldspar, the substitution of Na for K results in change of the number of nearest oxygen neighbors from 9 to 7 without change of the topology of the linkages of the aluminosilicate framework. Even for isotypes, there can be chemical differences: thus the fluorite and antifluorite structures are isotypes from the geometrical viewpoint, but the cations and anions are interchanged: the term ***antitypism*** can be used.

In case of doubt, it is safer to use the term ***homeotypism*** rather than isotypism. Homeotypism implies that the crystal structures are substantially similar, but some minor structural feature is different. Examples of such a difference are: (i) ***polytypes,*** in which a major structural unit (commonly a sheet) is linked together in more than one way (e.g., sphalerite versus wurtzite, tridymite versus cristobalite); (ii) ***parent-derivative pairs,*** in which a parent (e.g., diamond) is related to a derivative (e.g., blende) by splitting of a lattice complex; another example is the high–low transition in quartz; (iii) ***stuffed*** or ***nonstoichiometric structures*** in which either an extra lattice complex is used (e.g., nepheline, $NaAlSiO_4$, has an extra position for Na compared to tridymite $SiSiO_4$), or a lattice complex is not fully occupied (e.g., pyrrhotite $Fe_{1-x}S$ versus troilite FeS); (iv) ***cluster-replacement structures*** in which an atom in a lattice complex is replaced by a cluster of atoms to generate a lattice complex of higher multiplicity (e.g., replacement of Cl in halite by S_2 in pyrite). ***Heterotypes*** cannot be classified as isotypes or homeotypes. The old term ***isomorphism*** was based on similarity of external shape with the implication of similarity of internal structure. The term ***polymorphism*** describes chemical species that crystallize into several crystal structures, for example, quartz, tridymite, and cristobalite. Polymorphs can be homeotypic or heterotypic.

Lattice complexes (Fischer et al., 1973) are useful for classification of crystal structures, especially those of high symmetry with few atoms per cell. In addition, the German word *"Bauverband"* is needed to express the principal topologic connections of a crystal structure. Referring back to Section 8.3, the calcium carbide and pyrite structures belong to the same Bauverband as the halite structure because (i) the Ca atoms of CaC_2 and the Fe atoms of FeS_2 match directly with the Na atoms of halite, and (ii) the centers of the C_2 and S_2 doublets match with the Cl atoms. Furthermore, the AuCu and $AuCu_3$ ordered structures belong to the same Bauverband as the disordered

Au_xCu_{1-x} alloy because the atomic positions belong to the same lattice complex when the chemical distinction is removed and the tetragonal pseudo-isometric symmetry is converted to isometric symmetry. Thus the key feature is the **connectivity** of the structural components, and a connected unit may be either a single position or the center of a cluster. The cluster might be a doublet, a polygon, or a polyhedron. Perhaps the best translation of Bauverband is **structural connectivity**.

In many simple crystal structures, the structural connectivity can be specified by two symbols, one for a lattice complex, and one for a cluster. Table 8.1 lists the 16 invariant lattice complexes of the isometric system in standard positions. Complexes P, I, and F correspond, respectively, to the sphere centers for primitive, body-centered, and face-centered cubic packing, and D corresponds to the C atoms in diamond. The CsCl structure can be written $Cs(P) + Cl(P: \frac{1}{2}\frac{1}{2}\frac{1}{2})$ where $Cs(P)$ refers to occupancy of the P complex by Cs, and $Cl(P: \frac{1}{2}\frac{1}{2}\frac{1}{2})$ to occupancy by Cl of the P complex shifted by $x = y = z = \frac{1}{2}$. The NaCl structure can be written $Na(F) + Cl(F:$

Table 8.1 Standard settings of the 16 invariant isometric lattice complexes (Hellner, 1979)

Label	Space Group	Wyckoff Notation	Point Symmetry	Coordinates (in eighths)
P	Pm3m	*1(a)*	m3m	000
I	Im3m	*2(a)*	m3m	000; 444
J	Pm3m	*3(c)*	4/mmm	044; 404; 440
J^*	Im3m	*6(b)*	4/mmm	J; (J + 444)
F	Fm3m	*4(a)*	m3m	000; 044; 404; 440
D	Fd3m	*8(a)*	$\bar{4}$3m	F; (F + 222)
^+Y	P4$_3$32	*4(a)*	32	111; 537; 753; 375
$^+Y^*$	I4$_1$32	*8(a)*	32	^+Y; (^+Y + 444)
Y^{**}	Ia3d	*16(b)*	32	$\pm(^+Y^*)$
W	Pm3n	*6(c)*	$\bar{4}$m2	204; 420; 042; 604; 460; 046
W^*	Im3m	*12(d)*	$\bar{4}$m2	W; (W + 444)
T	Fd3m	*16(c)*	$\bar{3}$m	111; 133; 313; 331; 155; 515; 551; 177; 717; 771; 357; 735; 573; 375; 537; 753
S	I$\bar{4}$3d	*12(a)*	$\bar{4}$	023; 302; 230; 061; 106; 610; 254; 425; 542; 467; 746; 674
S^*	Ia3d	*24(d)*	$\bar{4}$	S; (S + 444)
^+V	I4$_1$32	*12(c)*	222	021; 102; 210; 063; 306; 630; 274; 427; 742; 465; 546; 654
V^*	Ia3d	*24(c)*	222	$\pm(^+V)$

$\frac{1}{2}$ 00). Because each lattice complex is occupied by a single atom, no symbol is needed for a cluster. A list of symbols for clusters is given in Hellner (1965; 1979, pp. 69–70): thus $2l$ stands for a doublet and $6o$ for an octahedron. The pyrite structure could be expressed as $Fe(F) + S(F: \frac{1}{2} 00; 2l)$. Also useful (Fischer, 1971, 1973, 1974) is the concept of **sphere packing,** defined as a 3D arrangement of spheres in which (i) the arrangement is periodic in three independent directions, and (ii) every two spheres may be connected by a sequence of spheres in contact. If the spheres are equivalent by space-group symmetry, the sphere packing is homogeneous. Readers should consult Hellner (1979) for systematic development of these concepts. This classification scheme will probably be used mostly for crystal structure of high symmetry.

An alternative approach is to classify crystal structures according to the connectivity of the strongest chemical bonds. Strictly speaking, this sentence implies understanding of chemical theory, but a similar conclusion is reached if emphasis is placed on the shorter interatomic distances. In graphite, the strong, short C–C bonds lie in a hexagonal sheet, and the sheets are held together only by weak long bonds. In MoS_2, double sandwiches of MoS_2 are only weakly bonded to each other, which property is used in lubrication. Calcite and aragonite contain strongly bonded CO_3 groups. Quartz contains a framework of corner-shared silicate tetrahedra. Hence, structures can be classified by the nature of the strongly bonded subunits and by the way in which these units are connected.

Mineralogists emphasize the tetrahedral $(Si,Al)O_4$ group in silicates, and use the following terms based on Greek words:

isolated tetrahedron, **nesosilicate**

corner-shared doublet, **sorosilicate**

corner-shared chain, **inosilicate**

corner-shared ring, **cyclosilicate**

corner-shared layer, **phyllosilicate**

corner-shared framework, **tectosilicate**

These broad groups can be further subdivided: thus, chains (Liebau, 1972) can be classified into single, double, or triple chains with various repeat distances (e.g., the 2-repeat double chain of amphibole). Unfortunately, this implicit emphasis on the silicate units tends to ignore the importance of the other structural units. For example, in the nesosilicate olivine $(Mg,Fe)_2 SiO_4$, the oxygen atoms lie close to the positions for hexagonal closest packing while the (Mg,Fe) atoms occupy the centers of a serrated chain of edge-shared octahedra. In clay minerals the octahedral sheet is as important structurally as the silicate sheet. Even weaker bonds such as Ca–O and K–O are important in determining the way in which a silicate unit is flexed. It seems best to use pragmatic classifications of complex crystal structures. In the remainder of this chapter, selected structures containing tetrahedral groups will be discussed, and in Chapter 9 some structures with octahedral groups are described. These structures are carefully selected to illustrate the relations between symmetry, packing, and connectivity. For brevity, the structures tend to be simple and have high symmetry, and readers should take note that many tetrahedral and octahedral structures are extremely complex.

Organic structures provide the prime example of weak bonding between strongly bonded clusters. The shape of most organic molecules is determined largely by the strong covalent bonds inside the molecule, and the intermolecular packing is determined mainly by van der Waals bonds. Many molecules have awkward shapes, and the crystal symmetry is usually low. Readers are referred to Kitaigorodskii (1961, 1973) for the simple principle that the protrusions of one molecule fit into the hollows of its neighbors. Awkward molecules will not fit neatly in space groups with axes of high rotation symmetry, and a zigzag pattern with glide or screw operations is preferred. Kitaigorodskii predicted that only 13 out of the 230 space groups would be mathematically favored for molecules of arbitrary shape with point symmetry **1** to **mmm**, and, indeed, most known organic structures fall into about 10 of these low-symmetry space groups; see also Fischer and Koch (1979).

At the opposite extreme to organic structures are metallic structures. The bonding in ideal metals is isotropic, and the packing of spheres provides a useful principle. When the spheres have different sizes, as in many alloys, the packing can become complex (e.g., Section 5.10). As the metallic bonding grades into covalent bonding, structural subunits can be recognized, such as the chains in Se and the sheets in As. Although hexagonal nets are extremely useful for classification of metals and alloys, additional structural features are needed.

To conclude this section, it seems clear that the geometry and symmetry of crystal structures must be studied case by case while bearing in mind the overall concepts of packing and connectivity. The interplay between general principles and specific details should become apparent in the remainder of Chapter 8 and Chapter 9.

8.5 Relation between Symmetry Elements of a Space Group and Cluster of Atoms

If all atoms of a cluster lie in a general equivalent position of a space group, there is no relation between the point symmetry of the cluster and the symmetry elements of the space group. However, the absolute position of the cluster may be restricted. In Figure 8.11*h*, doublet A is placed too close to the diad rotation axis to allow room for doublet B, which is required by the diad rotation symmetry. Space restrictions are less severe for screw axes and glide planes than for rotation axes and mirror planes, and, indeed, space group $P2_1/a$ is very common in crystal structures because of the freedom of placement of molecules. There are restrictions on the position and orientation of molecules that straddle a symmetry element of a space group. A doublet composed of identical atoms, for example, S_2, can lie with both atoms on any type of rotation axis R_n (Fig. 8.11*a*), but it can lie perpendicular to a rotation axis only for a diad (Fig. 8.11*b*). A doublet composed of different atoms, for example, a CO molecule, cannot lie perpendicular to any axis, even to a rotation diad axis (Fig. 8.11*c*). A symmetrical doublet can straddle a center of symmetry (8.11*d*), but an asymmetrical doublet cannot (8.11*e*). Both a symmetrical and an asymmetrical doublet can lie in a mirror plane (8.11*f*, 8.11*g*).

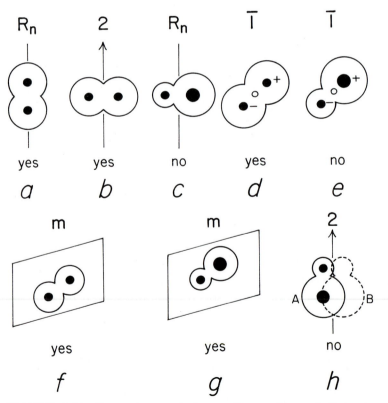

FIGURE 8.11 Symmetry controls on the position of diatomic molecules. In (*d*) and (*e*), positive and negative heights are needed to satisfy the center of symmetry (open circle).

The linear molecule CNCl lies at the intersection of two mirror planes in the ***CNCl crystal structure*** (Fig. 8.12): orthorhombic, a=5.68 Å, b=3.98 Å, c=5.74 Å, **Pmmn**, all atoms in (2a) $\frac{1}{4}\,\frac{1}{4}\,u$; $\frac{3}{4}\,\frac{3}{4}\,\bar{u}$ with $u(C)$ = 0.424, $u(N)$ = 0.626, $u(Cl)$ = 0.150 (Wyckoff, 1963, p. 173). Three unit cells are projected down the a axis. Symmetry elements are shown in the right-hand cell, and the atomic positions are given for the left-hand and middle cells. Each molecule lies parallel to the c axis either at height z = $\frac{1}{4}$ (filled symbols) or $\frac{3}{4}$ (open symbols). The symmetry of the molecule is **mm2**. If the molecule were not linear, it could not lie in this position in space group **Pmmn**, unless some kind of spatial disorder such as rotation gave an average symmetry of **mm2**. Representation of atoms in a complex molecule by spheres is somewhat unsatisfactory because of the big difference between the intra- and intermolecular bonding. The "light-bulb" shape in Figure 8.12 is a plausible interpretation. Because adjacent chlorine atoms are separated by either 3.98 Å (distance pq) or 3.87 Å ($pr = rq$), they are represented by a sphere of radius 1.9_5 Å. The Cl–N distance (ps) of 3.01 Å can then be obtained with a radius of 1.0_6 Å for the N atoms. For simplicity the C atom is given

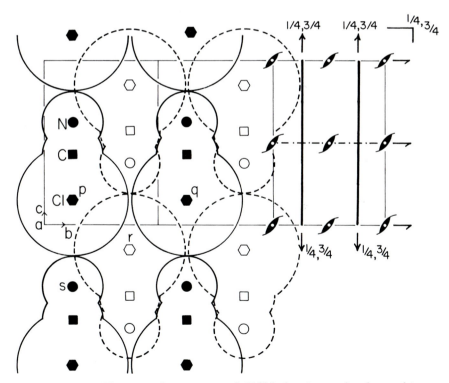

FIGURE 8.12 The crystal structure of ClCN showing molecular packing (left) and symmetry elements (right).

the same radius, and overlap with the N and Cl atoms gives the light-bulb shape. Indeed, actual light bulbs will pack together neatly as in Figure 8.12.

Each pair of N_2 atoms straddles a center of symmetry in the *cubic structure of N_2* (Exercise 8.11). Similarly in *solid CO_2, dry ice,* the CO_2 molecule is linear with the C atom lying on a center of symmetry and the two oxygen atoms symmetrically displaced. In *iodine* (Exercise 8.12, Fig. 8.24), each I_2 molecule is constrained by a rotation diad axis perpendicular to a mirror plane. The S_8 molecule in *rhombic sulfur* is nonplanar with four atoms up and four down like the top of a crown. The structure is tedious to draw because of the large unit cell (orthorhombic, a=10.5 Å, b=12.9 Å, c=24.5 Å) and difficult space group **Fddd**. Those readers who face the challenge of Exercise 8.13 will find that the S_8 molecules are tilted back and forth by the diamond glide planes. Each S_8 molecule consists of four S atoms in general positions, each repeated by a rotation diad axis. The S_6 molecule in *rhombohedral sulfur,* however, is highly symmetric (Fig. 7.36), and its six atoms are related by an inverse triad axis. Innumerable other examples of the relation between molecular geometry and the space group symmetry can be found, especially for organic compounds such as benzene and its derivatives. Interested readers can find many relevant structures in *Structure Reports* and journal articles. Biologically important molecules are particularly intriguing.

Turning to complex ions, there are many important types of which triangular, tetrahedral, and octahedral examples have already been described. The CO_3 cluster is strictly equilateral in *calcite* (Fig. 8.9) because of the passage of a triad axis through the C atom, but need not be strictly equilateral in *aragonite* (Fig. 8.10), where the cluster is bisected by a mirror plane. The elbow-shaped NO_2 anion in $NaNO_2$ (Fig. 7.25) has symmetry **m2m**, which enforces equality of the N—O distances.

8.6   Tetrahedra and Tetrahedral Complexes in Crystal Structures

Most textbooks of mineralogy and inorganic chemistry contain a figure showing some of the infinite number of tetrahedral complexes formed by starting with a single tetrahedron and then progressively sharing one or more vertices to form dumbbell-shaped dimers, rings, chains, layers, sheets, and 4-connected frameworks. Because chains may kink into various configurations, while layers can be bent or corrugated, and frameworks can be linked in an infinite number of ways, it is impractical to attempt a systematic enumeration. Silicate tetrahedra, SiO_4, are particularly important in Nature, and will ·be used primarily as examples here, following Liebau (1972).

8.6.1 STRUCTURES WITH ISOLATED TETRAHEDRA

Zircon, ZrSiO₄, has isolated SiO_4 tetrahedra separated by Zr atoms, each of which is bonded to eight oxygen atoms (Fig. 8.13). The crystal structure takes elegant advantage of the symmetry of tetragonal space group **I4₁/a2/m2/d**. Interaction between the symmetry elements produces an inverse tetrad

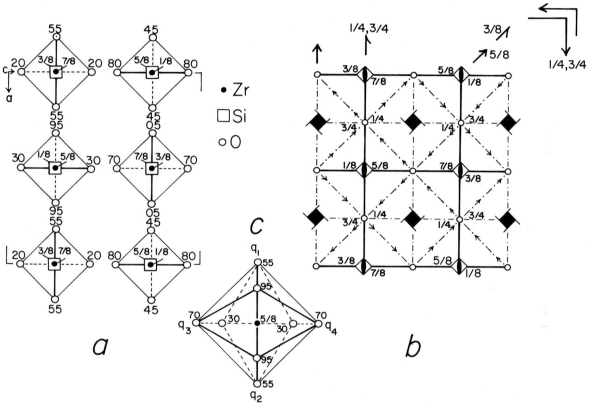

FIGURE 8.13 The crystal structure of zircon and symmetry elements of space group **I4₁/a2/m2/d**. Tetrahedra are outlined in (*a*). The coordination polyhedron for the Zr atom at ½,¼,⅝ is given in (*c*). For ease of comparison with the triangulated dodecahedron in Figure 3.16*c*, four vertices are labeled. Heights of oxygen atoms are in hundredths, and of Zr and Si in fractions. The space group **I4₁/amd** is referred to an origin at a center of symmetry whereas the diagram in International Tables for *X*-ray Crystallography, Vol. 1, p. 245, is referred to an origin with point symmetry $\bar{4}m2$. In diagram (*b*), it is necessary to specify the heights of the inversion points in the $\bar{4}$ axes, for example ⅜ and ⅞ for the $\bar{4}$ axis at $x = 0$ and $y = ¼$. In addition to the set of centers of symmetry at the corners and related positions, there is a second set of centers of symmetry displaced by ¼,¼,¼. Halfway between the mirror planes (heavy line) are diagonal glide planes (dot-dash) not listed in the space group symbol. The diamond glide planes (dot-dash-dot-arrow) lie in 45° positions with respect to the other vertical planes. In addition to **a** glide planes at height $z = ¼, ¾$ there are **b** glide planes at height $z = 0,½$. Four representative axes are shown in horizontal directions.

axis $\bar{4}$, which passes through the midpoints of opposing edges of a tetrahedron with Si–O = 1.62 Å. The tetrahedron also obeys the mirror plane. Screw tetrad axes 4_1 and 4_3 lie between the tetrahedra. A zirconium atom lies on each $\bar{4}$ axis displaced $c/2$ from the Si atom, and the actual values of the free coordinates of the oxygen atoms result in each Zr atom being bonded to four oxygen atoms in a tetrahedron at 2.13 Å and four more at 2.27 Å. These eight oxygen atoms lie at the vertices of a triangulated dodecahedron (diagram 8.13*c*), oriented differently from Figure 3.16*c*. The atomic coordinates referred to a unit cell with a=6.60 Å, c=5.98 Å with a center of symmetry at the origin are (Hazen and Finger, 1979):

Zr (*4a*) $\pm(0, \frac{3}{4}, \frac{1}{8})$

Si (*4b*) $\pm(0, \frac{3}{4}, \frac{5}{8})$

O (*16h*) $\pm(0, u, v; \ 0, \frac{1}{2} - u, v; \ \frac{1}{4} + u, \frac{1}{4}, \frac{3}{4} + v; \ \frac{3}{4} - u, \frac{1}{4}, \frac{3}{4} + v),$

$u = 0.066, v = 0.196$

(add $\frac{1}{2}, \frac{1}{2}, \frac{1}{2}$ to all coordinates for positions generated by the I-lattice).

From the viewpoint of connectivity of just the Zr and Si atoms, the zircon structure is rather simple. It is easy to show that the Zr and Si atoms lie on *D*-type lattice complexes (Table 8.1) displaced by $c/2$; of course, this diamond type of lattice complex undergoes tetragonal distortion in the actual zircon structure because of the bonding to oxygen atoms.

 The structures of **anhydrite, CaSO₄,** and **scheelite, CaWO₄,** resemble the zircon structure but are different. Readers will discover that the anhydrite structure (Exercise 8.14) is obtainable from the zircon structure by changing the z coordinates of alternate tetrahedra and cations by $\frac{1}{4}$. The Ca atom retains the triangulated dodecahedron with Ca–O distances ranging from 2.3 to 2.6 Å. The anhydrite and zircon structures are topologically distinct. The scheelite structure (Exercise 8.15) can be obtained from the zircon structure by rotating the tetrahedra by 30°. The space group **I4₁/a** of scheelite is a subgroup of the space group **I4₁/amd** of zircon. The Ca atom lies in a triangulated dodecahedron with four oxygens at 2.44 Å and four at 2.48 Å.

Some readers may like to look up the structures of **gypsum, CaSO₄ · 2 H₂O,** with space group **I2/c** (Wyckoff, 1965, p. 642), and **barite, BaSO₄,** with space group **Pnma** (Wyckoff, 1965, p. 45). The barite structure is straightforward, and it differs from that of anhydrite because the Ba²⁺ ion is larger than the Ca²⁺ ion with consequent need of a higher coordination number. The gypsum structure is complex because of the difficulty of fitting together sulfate tetrahedra, water molecules, and Ca²⁺ ions.

8.6.2   STRUCTURES WITH EDGE-SHARED TETRAHEDRA

Two tetrahedra can share a vertex, an edge, or a face. In ionic structures, cation repulsion is least for sharing of vertices. Edge sharing occurs in various molecules (e.g., Al_2Cl_6), and in chains of linked tetrahedra (e.g., $AlPS_4$, Fig. 7.7). In the **antifluorite** structure (Fig. 5.12*b*), each Li atom in Li_2O is surrounded by a tetrahedron of oxygen atoms, and each edge is shared to give an infinite framework. Readers should consult Wells (1975, Table 5.2) for further examples of edge sharing of tetrahedra. Sharing of vertices is now considered.

8.6.3   STRUCTURES WITH T_2O_7 DIMERS

The T_2O_7 dimer (T = tetrahedrally coordinated atom) is illustrated by the structure of **thortveitite, $Sc_2Si_2O_7$**. The monoclinic cell (Wyckoff, 1968, Fig. 216) has a = 6.54 Å, b = 8.52 Å, c = 4.67 Å, β = 102.55°, C12/m1, and positions:

O(1)	(*2a*)	$0,0,0; \frac{1}{2},\frac{1}{2},0$
Sc	(*4h*)	$\pm(0,u,\frac{1}{2}; \frac{1}{2},\frac{1}{2}+u,\frac{1}{2}), u = 0.308$
Si	(*4i*)	$\pm(v,0,w; \frac{1}{2}+v,\frac{1}{2},w), v = 0.223, w = 0.910$
O(2)	(*4i*)	$v = 0.391, w = 0.221$
O(3)	(*8j*)	$\pm(p,q,r; p,\bar{q},r; \frac{1}{2}+p,\frac{1}{2}+q,r; \frac{1}{2}+p,\frac{1}{2}-q,r), p = 0.235,$
		$q = 0.157, r = 0.714$

Each Si_2O_7 group (Fig. 8.14) consists of a shared oxygen of type O(1), lying at the intersection of a mirror plane and a rotation diad axis, two oxygens of type O(2) lying on a mirror plane, and four oxygens of type O(3) lying in general positions. Each Sc atom lies on a rotation diad axis, and is surrounded by six oxygen atoms at the corners of a distorted octahedron.

Why is the symmetry low for thortveitite? The answer is not clear, but it is probably difficult or perhaps impossible to achieve octahedral coordination of oxygen atoms around the Sc atoms when the symmetry is higher. Certainly, the thortveitite structure fits together rather nicely as explained in the figure legend, but there has been no attempt to investigate the theoretical possibilities of other arrangements.

The **melilite** family of structures does have high symmetry, as demonstrated by the specific structure of **hardystonite, $Ca_2ZnSi_2O_7$** (Fig. 8.15). In the tetragonal cell, a = 7.83 Å, c = 5.01 Å, $\mathbf{P\bar{4}2_1m}$, the coordinates (Louisnathan, 1969) are:

Zn (*2a*) $0,0,0;\ \tfrac{1}{2},\tfrac{1}{2},0$

O(1) (*2c*) $\tfrac{1}{2},0,u;\ 0,\tfrac{1}{2},\bar{u};\ u = 0.177$

Ca (*4e*) $v,\tfrac{1}{2}-v,w;\ \tfrac{1}{2}-v,\bar{v},\bar{w};\ \bar{v},\tfrac{1}{2}+v,w;\ \tfrac{1}{2}+v,v,\bar{w};\ v = 0.332,$
$w = 0.506$

Si (*4e*) $v = 0.139,\ w = 0.939$

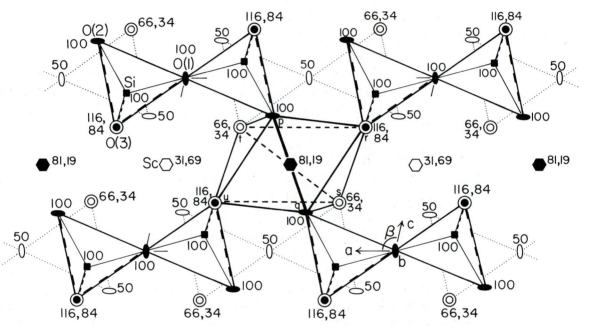

FIGURE 8.14 The crystal structure of thortveitite. In this b-axis projection, Si_2O_7 groups are shown in heavy line for O(1) and O(2) atoms at height 100 hundredths and O(3) atoms at heights 84 and 116 hundredths. The O(3) atoms superimpose in pairs and are shown by double circles. The Si-O bonds are shown by thin lines. Dotted lines show Si_2O_7 groups related by the **C** face-centering. Scandium atoms project in pairs at the hexagons, and an octahedron of oxygen atoms is shown for the Sc atom at $\tfrac{1}{2},0.81,\tfrac{1}{2}$. This octahedron is obtained from oxygen atoms at heights 66, 84, and 100 hundredths. It is somewhat distorted as shown by the lack of coincidence of oxygen atoms at heights 66 and 100 hundredths. An Sc atom at height $\tfrac{1}{2},1.19,\tfrac{1}{2}$ would be bonded to oxygen atoms at height 100, 116, and 134 hundredths, and the *pq* edge at height 100 would be shared between the two octahedra. The octahedron around the Sc atom at $0,0.69,\tfrac{1}{2}$ shares the edge *rs* between oxygen atoms at height 66 and 84 hundredths with the octahedron at $\tfrac{1}{2},0.81,\tfrac{1}{2}$, and similarly the octahedron around the Sc atom at $1,0.69,\tfrac{1}{2}$ shares the edge *tu*. Thus each octahedron shares three edges with other octahedra, in addition to sharing all six vertices with tetrahedra. Consequently, the thortveitite structure can be regarded as a 3D linkage of Si_2O_7 groups and sheets of edge-shared octahedra. A key feature of the structure is the tilting of the Si_2O_7 groups, which combined with the **C** face-centering allows the O(2) and O(3) oxygen atoms to lie in an octahedron around each Sc atom.

O(2)   (4e) $v = 0.140, w = 0.255$

O(3)   (8f) $x,y,z; y,\bar{x},\bar{z}; \frac{1}{2} - x, \frac{1}{2} + y, \bar{z}; \frac{1}{2} - y, \frac{1}{2} - x, z; \bar{x}, \bar{y}, z; \bar{y}, x, \bar{z};$
$\frac{1}{2} + x, \frac{1}{2} - y, \bar{z}; \frac{1}{2} + y, \frac{1}{2} + x, z; x = 0.082, y = 0.188,$
$z = 0.785$

Each Si_2O_7 group has its geometry controlled by two vertical mirror planes. Each Zn atom is surrounded by a tetrahedron of oxygen atoms, whose shape is controlled by the $\bar{4}$ axis. Each Ca atom lies inside a distorted square antiprism of oxygen atoms. Because the Si_2O_7 groups are cross linked by ZnO_4 tetrahedra, the hardystonite structure can be regarded geometrically as a layer structure in which crinkled sheets of vertex-shared tetrahedra of overall composition $ZnSi_2O_7$ are linked by Ca atoms (Zoltai, 1960).

The structure of *tilleyite, $Ca_5(Si_2O_7)(CO_3)_2$,* contains Si_2O_7 dimers and CO_3 triangles in general positions of space group $P2_1/a$ (Louisnathan and Smith, 1970). The Si_2O_7 group varies considerably in shape from one

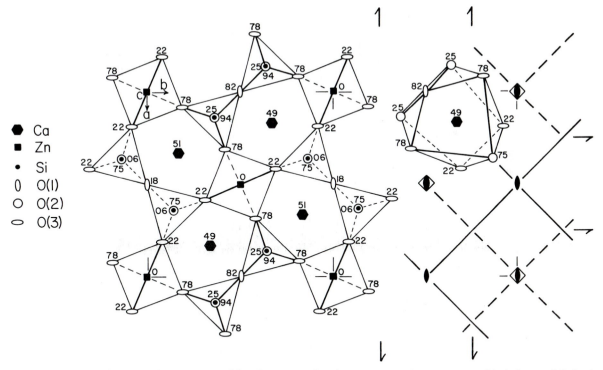

FIGURE 8.15 The crystal structure of hardystonite. In this *c*-axis projection, a crinkled sheet of linked ZnO_4 tetrahedra and Si_2O_7 dimers is shown in the left-hand unit cell, and a CaO_8 distorted square antiprism is shown in the right-hand unit cell along with symmetry elements. Each ZnO_4 tetrahedron is generated by a Zn atom at height 100 and 0 atoms at heights 78 and 122 hundredths. An Si atom at height 94 is surrounded by oxygen atoms at heights 78, 82, and 125 hundredths, while one at height 106 utilizes oxygens at 75, 118, and 122 hundredths.

structure to another, and the Si–O–Si angle between the two tetrahedra can change from 180° in thortveitite and hardystonite to 157° in tilleyite and 146° in kornerupine.

8.6.4   STRUCTURES WITH TO_3 RINGS

Further condensation of tetrahedra produces infinite chains and rings. The simplest ring, Si_3O_9, occurs in ***benitoite, BaTi(Si$_3$O$_9$)***. It is easy to guess that the symmetry is hexagonal. Figure 8.16 is drawn from the following coordinates (Fischer, 1969) in **P6̄c2**, $a = 6.64$ Å, $c = 9.76$ Å:

Ti	$(2c)$	$\frac{1}{3},\frac{2}{3},0; \frac{1}{3},\frac{2}{3},\frac{1}{2}$
Ba	$(2e)$	$\frac{2}{3},\frac{1}{3},0; \frac{2}{3},\frac{1}{3},\frac{1}{2}$
Si	$(6k)$	$u,v,\frac{1}{4}; v-u,\bar{u},\frac{1}{4}; \bar{v},u-v,\frac{1}{4}; \bar{v},\bar{u},\frac{3}{4}; u,u-v,\frac{3}{4}; v-u,v,\frac{3}{4};$
		$u = 0.0711, v = 0.2894$
O(1)	$(6k)$	$u = 0.2535, v = 0.1927$
O(2)	$(12l)$	$xyz; \bar{y},x-y,z; y-x,\bar{x},z; x,y,\frac{1}{2}-z; \bar{y},x-y,\frac{1}{2}-z;$
		$y-x,\bar{x},\frac{1}{2}-z; \bar{y},\bar{x},\frac{1}{2}+z; x,x-y,\frac{1}{2}+z; y-x,y,\frac{1}{2}+z;$
		$\bar{y},\bar{x},\bar{z}; x,x-y,\bar{z}; y-x,y,\bar{z}; x = 0.088, y = 0.4302, z = 0.1127$

The **c** glide planes pass through the origin of the cell and relate an Si_3O_9 ring at height $\frac{1}{4}$ to one above at height $\frac{3}{4}$. Oxygen atoms of type O(1) are bonded to two Si, whereas oxygen atoms of type O(2) are bonded to only one Si. Electrostatic neutrality is achieved according to Pauling's rules as follows. Each Si^{4+} ion supplies one charge unit to its four oxygen neighbors, and the two charges on the O(1) cation are immediately satisfied by the two Si neighbors. The O(2) cation receives only one charge unit from Si, and achieves balance by bonding to Ti^{4+} and Ba^{2+} ions. Each Ti^{4+} ion lies in a nearly regular octahedron of six O(2), and each O(2) is bonded to one Ti^{4+} ion, from which it receives $\frac{4}{6}$ of a charge unit. Each Ba^{2+} ion is bonded to six O(2) at the corners of a strongly distorted trigonal antiprism, and each O(2) is bonded to one Ba^{2+} from which it receives $\frac{2}{6}$ of a charge unit. The sum of $(1 + \frac{4}{6} + \frac{2}{6})$ balances the formal charge on the O(2) anion.

Turning to the packing features of benitoite, observe how the angular position of the Si_3O_9 rings is controlled by the octahedral coordination of O(2) atoms around the Ti atom. In order to obtain the octahedral coordination, Si_3O_9 rings at heights $\frac{1}{4}$ and $\frac{3}{4}$ must have a relative rotation near 35°. Barium–oxygen bonds are much weaker than Ti–O bonds, and the Ba atom is forced to "accept" the positions of the O(2) atoms enforced by the Ti–O bonding. The resulting polyhedron is a strongly distorted octahedron. Although the Si_3O_9 rings are so obvious in benitoite, the TiO_6 and BaO_6

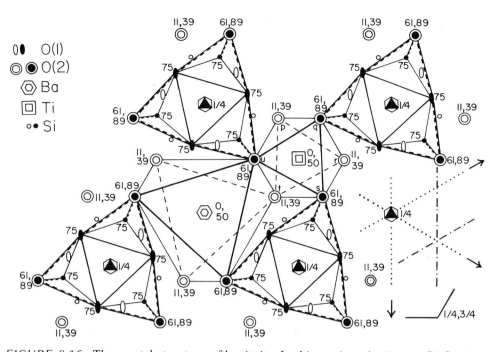

FIGURE 8.16 The crystal structure of benitoite. In this c-axis projection, an Si_3O_9 ring with Si and O(1) atoms at height 75 hundredths is centered in projection about each cell corner. These Si and O(1) atoms lie on a horizontal mirror plane at height $z = \frac{3}{4}$, and O(2) atoms at 61 and 89 hundredths lie symmetrically above and below. Each Si_3O_9 ring is related to another Si_3O_9 ring by an inverse hexad axis with inversion point at $z = \frac{1}{4}$ (or $\frac{3}{4}$), thereby giving Si and O(1) atoms at height 25 (open symbols not labeled with the height) and O(2) atoms at heights 11 and 39 hundredths (labeled with the height). Titanium atoms lie at heights 0 and 50 hundredths, and the octahedral coordination polyhedron is shown for an atom at height 50 hundredths. It utilizes three oxygen atoms at height 61 and three at height 39 hundredths. The octahedron, which uses O(2) atoms from height 89 and 11 hundredths, does not share a face with the first octahedron. Similarly, the Ba atoms (squares) at heights 0 and 50 hundredths each lie in an octahedron of O(2) atoms. Horizontal diad axes lie at heights $\frac{1}{4}$ and $\frac{3}{4}$ (lower right), and c glide planes (dotted lines) alternate with **n** glide planes (dot-dash lines).

octahedra must not be neglected, and, indeed, the benitoite structure can be regarded as a framework of linked tetrahedra and octahedra.

From the topological viewpoint, the connectivity of benitoite matches that of the AlB_2 structure (Fig. 5.10*b*). Collectively, the Ba and Ti atoms match the B atoms, and each center of an Si_3O_9 ring matches an Al atom.

 Actually this interpretation in terms of first neighbors is perhaps too simple. Sixfold coordination is rather low for the large Ba^{2+} ion, and there are six O(1) atoms lying at 3.43 Å, which compares with the Ba–O(2) distance of

2.77 Å. Although each O(1) atom is formally satisfied by the charge contribution from two Si atoms, some interaction with the adjacent Ba atom is expected. The six O(2) and six O(1) atoms would combine together to give a distorted dodecahedron.

For brevity, other structures with rings are not discussed here, but readers may wish to examine the structures of *beryl, Be₃Al₂Si₆O₁₈* (Hurlbut and Klein, 1977, Fig. 10.29; Gibbs et al., 1968), and *cordierite, (Mg,Fe)₂(Al₂) Si(Al₂Si₄O₁₆) · xH₂O* (Gibbs, 1966). The Si_6O_{18} rings of beryl are cross linked by Be in tetrahedral coordination and Al in octahedral coordination in a hexagonal structure with space group **P6/mcc**. Each 6-ring of cordierite contains the ordered sequence AlSiSiAlSiSi, and the symmetry is forced downward to orthorhombic **Cccm**. Actually, both beryl and cordierite contain a 3D framework of vertex-linked tetrahedra, whose centers are 3 Be and 6 Si in beryl and 4 Al and 5 Si in cordierite. Consequently, readers are cautioned that classification of some complex structures on the basis of subunits is a matter of convenience, and that the atomic packing depends on the interaction of all the structural units.

8.6.5   STRUCTURES WITH CHAINS OF VERTEX-SHARED TETRAHEDRA

Vertex sharing of tetrahedra produces many types of chains (Liebau, 1972), of which a few are selected in Figure 8.17. Chains 8.17*a*, 8.17*b*, and 8.17*c* are topologically identical but geometrically distinct. Each tetrahedron shares two vertices, one each with a neighboring tetrahedron, leaving two free vertices. The chain found in the pyroxene group of minerals can be idealized so that the shared vertices lie in a straight line and the tetrahedra are related by a mirror plane, a rotation diad axis, and a glide plane (dashes in diagram 8.17*a*). The chain found in $Ba_2Si_2O_6$ has a geometry in which the shared vertices lie in a zigzag, and the tetrahedra point alternately up and down to obey a screw diad. Whereas the pyroxene and $Ba_2Si_2O_6$ types of chains repeat every two tetrahedra, the wollastonite chain repeats every three tetrahedra. One tetrahedron is viewed down a triad axis, whereas the remaining two are viewed down a $\bar{4}$ axis. The chain in wollastonite, $CaSiO_3$, is depicted with a slight zigzag, as found in the actual crystal structure, but theoretically all shared vertices could lie in a straight line. Mirror symmetry would be retained. Using T for tetrahedral center and O for a vertex, all three chains have a simplified formula of TO_3.

Diagrams 8.17*d* and 8.17*e* show the double chains found in the amphibole group of minerals and in xonotlite. The amphibole chain is obtained by sharing vertices from two pyroxene chains, giving a repeat of T_4O_{11}. Five vertices are shared, and six are unshared. The diad axes of the pyroxene chains disappear, but new diad axes appear at the center of the 6-rings. The original mirror planes are retained, and a new one appears midway between the pyroxene chains. Two wollastonite chains share vertices to give the double chain T_6O_{17} of xonotlite, $Ca_6[Si_6O_{17}](OH)_2$, in which the wollastonite chains are related by a rotation diad and centers of symmetry.

In German, the terms *Einfachkette* and *Doppelkette* are used for single and double chains, and the prefixes *Zweier-* and *Dreier-* are used to show the repeat distance down the chain.

The symmetry of chains can be expressed formally using the notation for the symmetry of the 31 two-sided bands (Fig. 7.1). For axes labeled *a, b,* and *c* in the center of Figure 8.17, the symmetry symbols from Figure 7.1 are: pyroxene, **pbm2**; $BaSi_2O_6$, **p12$_1$1**; wollastonite, **p1m1**; amphibole, **pmm2**; xonotlite, **p12/m1**.

From the crystal–chemical viewpoint, T−O bonds are so strong that T−O distances and O−T−O angles deviate only slightly from the regular geometry of a tetrahedron. However, the T−O−T angles are only partly constrained by chemical bonding, and to a crude first approximation shared vertices can be regarded as universal joints. To understand the packing of crystal structures with tetrahedral chains, it is necessary to examine the ways in which the chains twist and flex to accommodate to the packing requirements of the other structural units. Unfortunately, all crystal structures with

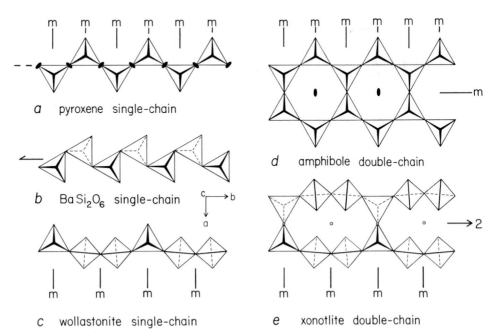

a   pyroxene single-chain

b $BaSi_2O_6$ single-chain

c   wollastonite single-chain

d   amphibole double-chain

e   xonotlite double-chain

FIGURE 8.17 Portions of vertex-linked tetrahedral chains. Each chain extends to infinity in the EW direction.

tetrahedral chains are complex, but the relatively simple structure of *diopside, CaMgSi$_2$O$_6$*, provides an excellent introduction.

This representative of the *pyroxene* family of minerals has the following structural parameters (Clark, Appleman, and Papike, 1969): monoclinic $a = 9.75$ Å, $b = 8.90$ Å, $c = 5.25$ Å, $\beta = 105.6°$, C$12/$c1;

Si	(*8f*)	$\pm(x,y,z; \bar{x},y,\frac{1}{2} - z)$, $x = 0.2862$, $y = 0.0933$, $z = 0.2293$
O(1)	(*8f*)	$x = 0.1156$, $y = 0.0873$, $z = 0.1422$
O(2)	(*8f*)	$x = 0.3611$, $y = 0.2500$, $z = 0.3180$
O(3)	(*8f*)	$x = 0.3505$, $y = 0.0176$, $z = 0.9953$
Mg	(*4e*)	$\pm(0,y,\frac{1}{4})$, $y = 0.9082$
Ca	(*4e*)	$y = 0.3015$

Add $x = \frac{1}{2}$, $y = \frac{1}{2}$ to obtain coordinates related by the C lattice. Interaction of the symmetry elements produces two sets of centers of symmetry, a set of diagonal glide planes halfway between the c glide planes, and a set of screw diad axes. The y coordinate of O(2) is only accidentally equal to $\frac{1}{4}$.

Because the (Si$_2$O$_6$)$_\infty$ chain lies along the c-axis, and because there is an interesting approximation to close packing of oxygen atoms in the bc plane, the diopside structure is usually projected down the a axis. Because the a and c axes are not perpendicular, the c axis cannot lie in the plane of the paper, and makes an angle of 15.6°. Lying in the plane of the paper is the $c^\star$ axis, which is mutually perpendicular to the a and b axes. The unit cell is $c \sin \beta$ wide (= 5.08 Å) along the $c^\star$ axis.

In Figure 8.18a, an infinite SiO$_3$ chain is represented by three tetrahedra containing seven oxygen atoms p–v with $x = 0.64$ or 0.65. Atoms p and t of type O(3) are shared by two tetrahedra, whereas atoms of type O(2) belong to only one tetrahedron. Each tetrahedron is completed by an O(1) atom at $x = 0.88$ and a silicon atom (filled triangle) at $x = 0.71$. The tapered solid lines show that each tetrahedron projects toward the reader. Almost hidden behind this chain is another chain composed of oxygen atoms at $x = 0.36$, 0.35, and 0.12 (open symbols) and silicon atoms at height 0.29 (inverted open triangle). Alternate tetrahedra of each chain are related by the 1c1 glide plane (three dashes), and the second chain (open symbols) is related to the first chain (filled symbols) by the rotation diad axes (arrows at extreme right). Two more chains occur in the unit cell, translated by $(\mathbf{a} + \mathbf{b})/2$ to satisfy the C face centering. Thus atoms v and p are translated to v' and p'. Tetrahedral vertices are shared only in a chain and not between chains. Furthermore, there is only one fundamental type of chain, which is merely translated and/or reflected to obtain the four different positions. This chain is twisted from the ideal shape in Figure 8.17a.

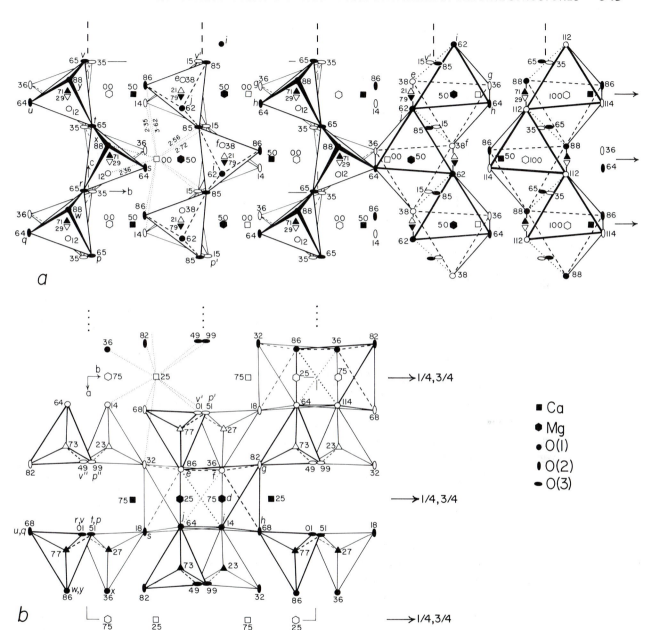

FIGURE 8.18 The crystal structure of the diopside variety of pyroxene in *a*-axis and *c*-axis projections. Two unit cells are shown for the *a*-axis projection [diagram (*a*)]. To avoid confusion, MgO$_6$ octahedra are shown only in the right-hand cell, and SiO$_4$ tetrahedra are shown in the left-hand cell. Symbols are either filled or left open to distinguish heights. The heights of some symbols are deliberately omitted to allow practice for the reader (exercise 8.23). Coloring the polyhedra is recommended. Positions of the *c* glide planes and the rotation diad axes are indicated by heavy dashes and arrows. Only one unit cell is given in the *c*-axis projection [diagram (*b*)]. Symbols and labels allow matching of the atomic positions. The *c* glide plane is now indicated by heavy dots.

Each Mg atom (hexagon) lies in a distorted octahedron of oxygen atoms, and each octahedron shares two edges to form a zigzag octahedral chain lying parallel to the silicate chain. To reduce confusion, octahedra are not drawn in the unit cell at the left. Two octahedral chains are related by C face centering in the right-hand unit cell. Each octahedral chain is obtained from one type of octahedron, which is translated and reflected by the **c** glide plane (three dashes) to give the zigzag shape. The chain at the extreme right is composed of oxygen atoms at heights 1.12 and 1.14 that form the upper triangular faces and oxygen atoms at heights 0.88 and 0.86 that form the lower triangular faces. Shared edges (dotted lines) join oxygen atoms at 0.88 and 1.12. The Mg atom lies at height 1.00.

Before proceeding to examine the linkage between the tetrahedral and octahedral chains, consider the c-axis projection in Figure 8.18b. Oxygen atoms labeled *p–y* outline the silicate chain mentioned two paragraphs back. All four Si—O bonds as well as the outer shape of each tetrahedron are shown. The characteristic trapezium-shaped cross section of the silicate chain of pyroxene maintains the same orientation after translation by the face-centering vector ($v \rightarrow v'$, $p \rightarrow p'$). Alternate tetrahedra of each chain are related by the **1c1** glide plane (heavy dots), which bisects the trapezium. Operation by the rotation diad axis (heavy arrows) inverts the trapezium and moves v to v'' and p to p''. In this projection, the zigzag octahedral chain projects as a rectangle with a confusing group of intersecting lines. The Mg atom d lies at the center of the distorted octahedron of six oxygen atoms e–j. Atom i is labeled with coordinate $z = 0.14$ to match the other oxygen atoms in the adjacent tetrahedron, but coordinate $z = 1.14$ is needed for the octahedron. Thus atoms e ($z = 0.86$), g (0.82), and i (1.14) form a triangular face nearest to the reader, and atoms f (0.36), h (0.68), and j (0.64) form a back face. Edges from e (0.86) to i (1.14) and from f (0.36) to j (0.64) are shared between octahedra centered on Mg atoms at $z = 0.75$ and 0.25, respectively, and do not intersect as appears in projection.

Each Ca atom (square) lies at the center of a distorted pentagonal prism of oxygen atoms, and the 10 Ca—O vectors are shown by the finely dotted lines in both projections.

To understand the geometrical relations between the two types of chain and the pentagonal prism, suppose that you were asked to invent a crystal structure incorporating these three units. Obviously this is a very difficult problem if considered *ab initio*. Consider first the problem of sharing vertices between ideal tetrahedral and zigzag octahedral chains with the appropriate sizes for Si—O = 1.63 Å and Mg—O = 2.08 Å. You will find that the chains can fit together with little distortion if vertex k is shared with vertex m, and l with n', for the fragments of ideal chains at the upper right of Figure 8.19. But you will then find difficulty in proceeding further to make a complete crystal structure. For complex structures, it is usually better to take

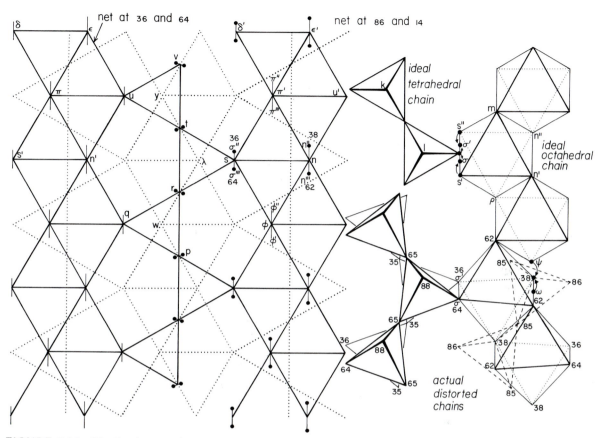

FIGURE 8.19 Idealized nets of oxygen atoms in the pyroxene structure (left) and the distortions required in order that the tetrahedral and octahedral chains can share vertices (right). To understand the figure, first compare the nodes of the idealized nets at heights 0.14 and 0.86 (dotted line) and at 0.36 and 0.64 (continuous line) with the oxygen positions in Figure 8.18*a*. Then study how an ideal octahedral chain (upper right) is distorted to give the two octahedra at the lower right. The distortion results from two interactions with the silicate chains. Vertices s' and s'' move to σ and σ', as shown by the arrows, so that each becomes shared with a tetrahedral node. This causes rotation of the tetrahedra as shown below. Thus movement of vertex σ at height 0.64 caused clockwise rotation of its tetrahedron while movement of σ' at height 0.36 caused anti-clockwise rotation. The second interaction is shown at the lower right. When the tetrahedral and octahedral chains are regular in shape, vertices ω and ψ cannot meet. The arrows show the movements that allow shar-ing of the vertices. Returning to the nets at the left, the cross marks show the displacements of the nodes. In addition to these displacements, there are small changes of the x-coordinates, and a 15.6° change of the β angle so that the orthorhombic symmetry of the idealized nets gives the monoclinic cell of diopside.

the actual structure and find out how the pieces fit together. Invariably there is a simple underlying pattern such that the pieces fit together in an elegant way.

The key to the pyroxene structure is a type of 2D net that is repeated four times by symmetry operators. All the oxygen atoms in the pyroxene structure lie close to four planes with $x \sim 0.14$, 0.36, 0.64, and 0.86. Furthermore, the oxygen atoms in Figure 8.18*a* fall at the corners of distorted triangles. The positions can be idealized to give the 2D nets at the left of figure 8.19. Atoms *p–v* are idealized to give a regular zigzag of *corner-sharing* equilateral triangles that matches the outline of the ideal pyroxene chain in Figure 8.17*a*. The 2D net (heavy line) is completed with the zigzag chain of *side-sharing* triangles $\delta\epsilon\pi un's'$, which repeats as $\delta'\epsilon'\pi'u'ns$. Pentagons (e.g., $utrqn'$) are formed between the two types of equilateral triangles. The two types of chain will fit together only if the first type of equilateral triangle (e.g., *tuv*) is smaller than the second type (e.g., $\delta\epsilon\pi$) with $\delta\epsilon/ut$ = 1.16.

The nets in heavy line lie at heights $x = 0.36$ and 0.64, whereas those in dotted line lie at 0.14 and 0.86. They obey orthorhombic symmetry. Using the axial orientation for pyroxene (after placing $\beta = 90°$), the space–group symmetry of the four nets is **Cmcm**.

Small distortions of the ideal nets yield the pyroxene structure. Oxygen atoms projecting onto *r, s,* and *t* from $x = 0.36$ and onto λ from $x = 0.14$ generate a tetrahedron with centroid projecting onto λ with $x \sim 0.3$. A second tetrahedron is produced by oxygen atoms projecting onto *r, s,* and *t* from $x = 0.64$ and onto λ from $x = 0.86$ to give a centroid at $x \sim 0.8$. Consequently, it is easy to get the SiO_3 chains by placing a silicon atom at both centroids, and by repeating the symmetry elements.

But how to get the octahedral chain? The lozenge $s\pi'n\phi$ does not match the ideal octahedron $s''mn''n'\rho s'$. However, two distortions provide the answer. Move oxygen atom *s* at $x = 0.36$ to σ'' and oxygen atom *s* at $x = 0.64$ to σ'''. Move oxygen atom *n* at $x = 0.36$ to n'' at $x = 0.38$, and oxygen atom *n* at $x = 0.64$ to n''' at $x = 0.62$. Symmetrically related movements of oxygen atoms at π' and ϕ allow construction of a distorted octahedron whose shape is shown at the bottom right. These movements cause two rotations of each tetrahedron as shown by the arrows.

Putting all the ideas together, the complex geometry of the pyroxene structure results from the difficulty of fitting together a silicate tetrahedral chain with an MgO_6 octahedral chain to give a crystal structure linked regularly in 3D. The final structure is a compromise in which the two chains are distorted from their regular shape in order to allow sharing of vertices. Thereby the chains link up to give a compact framework. Underlying this framework is an idealized 2D net composed of two types of equilateral triangles and one type of pentagon. The orthorhombic symmetry of the net is destroyed by the distortions, but the new space group **C12/c1** inherits some of the symmetry elements.

Each pentagonal prism of the idealized net is occupied by a Ca atom in the diopside structure, but the Ca–O distance ranges widely because of the distortions. Calculated distances for the above-listed parameters are Ca–O(1)

= 2.36 Å, Ca–O(2)$'$ = 2.35 Å, Ca–O(3) = 2.56 Å, Ca–O(3)$'$ = 2.72 Å, and Ca–O(2)$'$ = 3.62 Å (Fig. 8.18a). The sum of the ionic radii from Table 1.1 is ~2.4 Å, with the exact value depending on the coordination numbers of the ions. For Ca at the center of the idealized pentagonal prism the mean Ca–O distance of ~2.7 Å would be too great for ionic bonding. Removal of the two O(2)$'$ atoms to 3.6 Å allows the other eight atoms to move closer to the Ca atom. Regularity of distances is not obtained because the bonding requirements of the silicate tetrahedron and the MgO_6 octahedron are stricter than for the calcium polyhedron. Nevertheless, the Ca atom is an important component of the diopside structure, not only for balancing the charge in the $CaMgSi_2O_6$ formula, but also for providing a significant contribution to the chemical bonding and internal energy.

Because of the flexibility provided by the distortions of the subunits, it is not surprising that the diopside structure is adopted by a wide range of compositions including $LiAlSi_2O_6$ (spodumene) and $NaAlSi_2O_6$ (jadeite). There is a complex literature on the relation between chemistry and structural geometry. The distortions are so severe in some varieties that the space–group symmetry is lowered (e.g., **P12$_1$/c1** for clinoenstatite $Mg_2Si_2O_6$). Finally, topologically different ways of stacking together the chains give rise to two types of orthorhombic varieties of pyroxenes (references listed in Deer, Howie, and Zussman, 1978; Pannhorst, 1978).

8.6.6 STRUCTURES WITH LAYERS OF VERTEX-SHARED TETRAHEDRA

Vertex sharing of tetrahedra produces several types of layers (Liebau, 1972), of which four are selected in Figure 8.20. Two pyroxene chains TO_3 (Fig. 8.17a) are cross linked to give the double chain T_4O_{11} of amphibole (Fig. 8.17d), and further cross linking gives the T_2O_5 layer of *mica* (Fig. 8.20a). The lines show the projections of the T–O bonds and O–O edges in a plan and an elevation of the layer. In its most regular shape, the layer has hexagonal symmetry. Each tetrahedron shares an oxygen atom with three adjacent tetrahedra, and these three oxygen atoms are called the ***basal*** oxygens because they form the base of the tetrahedron when viewed in plan. The remaining oxygen atom (small filled circle) is called the ***apical*** oxygen because it lies at the apex of the tetrahedron. Ignore the other atoms for the time being.

The high-temperature ***hexagonal polymorph*** of ***BaAl$_2$Si$_2$O$_8$*** (Takéuchi, 1958) contains a double layer T_4O_8 obtained by sharing the apical oxygens of two mica layers (Fig. 8.20b). The T sites are occupied alternately by Al and Si, and the double layers are cross linked by weak Ba–O bonds (Exercise 8.28). In the ***hexagonal polymorph*** of ***CaAl$_2$Si$_2$O$_8$***, the double layers are

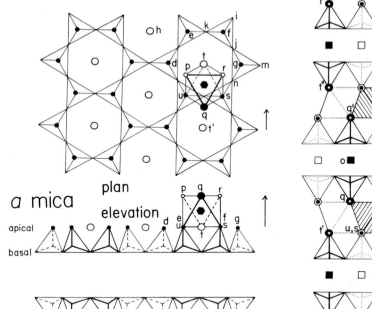

a mica

plan

elevation

apical

basal

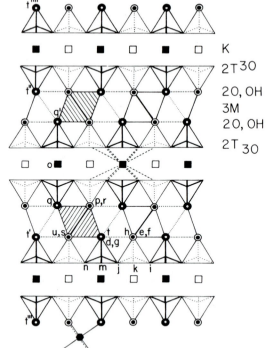

K

2T 3O

2O, OH

3M

2O, OH

2T 3O

e 2-layer orthorhombic mica

b hexagonal BaAl₂Si₂O₈

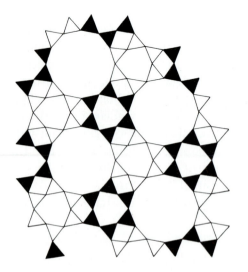

c manganpyrosmalite

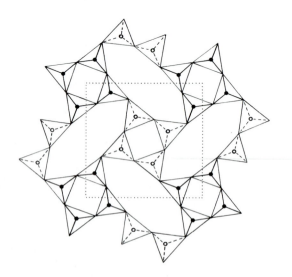

d apophyllite

distorted considerably with respect to those in $BaAl_2Si_2O_8$ in order to accommodate to the smaller Ca^{2+} ion (Takéuchi and Donnay, 1959).

In *manganpyrosmalite, $Mn_8(Si_6O_{15})(OH)_9Cl$,* a single layer is composed of 4-, 6-, and 12-rings of tetrahedra. Each 6-ring is attached to three 6-rings whose apical oxygen atoms point the opposite direction (open versus filled triangles in diagram 8.20c). The single layers are cross linked by Mn, OH, and Cl species (Takéuchi et al., 1969).

In *apophyllite, $Ca_4K(Si_8O_{20})F \cdot 8H_2O$,* each single layer is composed of 4- and 8-rings (Colville et al., 1971). Although the rings are distorted, tetragonal symmetry is retained.

Upon referring to Figure 1.2, readers will find that the silicate layers in mica, manganpyrosmalite, and apophyllite can be obtained by replacing each close-packed circle with a tetrahedron in diagrams 1.2d, 1.2g, and 1.2e, respectively. For the mica layer all tetrahedra point upward, whereas in manganpyrosmalite half point upward and half downward. To obtain the apophyllite layer from 1.2e, considerable distortion is needed, but the topological relationship remains.

There are 80 symmetry groups for two-sided layers (Fig. 7.4). The symmetry groups for the layers in Figure 8.20 are: mica, **p6mm**; $BaAl_2Si_2O_8$, **p6/mmm**; manganpyrosmalite, **p$\bar{3}$m1**; apophyllite, **p42$_1$2**.

Returning now to the tetrahedral layer in mica, consider how two tetrahedral layers share apices with an octahedral layer to give a sandwich.

First, try to fill space just with tetrahedral layers. You will find that the layers will not fit together nicely. Furthermore, for a T_2O_5 formula, the T atom would need to be pentavalent, and no obvious candidate is available except, perhaps, phosphorus.

Second, observe that apical oxygens *defgsu* form a hexagon about position t (Fig. 8.20a). Filling space would be easier if this position were occupied, and, indeed, position t in mica is occupied either by hydroxyl, OH, or fluorine, F (open circle). The apical oxygens and hydroxyls collectively generate a close-packed hexagonal layer of spheres (ignoring a slight difference in radius).

FIGURE 8.20 (*Opposite*) Portions of vertex-linked tetrahedral layers. (a) The mica layer in plan and elevation. Open circles show hydroxyls that fill the voids between the apical oxygens (small filled circles). An octahedron of type M1 is also shown. (b) The condensation of two tetrahedral layers of mica type into the double layer found in hexagonal $BaAl_2Si_2O_8$. (c) A layer in manganpyrosmalite, with black tetrahedra pointing in the opposite direction to the open tetrahedra. (d) A layer in apophyllite with 4-rings alternately pointing up and down. Dots show the unit cell. (e) An elevation of the 2-layer orthorhombic polytype of mica. The double circles show an overlap of two apical oxygens and one hydroxyl. Heavy and light lines show different relative depths. Bonds to apical oxygen atoms are shown for one K atom (square).

Third, place a small atom (filled hexagon) in the dimple between two oxygens *u* and *s* and hydroxyl *t*. An octahedron is completed by adding two more oxygen atoms *p* and *r* and hydroxyl *q*. In plan, *pqr* and *stu,* respectively, form the top and bottom faces of the octahedron. In elevation, triangle *pqr* degenerates into a line segment, as does triangle *stu*. Just as positions *u, s,* and *t* were derived from the upper apices of a tetrahedral layer, so can positions *p* and *r* provide apices for a second tetrahedral layer that is inverted with respect to the first layer. Position *q* is the prototype for a second set of hydroxyls.

Because diagram 8.20*a* would become rather confused if the octahedral and second tetrahedral layers were added, the fourth step is construction of the elevation in diagram 8.20*e*. Instead of constructing a north–south elevation as in diagram 8.20*a*, construct an east–west elevation. The initial tetrahedral layer is represented by basal oxygens *ikjmn* and apical oxygens that superimpose to give three positions *ef, dg,* and *us* in diagram 8.20*e*. Hydroxyl *t* superimposes onto oxygens *d* and *g,* and hydroxyl *h* onto oxygens *e* and *f*. Thus, in diagram 8.20*e*, there is one hydroxyl per two apical oxygens for each double circle. The ornament changes to show a difference in relative depth. The shaded lozenge shows the octahedron composed of oxygens *prsu* and hydroxyls *q* and *t*. This octahedron is repeated to give an octahedral layer.

A 3D model should be useful here. Many universities and colleges possess a ball-and-spoke model of mica from which a student can identify tetrahedra and octahedra. Joining oxygen positions by colored wool or cotton is helpful. Two more types of model are useful. First, take a sheet of cardboard and glue tetrahedra to it as in diagram 8.20*a*. Then construct octahedra with appropriate edge length. Show from diagram 8.20*a* that the ratio of octahedral to tetrahedral edge length is $2/\sqrt{3}$. Take another sheet of cardboard and glue octahedra together to form a layer as in Figure 5.18*b*. Each octahedron shares an edge with six adjacent octahedra. Place the tetrahedral and octahedral sheets together so that vertices are shared. Second, make a ball model. Glue together balls in an open hexagonal array to represent the basal oxygens of the tetrahedral layer, and glue a ball into each dimple to give the apical oxygens. Insert a ball of the same size but different color into each hole between the apical oxygens to represent hydroxyl positions. Select two oxygen balls *u* and *s* and a hydroxyl ball *t,* as in Figure 8.20*a*. Place a small ball in the dimple to represent an octahedrally coordinated atom. Complete the octahedron with two balls of oxygen type at *p* and *r* and a ball of hydroxyl type at *q*.

Now study Figure 8.21*a*. There are two types of octahedra in mica, labeled M1 and M2, whose upper faces are distinguished by the heavy and light stripes. In an M1 octahedron, the hydroxyl groups (large circles) are diametrically opposed, and project at the top and bottom. In an M2 octa-

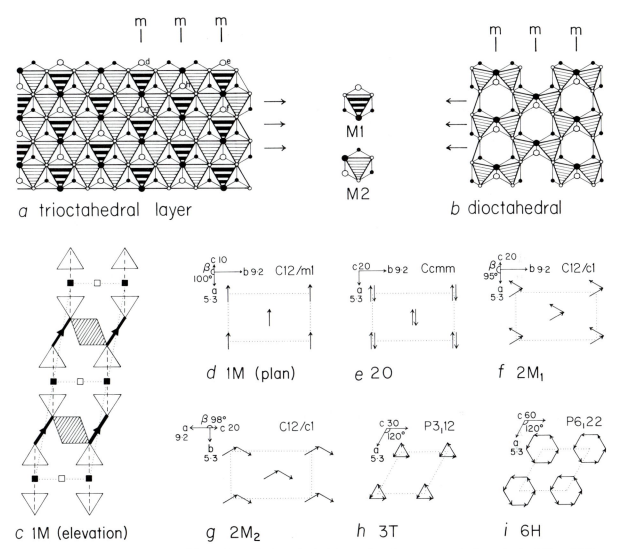

FIGURE 8.21 Mica structure (*a*) A plan of trioctahedral layer showing two types of octahedra M1 and M2. Thin lines show the six shared edges of each octahedron. Large and small circles, respectively, represent hydroxyl and apical oxygen. (*b*) A plan of dioctahedral layer composed of distorted M2 octahedra. (*c*) A simplified elevation of 1M polytype showing K (square) and outlines of tetrahedra and octahedra. The tilt vector is shown by the heavy arrow. (*d* to *i*) The six simple polytypes. Each diagram shows the outline of the unit cell (dots), the projection of the tilt vector, and the cell dimensions and space group.

hedron, the hydroxyl groups share an edge. In an octahedral layer there are two M2 octahedra for each M1 octahedron. Although the *positions* in the octahedral layer fall into three hexagonal arrays, the overall layer symmetry is not hexagonal, because arrays are staggered in projection. The smallest unit with highest symmetry is represented by rectangle *defg*. The unit is centered by position h, and obeys mirror symmetry (vertical planes marked m) and 2-fold rotation symmetry (arrows). Note that labels do not match between Figure 8.20 and Figure 8.21.

Returning now to the ball model, the octahedron already constructed is of type M1. Add appropriate balls to give octahedra of type M2. Note that hydroxyl positions switch from left to right in alternate M2 octahedra in response to the mirror planes, but that there is only one orientation for the M1 octahedron because the hydroxyl positions lie on a mirror plane. At this stage, the ball model consists of one tetrahedral layer sharing apices with one octahedral layer. To complete a model of mica, add a tetrahedral layer to the other side of the octahedral layer to give a closed sandwich (Fig. 8.20e). Make another sandwich (or club together with fellow students). Adjacent sandwiches are joined by large balls representing potassium atoms. Each potassium atom nestles into a dimple of six basal oxygens from one tetrahedral layer and six basal oxygens from a tetrahedral layer of the next sandwich. In plan, each potassium atom would superimpose on two hydroxyls, and in elevation all three lie in a vertical line (q, o, and q' in Fig. 8.20e).

Mica has the general formula $A_1 M1_1 M2_2 T_4 O_{10} B_2$. In hydroxyl–phlogopite the composition is $KMg_3(AlSi_3)O_{10}(OH)_2$. The K atom occupies the A site, which can also be occupied in mica by Na and Ca. Both the M1 and M2 types of octahedral sites are occupied by Mg or by other octahedrally coordinated atoms. The tetrahedral sites in phlogopite are randomly occupied by 1 Al and 3 Si atoms. The B site is occupied by hydroxyl. Phlogopite is known as a trioctahedral variety of mica because all three octahedral sites are occupied. Muscovite is a dioctahedral mica because only the two M2 sites are occupied by octahedral atoms. Its formula is $K\square Al_2(AlSi_3)O_{10}(OH)_2$, where $\square$ represents the void. Aluminum occurs in both octahedral and tetrahedral coordination. Although this book is concerned principally with the geometry of crystal structures, it should be emphasized again that chemical bonding is also important in determining the atomic arrangement of crystal structures. Mica is an excellent example. For phlogopite, there is a balance of the formal charges: thus K + 1, 3 Mg + 6, Al + 3, 3 Si +12 gives 22 charges, which are balanced by –20 for 10 oxygens and –2 for 2 hydroxyls. In muscovite, 2 Al^{3+} and a void replace 3 Mg^{2+}. Charge balance would not be achieved for phlogopite unless one tetrahedral position were occupied by Al. The structural details of mica fit well with the ionic model (Brindley and Brown, 1980), but some covalency should occur.

So far, only the general geometrical features of the mica structure have been discussed. Three geometrical details are particularly instructive. First, what happens to the shape of the octahedral layer when the M1 site is unoccupied? Does the layer tend to collapse around the hole? Actually, the hole expands as in Figure 8.21b! This surprising result can be explained simply by electrostatic repulsion between cations in M2 sites. When M1 sites are occupied as well as M2 sites, all cations are repelled symmetrically, but when the M1 sites are unoccupied, the M2 sites move away from each other. The arrays of basal oxygen atoms then lose hexagonal symmetry and adopt di-trigonal symmetry. Each M2 octahedron becomes distorted but the tetrahedral-octahedral sandwich retains the **C**, **2**, and **m** symmetry operations.

The second geometrical feature is the relationship between the dimensions of the tetrahedral and octahedral layers. A perfect fit requires a ratio of $2/\sqrt{3}$ between the octahedral and tetrahedral edges. In most micas such a perfect fit does not occur, and the tetrahedral and octahedral layers accommodate by distortion. However, the **C**, **2**, and **m** operations are retained in the tetrahedral–octahedral sandwich.

The third geometrical factor gives polytypism. Like the leaning tower of Pisa, the upper face of an octahedron (*prq,* Fig. 8.20*a*) does not lie over the bottom face (*stu*). An octahedron in the lower half of Figure 8.20*e* is deliberately hatched to emphasize this tilt. To make the tilt more obvious, Figure 8.21*c* shows only the outlines of one octahedron and the four adjacent tetrahedra. The heavy arrow shows the tilt across the mica sandwich. There is no tilt across the K atoms, and the tilt of a sandwich arises solely at the octahedron. The simplest polytype arises when all sandwiches tilt the same way. Two sandwiches are shown in Figure 8.21*c,* and two unit cells are outlined by dots. This polytype is denoted 1M because the repeat is one layer and the symmetry is monoclinic. The basic unit *defg* in Figure 8.21*a* can be represented symbolically by Figure 8.21*d.* Each arrow represents the projection of the tilt, and it corresponds to the displacement between hydroxyl positions in Figure 8.21*a* and 8.21*c.* The operations of the space group **C12/m1** of the 1M polytype are simply those of the octahedral layer. Returning to Figure 8.20*a,* the base of the octahedron can be chosen in six ways: *tus, tsg, tgf, tfe, ted,* and *tdu.* For tetrahedral layers with hexagonal geometry, there is no structural control over the choice of these six ways. The most obvious mica polytype is one with a completely random choice of the six orientations of the sandwiches, and natural phlogopites show considerable stacking disorder.

Theoretically, there are six simple polytypes with completely regular stacking of sandwiches (Smith and Yoder, 1956). For the 1M type, the tilt vector points the same way for all sandwiches (Figs. 8.21*c* and 8.21*d*). Rotation by 180° between adjacent vectors give a two-layer repeat and orthorhombic symmetry (Figs. 8.21*e,* 8.20*e*). This 2O polytype retains the **C**, **121**, and **1m1** operations of the mica sandwich, and the zigzag pattern of adjacent layers generates a **c11** glide plane and a **11m** mirror plane to yield space group **Ccmm**. In Figure 8.20*e,* the positions labeled *t* and *q* show how the tilts are compensated. Positions *t″* and *t″″* obey the **11m** mirror operation, as do *q* and *q′.* Positions *t′* and *q′* demonstrate the **c11** glide operation.

Rotation by 120° gives the 3T polytype (Fig. 8.21*h*). A screw triad axis 3_1 is generated, and the unit cell contains three layers and has trigonal symmetry. Clockwise rotation gives a 3_2 axis and the enantiomorphic space group **P3$_2$12**.

Alternate clockwise and anticlockwise rotation of 120° gives the 2M$_1$ polytype (Fig. 8.21*f*). This two-layer monoclinic polytype loses the mirror operation, and has space group **C12/c1**.

Anticlockwise rotation by 60° gives the six-layer hexagonal polytype 6H with space group **P6$_1$22** (Fig. 8.21*i*). Its enantiomorph **P6$_5$22** has clockwise rotation. Alternation of the clockwise and anticlockwise rotations gives a second monoclinic polytype 2M$_2$ also with space group **C12/c** (Fig. 8.21*g*).

Distortions of the tetrahedral and octahedral layers complicate this simple development. Thus the ditrigonal geometry of dioctahedral micas favors the 2M$_1$ polytype. Nevertheless, the crystal structure and polytypism of mica provide a beautiful example of the way in which symmetry operations result from the packing of structural units. The relation of mica to other layer structures is discussed in the next chapter.

8.6.7   STRUCTURES WITH FRAMEWORKS OF VERTEX-SHARED TETRAHEDRA

There is an infinity of topologically different frameworks obtained by sharing all the vertices of tetrahedra. Systematic enumeration according to various algorithms is yielding a useful structural classification, and a few simple structures are described here to illustrate the relation between topology and symmetry.

The framework of tridymite, SiO_2, can be obtained from the mica sheet (Fig. 8.20*a*) by reversing alternate tetrahedra to give the sheet in Figure 5.21*a*. Apical oxygens are then shared to give the framework. The high-temperature variety belongs to space group **P6$_3$/mmc**. Alternation of direction of the apical oxygens results in the 6_3 axis. The mirror planes are easy to locate in the crystal structure, and the **c** glide planes lie EW in Figure 5.21*a* passing through the middle of the 6-rings. At low temperatures, tridymite collapses to several structures with low symmetry. The topology of the framework is retained, but Si—O—Si angles are reduced. The cristobalite polytype belongs ideally to space group **Fd3m**, but complications arise because of displacements of the oxygen atoms. The tetrahedral nodes of cristobalite and tridymite match the carbon positions of diamond and lonsdaleite, respectively, and the *D* lattice complex can be used for description of cristobalite. The quartz polymorph has been described in detail (Figs. 7.16, 7.17).

Sodalite, $Na_4Al_3Si_3O_{12}Cl$, has a tetrahedral framework whose nodes lie at the vertices of close-packed truncated octahedra (Fig. 5.23*a*). If the Al and Si atoms are assumed to occupy the tetrahedral sites at random, the crystal structure can be idealized as follows: isometric, $a = 8.87$ Å, $I\bar{4}3m$,

Na (8c) $uuu; u\bar{u}\bar{u}; \bar{u}u\bar{u}; \bar{u}\bar{u}u; u = 0.178$ Add $\frac{1}{2}\frac{1}{2}\frac{1}{2}$ for I vector.

T   (12d) $\frac{1}{4}\frac{1}{2}0; 0\frac{1}{4}\frac{1}{2}; \frac{1}{2}0\frac{1}{4}; \frac{1}{4}0\frac{1}{2}; \frac{1}{2}\frac{1}{4}0; 0\frac{1}{2}\frac{1}{4}$

O   (24g) $sst; tss; sts; \bar{s}s\bar{t}; \bar{t}s\bar{s}; \bar{s}t\bar{s}; s\bar{s}\bar{t}; t\bar{s}\bar{s}; s\bar{t}\bar{s}; \bar{s}\bar{s}t; \bar{t}\bar{s}s;$
            $\bar{s}\bar{t}s, s = 0.144, t = 0.438.$

Cl (2a)   000

The easiest way to understand the sodalite structure is to first study Figure 5.23a. Ignore the difference between Al and Si, and place the 6 Al and 6 Si randomly on the 12 tetrahedral nodes of the cubic unit cell. These nodes belong to lattice complex W^* in Table 8.1. Place each of the 24 oxygen atoms at the midpoint of a branch. The clinographic projection of the cubic unit cell (Fig. 8.22a) shows two truncated octahedra sharing the 6-ring of T atoms labeled d–i. One is centered about the hexagon j that lies at a cell corner, and the other is centered about the body-center k. The chlorine anion is so large (1.8 Å radius) that it must be placed at the center of the two truncated octahedra, namely, j and k. Each of the eight sodium atoms is placed temporarily at the center of a 6-ring. Position n is shown with six dotted lines radiating out to oxygen atoms and two lines to chlorine atoms k and y. Sodium atoms m, n, o, and t each lie on a front-facing hexagon of the truncated octahedron centered about k, whereas positions p, q, r, and s each lie on a back-facing hexagon. Can this simple arrangement be correct? Not quite because (i) the Na–Cl distance is too long, (ii) the O atoms do not form regular tetrahedra about the T atoms, and (iii) the Al and Si atoms alternate regularly over the tetrahedral nodes. However, the actual structure (Loens and Schulz, 1967) has the same topology as the simple arrangement, and differs by only small displacements. Detailed geometrical calculations (Taylor and Henderson, 1978) incorporate the following features.

The cell dimension is controlled by the mean T–O distance ($\sim$1.68 Å for Si–O = 1.61 Å and Al–O = 1.75 Å), and must be near 8.9 Å. In the idealized structure, the two Cl atoms lie at the corner and body center while the eight Na atoms lie at $(\pm\frac{1}{4}, \pm\frac{1}{4}, \pm\frac{1}{4})$. Each Cl atom (e.g., position u in Fig. 8.22a) lies at the center of a cube of Na atoms at $8.9 \times \sqrt{3}/4 = 3.9$ Å, and each Na atom lies midway between two Cl atoms. In order to attain an Na–Cl distance matching the sum of the ionic radii (2.8 Å, Table 1.1), four Na atoms at the corners of a tetrahedron (v, w, x, m) move toward the Cl atom at u, and the other four (unlabeled) move away. Appropriate arrows are attached to all the Na atoms in Figure 8.22a: thus atom n moves away from y and toward k, and atom q moves away from k and toward j. The predicted position of Na is at uuu, where $u = 2.8/3.9 \times 0.25$. This value of 0.179 agrees with the observed value of 0.178.

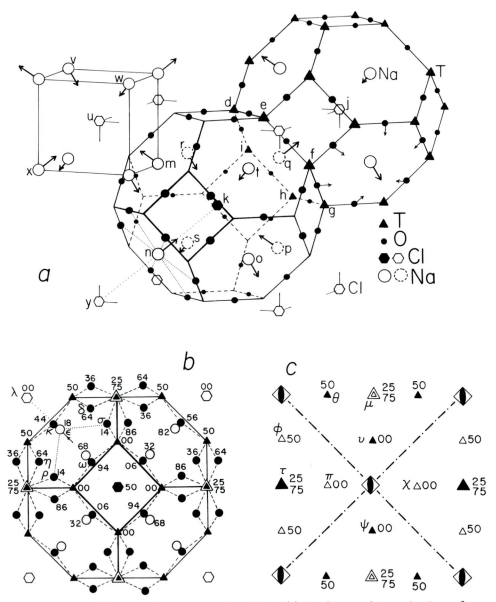

FIGURE 8.22 The crystal structure of sodalite. (a) A clinographic projection of two truncated octahedra in a cubic unit cell. Chlorine atoms lie at each corner (open hexagon) and body-center (filled hexagon) of the unit cell. A tetrahedrally-coordinated atom lies at each vertex of a truncated octahedron, and is shown by a filled triangle in one of the polyhedra. Each midpoint of an edge (filled circle) is a tentative position for an oxygen atom. The initial placement of an Na atom at the midpoint of a 6-ring (open circle) is changed by an arrow. (b) A plan of idealized structure in $\mathbf{I\bar{4}3m}$. Continuous lines join tetrahedral nodes whereas dashed lines show T-O bonds. The Na atom ϵ is bonded to three oxygens at σ, ρ, and κ and a Cl atom at the origin, but the $\kappa\epsilon$ bond is hidden in this projection. (c) A plan of Si and Al positions for the actual ordered structure. Two inverse tetrad axes and two diagonal glide planes are shown, but other symmetry elements are omitted.

Whereas a regular tetrahedron has six equal O–T–O angles of 109.47°, the simple sodalite model has four angles of 120° and two of 90°. Oxygen atoms must move away from the center of each 4-ring to increase the O–T–O angles of 90° by ~20°, and toward the center of each 6-ring to decrease the 120° angles by ~10°. Furthermore, the oxygen positions are constrained by the need to obtain a Na–O distance near 2.4 Å for ionic bonding. The overall result of these two factors gives the atomic positions shown in a plan of the unit cell (Fig. 8.22b). Heavy lines show the edges of a truncated octahedron. Oxygen atoms move away from the centers of the 4-rings and toward the centers of the 6-rings, and the T–O bonds are shown by dashed lines. An arrow shows the displacement of an Na atom ϵ from the center of a 6-ring toward a Cl atom λ to which it is linked by a dotted line. The Na atom is also bonded to three oxygen atoms (ρ, σ, κ) of type O(1) at 2.4 Å but is too far (2.9 Å) from the other three oxygen atoms (η, δ, ω) for the bonding to be strong. Such an offset of a cation from the center of a 6-ring is common in framework silicates, especially in zeolites.

The third stage is to consider the effects of ordering of the Si and Al atoms on the tetrahedral nodes. To a first approximation, the geometry of the framework is unchanged, and Figure 8.22c shows the tetrahedral positions. In diagrams 8.22a and 8.22b, all tetrahedral nodes are shown by filled triangles except for the overlapping nodes in 8.22b. In diagram 8.22c, filled symbols represent Si and open ones Al, or vice versa. In the 6-ring $\theta\mu\nu\pi\tau\phi$ and the 4-ring $\pi\nu\chi\psi$, the atoms alternate regularly, and indeed there is only one way to assign atoms once the first atom is placed. In diagram 8.22b, there are diagonal mirror planes (e.g., $\eta \rightarrow \delta$, $\omega \rightarrow \omega$), but in diagram 8.22c the mirror operation is lost (e.g., ϕ and θ). However, a diagonal glide operation remains ($\theta \rightarrow \psi$; $\phi \rightarrow \chi$). The inverse tetrad axes also remain, but the **I** body centering is lost (e.g., θ and χ). Thus the regular alternation of Si and Al changes the space group from **I$\bar{4}$3m** (diagram 8.22b) to **P$\bar{4}$3n** (diagram 8.22c).

The term *topologic symmetry* describes the symmetry of a framework when idealized into its most regular shape by movements that leave intact the topologic relationships between nodes. Interatomic angles can be changed, but not contacts between first neighbors. *Topochemical symmetry* describes the symmetry of an idealized framework when the chemical occupancy of the nodes is considered (P.B. Moore in Smith, 1974). The tetrahedral nodes of the sodalite framework fit space group **I4/m3m**, but the requirement of tetrahedral geometry forces the oxygen atoms into the lower space group **I$\bar{4}$3m** (diagram 8.22b). This topologic symmetry is degraded into topochemical symmetry when Si,Al alternation gives **P$\bar{4}$3n**. Note that the apparent mirror planes along $\phi\tau$ and $\tau\pi\chi$ in diagram 8.22c are destroyed when oxygen positions from diagram 8.22b are considered.

Writing the sodalite composition as $Na_3Si_3Al_3O_{12}$ · NaCl is structurally misleading because all sodium atoms are equivalent.

When the symmetry of a framework is degraded by ordering of the T atoms, the T atoms stay exactly on or very close to their ideal positions, and the O atoms move away from their ideal positions to provide for differences in T–O distances.

 All members of the *feldspar* family of minerals (Smith, 1974) contain an infinite framework of vertex-linked $(Si,Al)O_4$ tetrahedra, whose irregular interstices contain various large cations. The ideal structure is exemplified by the *high sanidine* polymorph of $KAlSi_3O_8$ whose Al and Si atoms occupy the tetrahedral nodes randomly, and a superstructure is represented by the *low microcline* polymorph in which each of the Al and 3 Si atoms occupies a distinct crystallographic site.

High sanidine has the following parameters: $a = 0.8564$ nm, $b = 1.3030$ nm, $c = 0.7175$ nm, $\beta = 115.99°$, **C12/m1**,

	M	T_1	T_2	O_{A1}	O_{A2}	O_B	O_C	O_D
x	0.2840	0.0097	0.7089	0	0.6343	0.8273	0.0347	0.1793
y	0	0.1850	0.1178	0.1472	0	0.1469	0.3100	0.1269
z	0.1352	0.2233	0.3444	0	0.2858	0.2253	0.2579	0.4024

In natural sanidine, the M site is occupied mainly by K but with some Na and other large ions with low ionic potential. The two (Si,Al) sites are denoted by T, and along with the five oxygen sites are labeled with subscripts assigned by W.H. Taylor in 1933.

Because there are mirror planes at $y = 0$ and $\frac{1}{2}$, it is sufficient to show atoms with $0 < y < \frac{1}{2}$ in the b-axis projection (Fig. 8.23a). Filled and open symbols are used to distinguish atoms at different levels.

Diagram 8.23b shows a b-axis projection of the general positions and symmetry elements for **C12/m1**. The coordinates of the general positions can be developed as follows (see also Section 7.13):

$xyz; \frac{1}{2} + x, \frac{1}{2} + y, z$ operation of C face centering on starting position
$\bar{x}y\bar{z}; \frac{1}{2} - x, \frac{1}{2} + y, \bar{z}$ operation of 2-fold axis at $x = z = 0$ on first two
positions

FIGURE 8.23 (*Opposite*) The crystal structure of sanidine. (*a*) The *b*-axis projection showing T-O bonds (continuous lines) and K-O bonds (dashed lines) in a unit cell outlined by dots. (*b*) The general positions and symmetry elements for **C12/m1**. (*c*) The T-O linkages of the framework with emphasis placed on the way in which the 4-rings can be linked into either double-crankshaft or bifurcated chains. Diad axes and centers of symmetry are also shown. Heights given either in fractions or hundredths of *b*.

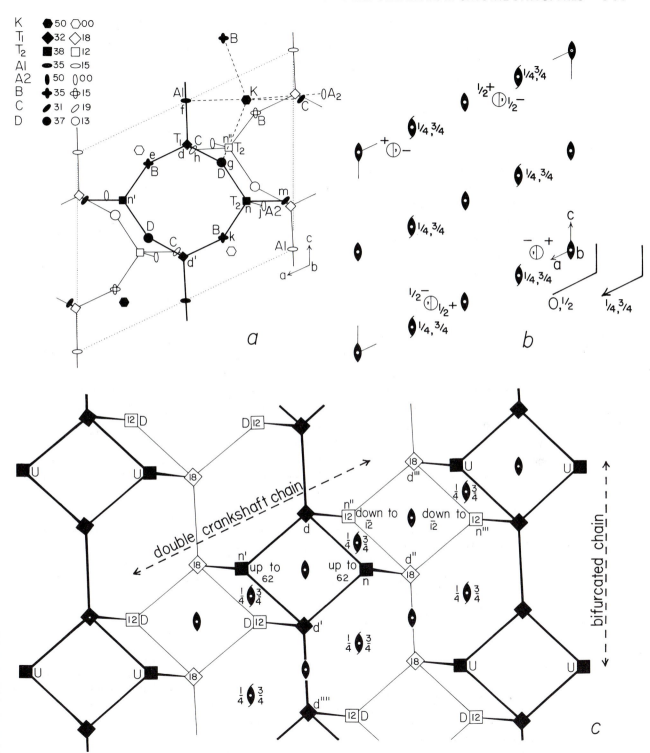

$x\bar{y}z;\ \frac{1}{2} + x, \frac{1}{2} - y, z$ operation of mirror plane at $y = 0$ on first four positions

$\bar{x}\bar{y}\bar{z};\ \frac{1}{2} - x, \frac{1}{2} - y, \bar{z}$

Special positions are:

point symmetry $\mathbf{m}$	multiplicity 4; $x0z$; $\bar{x}0\bar{z}$; $\frac{1}{2} + x, \frac{1}{2}, z$; $\frac{1}{2} - x, \frac{1}{2}, \bar{z}$
point symmetry $\mathbf{2}$	multiplicity 4; $0y0$; $0\bar{y}0$; $\frac{1}{2}, \frac{1}{2} + y, 0$; $\frac{1}{2}, \frac{1}{2} - y, 0$ and $0y\frac{1}{2}$; $0\bar{y}\frac{1}{2}$; $\frac{1}{2}, \frac{1}{2} + y, \frac{1}{2}$; $\frac{1}{2}, \frac{1}{2} - y, \frac{1}{2}$
point symmetry $\bar{\mathbf{1}}$	multiplicity 4; $\frac{1}{4}\frac{1}{4}0$; $\frac{1}{4}\frac{3}{4}0$; $\frac{3}{4}\frac{3}{4}0$; $\frac{3}{4}\frac{1}{4}0$ and $\frac{1}{4}\frac{1}{4}\frac{1}{2}$; $\frac{1}{4}\frac{3}{4}\frac{1}{2}$; $\frac{3}{4}\frac{3}{4}\frac{1}{2}$; $\frac{3}{4}\frac{1}{4}\frac{1}{2}$
point symmetry $\mathbf{2/m}$ (also a center of symmetry)	multiplicity 2; 000; $\frac{1}{2}\frac{1}{2}0$ and $0\frac{1}{2}0$; $\frac{1}{2}00$ and $00\frac{1}{2}$; $\frac{1}{2}\frac{1}{2}\frac{1}{2}$ and $0\frac{1}{2}\frac{1}{2}$; $\frac{1}{2}0\frac{1}{2}$

The M and O_{A2} atoms lie on a mirror plane, and the O_{A1} atom on a rotation diad axis.

Diagram 8.23c shows only the T atoms, and diagram 8.23a shows how adjacent T and O atoms are linked. A T_1 atom (d) is linked to four oxygen atoms, O_{A1} at f, O_B at e, O_C at h, and O_D at g. The link to O_C hardly shows because the ellipse for O_C is close to the square for T_1. A T_2 atom (n) is linked to O_{A2} at j, O_B at k, O_C at m, and O_D at g. Each oxygen atom is linked to two T atoms: O_{A1} to two T_1 atoms (remember that O_{A1} lies on a vertical rotation diad), O_{A2} to two T_2 atoms (remember that O_{A2} lies on a horizontal mirror plane), and O_B, O_C, and O_D to one T_1 and one T_2. The T–O linkages form an infinite framework with T_1–O distances calculated in Exercise 4.5 and the following T_2–O distances: $O_{A2} = 0.1645$ nm, $O_B = 0.1631$ nm, $O_C = 0.1645$ nm, $O_D = 0.1638$ nm.

The coordination of the K atom is quite irregular, but nine oxygen atoms can be regarded as first neighbors (dashed lines in diagram 8.23a): one $O_{A2} = 0.271$ nm, one O_{A1} at $f = 0.291$ nm, two O_D at $g = 0.295$ nm, two $O_B = 0.302$ nm, and two $O_C = 0.314$ nm.

Ignoring the O and K atoms, the T atoms can be considered as the nodes of a 4-connected net. Each pair of T atoms that share an oxygen atom is joined by a line, which becomes a branch of the net. The net is 4-connected because four branches meet at each node. Nodes n', d, n, and d' form a 4-ring, whose opposing nodes are related by a rotation diad axis. Nodes d, n'', d'', and n form another type of 4-ring, whose opposing nodes are related by a center of symmetry at height $z = \frac{1}{4}$. These two types of 4-rings are repeated by the symmetry elements so that they generate an infinite chain, which is known as the "double crankshaft chain." To see the crankshaft, observe that the 4-rings of type $n'dnd'$ (heavy line) are higher than the 4-rings of type $n''d''' n''' d''$ (thin line), and that they are connected by tilted rings of type $dn''d''n$. The chain is considerably distorted, but it has the zig-zag shape of a crankshaft. In the c direction, double crankshaft chains are connected by horizontal linkages of type $d'd''''$ that are related by vertical diad axes. Alternatively, the feldspar net can be considered as the cross-linkage of "bifurcated chains" lying along the c axis. In the b direction, link-ages occur through mirror planes at $y = 0$ and $\frac{1}{2}$. The T_2 node n at $y = 0.3822$ is bonded through O_{A2} at $y = \frac{1}{2}$ (ellipse j in diagram 8.23a) on a mirror plane, and this oxygen is then bonded to another T_2 node at $y = 0.6178$. Similarly, T_2 node n'' at height $y = 0.1178$ is bonded to the nearby O_{A2} atom at $y = 0$. Symbolically the linkages in the b direction can be repre-sented in diagram 8.23c by using U and D to represent an upward and a downward linkage, respectively.

It is not known why $KAlSi_3O_8$ adopts the feldspar structure rather than one of the infinite number of other framework structures that can be in-vented. Although the oxygen atoms are not close packed in all directions in feldspar, they do approximate rather crudely to cubic closest packing, as explained by G.O. Brunner in Smith (1974, p. 41), and perhaps this relative compactness contributes to the stability.

Because the 4 Al and 12 Si atoms of high sanidine fall into two 8-fold sites of **C12/m1**, the space group symmetry must drop when the Al and Si atoms are fully ordered. Indeed the **2** and **m** operations are lost resulting in triclinic space group **C$\bar{1}$**. Strictly speaking, new axes should be chosen to convert **C$\bar{1}$** to **P$\bar{1}$**, but it is convenient to retain the C-centered cell with a 4-fold general position. The 8-fold T_1 position of **C12/m1** splits into two 4-fold positions labeled $T_1 0$ and $T_1 m$ by H.D. Megaw, and the T_2 position goes to $T_2 0$ and $T_2 m$. The four Al atoms of low albite go into the $T_1 0$ position.

In *celsian, BaAl$_2$Si$_2$O$_8$,* the Al and Si atoms alternate on the tetrahedral sites. Whereas in sodalite such alternation did not change the cell repeat, in celsian the c dimension must be doubled. Thus in Figure 8.23c, put Al at d'''', Si at d', Al at both n and n', and Si at d. Positions d'''' and d are no longer equivalent, and it is necessary to double c. The resulting unit cell has space group **I12/c1**.

This completes the development of structures containing progressively greater polymerization of tetrahedra. Further structures containing tetra-hedra will appear in the next chapter, which is devoted to the polymeriza-tion of octahedra.

EXERCISES   (*More Difficult)

8.1 Make several cubes out of cardboard. Mark each corner with a felt pen. Stack the cubes together to make a model of simple cubic packing. Locate the positions of 4, $\bar{3}$, and 2 axes and of the two types of mirror planes. Then mark the center of each face to make a model of face-centered cubic packing, and show that all the symmetry elements are retained. *Identify new types of symmetry planes produced by the interaction of the F lattice with the symmetry elements found in simple cubic packing.

8.2* If ball-and-spoke models are available for any of the structures shown in Figure 8.2, identify the symmetry operations.

8.3* Make several 120° prisms out of 120° rhombuses and rectangles of cardboard. Mark each corner with a felt pen. Stack the prisms together to make a model of simple hexagonal packing. Locate the positions of the symmetry elements in Figure 8.3.

8.4* If a model of graphite is available, identify the symmetry operations. Alternatively, show how the positions in Figure 8.4b obey the symmetry elements in diagram 8.3g.

8.5* Build layers of MoS_2 type (Fig. 8.5) using balls of two sizes. Stack them in the appropriate positions. Identify the symmetry operations in space group $P6_3/mmc$ (Fig. 8.3g).

8.6* Several metals (Wyckoff, 1963, p. 14), including lanthanum, crystallize with a hexagonal unit cell and coordinates $0,0,0; 0,0,\frac{1}{2}; \frac{1}{3},\frac{2}{3},\frac{3}{4}; \frac{2}{3},\frac{1}{3},\frac{1}{4}$. Draw a c-axis projection of the unit cell of La metal (a = 3.77 Å, c = 12.2 Å). Determine the space group. Show that the atomic packing can be described as **ABAC** in contrast to **AB** for hcp and **ABC** for ccp.

8.7* The crystal structure of NbO (Wyckoff, 1963, p. 94), has the following atomic positions in a cubic unit cell with a = 4.21 Å:

Nb $0\frac{1}{2}\frac{1}{2}; \frac{1}{2}0\frac{1}{2}; \frac{1}{2}\frac{1}{2}0$ O $\frac{1}{2}00; 0\frac{1}{2}0; 00\frac{1}{2}.$

Draw a plan of the unit cell. Determine the space group, the atomic coordination, and Nb–O distance.

8.8 Examine a model of the halite structure and identify the symmetry operations in the symbol **Fm3m**.

8.9* Build a model of CaC_2 type using balls of two sizes. Cork balls are easy to truncate for assembly into C_2 doublets. The relative size of the balls is not critical because the axial ratio c/a of the tetragonal unit cell will adjust automatically.

8.10* Study Figures 8.7 and 8.8, and then draw Figure 8.7 from memory.

8.11* The crystal structure of the cubic form of N_2 (Wyckoff, 1963, p. 30) has a = 5.64 Å and coordinates $\pm(uuu;$ $\frac{1}{2} + u, \frac{1}{2} - u, \bar{u}; \bar{u}, \frac{1}{2} + u, \frac{1}{2} - u;$ $\frac{1}{2} - u, \bar{u}, \frac{1}{2} + u)$ where $u \cong 0.054$. Draw a plan of the unit cell and determine the space group by analogy with the structure of pyrite. Determine intra- and intermolecular N_2 distances. The crystal structure (Wyckoff, 1963, p. 368) of dry ice, solid CO_2, can be obtained from that of cubic N_2 by replacing each N atom by an O atom, but with $u \cong 0.11$, and by placing carbon atoms at 000, $0\frac{1}{2}\frac{1}{2}$, $\frac{1}{2}0\frac{1}{2}$ and $\frac{1}{2}\frac{1}{2}0$. Using a = 5.58 Å, draw a plan of the unit cell. Show that there are four linear CO_2 molecules per cell. Determine the intra- and intermolecular distances.

8.12* The crystal structure of iodine (Wyckoff, 1963, p. 52) has a = 7.27 Å, b = 9.79 Å, c = 4.79 Å and coordinates $\pm(0,u,v;$ $\frac{1}{2},\frac{1}{2} + u, \bar{v};$ $\frac{1}{2}, u, \frac{1}{2} + v;$ $0, \frac{1}{2} + u, \frac{1}{2} - v)$, u = 0.116, v = 0.149. Draw the a-axis projection and determine the space group. Calculate the intramolecular distances and identify the closest neighbors from adjacent molecules.

8.13** Look up the crystal structure of rhombic sulfur in Wyckoff (1963, p. 33), and plot the atomic coordinates from the listed data. Identify the symmetry operations in space group **Fddd**. Show that the 128 atoms per unit cell occur as 16 identical S_8 molecules, each shaped like the top of a crown.

8.14* The crystal structure of anhydrite, $CaSO_4$, has a simple relation to the zircon structure, which is easily seen by choosing a new set of axes from those in Wyckoff (1965, p. 18). In an orthorhombic cell with a = 6.99 Å, b = 7.00 Å, c = 6.24 Å, **Amma**, the atoms utilize the following special positions.

$(4c)$ $\pm(\frac{1}{4}, 0, u;$ $\frac{1}{4}, \frac{1}{2}, \frac{1}{2} + u)$

$(8f)$ $\pm(v0w;$ $\frac{1}{2} - v, 0, w;$ $v, \frac{1}{2}, \frac{1}{2} + w;$ $\frac{1}{2} - v, \frac{1}{2}, \frac{1}{2} + w)$

$(8g)$ $\pm(\frac{1}{4}, p, q;$ $\frac{1}{4}, \bar{p}, q;$ $\frac{1}{4}, \frac{1}{2} + p, \frac{1}{2} + q;$ $\frac{1}{4}, \frac{1}{2} - p, \frac{1}{2} + q)$

Draw the c-axis projection using the coordinates: Ca, u = 0.650; S, u = 0.155; O(1), v = 0.08, w = 0.297; O(2), p = 0.172, q = 0.017. Calculate S–O and Ca–O distances and show that each S atom lies in a tetrahedron and each Ca atom in a triangulated dodecahedron. Determine the relation between the anhydrite and zircon types of structure, and show why the anhydrite structure does not have tetragonal symmetry. Locate all six types of symmetry planes in the anhydrite structure.

8.15* The crystal structure of scheelite, $CaWO_4$, has a tetragonal unit cell (Kay et al., 1964), with a = 5.243 Å, c = 11.38 Å, **14_1/a**. Draw the c-axis projection using the atomic coordinates:

Ca $(4b)$ $0,0,\frac{1}{2}$; $\frac{1}{2},0,\frac{1}{4}$; bc

W   (4a) $0,0,0; 0,\frac{1}{2},\frac{1}{4}$; bc

O   (16f) $x,y,z; \bar{x},\bar{y},z; x, \frac{1}{2} + y, \frac{1}{4} - z; \bar{x}, \frac{1}{2} - y, \frac{1}{4} - z; \bar{y},x,\bar{z};$

$y,\bar{x},\bar{z}; \bar{y}, \frac{1}{2} + x, \frac{1}{4} + z; y, \frac{1}{2} - x, \frac{1}{4} + z;$ bc; $x =$

$0.241, y = 0.151, z = 0.086$

bc means add $\frac{1}{2},\frac{1}{2},\frac{1}{2}$ for the body centering. Calculate W–O and Ca–O distances, and show that each W lies in a tetrahedron of oxygen atoms, and each Ca atom lies in a triangulated dodecahedron of oxygen atoms. Determine the relation between the zircon (Fig. 8.13) and scheelite structures. Draw the space group of symmetry elements for $I4_1/a$, and compare with $I4_1/amd$.

8.16 Locate the edge-sharing chains of tetrahedra in the $AlPS_4$ structure (Fig. 7.7).

8.17 Do edge-shared tetrahedra occur in the fluorite structure (Fig. 5.12a)?

8.18* Using three felt pens, color the two levels of Si_2O_7 groups and the ScO_6 octahedron in Figure 8.14. Construct the ScO_6 octahedra centered on Sc atoms at $y = 0.69$, and color them with a fourth pen. Show that ScO_6 octahedra at $y = 0.31$ and 0.81 do not share edges. Calculate the Si–O and Sc–O distances.

8.19* Using three felt pens, color the two levels of Si_2O_7 groups and the ZnO_4 tetrahedra in Figure 8.15. Extend the symmetry elements from the right-hand into the left-hand unit cell. Draw the square antiprism for a Ca atom at height $51/100$. Calculate the Ca–O, Zn–O, and Si–O distances.

8.20* Using three felt pens, color the Si_3O_9 rings and the TiO_6 and BaO_6 octahedra in Figure 8.16. Using tracing paper, link the open symbols to find the second level of Si_3O_9 rings. Extend the symmetry elements to cover the entire diagram. In particular, show that the Ba and Ti atoms lie on inverse hexad axes. Check your answer with Volume 1, page 295, of *International Tables for X-Ray Crystallography*. Calculate the Si–O, Ti–O, and Ba–O distances. Show that the TiO_6 and BaO_6 octahedra share edges, and count the number.

8.21 Glue balls together to make models of the pyroxene and amphibole chains. Alternatively, draw circles of appropriate radius in Figure 8.17. *Glue balls together to make models of the other three chains. Invent your own chains.

8.22* What would be the band symmetry group if (i) alternate tetrahedra of the pyroxene chain pointed downward into the paper instead of upward, and (ii) all tetrahedra pointed upward in the $BaSi_2O_6$ chain.

8.23* Determine the heights omitted in Figure 8.18a, and check your answer by comparing with listed heights in the adjacent unit cell (e.g., the three atoms labeled with i should have the same height). Color the chains of silicate and MgO_6 polyhedra. Using tracing paper, copy the positions of the zigzag octahedral chains in the right-hand unit cell. Then place the tracing paper over the left-hand unit cell, and determine the complex interaction between the silicate tetrahedra and MgO_6 octahedra. Use Figure 8.18b to check the third dimension. Locate the atoms in diagram 8.18a that correspond to atoms labeled p'' and v'' in diagram 8.18b. Color the polyhedra in Figure 8.19, and compare the net and the idealized and distorted polyhedra with the atomic positions in Figure 8.18a.

8.24** Study the crystal structures of the pyroxenoid group of chain silicates using the references listed in Deer, Howie, and Zussman (1978) and Ohashi and Finger (1978). Note particularly the linkages between tetrahedra and octahedra.

8.25 Study the four types of tetrahedral layers in Figure 8.20, close the book, and draw the layers from memory.

8.26* (i) If adjacent tetrahedra of the mica layer pointed alternately up and down what would be the symmetry group? (ii) What would be the symmetry groups of the manganpyrosmalite and apophyllite layers if all tetrahedra pointed upward?

8.27* Build models of the tetrahedral and octahedral layers of mica as described in the text. Study the diagrams of the mica structure in Figures 8.20 and 8.21, close the book, and sketch the key features of the diagrams.

8.28* Draw the c-axis projection of the idealized structure of hexagonal $BaAl_2Si_2O_8$ using the following parameters: $a = 5.3$ Å, $c = 7.8$ Å; **P6/mmm**;

Ba	(*1a*)	$0,0,0$
(Al,Si)	(*4h*)	$\pm(\frac{1}{3},\frac{2}{3},u;\ \frac{2}{3},\frac{1}{3},u)$, $\quad u = 0.29$
O(1)	(*2d*)	$\frac{1}{3},\frac{2}{3},\frac{1}{2};\ \frac{2}{3},\frac{1}{3},\frac{1}{2}$
O(2)†	(*12n*)	$\pm(u,0,v;\ 0,u,v;\ u,u,v;\ u,0,\bar{v};\ 0,u,\bar{v};$
		$u,u,\bar{v})$, $\quad u = 0.45, v = 0.21$

(†50% random occupancy)

Show that the (Al,Si) and O atoms generate a double layer of vertex-linked tetrahedra as in Figure 8.20b, and that the layers are cross-linked by Ba atoms coordinated to 12 oxygens of type O(2). Note the orientational disorder. **Suppose that the Al and Si atoms occupy alternate tetrahedral nodes, instead of being randomly placed as in the idealized structure. Determine the space group. **Read Takéuchi (1958) and Takéuchi and Donnay (1959) to discover the structural complexities caused by stacking disorder and replacement of the large Ba^{2+} cation by the small Ca^{2+} cation.

8.29* Take a cardboard model of a truncated octahedron (Chapter 3), and mark the atomic positions shown in Figure 8.22a. The Cl position must be imagined at the body center. Stack together several truncated octahedra.

8.30** Examine Figure 8.22b. Then plot a similar plan using the following data for sodalite (Loens and Schulz, 1967) in which distinction was made between Si and Al positions: a = 8.870 Å, **P$\bar{4}$3n,**

Na	8e	uuu, u = 0.1777
Cl	2a	000
Al	6c	¼ ½ 0
Si	6d	¼ 0 ½
O	24i	xyz, x = 0.1401, y = 0.4385, z = 0.1487

Try to deduce all the positions just from the space group operations, and check with *International Tables for X-Ray Crystallography,* Volume 1, page 327. Calculate distances between first neighbors, and O—Si—O and Si—O—Al angles. Read Taylor and Henderson (1978) to find out how the geometry of the sodalite structure changes as different cations and anions are substituted.

8.31** Examine Figure 8.23. Then plot a projection similar to diagram 8.23a using the following data for synthetic Rb-feldspar: a = 0.8839 nm, b = 1.3034 nm, c = 0.7182 nm, β = 116.29°,

	Rb	T_1	T_2	O_{A1}	O_{A2}	O_B	O_C	O_D
x	0.2958	0.0104	0.7227	0	0.664	0.832	0.0464	0.1682
y	0	0.1904	0.1195	0.1525	0	0.1596	0.3145	0.1283
z	0.1468	0.2227	0.3440	0	0.283	0.226	0.263	0.403

Calculate the T—O and·Rb—O distances.

8.32** Read Chapters 3 and 4 of Smith (1974) to discover the complex literature on the order–disorder and geometry of feldspar minerals.

ANSWERS

8.1 In **F4/m$\bar{3}$2/m** there are simple glide planes halfway between the first set of mirror planes, and diagonal glide planes halfway between the second set of mirror planes. See Buerger (1956, p. 436) for a drawing of all the symmetry elements.

8.6 **P6$_3$/mmc.** See Figure 5.3 for comparison with ccp and hcp.

8.7 The Nb and O atoms lie, respectively, at the midpoints of faces and edges. Each Nb atom is at the center of a square of O atoms at 2.10_5 Å, and vice versa. The Nb—O bonds generate a 4-connected 3D net (Fig. 5.27*e*). **Pm3m.**

8.11 **Pa3.** To obtain the same tilts of the N_2 molecules as for the S_2 doublets translate by $(\mathbf{a} + \mathbf{b})/2$. Intramolecular N—N = 1.06 Å. Each N atom has six neighbors from adjacent molecules, three at 3.59 Å and three at 3.64 Å. The centers of the N_2 molecules are in a fcc array, for which the coordination is 12-fold. The next six closest N atoms are at 4.40 and 4.44 Å, which distances are so much larger than 3.59 and 3.64 that they are ignored in considering the packing of the N_2 molecules. In solid CO_2, the C—O distance is 1.06 Å. Each O atom has six near neighbors, three at 3.20 Å and three at 3.43 Å.

8.12 See Figure 8.24. The space group is **B(m** or **c)(a** or **c)(b** or **n)**, and **Bmab** is the conventional choice. Note the overlap of **c** and **a** glide planes at $y = \frac{1}{4}$ and $\frac{3}{4}$. Axes and centers of symmetry are omitted. The intramolecular distance is 2.68 Å. Each I atom lies at 3.55 Å to two I atoms from ad-

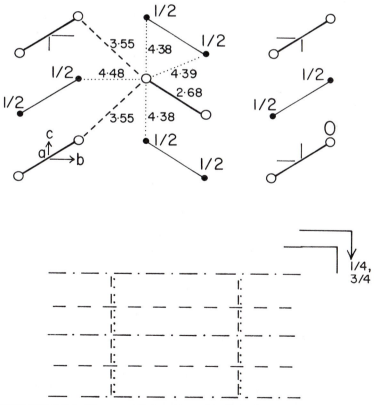

FIGURE 8.24 The crystal structure of iodine. Answer to exercise 8.12.

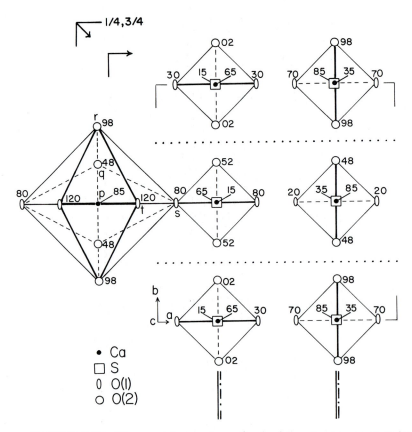

FIGURE 8.25 The crystal structure of anhydrite. Answer to exercise 8.14.

jacent molecules with the same x coordinate and at 4.38–4.48 Å to eight I atoms displaced by $x = \frac{1}{2}$. Thus there are sheets of I_2 molecules in a tweed configuration separated from adjacent sheets by weak bonds.

8.14 See Figure 8.25. Both S–O(1) and S–O(2) are 1.48 Å. In the triangulated dodecahedron, $pq = 2.59$, $pr = 2.44$, $ps = 2.33$, and $pt = 2.50$ Å. Alternate vertical chains of SO_4 tetrahedra and Ca atoms are displaced by $z = \frac{1}{4}$ with respect to the zircon structure, thereby destroying the 4-fold symmetry. Five of the symmetry planes are shown. Mirror and diagonal glide planes overlap at $x = \frac{1}{4}, \frac{3}{4}$. Mirror planes occur at $y = 0, \frac{1}{2}$ but are not shown, to avoid confusion.

8.15 See Figure 8.26. W–O = 1.78_3 Å; Ca–O(1) = 2.43_9 Å (pr) and 2.47_8 Å (pq). The tetrahedra in the scheelite structure are rotated 31° with respect to the zircon structure, with consequent loss of the **m** and **d** operations. All the symmetry elements of $I4_1/a$ can be found in $I4_1/amd$, and the former is

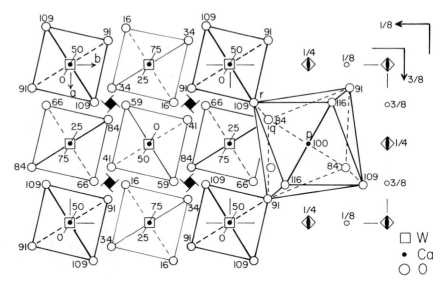

FIGURE 8.26   The crystal structure of scheelite. Answer to exercise 8.15.

a cell-equivalent subgroup, but not a class-equivalent subgroup, of the latter. See Bayer (1978).

8.16   Atoms *l* and *m* share edges *hj* and *ik* to form chains parallel to the *a*-axis. Similarly, atoms *p* and *q* form chains parallel to the *b-axis.*

8.17   From the mathematical viewpoint, each F atom is surrounded by four Ca atoms at the vertices of a tetrahedron, but from a crystal–chemical viewpoint the coordination polyhedron is constructed from the eight large F^- anions that surround each Ca^{2+} cation. These eight anions generate a cube, each of whose eight edges is shared with adjacent cubes to form an edge-shared framework. In the antifluorite structure, four large O^{2-} anions form a tetrahedron about each Li^{2+} cation.

8.18   Si–O(1) = 1.60_3 Å, Si–O(2) = 1.62_0 Å, Si–O(3) = 1.63_2 Å. Sc–O(2) = 2.11_6 Å, Sc–O(3) = 2.08_5 and 2.20_1 Å.

8.19   Ca–O(1) = 2.48_5 Å, Ca–O(2) = 2.47_0 Å, Ca–O(3) = 2.41_0 Å. Zn–O(3) = 1.93_4 Å. Si–O(1) = 1.64_5 Å, Si–O(2) = 1.58_3 Å, Si–O(3) = 1.62_2 Å.

8.20   Si–O(1) = 1.630 and 1.648 Å (two symmetry-independent values), Si–O(2) = 1.605 Å; Ti–O(2) = 1.942 Å; Ba–O(2) = 2.767 Å [note Ba–O(1) = 3.434 Å]. Each octahedron shares three edges with its neighbors. Thus the TiO_6 octahedron shares edges *pq, rs,* and *tu* with BaO_6 octahedra.

8.22   (i) **p12₁/m1**, (ii) **pb11**.

8.26   (i) **p3̄m1**, (ii) **p6mm** and **p4bm**.

8.28   If an Si atom is placed at ⅓,⅔,*u*, Al atoms must lie at ⅓,⅔,*ū*; ⅔,½,*u*, and the second Si atom at $\frac{2}{3}, \frac{1}{3}, \bar{u}$. These positions correspond to special position (*2d*) in **P$\bar{3}$m1** (*International Tables for X-Ray Crystallography,* p. 270).

8.30 Silicate tetrahedron: Si–O = 1.628 Å, O–O = 2.629, 2.714 Å, O–Si–O = 107.7°, 113.0°; aluminate tetrahedron: Al–O = 1.728 Å, O–O = 2.806, 2.855 Å; Si–O–Al = 138.3°; Na–Cl = 2.73 Å; Na–O = 2.35, 3.09 Å.

8.31 T_1 tetrahedron: O_{A1} 0.1633, O_B 0.1629, O_C 0.1644, O_D 0.1634 nm; T_2 tetrahedron: O_{A2} 0.1633, O_B 0.1627, O_C 0.1634, O_D 0.1635; distances to Rb: O_{A1} 0.307, O_{A2} 0.295, O_B 0.317, O_C 0.312, O_D 0.304.

9

Atomic Packing and Symmetry II. Octahedral Structures

Preview This chapter continues the development of the last chapter, and pays particular attention to structures containing octahedra. Because atoms found in octahedral coordination tend to have lower ionic potential than atoms in tetrahedral coordination, the ionic model leads to the prediction that edge sharing of octahedra should be more common than for tetrahedra. This is true, and, furthermore, even faces can be shared by octahedra. This extra complexity is accentuated by the number of ways in which (say) two shared edges can be selected from the 12 edges of an octahedron. Only a few representative structures can be given here.

9.1 Structures with Isolated Octahedra

Because an octahedron approximates to a sphere, it is not surprising that an octahedral group can replace the Cl of the halite (NaCl) structure type. In the *NaSbF$_6$ structure,* each Sb atom is surrounded by an octahedron of six F atoms, and the Na and Sb atoms correspond to the Na and Cl positions of the halite structure. The cubic unit cell has $a = 8.18$ Å and retains the space group **Fm3m** of halite. Atomic coordinates are (Wyckoff, 1965, p. 324):

$$\text{Na} \quad 000 \qquad \text{Sb} \quad \tfrac{1}{2}\tfrac{1}{2}\tfrac{1}{2}$$

$$\text{F} \quad \pm(u00; 0u0; 00u), \qquad u = 0.283$$

Add $\tfrac{1}{2}\tfrac{1}{2}0; 0\tfrac{1}{2}\tfrac{1}{2}; \tfrac{1}{2}0\tfrac{1}{2}$ to obtain positions related by the face centering. In the clinographic projection of the unit cell (Fig. 9.1), SbF$_6$ octahedra are centered about the midpoints of the cell edges and about the body center. A sodium atom lies at each cell corner and face center, where it is surrounded by an octahedron of F atoms. For the ionic model, the strongly bonded $(SbF_6)^{-1}$ anionic complex is weakly bonded to Na$^+$ cations, as indicated by the shortness of the Sb–F distance (1.78 Å) compared with the Na–F

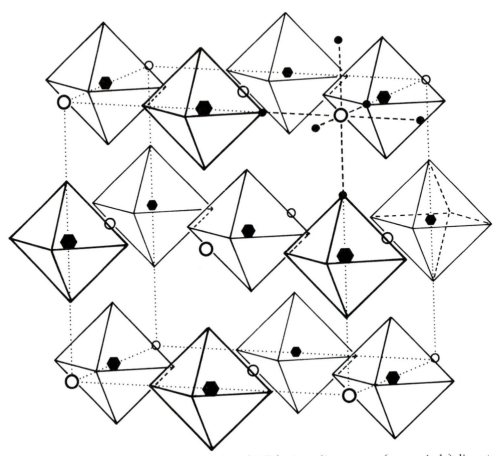

FIGURE 9.1   The crystal structure of $Na(SbF_6)$. A sodium atom (open circle) lies at each corner of the unit cell (dotted lines) and each face-center. An Sb atom (hexagon) lies at the midpoint of each edge and at the body-center, and is surrounded by an octahedron of F atoms. Six F atoms (filled circle) are shown bonded (dashed lines) to an Na atom, but other Na-F bonds are omitted to reduce overlap.

distance (2.31 Å). To emphasize the anionic complex, octahedra are not drawn about the Na atoms in Figure 9.1.

From a purely topologic viewpoint, ignoring the distinction between the Na and Sb atoms gives a framework of vertex-shared octahedra. Indeed, the $NaSbF_6$ structure can be regarded as a superstructure of the parent ReO_3 structure (Fig. 5.27*f*). Furthermore, the $NaSbF_6$ structure can be described compactly using the *F* lattice complexes of Table 8.1 and the symbol 6*o* for an octahedral group (Hellner, 1979, p. 70).

 The **$TlSbF_6$ structure** is obtained from the CsCl structure by placing a Tl atom at each Cs site and an Sb-centered octahedron of F atoms at each Cl site. The cubic unit cell of CsCl is degraded to a rhombohedral unit cell

(Wyckoff, 1965, p. 331). The symmetry is degraded from **Im3m** to **R3̄m**, and each Tl atom is surrounded by a trigonal antiprism of F atoms at 3.14 Å.

The *K₂PtCl₆* structure is obtained from the fluorite structure by placing a K atom at each Ca site and a Pt-centered octahedron of Cl atoms at each F site. Each K atom has 12 Cl atoms as first neighbors. See Exercise 9.2.

The crystal structure of *metavariscite, AlPO₄ · 2H₂O,* has isolated $AlO_4(H_2O)_2$ octahedra (Exercise 9.3). However, each of the four oxygen atoms is shared between an octahedron and a PO_4 tetrahedron to give an infinite framework of vertex-shared octahedra and tetrahedra. It is rather artificial to select either the isolated octahedron or the isolated tetrahedron for classification of the metavariscite structure, and the infinite framework is more important for the overall stability of the crystal structure.

9.2   Structures with Vertex-Linked Octahedra

Sharing of vertices of octahedra leads theoretically to dimers, clusters, chains, sheets, and frameworks, and only a few illustrative structures are given here. Readers are referred to Moore (1970a) for a description of stereoisomerism among corner-sharing octahedral and tetrahedral chains.

Pseudobrookite, Fe₂TiO₅, contains a chain of Ti-centered oxygen octahedra, each of which shares two diametrically opposite vertices with adjacent octahedra to give the composition TiO_5 (Fig. 9.2*a*). However, the structure should really be regarded as a framework, because the unshared oxygen atoms of adjacent TiO_5 chains are cross-linked by Fe atoms, each of which lies at the center of an octahedron of oxygen atoms.

 The actual structure is difficult to understand because of irregular distortion of the polyhedra, and is best approached from an idealized structure (Fig. 9.2*b*). Start with a 4.4.3.3.3 2D net (Fig. 2.1*d*). Draw diagonals across alternate squares. Imagine that each of these diagonalized squares is the projection of a vertex-sharing chain of octahedra (diagram 9.2*a*) viewed down the axis of the chain. Place oxygen atoms *p, q, r,* and *s* at height ½ and represent them by filled circles in diagram 9.2*b*. Place a Ti atom (filled hexagon) at the center of these four atoms (position *t*). It will also lie at height ½. Place an oxygen atom (open circle inscribed inside the hexagon) at height 0. This oxygen atom is shared by two octahedra centered about Ti atoms at heights ½ and (½ – 1). Similarly, place oxygen atoms at *d, e, f,* and *g,* but at height 0 instead of height ½ for atoms *p–s*. The Ti atom will also lie at height 0, and the shared oxygen at ½. Use open symbols for atoms at height 0 and filled ones for height ½. Place another octahedral chain at *ijklm*. The oxygen atoms at *g, j,* and *r* define the vertices of a tetrahedron. Atoms *g* and *j* are at height 0, and atom *r* at ½ is repeated at (½ – 1). Place an Fe atom (square)

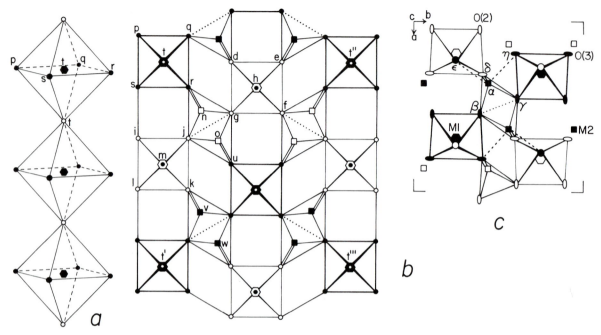

FIGURE 9.2 The crystal structure of pseudobrookite. (*a*) The clinographic drawing of an idealized chain of vertex-shared octahedra. (*b*) The idealized structure based on 4.4.3.3.3 2D net. All atoms occur either at height 0 (open symbols) or ½ (filled symbols). Circles represent oxygen atoms, and squares and hexagons, respectively, represent tetrahedrally- and octahedrally-coordinated atoms. (*c*) The actual structure of pseudobrookite. The tetrahedral site from diagram (*b*) has now become octahedrally coordinated: thus, the M2 atom at α is bonded to oxygen atoms ϵ and η (dashed lines) as well as the original four atoms (continuous lines). The octahedron around the M1 site is distorted so that oxygen atoms ϵ and η can move closer to the M2 atom. Oxygen atoms are differentiated as 0(1) circle, 0(2) vertical ellipse, and 0(3) horizontal ellipse.

at the center (*n*) of the tetrahedron, and draw the Fe–O bonds as well as the tetrahedral edges. Place a second Fe atom at position *o* and complete the tetrahedron with position *u*. The two tetrahedra share edge *jg*. Continuation of this procedure gives a unit cell whose corners project at *t–t'''*. Iron atoms lie either at height 0 (e.g., *n* and *o*) or ½ (e.g., *v* and *w*).

Edge sharing of Fe-centered oxygen tetrahedra is energetically unfavorable for the ionic model, and indeed the actual structure of pseudobrookite (Wells, 1965, p. 297) is distorted so that each Fe atom has six near neighbors instead of four. The crystallographic data are: $a = 9.79$ Å, $b = 9.93$ Å, $c = 3.72$ Å, **Bbmm** (listed in *International Tables for X-ray Crystallography* as **Cmcm** with relabeled axes). All atoms in pseudobrookite lie in the following special positions:

$$(4c) \quad \pm(u,\tfrac{1}{4},0; \tfrac{1}{2}+u,\tfrac{1}{4},\tfrac{1}{2})$$

$$(8f) \quad \pm(vw0; v,\tfrac{1}{2}-w,0; \tfrac{1}{2}+v,w,\tfrac{1}{2}; \tfrac{1}{2}+v,\tfrac{1}{2}-w,\tfrac{1}{2})$$

From the intensities of the X-ray diffraction pattern, the following coordinates were obtained:

M1	$u = 0.190$	O(1)	$u = 0.730$
M2	$v = 0.135; w = 0.560$	O(2)	$v = 0.045; w = 0.110$
		O(3)	$v = 0.310; w = 0.095$

The c-axis projection (Fig. 9.2c) can be compared with the idealized structure of diagram 9.2b. Each octahedral chain is kinked so that the shared oxygen [type O(1): circle] does not project directly over the M1 octahedral site. Each M2 atom (square) is surrounded by six oxygens: thus atom α is surrounded by two O(2) at β and γ, two O(3) at δ, one O(1) at ϵ and one O(3) at η. The first four oxygen atoms lie at ~1.9 Å and the second two at ~2.2 Å. From the crystal–chemical viewpoint, it is desirable to regard each M2 atom as being bonded to all six oxygens, even though the distances are so different. The M1 octahedron is also distorted. In diagram 9.2c, the octahedron around each M2 atom is not outlined, and only the M2–oxygen bonds are shown. Adjacent M2 octahedra share an edge $\beta\gamma$, and adjacent M1 and M2 octahedra share an edge (e.g., $\gamma\eta$ and $\delta\epsilon$). In the first description of the pseudobrookite structure, 4 Ti^{4+} were placed in M1 and 8 Fe^{3+} in M2. Later studies listed in Smyth (1974) showed that M1 is actually occupied by 4 Fe, and M2 by a disordered arrangement of 4 Ti and 4 Fe.

The mineral ***armalcolite*** was collected by the first three astronauts to land on the Moon (*Arm*strong, *Al*drin, *Col*lins). Because lunar rocks are more reduced than terrestrial rocks containing pseudobrookite, the iron in armalcolite is divalent. The resulting formula $(Fe,Mg)^{2+}Ti^{4+}_2O_5$ is accommodated in a crystal structure of pseudobrookite type, in which the Fe and Mg atoms occupy M1, and the Ti atoms occupy M2. Details are given in Wechsler et al. (1976), and used in Exercise 9.4. Readers may recall that similar differences of cation assignments were found in normal and inverse spinel (Section 5.6). Polyhedral drawings are given in Smyth (1974, Fig. 1) and Wechsler et al. (1976, Fig. 4).

Vertex-shared octahedral sheets occur in the K_2NiF_4 ***structure*** (Fig. 9.3). A clinographic projection (diagram 9.5a) shows an NiF_6 octahedron centered about each corner and body center of a tetragonal unit cell (dotted line). Each octahedron shares a vertex with four neighbors to give an infinite sheet of formula NiF_4. Each K atom is bonded to eight F atoms of one sheet and one F atom of the next sheet: thus atom p is bonded to four F atoms of type q, four F atoms of type r, and one F atom of type s.

The K_2NiF_4 structure has space group **I4/mmm** and an elongated tetragonal cell with $a = 3.95$ Å, $c = 13.71$ Å (Wyckoff, 1965, p. 68). The atomic coordinates are:

Mg	(*2a*)	000	K	(*4e*)	± 00*u*	*u* = 0.352
F(1)	(*4c*)	0½0; ½00	F(2)	(*4e*)	± 00*u*,	*u* = 0.151

Add ½½½ for body-centering vector.

Figure 9.3*b* shows the *a*-axis projection. Note how the K atom (*p*) lies almost at the center of the square formed by four F atoms (*q*), and below F atom *s*. Diagram 9.3*c* shows the elegant packing viewed down the tetrad axis. Six octahedra are shown at the bottom as squares with crossing diagonals. A fluorine atom of type *q* is shown as a heavy circle inscribed inside one of the diagonalized octahedra. Going upward in the diagram, the octahedral outlines are omitted, and the circles are emphasized. Potassium atoms (dashed circles) fit neatly between the heavy circles. At the top of diagram 9.3*c*, thin circles represent F atoms of type *r*. This diagram assumes that

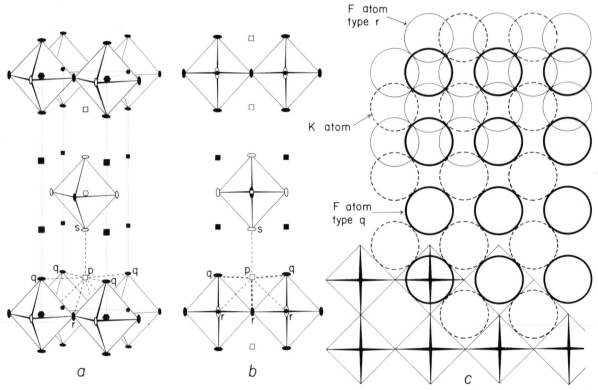

FIGURE 9.3 The crystal structure of K$_2$NiF$_4$. (*a*) The clinographic projection of tetragonal unit cell dotted line). Each Ni atom (hexagon) is surrounded by six F atoms (ellipses). The bonds (dashed line) are shown for one of the K atoms (square). (*b*) The *a*-axis projection. Because of overlap, the hexagon for each Ni atom is reduced in size to fit inside the ellipse for the superimposed K atom. (*c*) The packing diagram of layers viewed down the tetrad axis.

K and F atoms are spheres with identical radii—this is approximately true for the ionic model ($K^+ \sim 1.5$ Å; $F^- \sim 1.4$ Å; note that the radius in Table 1.1 depends on coordination number). Indeed, the K_2NiF_4 structure is the prototype for many structures (e.g., Rb_2CoF_4, Ca_2MnO_4, Sr_2TiO_4), and in all these structures the ionic radii of the first and third species are nearly equal. Slight changes of axial ratio and u coordinates accommodate the differences in ionic radii. The second requirement is that the ionic radius of the second species is suitable for octahedral coordination to the third species. A third requirement is that the charges of the cations and anions are equal and opposite to give electrostatic neutrality for the ionic model. Thus when O^{2-} replaces F^- in the above formula, a divalent cation replaces the univalent K. The square packing of atoms (diagram 9.3c) of K_2NiF_4 is not as efficient as *closest* packing of equal spheres, for which hexagonal layers would be needed. Nevertheless, the K_2NiF_4 structure is a neat way of obtaining close (*not closest*) packing of a vertex-shared octahedral sheet with intervening large atoms.

The crystal structure of *alunite, $KAl_3(OH)_6(SO_4)_2$* contains a layer of vertex-shared octahedra whose centers lie at the nodes of a kagomé net (Fig. 9.4). Each octahedron shares a vertex with each of four adjacent octahedra, leaving two unshared vertices. In turn, each of these two vertices is shared with a tetrahedron. The resulting octahedral–tetrahedral layers are linked by K atoms in icosahedral coordination and by hydrogen bonding.

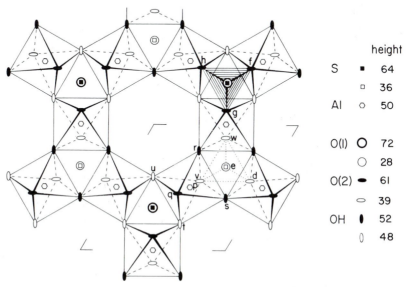

FIGURE 9.4 The vertex sharing between octahedra and tetrahedra in alunite. Projection down c-axis of triple-volume hexagonal cell.

The details of the crystal structure are complex (Exercise 9.5). The unit cell is rhombohedral with space group $\mathbf{R\bar{3}m}$, but it is easier to use the triple-volume hexagonal cell with $a = 6.97$ Å, $c = 17.27$ Å. Figure 9.4 shows only atoms lying between heights $z = 0.28$ and 0.72. Each Al atom (e.g., p) at height 0.50 (hexagon) is surrounded by one O(2) at 0.61 (q), two OH at 0.52 (r,s), two OH at 0.48 (t,u), and one O(2) at 0.39 (v). Each Al-centered octahedron shares an OH at height 0.52 (e.g., r,s) with each of two adjacent octahedra, and an OH at height 0.48 (t,u) with each of two more octahedra. The Al atoms lie on the nodes of a kagomé net (Fig. 2.1h), and the sharing of octahedral vertices produces an infinite layer of octahedra lying in 3- and 6-rings. The octahedra are tilted such that q is not directly over v and p. This results from vertex sharing with tetrahedra. Three O(2) atoms (w, v, d) form an equilateral triangle at height 0.39. An S atom (open square) lies at height 0.36, and an inverted tetrahedron (dotted line) is completed by an O(1) atom at height 0.28 (e). Similarly three O(2) atoms (f, g, h) at height 0.61 form the base of an upright tetrahedron (hatched) centered on an S atom (filled square) at height 0.64. The angle of tilt of the octahedra is governed by the relative dimensions of the tetrahedron and octahedron. In *crandallite,* $CaAl_3(OH)_6[PO_{3.5}(OH)_{0.5}]_2$, the SO_4 tetrahedron of alunite is replaced by a $PO_{3.5}(OH)_{0.5}$ tetrahedron for which there is substitutional disorder between O and OH. Readers who complete Exercise 9.5 will find that the layer centered on Al atoms at height 0.5 is repeated by the rhombohedral lattice according to the translations $(\frac{1}{3},\frac{2}{3},\frac{2}{3})$ and $(\frac{2}{3},\frac{1}{3},\frac{1}{3})$. The K atom of alunite is surrounded by six O(2) and six OH in a distorted icosahedron. Adjacent layers are also joined by hydrogen bonding. Structural details are given in Blount (1974) and Wang et al. (1965), and the relationship of the alunite layer to other layers of linked octahedra is discussed in Moore and Araki (1977).

The *ReO₃ structure* type is obtained by sharing all six vertices of an octahedron (Fig. 5.27f). The *perovskite structure* type (Fig. 5.29) is obtained by placing a large cation, for example, Ca, in each void of the octahedral framework as in the mineral $CaTiO_3$ (Fig. 5.29a).

As mentioned in Section 5.9, there are many distorted varieties of the perovskite structure, and the orthorhombic type of $MgSiO_3$ was depicted in Figure 5.29b. An Si atom occupies the center of each octahedron of six oxygen atoms, and an Mg atom lies at the center of each void. Twisting of the octahedral framework allows the Mg atom to lie closer to six oxygen atoms (2.1–2.2 Å) than the next six oxygen atoms (2.5–3.2 Å), and these six oxygen atoms lie at the corners of a distorted trigonal prism. This polyhedron shares an edge with each of three adjacent SiO_6 octahedra.

The *CaB₆ structure* (Fig. 5.28a) has a branch between each pair of vertices of adjacent octahedra. The **pyrite** (Figs. 8.7, 8.8) and **calcite** structures (Fig. 8.9) can be described in terms of vertex-linked octahedra, but this type of description should not be allowed to overshadow the presence of tightly bonded S_2 doublets and CO_3 complex ions.

 The **PdS$_2$ structure** is based on an orthorhombic distortion of the crystal structure of pyrite. It does not contain regular octahedra, and some readers may wish to do Exercise 9.6.

9.3   Structures with Clusters of Edge-Shared Octahedra: Enumeration of Clusters

An octahedron can share an edge with an adjacent octahedron to give a dimer, and additional sharing can lead to clusters, chains, sheets, and frameworks.

Moore (1970*b*) obtained a systematic retrieval and classification of **edge-shared clusters** of octahedra. Consider first only those clusters *whose octahedral centers lie in a plane.* Each cluster with r octahedral centers (M) and s vertices (ϕ) can be written $M_r\phi_s$. The problem is to find how many topologically different clusters occur for each r and s, and what is the ideal geometry and symmetry of each cluster.

The solution is obtained from the intersection of a triangular net with a hexagonal net from Figure 2.1. An octahedron (Fig. 9.5) can be obtained in the orientation of Figure 3.3*f* by using a triangle as the upper face of an octahedron and by turning six horizontal branches of the hexagonal net into inclined edges. The octahedron is completed by using a second triangle (dashed lines). An infinite planar cluster (i.e., a layer) is obtained by using all edges of the hexagonal net, and is represented by the $Mg(OH)_2$ layer of **brucite** (Fig. 5.18*b*). Because this layer has the greatest fraction possible of shared edges for a *planar* cluster, s/r cannot be less than 2 in *planar* clusters.

In order to enumerate all clusters, it is desirable to simplify the problem. The dimer $M_2\phi_{10}$ (Fig. 9.6*a*) can be represented merely by a line joining the M positions because the two octahedra can be reconstructed by drawing hexagons and triangles. The three possible trimers can also be represented simply by straight lines linking the M positions. Two of these trimers (9.6*b*, 9.6*c*) have composition $M_3\phi_{14}$ but the third one (9.6*d*) has composition $M_3\phi_{13}$. This third trimer is triangular in shape and has three shared edges instead of the two shared edges on the linear and angled trimers. Further progress utilizes **graph theory,** whose mathematical basis is given in Ore (1963). Each M position is called a **node** and each link between *adjacent* M positions (i.e., M positions sharing an edge) is called a **branch**. A **graph** is obtained by joining branches. A unique graph is obtained when the branches do not form one or more circuits. The two trimers $M_3\phi_{14}$ have

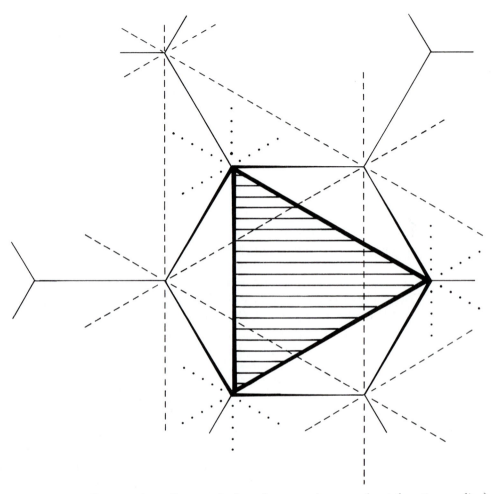

FIGURE 9.5 Construction of an octahedron from one hexagonal net (continuous line) and two triangular nets (dashed and dotted lines).

unique graphs, but the $M_3\phi_{13}$ trimer has a circuit. Only two branches are needed to link the three nodes (continuous lines), and the third branch (dashed line) is redundant. Although there are three ways of choosing two branches out of three, these three ways are topologically equivalent and a unique graph is obtained with only two branches. The chosen and rejected ones are called ***primary*** and ***secondary branches,*** respectively. Each triangular circuit results in a tetrahedral hole between the three adjacent octahedra. For each tetrahedral hole there is a loss of an octahedral vertex. For p circuits or secondary branches, the general stoichiometry is $M_r\phi_{4r + 2 - p}$.

For $r = 4$, there are seven clusters (diagram 9.6e). Five have no secondary branches and formula $M_4\phi_{18}$, one has one secondary branch and formula $M_4\phi_{17}$, and the remaining one with $p = 2$ has formula $M_4\phi_{16}$. As r increases,

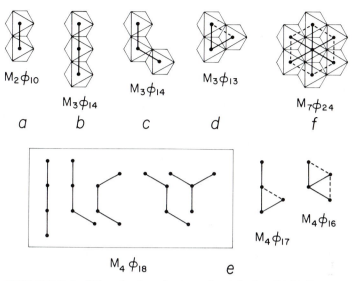

FIGURE 9.6 Edge-sharing clusters of octahedra. Light lines are octahedral edges, and heavy lines are branches of graphs connecting the *M* positions (dot). Primary and secondary branches are shown by continuous and dashed lines, respectively.

the number of possibilities grows, and a systematic counting procedure is needed (P. B. Moore, unpublished). Particularly interesting theoretically is the closed cluster $M_7\phi_{24}$ (diagram 9.6*f*). Clusters with identical (*s, r*) are called *isomers,* and those with identical (*r, p*) are called *topological isomers.*

For $r > 3$, additional clusters occur for which the octahedral centers are not confined to a single plane. The simplest nonplanar cluster, $M_4\phi_{16}$, is obtained from a planar cluster, $M_3\phi_{13}$, by utilizing three vertices *p, q,* and *r* (Fig. 9.7) for the base of the fourth octahedron and adding three unshared vertices to form the upper triangular face. For octahedra with regular shape, this $M_4\phi_{16}$ cluster has point symmetry $\bar{4}3m$. Each M position lies at the vertex of an imaginary regular tetrahedron. A triad axis projects upward in Figure 9.7, and an inverse tetrad axis passes through the midpoints of each pair of nonadjacent edges of the imaginary tetrahedron. There is a real tetrahedral void between the inner faces of the four octahedra. P. B. Moore enumerated all of the 144 geometrically distinct clusters in which a central octahedron shares from 1 to 12 of its edges with one or more adjacent octahedra. Sharing of all 12 edges gives the cluster $M_{13}\phi_{38}$, whose point symmetry is the same as that of the central octahedron, namely, $4/m\bar{3}2/m$. This cluster can be expanded to infinity giving the *halite* structure (Fig. 5.8), previously described in terms of closest packing of two types of spheres with radius ratio ($\sqrt{2} - 1$). In the halite structure, each Na atom lies at the center of an octahedron of Cl atoms, and each octahedral vertex is shared between six octahedra to give the stoichiometry $M_1\phi_1$.

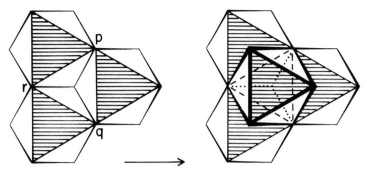

FIGURE 9.7 Conversion of planar $M_3\phi_{13}$ cluster into tetrahedral $M_4\phi_{16}$ cluster of edge-shared octahedra by addition of octahedron (heavy outline). The triangle pqr of unshared edges forms the base of the fourth octahedron. The tetrahedral void in the $M_4\phi_{16}$ cluster is outlined by dashed and dotted lines, all of which represent shared edges.

Clusters can be classified as **rigid** and **floppy**. Whereas the $M_2\phi_{10}$ cluster can be bent around the shared edge, the $M_7\phi_{24}$ cluster is rigidly locked into an hexagonal configuration. This statement is true for both cardboard and sphere models. Mathematical limits of floppiness are given by Wells (1975, p. 157).

The edge-sharing dimer $M_2\phi_{10}$ occurs as the molecule Nb_2Cl_{10} in a monoclinic crystal structure with space group **C12/m1** and cell dimensions $a = 18.30$ Å, $b = 17.96$ Å, $c = 5.89$ Å, $\beta = 90.6°$ (Zalkin and Sands, 1958). In spite of the large unit cell and low symmetry, the *Nb_2Cl_{10} structure* is easy to understand because the unit cell is almost orthogonal and the atoms lie in layers with z almost in multiples of ¼ (Fig. 9.8):

Nb(1)	*4g*	$x = 0$	$y = 0.111$	$z = 0$
Nb(2)	*8j*	0.333	0.111	0.525
Cl(1)	*4i*	0.053	0	0.225
Cl(2)	*8j*	0.056	0.191	0.240
Cl(3)	*8j*	0.103	0.097	0.782
Cl(4)	*4i*	0.280	0	0.744
Cl(5)	*8j*	0.279	0.189	0.770
Cl(6)	*8j*	0.232	0.098	0.293
Cl(7)	*4i*	0.381	0	0.298
Cl(8)	*8j*	0.389	0.190	0.284
Cl(9)	*8j*	0.434	0.098	0.760

The symmetry elements and general positions of **C12/m1** are given in Figure

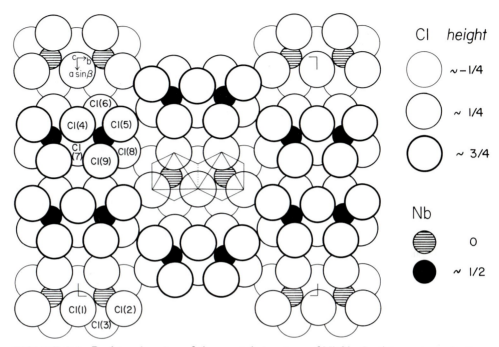

FIGURE 9.8 Packing drawing of the crystal structure of $NbCl_5$. In this *c*-axis projection, Cl and Nb atoms are drawn as circles of radius 1.4 and 0.9 Å, respectively. These values were chosen merely for convenience of drawing. Actual Cl-Cl and Nb-Cl distances are 3.3–3.5 and 2.3–2.6 Å, respectively. Molecules are drawn for Nb(1) atoms (hatched) at height 0 and Nb(2) atoms (filled circle) at heights 0.475 and 0.525. Chlorine atoms attached to Nb(1) lie at heights $\sim -\frac{1}{4}$ and $\sim \frac{1}{4}$, and those attached to Nb(2) lie at $\sim \frac{1}{4}$ and $\sim \frac{3}{4}$. The octahedral representation of a molecule is shown at the center. The Cl atoms lie close to two-thirds of the positions for hexagonal closest packing.

7.29. There are two crystallographic types of molecules. The four Nb(1) and Cl(1) and the eight Cl(2) and Cl(3) yield two Nb_2Cl_{10} dimers, each of which has a center of symmetry at the middle of the molecule where a diad rotation axis intersects a mirror plane. The eight Nb(2) and remaining Cl atoms yield four Nb_2Cl_{10} dimers, each of which has only a mirror plane of symmetry passing through Cl(4) and Cl(7) atoms. Neither molecule has the ideal geometry of the dimer in Figure 9.6*a*, but the topological relationships between the octahedra are the same. Electrostatic repulsion between each pair of Nb atoms explains the shortening of each shared octahedral edge.

Although the molecules tend to be ellipsoidal, they do not fill space with the greatest possible efficiency. The molecules form a distorted hexagonal arrangement in each layer, and there is a large hole at the center of each six molecules of a layer. It is not always easy to explain the details of packing of molecules linked by weak van der Waals forces. For Nb_2Cl_{10}, the key feature appears to be the positions of the Cl atoms that fall near the positions of hexagonal closest packing. Although one third of the positions for hcp are unoccupied, the neatness of this packing apparently overcomes the weakness caused by the large holes. Each hole is large enough to hold 5 water molecules, and the possibility of hydration might be worth testing.

The ***pharmacosiderite*** structure (Buerger et al., 1967) contains an $M_4\phi_{16}$ type of cluster with composition $Fe^{3+}_4(OH)_4O^{2-}_{12}$ centered about each corner of a cubic unit cell with $a = 7.98$ Å and space group **P$\bar{4}$3m** (Fig. 9.9). Indeed, this is not surprising, because the point symmetry of the isolated cluster is $\bar{4}$3m. However, the space group can assume the symmetry of the cluster only for one special orientation in which each $\bar{4}$ axis is parallel to a cell edge and each **3** axis is parallel to a body diagonal. The clusters are not isolated as were the Nb_2Cl_{10} dimers, and each O^{2-} anion also belongs to an As^{5+}-centered tetrahedron. Such a sharing of an octahedral and a tetrahedral vertex produces a framework containing one large hole per unit cell. Also present are seven H_2O and one K whose positions are poorly defined.

The general and special positions of **P$\bar{4}$3m** can be handled algebraically. The 24-fold general position has the following coordinates:

$xyz; zxy; yzx$ Cyclic rotation for triad axis

$\left.\begin{array}{l} y\bar{x}\bar{z}; x\bar{z}\bar{y}; z\bar{y}\bar{x} \\ \bar{x}\bar{y}z; \bar{z}\bar{x}y; \bar{y}\bar{z}x \\ \bar{y}x\bar{z}; \bar{x}z\bar{y}; \bar{z}y\bar{x} \end{array}\right\}$ Operate on first three with $\bar{4}$ along [**001**] giving $xyz \rightarrow y\bar{x}\bar{z}; \bar{x}\bar{y}z; \bar{y}x\bar{z}$

$\left.\begin{array}{l} yxz; xzy; zyx \\ \bar{x}y\bar{z}; \bar{z}x\bar{y}; \bar{y}z\bar{x} \\ \bar{y}\bar{x}z; \bar{x}\bar{z}y; \bar{z}\bar{y}x \\ x\bar{y}\bar{z}; z\bar{x}\bar{y}; y\bar{z}\bar{x} \end{array}\right\}$ Operate on first eight with mirror plane at $x = y$ giving $xyz \rightarrow yxz$

The 12-fold special position (Wyckoff type *i*) with **m** point symmetry is obtained by $xyz \rightarrow xxz$ giving:

$$xxz; zxx; xzx; x\bar{x}\bar{z}; x\bar{z}\bar{x}; z\bar{x}\bar{x}; \bar{x}\bar{x}z; \bar{z}\bar{x}x; \bar{x}\bar{z}x; \bar{x}x\bar{z}; \bar{z}x\bar{x}; \bar{x}z\bar{x}$$

The 4-fold special position (*e*) with **3m** point symmetry is obtained by setting $x = y = z$ giving: $xxx; x\bar{x}\bar{x}; \bar{x}\bar{x}x; \bar{x}x\bar{x}$. There are two 3-fold special positions (*d*) and (*e*) with $\bar{4}$2m point symmetry: $\frac{1}{2}00; 0\frac{1}{2}0; 00\frac{1}{2}$ and $0\frac{1}{2}\frac{1}{2}; \frac{1}{2}0\frac{1}{2}; \frac{1}{2}\frac{1}{2}0$. Readers can discover the other special positions, and check with page 324 of Volume 1 of *International Tables for X-Ray Crystallography*.

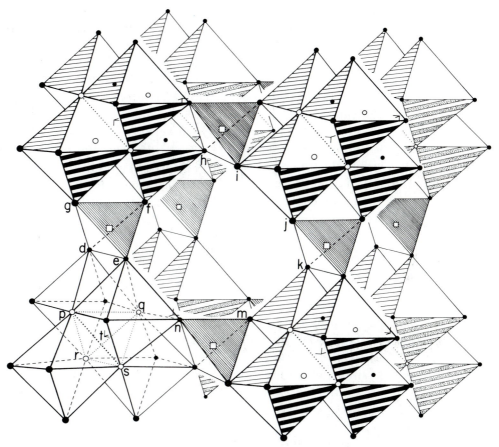

FIGURE 9.9 Clinographic projection of polyhedral framework of pharmacosiderite. A cluster of four octahedra is shown about the tetrahedral cavity (dotted lines joining *p, q, r,* and *s*) at the lower left. All other octahedra are shown only by the front faces, two sets of which are hatched. An Fe atom (not shown) lies near the center of each octahedron, and each octahedral cluster is centered about a cell corner (e.g., *t*). An As atom (square) lies at the center of a tetrahedron which shares two vertices (e.g., *d* and *e*, and *f* and *g*) with adjacent octahedra. Open and filled circles, respectively, represent OH and 0(1).

The framework coordinates for pharmacosiderite are:

Fe   (*4e*) $x = 0.143$ O(1)   (*12i*) $x = 0.122; z = 0.379$

As   (*3d*)   OH   (*4e*) $x = 0.885$

Probably the best way to understand the framework of pharmacosiderite is to build a model using the clinographic projection of Figure 9.9 as a guide.

Make an $M_4\phi_{16}$ cluster from four octahedra and one tetrahedron. Glue one face of each octahedron accurately onto a face of the tetrahedron. Orient the cluster so that the vertices of the tetrahedron match *p–s*. Seven more such clusters are needed to complete a unit cell. Each octahedron has three shared vertices (OH; open circle) and three unshared vertices [O(1); filled circle]. Each cluster occurs with its center (e.g., *t*) at a corner of the cubic unit cell (marked by a corner symbol), and its triad axes along body diagonals. Observe that a pair of unshared vertices (e.g., *d* and *e*) form a tetrahedron with unshared vertices (e.g., *f* and *g*) from the next cluster. For pharmacosiderite, this tetrahedron is 10 percent larger than the tetrahedron inside the $M_4\phi_{16}$ cluster, but in an idealized model the two types of tetrahedra would be equal. It is not easy to link the clusters with tetrahedra. Narrow strips of tape can be used. Short lengths of bent wire can be fastened to the polyhedral surfaces with strong glue. The final model can be stiffened by using a needle that passes through the midpoints of *rs, pq, de,* and *fg,* plus similar needles related by the lattice symmetry. Each such needle defines an edge of the unit cell. Whereas the first type of tetrahedron does not contain an atom, the second type contains an As atom (square). Each octahedron is centered by an Fe atom. Each unit cell of the pharmacosiderite framework contains a large cavity, and each cavity is linked to six adjacent cavities by a nonplanar window defined by eight O(1) atoms (e.g., *efhijkmn*). The positions of atoms in large cavities of frameworks of many zeolite minerals are poorly defined, and this is also true for the seven H_2O and one K of pharmacosiderite.

Heteropoly and *isopoly complexes* of the transition metals of groups 5 and 6 form a remarkable series of crystal structures reviewed by Evans (1971). Each complex consists of metal-centered oxygen polyhedra that share vertices or edges or both. An isopoly complex (e.g., heptamolybdate ion) involves only one metal, and a heteropoly complex (e.g., hexamolybdotellurate ion) involves more than one metal. Most polyhedra are octahedral, but tetrahedra occur.

The *heptamolybdate ion* $(Mo_7O_{24})^{6-}$ occurs in $K_6Mo_7O_{24} \cdot 4H_2O$ (Evans et al., 1975). Unfortunately, the complete crystal structure is too complicated to give here, but a stereo view of the Mo—O bonds and a polyhedral representation (Figs. 9.10a, 9.10b) are given from Evans (1971, p. 34, 35). Each Mo^{6+} ion lies inside an octahedron of O^{2-} ions. The central Mo^{6+} ion shares an octahedral edge with four octahedra and a vertex with two octahedra. Its maximum symmetry is **2mm**, but interaction with cross-linking K and H_2O yields space group symmetry $\mathbf{P2_1/c}$.

The *hexamolybdotellurate ion* $(TeMo_6O_{24})^{6-}$ occurs in the crystal structure of $K_6[TeMo_6O_{24}] \cdot 7H_2O$ and other complex structures (Evans, 1974).

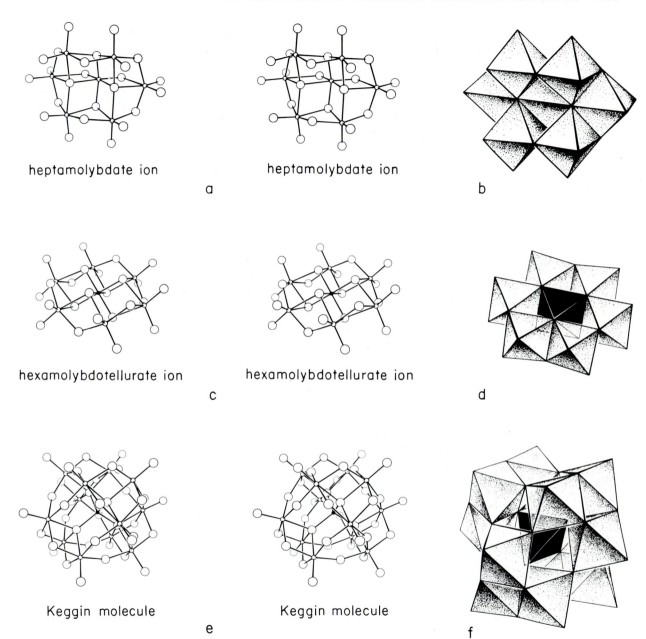

heptamolybdate ion

a

heptamolybdate ion

b

hexamolybdotellurate ion

c

hexamolybdotellurate ion

d

Keggin molecule

e

Keggin molecule

f

FIGURE 9.10 A stereo diagram of metal-oxygen bonds and polyhedral representation of heptamolybdate ion (a) small circle, Mo^{6+}, large circle, O^{2-}, (b); hexamolybdotellurate ion (c) small filled circle, Te^{6+}, small open circle, Mo^{6+}, (d); Keggin molecule (e) and (f). From Evans (1971). Reprinted by permission.

A Te-centered octahedron shares an edge with each of six surrounding Mo-centered octahedra, and each Mo-centered octahedron shares an edge with its two neighboring Mo-centered octahedra (Figs. 9.10c, 9.10d). Indeed, the hexamolybdotellurate ion has the topology of the $M_7\phi_{24}$ cluster in Figure 9.6f. Although the maximum symmetry of the ion is $\bar{3}m$, the linkage to K and H_2O results in space-group symmetry **Pbca**.

The first heteropoly complex to be discovered is commonly known as the ***Keggin molecule,*** and is typified by the precipitation agent for analysis of phosphate ions. The general formula (Evans, 1971, p. 10) is

$$[T^x M_{12} O_{40}]^{x-8}$$

where T is a tetrahedrally coordinated ion at the center of the complex and M is an octahedrally coordinated ion: x is the charge on the T ion, and the M ion is hexavalent. The twelve M atoms (Figs. 9.10e, 9.10f) lie at the vertices of a distorted cuboctahedron. Each octahedron shares one vertex with the central tetrahedron, an edge with each of two octahedra, and a vertex with each of two octahedra; one vertex is unshared. Each vertex of the tetrahedron is shared with three octahedra, and this sharing is the cause of the distortion of the cuboctahedron of M sites. The T site can be occupied by many ions including B^{3+}, Si^{4+}, P^{5+}, Fe^{3+}, Cu^{2+}, and As^{5+}, while the M site is usually occupied by Mo^{6+} or W^{6+}. The Keggin molecule changes shape for the different ions, and has a maximum point symmetry of $\bar{4}3m$ inherited from the central tetrahedron. Keggin molecules are cross-linked by large cations and water molecules to give several types of crystal structures. In type *A*, typified by $K_3(PMo_{12}O_{40}) \cdot 4H_2O$, two molecules lie in a cubic cell ($a = 12$ Å) centered on 000 and ½ ½ ½. Because the molecule at the body center is turned 90° with respect to its eight neighbors at 000 and related lattice positions, the space-group symmetry is degraded from **I43m** for parallel molecules to **Pn3m**. In type *B,* the molecular centers lie in a cubic cell ($a = 22$ Å) centered at the coordinates of carbon atoms in diamond. Because each molecule is rotated 90° with respect to its four neighbors, the space-group symmetry becomes **F43m**. Other types have lower symmetry. Some readers might wish to study the systematic theory of the relation of point symmetry, position, and angular orientation of molecules to the resultant space-group symmetry.

9.4   Structures with Chains of Edge-Sharing Octahedra

Figure 9.11 shows the three simplest chains of edge-shared octahedra. The chain found in $TcCl_4$ is obtained by sharing opposing edges (dots) with adjacent octahedra. Sharing of two nonopposing edges gives the zigzag chain

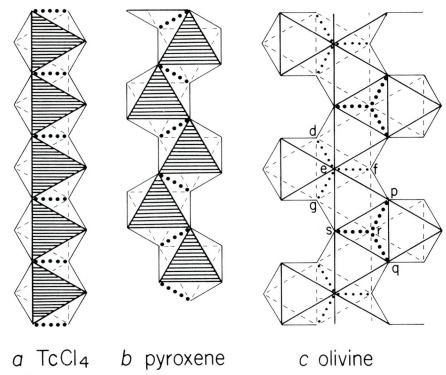

a TcCl₄   b pyroxene   c olivine

FIGURE 9.11   The three simplest chains of edge-shared octahedra. Dots show shared edges.

found with a distorted shape in pyroxene (Fig. 8.19). The serrated chain found in olivine involves sharing of adjacent edges. Vertices p, r, and s form an inverted tetrahedral hole between three adjacent octahedra, and vertices d, e, f, and g form an upright tetrahedral hole.

Technetium tetrachloride (Elder and Penfold, 1966) has the following atomic coordinates in an orthorhombic unit cell with $a = 11.65$ Å, $b = 14.06$ Å, $c = 6.03$ Å, **Pbca**. All atoms are in an 8-fold general position.

$$\pm(xyz;\ \tfrac{1}{2} + x, \tfrac{1}{2} - y, \bar{z};\ \bar{x}, \tfrac{1}{2} + y, \tfrac{1}{2} - z;\ \tfrac{1}{2} - x, \bar{y}, \tfrac{1}{2} + z)$$

Atomic coordinates are: Tc 0.628, 0.179, 0.398; Cl (1) 0.743, 0.308, 0.574; Cl (2) 0.512, 0.308, 0.224; Cl (3) 0.746, 0.071, 0.547; Cl (4) 0.510, 0.070, 0.250. The crystal structure (Fig. 9.12) is projected down the a axis to display the TcCl₄ chain in the same orientation as in Figure 9.11a. Technetium atoms occur at heights $x = 0.13$, 0.37, 0.63, and 0.87, but only one TcCl₄ chain is shown. This is composed of Tc atoms at height 0.13 and Cl atoms at heights 0.01 and either 0.24 or 0.25. Readers may like to color the chain, and then proceed to identify the three other chains that could be distin-

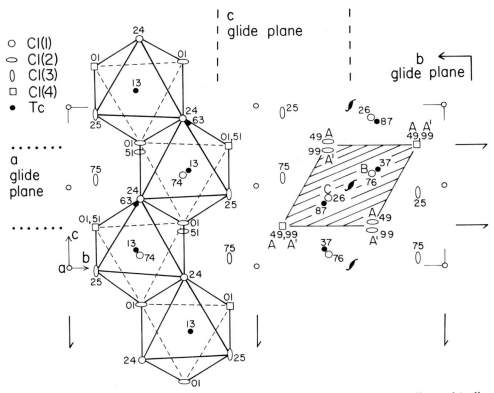

FIGURE 9.12 Crystal structure of TcCl₄ projected down *a*. Four crystallographically distinct types of Cl atoms are shown by different symbols. Small circles are centers of symmetry. Only one out of the four TcCl₄ chains is shown. The pseudohexagonal cell is hatched, and *A*, *B*, and *C* refer to closest-packing of equal spheres.

guished by separate colors. A second structural feature is the packing of Cl atoms. These lie approximately at heights $x = 0$, $\frac{1}{4}$, $\frac{1}{2}$, and $\frac{3}{4}$, and fall approximately in a hexagonal pattern. A pseudohexagonal cell can be chosen (hatched 120° rhombus), and the Cl atoms lie close to the positions of closest-packed spheres (Fig. 5.3). Indeed, the stacking can be denoted symbolically as *ABA'C*, sometimes called *"double hexagonal" closest packing.* The positions of the glide planes, screw axes, and centers of symmetry are given.

The TcCl₄ type of edge-shared octahedral chain is represented in minerals by cross-linked (not isolated) AlO₄ and TiO₄ chains. Figure 9.13 shows a *c*-axis projection of an idealized version of the crystal structure of *sillimanite,* one of the three polymorphs of Al_2SiO_5. In this theoretical structure $Al^{VI}(Al + Si)^{IV}O_5$, half the Al atoms are in edge-shared octahedral chains, and half randomly occupy tetrahedrally coordinated sites with an equal number of Si atoms. The crystallographic data are: orthorhombic $a = 7.49$ Å,

$b = 7.68$ Å, $c = 2.89$ Å, space group **Pbam**, and atomic coordinates:

Al   000; ½½0

T $pq0; \bar{p}\bar{q}0; ½ + p, ½ - q, 0; ½ - p, ½ + q, 0$ with $p = 0.147$
and $q = 0.342$

O(1) $p = 0.359, q = 0.422$

O(2)   ½0½; 0½½

O(3) $uv½; \bar{u}\bar{v}½; ½ + u, ½ - v, ½; ½ - u, ½ + v, 0$ with
$u = 0.125$ and $v = 0.223$

To understand the structure, first note that all atoms are in special positions lying on mirror planes at either $z = 0.00$ (open symbols) or 0.50 (filled symbols). Each Al atom (hexagon) lies at the center of an octahedron composed of two O(3) at the same height and four O(1) displaced by ±0.50. Each octahedron shares a horizontal edge O(1)–O(1) with an overlying octahedron and with an underlying octahedron to give an infinite AlO_4 chain parallel to the c axis. The octahedron is viewed down a 2-fold axis (Fig. 3.3*h*). Each T site (square) lies in the center of a tetrahedron composed of one O(1) and O(2) at the same height and two O(3) displaced by ±0.50. Adjacent tetrahedra share an O(2) atom to form a T_2O_7 group.

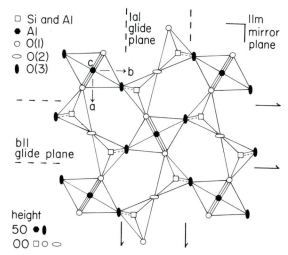

FIGURE 9.13 Hypothetical structure (*c*-axis projection) obtained by removing the distinction between tetrahedral sites in sillimanite. A center of symmetry (not shown) occurs at the corners and midpoint of edges of the unit cell.

Whereas the edge-shared octahedral chains in $TcCl_4$ are tightly bonded in the chain and very weakly bonded between the chains (Cl–Cl van der Waals bond), the edge-shared octahedral chains in sillimanite should really be regarded as a component of a framework structure: indeed the Si–O and Al–O tetrahedral bonds are stronger than the Al–O octahedral bond.

The actual structure of sillimanite has ordering of Si and Al atoms among the T sites. In each T_2O_7 group there is one Si and one Al atom. As shown in Exercise 9.15, this results in doubling of the c axis and change of space group from **Pbam** to **Pbnm**. Some readers who studied Section 7.13 may wish to show how **Pbnm** is a subgroup of **Pbam** when the c repeat is doubled: the procedure is analogous to that given in Figure 7.31.

Those readers interested in examining other minerals with AlO_4 chains cross-linked by other structural units should consult Bragg et al. (1965) and Papike and Cameron (1976) for the structures of *andalusite, kyanite, epidote group,* and *staurolite.* Particularly interesting is the structural relation of sillimanite to andalusite and kyanite, the other two polymorphs of Al_2SiO_5. In andalusite, the AlO_4 chains are cross-linked by SiO_4 tetrahedra and AlO_5 trigonal bipyramids and in kyanite the cross-linking is by SiO_4 tetrahedra and AlO_6 octahedra. Kyanite is also interesting because the oxygen atoms are close to the positions of cubic closest packing of equal spheres. The polyhedral diagram (Moore and Smith, 1970, Fig. 3c) is complex.

The *olivine* structure was briefly mentioned in Chapter 5 as an example of a close-packed array of oxygen atoms in hcp positions, one eighth of whose tetrahedral voids are occupied by Si, and one half of whose octahedral voids are occupied by Mg. The serrated chain of edge-shared octahedra (Fig. 9.11c) was shown in a sphere drawing (Fig. 5.20e).

Olivine occurs in the Earth as a mineral series ranging from Mg_2SiO_4 (forsterite) to Fe_2SiO_4 (fayalite) with minor substitution of Ni, Ca, and so. The crystal structure of forsterite (Birle et al., 1968) is now given (Fig. 9.14): orthorhombic $a = 4.762$ Å, $b = 10.225$ Å, $c = 5.994$ Å, **Pbnm**; atomic coordinates referred to a cell with center of symmetry at 000:

Mg(1) 000	Mg(2) 0.990,0.277,¼	Si 0.427,0.094,¼
O(1) 0.766,0.092,¼	O(2) 0.220,0.448,¼	O(3) 0.278,0.163,0.034

For **Pbnm** (Fig. 9.14b) a center of symmetry is produced at the origin by placing the **b11** glide plane at $x = ¼, ¾$, the **1n1** glide plane at $y = ¼, ¾$, and the **11m** mirror plane at $z = ¼, ¾$. Only the O(3) atoms are in an 8-fold general position with xyz transformed as follows:

$$xyz \rightarrow ½ - x, ½ + y, z \qquad \textbf{b11 } \text{glide plane at } x = ¼$$

$$½ + x, ½ - y, ½ + z; \bar{x}, \bar{y}, ½ + z \qquad \textbf{1n1 } \text{glide plane at } y = ¼$$

$$x, y, \tfrac{1}{2} - z;\ \tfrac{1}{2} - x, \tfrac{1}{2} + y, \tfrac{1}{2} - z \quad \textbf{11m} \text{ mirror plane at } z = \tfrac{1}{4}$$
$$\tfrac{1}{2} + x, \tfrac{1}{2} - y, \bar{z};\ \bar{x}\bar{y}\bar{z}$$

Special positions with 4-fold multiplicity lie on the mirror plane giving the following coordinates for $z = \tfrac{1}{4}$:

$$x y \tfrac{1}{4};\ \tfrac{1}{2} - x, \tfrac{1}{2} + y, \tfrac{1}{4};\ \tfrac{1}{2} + x, \tfrac{1}{2} - y, \tfrac{3}{4};\ \bar{x}\bar{y}\,\tfrac{3}{4}$$

This position is occupied by Mg(2), Si, O(1) and O(2). Setting $x = y = z = 0$ gives the 2-fold position $(000;\ \tfrac{1}{2}\,\tfrac{1}{2}\,\tfrac{1}{2})$ on a center of symmetry that is occu-

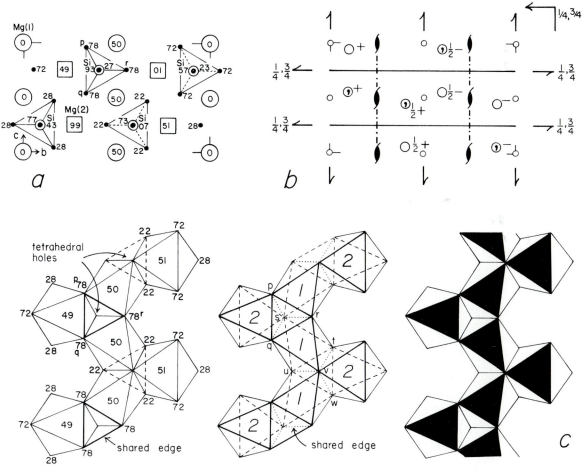

FIGURE 9.14 Olivine structure projected down a-axis. (a) Atomic positions: Mg(1) large circle; Mg(2) square; Si, small circle; O dot; x-coordinate in hundredths. (b) Symmetry elements. (c) Three ways of depicting an octahedral chain; left-hand drawing, x-coordinate in hundredths; central drawing, 1 and 2 refer to Mg atom.

pied by Mg(1). Space group **Pbnm** is obtained from **Pnma** (62) in Table 7.5 by relabeling axes.

The a-axis projection (Fig. 9.14a) shows the near-hexagonal array of oxygen atoms and Mg atoms (larger circle and square) and Si atoms (smaller circle). The oxygen atoms have coordinates 0.22 to 0.28 and 0.72 to 0.78 rather than falling exactly at $z = 0.25$ and 0.75 for hexagonal closest packing.

There are several ways of choosing half the octahedral voids and one eighth of the tetrahedral voids in hcp (Birle et al., 1968). The choice in olivine is an elegant way of reducing electrostatic repulsion between cations. No faces are shared between MgO_6 and SiO_4 polyhedra, but some edges are shared. The idealized serrated chain of edge-shared MgO_6 octahedra (Fig. 9.11c) is distorted in olivine. Three identical chains parallel to the c axis are depicted in different ways (Fig. 9.14c). Only the visible faces of the octahedra are shown in the right-hand chain. Octahedra of the central strip are centered on Mg(1), and projecting ones are centered on Mg(2). In the left-hand drawing, the height of each Mg atom is shown inside its octahedron, and the height of most O atoms is shown outside the octahedron. In the middle drawing, all octahedral edges are shown. A tetrahedral hole pointing downward (p–s) alternates with a tetrahedral hole pointing upward (t–w). Octahedra share edges marked by dotted lines. In the left-hand drawing, the horizontal face of each tetrahedral hole is marked either by a heavy line or a dashed line. From the geometrical viewpoint, each tetrahedral hole could be occupied by an Si atom, but this would involve sharing of octahedral and tetrahedral faces with consequent strong electrostatic repulsion. Instead, each Si atom occupies a tetrahedron that is directly opposite a tetrahedral void. Thus oxygen atoms p, q, and r at height $x = 0.78$ form a tetrahedron with an oxygen atom at $x = 1.27$ and an Si atom at 0.93 (cf. diagrams 9.14a and 9.14c). The adjacent tetrahedral void is formed by atoms p, q, r, and an oxygen at 0.27. Each Si-centered tetrahedron shares an edge with three octahedra while each *unoccupied* tetrahedron shares three edges with an occupied tetrahedron and the remaining three with octahedra. In accordance with ionic theory, unshared edges are longer than shared edges.

The three serrated chains in diagram 9.14c are correctly spaced for three unit cells along the b axis, and the intervening gaps have exactly the same shape as the chains, as is required by the **b11** glide plane. Each gap is filled by a serrated chain displaced by $a/2$, and its Mg(1) and Mg(2) atoms lie at heights 0.01 to –0.01, as can be found in diagram 9.14a. Adjacent chains of different heights share only vertices.

The olivine structure of Mg_2SiO_4 becomes unstable at high pressure, and is replaced by the *β-structure* and then the *spinel structure,* both of which are based on cubic closest packing of oxygens (Moore and Smith, 1971).

Like olivine, the *humite* group of minerals has anions near hcp positions, and the Mg and Si atoms also occupy octahedral and tetrahedral sites.

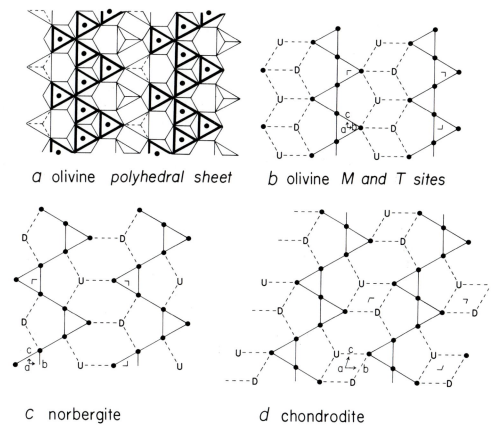

a olivine polyhedral sheet   b olivine M and T sites

c norbergite   d chondrodite

FIGURE 9.15 Idealized abstract drawings of the olivine, norbergite, and chondrodite structures. (*a* and *b*) Sheet of linked octahedra and tetrahedra, and corresponding sheet of linked Mg (dot) and Si atoms (*U*, upward-pointing tetrahedron; *D*, downward-pointing). (*c* and *d*) Sheets of linked Mg and Si in norbergite and chondrodite.

However, the topologic relations are complex in the humite minerals, as represented in Figure 9.15 for *norbergite*, $Mg_3SiO_4(OH,F)_2$, and *chondrodite*, $Mg_5Si_2O_8(OH,F)_2$. References to *humite*, $Mg_7Si_3O_{12}(OH,F)_2$, and *clinohumite*, $Mg_9Si_4O_{16}(OH,F)_2$, are given in Exercise 9.17. The easiest way to understand the humite structures is to note that all four have $a = 4.7$ Å like olivine. The olivine structure can be obtained by stacking polyhedral sheets of linked octahedra and tetrahedra along the *a* axis. Figure 9.15*a* is obtained from Figure 9.14*c* by adding tetrahedra between the serrated chains of octahedra. Alternate tetrahedra point upward and downward, and Mg atoms are shown by large dots. Figure 9.15*b* is a simplified version obtained by linking adjacent Mg and Si atoms. The latter are denoted U and D to differentiate between tetrahedra pointing upward and downward. Continuous lines show serrated chains, and dashed lines link adjacent octahedra and tetrahedra. Taken together, the continuous and dashed lines form a tessellation of triangles, rhombuses, and pentagons. Corner marks show the unit cell. In norbergite, the Mg atoms form another type of serrated chain, and the tessellation is composed of triangles, pentagons, and hexagons. In chondrodite, the octahedral chain becomes even more complex, and the tessellation contains triangles, rhombuses, pentagons, and hexagons. Although the chemical formulas of the humite minerals can be written $n\text{Mg}_2\text{SiO}_4 \cdot \text{Mg(OH,F)}_2$ with n = 1, 2, 3, and 4, respectively, for norbergite, chondrodite, humite and clinohumite, all four minerals are based on a single array of hcp anions, and are not composed of layers of olivine and brucite. Each humite mineral has a distinctive range of chemical composition (refs. in Exercise 9.17).

9.5   Structures with Layers of Edge-Sharing Octahedra

The two simplest structures with edge-shared octahedral layers are named the CdI_2 and CdCl_2 types. In the idealized version of the **CdI_2 structure**, large anions occupy positions of hexagonal closest packing of spheres, and small cations just fit into half the octahedral holes of each alternate layer, leaving empty each intervening layer of octahedral holes (Fig. 5.18). The axial ratio of the unit cell would be $c/a \simeq 1.63$, and the atomic coordinates for a generalized formula $\text{M}\phi_2$ would be:

M   (*1a*)   000
ϕ (*2d*) $\pm(\frac{1}{3}, \frac{2}{3}, u); u = \frac{1}{4}$

The space group would be **P$\bar{3}$m1** (Fig. 7.23) and both atom types would be in special positions. Actual crystal structures have u varying from ~0.22 to ~0.26 as the ionic radii change, and the axial ratios lie in the range 1.4 to 1.6 because cation repulsion causes thinning of the octahedral layer. Note that the AlB_2 structure has $u = \frac{1}{2}$ instead of $u = \frac{1}{4}$ for the CdI_2 structure.

Particularly interesting from the viewpoint of packing is the distorted variant assumed by **brucite, Mg(OH)$_2$**, and **portlandite, Ca(OH)$_2$**. From a neutron-diffraction study of portlandite, Busing and Levy (1957) determined the following coordinates for a hexagonal cell, a = 3.592 Å, c = 4.906 Å, **P$\bar{3}$m1**:

Ca   (*1a*)   000
O   (*2d*) $\frac{1}{3}, \frac{2}{3}, u; \frac{2}{3}, \frac{1}{3}, \bar{u}, u = 0.234$
H   (*2d*) $u = 0.425$

Figure 9.16*a* is a plan of an octahedral layer. Oxygen atoms at height 0.23 (filled circle) provide the upper face of each octahedron (shaded) and ones

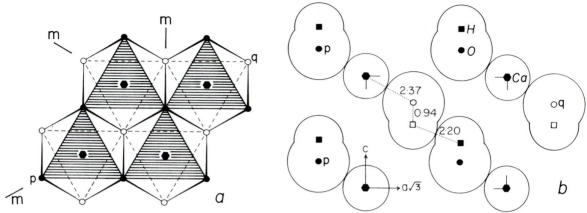

FIGURE 9.16 Crystal structure of portlandite, $Ca(OH)_2$. (*a*) *c*-axis projection showing a Ca atom (hexagon) at the center of each octahedron of O atoms (filled and open circles); H not shown. (*b*) Packing drawing through vertical section *pq* of diagram (*a*). Each OH group is represented by truncated spheres centered on O (circle) and H (square).

at height –0.23 (open circle) provide the lower face (dashed line). A calcium atom (hexagon) occurs at the center of each octahedron, and defines a corner of the hexagonal unit cell. Between each three adjacent octahedra is a tetrahedral dimple. A hydrogen atom (not shown) lies directly under each oxygen atom projecting 0.94 Å away from the median plane of the octahedral layer. Diagram 9.16*b* is a vertical section through *pq* showing how the atoms can be represented by touching spheres. The observed Ca–O, O–H and H–H distances of 2.37, 0.94, and 2.20 Å can be interpreted in terms of ionic radii of 1.0 and 1.4 Å for Ca^{2+} and O^{2-} ions and a van der Waals radius of 1.1 Å for H. Such an interpretation is consistent with the excellent cleavage between octahedral layers and the softness of portlandite crystals. An inverse triad axis passes vertically through each Mg atom, and a mirror plane (**m**) is perpendicular to each horizontal axis of the unit cell.

The ideal **$CdCl_2$ structure** has the large anions in the positions of cubic closest packing and the small cations in half the octahedral holes of each alternate layer. Symbolically the CdI_2 and $CdCl_2$ structures can be described with the *ABCabc* notation for closest-packing:

CdI_2 structure

A	c	B	c
I	Cd	I	void

$CdCl_2$ structure

A	c	B	a	C	b	A	c	B	a	C	b
Cl	Cd	Cl	void	Cl	Cd	Cl	void	Cl	Cd	Cl	void

The CdCl$_2$ structure has a rhombohedral unit cell and space group $\mathbf{R\bar{3}m}$, and the atomic coordinates for a generalized formula are

M   (*1a*)   000
ϕ (*2c*) $\pm(uuu)$

It is easier to work with the triple-volume hexagonal cell and the coordinates:

M   000; $\frac{1}{3}, \frac{2}{3}, \frac{2}{3}$; $\frac{2}{3}, \frac{1}{3}, \frac{1}{3}$
ϕ $\pm(uuu;\ \frac{1}{3}+u, \frac{2}{3}+u, \frac{2}{3}+u;\ \frac{2}{3}+u, \frac{1}{3}+u, \frac{1}{3}+u)$

For cations in the ideal positions of cubic closest packing, the parameter u is exactly $\frac{1}{4}$, and indeed all known structures of CdCl$_2$ type have u between 0.25 and 0.26. The axial ratio c/a for the hexagonal cell of the idealized CdCl$_2$ structure is $\sim$4.9, and the observed range for actual structures is 4.5 to 5.5. From the viewpoint of simple packing, adjacent layers are identical in the CdI$_2$ and CdCl$_2$ structures, and random stacking would be expected. Polytypism has been observed for CdI$_2$, but most materials apparently adopt just one of the ordered structures.

Sheets of edge-shared octahedra are extremely important in layer silicates, of which mica was already described in Section 8.6.6, devoted to layers of vertex-shared tetrahedra. In Figure 8.20e, two tetrahedral layers are cross-linked by a sheet of edge-shared octahedra. When all octahedral sites are filled (Fig. 8.21a), the octahedral layer is geometrically regular, but when only two out of three octahedral sites are occupied (Fig. 8.21b), the octahedral layer is distorted. Bragg et al. (1965, Chapter 13) and Brindley and Brown (1980) give descriptions of structures containing octahedral sheets. Symbolically they can be classified as follows: M octahedral; T tetrahedral; A alkali or alkaline-earth; aq water;

	Trioctahedral	Dioctahedral
M	*Brucite*	*Gibbsite*
MT	*Serpentine group*	*Kaolinite group*
TMT	*Talc group*	*Pyrophyllite*
A. TMT.	*Mica group*	*Mica group*
M. TMT.	*Chlorite group*	*Chlorite group*
aq. TMT.	*Smectite and vermiculite groups*	

Many complexities occur from different stacking sequences (cf. Fig. 8.21 for mica), and from distortion caused by misfit between tetrahedral and octahedral sheets, including formation of curved sheets.

9.6 Structures with Frameworks of Edge-Sharing Octahedra

The *halite* structure (Fig. 5.8, 8.9*a*) can be regarded as a framework of edge-shared octahedra. In Figure 5.8, each Na lies at the center of an octahedron of six Cl atoms. Each edge is shared between two octahedra, and the whole structure can be regarded as a framework of edge-sharing octahedra. However, the halite structure is better envisaged as the filling by Na atoms of all octahedral holes between ccp of Cl atoms.

The *spinel* structure (Fig. 5.19) can also be regarded as a framework of edge-shared octahedra cross-linked by tetrahedra, but again is better envisaged as the partial filling of octahedral and tetrahedral holes in cubic closest packing.

9.7 Structures with Frameworks of Edge- and Vertex-Sharing Octahedra

Rutile, one of the three polymorphs of TiO_2, contains a framework of edge- and vertex-sharing octahedra (Fig. 5.17*d*). The crystallographic data (see Exercise 4.2) are: tetragonal, $a = 4.59$ Å, $c = 2.96$ Å, **P4$_2$/mnm**, atomic coordinates:

Ti (*2a*) $000; \frac{1}{2}\frac{1}{2}\frac{1}{2}$
O (*4f*) $\pm(uu0; \frac{1}{2} + u, \frac{1}{2} - u, \frac{1}{2})$ with $u = 0.305$

A clinographic projection (Fig. 9.17) shows chains of edge-sharing octahedra lying parallel to the *c* axis, each one projecting in Figure 5.17*d* as a rhombus crossed by a heavy line. Figure 9.17 shows three unit cells. A Ti atom lies at each corner (e.g., *p–s, p'–s'*) and each body center (e.g., *t*). Oxygen atoms *d* to *i* form an octahedron about Ti atom *t,* and edges *gh* and *de* are shared with octahedra lying below and above (not shown) to form a chain of TcCl$_4$ type (Fig. 9.11*a*). Similarly oxygen atoms *i* to *n* form an octahedron about Ti atom *s,* and share edges *ij* and *mn* to form a chain that is rotated 90° with respect to the first chain. Atom *i* is shared between octahedra centered on *t, s,* and *s',* and atom *f* is shared between octahedra centered on *t, q,* and *q'.* Such vertex sharing leads to an infinite framework of edge- and vertex-sharing octahedra. The rutile structure can be also described (Fig. 5.17) as a distortion of an hcp array of oxygen atoms, half of whose octahedral interstices are occupied by Ti.

The *hollandite* structure type has an ideal formula $A_2M_8X_{16}$, typified by $Ba_2Mn_8O_{16}$. Crystallographic data (Byström and Byström, 1950) are: tetragonal, $a = 9.8$ Å, $c = 2.86$ Å, **I4/m** (Fig. 7.21), atomic coordinates:

Ba (*2b*) $00\frac{1}{2}$ Add $\frac{1}{2}\frac{1}{2}\frac{1}{2}$ for **I** centering
Mn (*8h*) $uv0; \bar{u}\bar{v}0; \bar{v}u0; v\bar{u}0$ with $u = 0.348, v = 0.167$

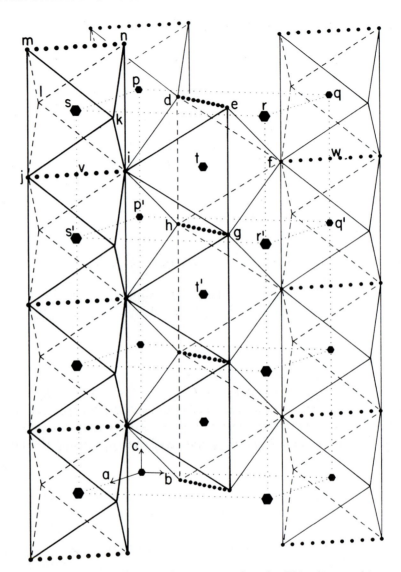

FIGURE 9.17 The crystal structure of rutile. This clinographic projection shows three unit cells (faint dots). Octahedral chains are shown for Ti atoms at *ss'*, *tt'*, and *qq'*, but the chain for Ti atoms *rr'* is omitted for clarity. Part of the octahedron about *p* is shown. Exercise 9.20 explains the method of construction.

O(1) (*8h*) $u = 0.153, v = 0.180$
O(2) (*8h*) $u = 0.542, v = 0.167$

The *c*-axis projection (Fig. 9.18) shows rhombus-shaped cross sections of octahedral chains. In rutile, octahedral chains are cross-linked only by vertices, but in hollandite each octahedral chain (e.g., *p–t*) shares edges (e.g., *st*) with an adjacent chain (e.g., *s–w*) to give a double chain. The rotation tetrad axis repeats each double chain to produce a channel (e.g., *r, d–h, u,s*) with a square cross section. A Ba atom at *i* lies at height *z* = 0, and is bonded to eight oxygen atoms (*d, f, h, s*) at height ±½ to give cubic coordination.

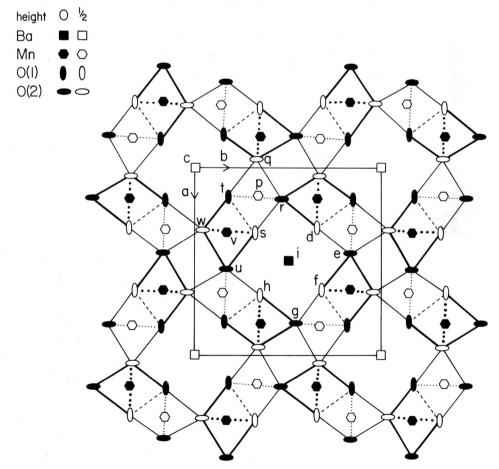

FIGURE 9.18 The crystal structure of ideal hollandite. In this *c*-axis projection, atoms at height 0 and ½ are shown by filled and open symbols, respectively. Unshared and shared octahedral edges are shown respectively by continuous and dotted lines. Thick and thin lines are used to differentiate the two heights of octahedra.

Most observed structures of hollandite type have some unoccupied A and M sites giving a general formula $A_{2-n}M_{8-m}X_{16}$, where $0.8 < n < 1.3$ and $0.1 < m < 0.5$. Site A is occupied by a large cation such as K^+ and Ba^{2+} and site B by Mn^{4+}, Fe^{3+}, or Mn^{2+}. Site X is occupied by oxygen and hydroxyl. Structural disorder is common, but the tendency for n to be near unity probably results from near-regular alternation of an A atom with a void in the square tunnels. For regular alternation, the space-group symmetry must drop from $I4/m$ to $P4/m$ (or lower) if c remains at 2.86 Å. If the c repeat is doubled, the space group can remain at $I4/m$. Natural hollandites are only pseudotetragonal.

Are other structure types possible? Indeed there is an infinity of ways of cross-linking $TcCl_4$-type octahedral chains, of which rutile and hollandite are merely two simple examples (Bursill, 1979). Figure 9.19 shows how the double chain of hollandite can be cross-linked in a different way to give the crystal structure of *ramsdellite,* whose chanels are rectangular instead of square. Whereas hollandite was obtained from rutile by edge sharing of two

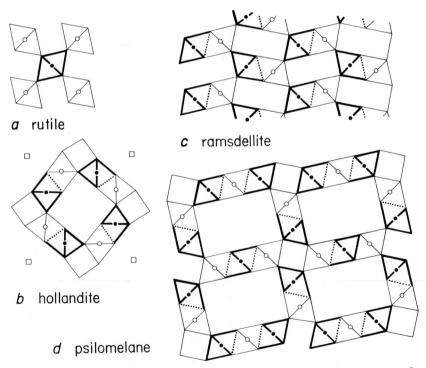

a rutile

c ramsdellite

b hollandite

d psilomelane

FIGURE 9.19 The linkage of octahedral chains in idealized structures of rutile, hollandite, ramsdellite, and psilomelane. In these projections down the chain axis, octahedral centers are shown by open and closed circles to distinguish the two levels. Edges shared between two chains are shown by dotted lines. The *A* positions of hollandite are shown by squares.

chains, **psilomelane** is obtained by condensation of both two chains and three chains.

The octahedral linkages in **diaspore, AlO(OH)**, are topologically the same as those in the ramsdellite structure type of MnO_2. The diaspore structure was treated in Figure 5.16 as a distorted hcp array of equal spheres, half of whose octahedral interstices are occupied by Al atoms. Figure 9.20 is an accurate c-axis projection of all atoms using the following crystallographic data

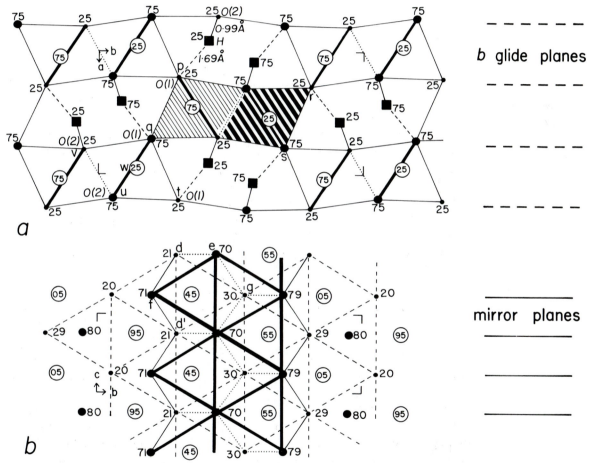

FIGURE 9.20 The crystal structure of diaspore. (*a*) The c-axis projection giving an end-on view of the double chains of edge-shared octahedra of oxygen atoms (dot) surrounding Al atoms (circles). The double chain at the center of the cell is shaded, and others are centered at the corners of the cell. Each H atom (solid square) is shown bonded to the nearest O(2) atom at 0.99 Å, and weakly linked to an O(1) atom at 1.69 Å. Heights given to the nearest hundredth of c. (*b*) The a-axis projection giving a lengthways view of a double-chain of octahedra. The uppermost face of each octahedron is emphasized, and shared edges are shown by dotted lines. Dashed lines show one set of oxygen atoms in positions related to hexagonal closest-packing when taken in conjunction with oxygen atoms at height $x = 70,71,79$ and 80 hundredths. The H atoms are omitted for clarity.

(Busing and Levy, 1958): orthorhombic, $a = 4.396$ Å, $b = 9.426$ Å, $c = 2.844$ Å, **Pbnm**. This space group was considered in detail for the olivine structure. All atoms of diaspore lie in special positions $\pm(xy\frac{1}{4}; \frac{1}{2} - x, \frac{1}{2} + y, \frac{1}{4})$ on a mirror plane with coordinates:

Atom	Al	O(1)	O(2)	H
x	0.9549	0.2880	0.8030	0.5905
y	0.1446	0.8011	0.9468	0.9124

In *a*-axis projection, oxygen atoms occur in a hexagonal array (dashed lines) at heights $\pm(0.20, 0.21, 0.29,$ and $0.30)$. A hexagonal closest-packed array would result if these heights were adjusted to 0.25 and 0.75, and if minor changes were made in horizontal coordinates and axial ratios. Aluminum atoms lie at $\pm(0.05$ and $0.45)$, and each is near the center of an octahedron of oxygen atoms. Those at ±0.45 form an infinite double column of edge-sharing octahedra parallel to the *c* axis. The uppermost face of each octahedron is shaded, and shared edges are shown by dots. Aluminum atoms at ±0.05 also form a double column (not drawn), and adjacent double columns share vertices to give an infinite framework.

The cross-linking of double columns is seen best in the *c*-axis projection. The central column is linked at vertices *p* to *s* to four adjacent columns in the same manner as in ramsdellite (Fig. 9.19).

A hydrogen nucleus lies at 0.99 Å from each O(2) atom of diaspore, converting it into a hydroxyl anion OH^-. Each H is also weakly bonded to an O(1) atom at 1.69 Å. To avoid confusion, the H nuclei (filled square), whose positions were determined by neutron diffraction, are shown only in the *c*-axis projection. As expected from electrostatic arguments, the H nuclei project away from the Al^{3+} ions and each other. The Al—O(1) distances ($wq = 1.85$ Å, $wt = 1.86$ Å) are shorter than the Al—O(2) distances ($wv = 1.98$ Å, $wu = 1.97$ Å) because O(2) is bonded to H. Shared octahedral edges ($de = 2.54$ Å, $eg = 2.46$ Å) are shorter than unshared ones ($df = 2.79$ Å, $fe = 2.74$ Å, $dd' = 2.84$ Å).

The α-*PbO₂* **structure** (Fig. 5.15) is also based on occupancy of half the octahedral holes in hcp of equal spheres, but the selection of holes is different from the diaspore structure. Indeed the α-PbO_2 structure can be constructed from the pyroxene type of octahedral chain (Fig. 9.11*b*). Crystallographic data (Wyckoff, 1963, p. 259) are orthorhombic, $a = 4.95$ Å, $b = 5.95$ Å, $c = 5.50$ Å, **Pbcn**, atomic coordinates:

Pb (*4c*) $\pm(0u\frac{1}{4}; \frac{1}{2}, \frac{1}{2} + u, \frac{1}{4})$ with $u = 0.178$

O (*8d*) $\pm(xyz; \frac{1}{2} - x, \frac{1}{2} - y, \frac{1}{2} + z; \frac{1}{2} + x, \frac{1}{2} - y, \bar{z}; \bar{x}, y, \frac{1}{2} - z)$ with $x = 0.178, y = 0.410, z = 0.425$

The *a*-axis projection (Fig. 9.21) reveals oxygen atoms near the ideal positions of hcp. Those at heights 0.72 and 0.78 are connected by a triangular array of dashed lines, and those at heights 0.28 and 0.22 lie near each center of half the triangles. Addition of 0.25 to the heights gives a close approximation to the positions of hcp in Figure 5.3*b*. Each Pb atom (e.g., *d*) is sur-

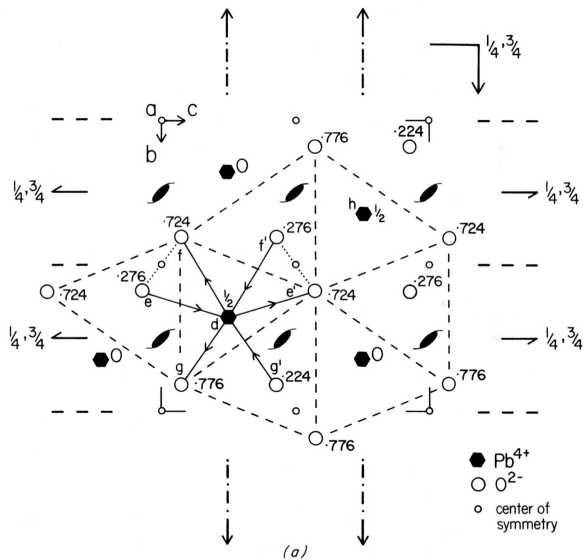

(a)

FIGURE 9.21 The crystal structure of α-PbO$_2$. (*a*) The *a*-axis projection showing symmetry elements and atomic positions. Dashed lines show a triangular array of oxygen atoms, and arrowed lines show octahedral coordination of *Pb*. (*b*) Pyroxene-type, edge-shared, octahedral chains. Dotted lines *ef* and *e'f'* are shared edges. Numbers show the *x*-coordinate of Pb atoms. Move the chains along the arrows to obtain the structure of α-PbO$_2$.

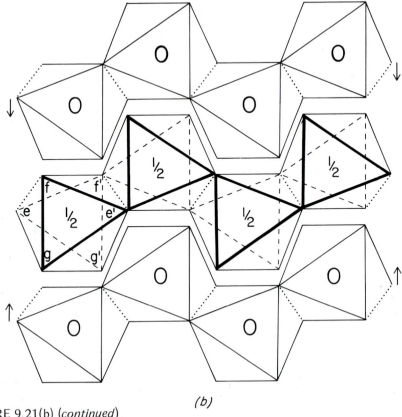

(b)

FIGURE 9.21(b) *(continued)*

rounded by an octahedron of six oxygen atoms, and three bonds go upward (e.g., to *f, e′,* and *g*) and three go downward (to *e, f′* and *g*). All Pb—O distances are 2.168 Å within experimental error.

Figure 9.21*b* shows how the structure can be constructed from infinite edge-shared octahedral chains of pyroxene type. Each chain lies along the *c* axis, and a chain at height $x = \frac{1}{2}$ is linked by octahedral vertices to four chains displaced by $\pm a/2 \pm b/2$. Adjacent chains are pulled apart for clarity, and the structure can be assembled by moving outer chains along the arrows.

9.8   Structures with Face-Sharing Octahedra

Two octahedra share a face to give Al_2O_9 dimers in the structure of *corundum,* and the dimers share edges and vertices. Corundum, however, is better described in terms of partial void filling in hcp (Fig. 5.14).

Wells (1975, pp. 186–188) described several interesting structures with face-sharing octahedra, and Moore (1970b) has found interesting face-sharing octahedral triplets in basic iron phosphate minerals.

EXERCISES   (*More Difficult)

9.1 Draw a plan of the unit cell of the $NaSbF_6$ structure using the atomic coordinates given in Section 9.1. Draw edges for at least one SbF_6 octahedron. Compare with the clinographic drawing (Fig. 9.1). To help visualization, color the octahedra to show relative depth. For isostructural $NaMoF_6$ ($a = 8.19$ Å, $u = 0.212$) calculate Na–F and Mo–F distances.

9.2* Draw a plan of the unit cell of the K_2PtCl_6 structure using $a = 9.75$ Å, $Fm3m$,

   Pt 000

   K   ¼ ¼ ¼; ¾ ¾ ¾

   Cl $\pm(u00; 0u0; 00u)$, $u = 0.24$

   Add ½ ½ 0; 0½ ½; ½ 0½ for the face-centering.

Show that the Pt and K atoms are related to the Ca and F atoms of the fluorite structure. Show that each Pt atom lies in an octahedron of Cl atoms, and that each K atom has 12 Cl atoms as first neighbors. Calculate the Pt–Cl and K–Cl distances, and determine what type of Archimedean polyhedron is formed around a K atom by its first neighbors.

9.3** Draw the a-axis projection of a unit cell of metavariscite (Kniep and Mootz, 1973) using: $a = 5.178$ Å, $b = 9.514$ Å, $c = 8.454$ Å, $\beta = 90.35°$, $P12_1/n1$. This space group was obtained from $P12_1/c1$ by change of axes in order to obtain a pseudo-orthorhombic cell. Draw the two space-group diagrams, and if necessary choose the positions of the screw axis and diagonal glide plane so that there is a center of symmetry at the origin of the unit cell. Show that the coordinates of a general position are $\pm(xyz; \frac{1}{2} - x, \frac{1}{2} - y, \frac{1}{2} + z)$. The atomic coordinates are:

	x	y	z		x	y	z
Al	0.4031	0.3254	0.3063	O(5)	0.1162	0.4477	0.3220
P	0.9090	0.1469	0.1837	O(6)	0.4041	0.3624	0.0790
O(1)	0.1650	0.1790	0.2704	H(1)	0.139	0.530	0.288
O(2)	0.9071	0.2168	0.0209	H(2)	0.963	0.432	0.331
O(3)	0.6852	0.2044	0.2813	H(3)	0.274	0.334	0.023
O(4)	0.8854	0.9861	0.1723	H(4)	0.539	0.369	0.023

The chemical formula of metavariscite is $AlPO_4 \cdot 2 H_2O$. Calculate interatomic distances and show (i) that six oxygen atoms form an octahedron about each Al atom; (ii) that the first four oxygen atoms form a tetrahedron around each P atom; (iii) that H(1) and H(2) form a water molecule with O(5), and similarly for H(3), H(4), and O(6); and (iv) that vertex sharing of octahedra and tetrahedra yields an infinite framework. Calculate P–O, Al–O and H–O distances, and H–O–H angles.

9.4* Draw the c-axis projection of a unit cell of armalcolite (Wechsler et al., 1976) using: $a = 9.776$ Å, $b = 10.021$ Å, $c = 3.748$ Å, **Bbmm**. Atomic coordinates are:

M1 (occupied by Fe, Mg)   (*4c*) $x = 0.1922$

M2 (occupied by Ti)   (*8f*) $x = 0.1348, y = 0.5645$

O1   (*4c*) $x = 0.7264$

O2   (*8f*) $x = 0.0468, y = 0.1147$

O3   (*8f*) $x = 0.3130, y = 0.0656$

Calculate the distances to the six nearest oxygen atoms from M1 and M2 atoms. Identify the positions of the planes of symmetry in the space-group symbol, and determine the types and positions of the other planes of symmetry.

9.5** Draw the c-axis projection of a unit cell of alunite (Wang et al., 1965) using: $a = 6.97$ Å, $c = 17.27$ Å, **R$\bar{3}$m**. Atomic coordinates in the triple-volume hexagonal cell are:

K   (*3a*)   000   Add $\frac{1}{3}, \frac{2}{3}, \frac{2}{3}$ and $\frac{2}{3}, \frac{1}{3}, \frac{1}{3}$ to all coordinates

S   (*6c*) $\pm(00z)$, $z = 0.303$

Al   (*9d*) $\frac{1}{2}0\frac{1}{2}$; $0\frac{1}{2}\frac{1}{2}$; $\frac{1}{2}\frac{1}{2}0$

O(1)   (*6c*) $z = 0.384$

O(2)   (*18h*) $\pm(x\bar{x}z; x, 2x, z; 2\bar{x}, \bar{x}, z)$, $x = 0.218, z = 0.941$

OH   (*18h*) $x = 0.125, z = 0.142$

Show that the Al atoms at height $z = \frac{1}{2}$ lie on a kagomé net, and that those at heights $\frac{1}{6}$ and $\frac{5}{6}$ lie on two more kagomé nets that are related to the first net by rhombohedral symmetry. Show that each Al atom at height $\frac{1}{2}$ is surrounded by two O(2) atoms at heights 0.392 and 0.608 and four OH at 0.475 and 0.525. Draw the edges of the octahedra, and show that the octahedra form a vertex-shared sheet as in Figure 9.4. Show that the S atom at $\frac{1}{3}, \frac{2}{3}, 0.364$ lies in a tetrahedron of three O(2) at height 0.392 and one O(1) at height 0.283. Check that this tetrahedron shares a vertex but not an edge with three adjacent octahedra. Show that each O(1) is 2.96 Å away from three OH. Determine that each K atom is surrounded by six O(2) and six

OH that lie at the vertices of a distorted icosahedron. Draw the space groups of equivalent general positions and symmetry elements for $R\bar{3}m$, and check that all the atomic coordinates of alunite correspond to special positions.

9.6** The crystal structure of PdS_2 contains 8S atoms and 4Pd atoms in an orthorhombic unit cell with $a = 5.46$ Å, $b = 5.54$ Å, $c = 7.53$ Å. Draw the c-axis projection using the coordinates:

Pd $000; \frac{1}{2}\frac{1}{2}0; 0\frac{1}{2}\frac{1}{2}; \frac{1}{2}0\frac{1}{2}$

S $\pm(xyz; \frac{1}{2} + x, \frac{1}{2} - y, \bar{z}; \bar{x}, \frac{1}{2} + y, \frac{1}{2} - z; \frac{1}{2} - x, \bar{y}, \frac{1}{2} + z)$ with
$x = 0.107, y = 0.112$, and $z = 0.425$

Determine the space-group symbol, and locate all the symmetry elements. Calculate interatomic distances, and determine the coordination. What is the relation to the pyrite and halite structures? Give a crystal–chemical explanation why the PdS_2 structure differs from the pyrite structure.

9.7 Build cardboard and sphere models of clusters in Figure 9.6. (This could be a group effort for a class.) After studying the models, close the book and reconstruct Figure 9.6.

9.8 Build both cardboard and sphere models of the cluster in Figure 9.7. Check that the ϕ spheres constitute a fragment of cubic closest packing of equal spheres. Show that the point symmetry is $\bar{4}3m$ for the cluster, and check that there is no center of symmetry.

9.9** The crystal structure of $MoCl_5$ is based on a monoclinic cell $a = 17.31$ Å, $b = 17.81$ Å, $c = 6.08$ Å, $\beta = 95.7°$, $C2/m$. Draw the c-axis projection using the following coordinates (Sands and Zalkin, 1959):

Mo(1)	4g	$x = 0$	$y = 0.108$	$z = 0$
Mo(2)	8j	0.333	0.108	0.435
Cl(1)	4i	0.079	0	0.867
Cl(2)	8j	0.077	0.192	0.855
Cl(3)	8j	0.075	0.094	0.323
Cl(4)	4i	0.260	0	0.243
Cl(5)	8j	0.259	0.193	0.238
Cl(6)	8j	0.253	0.093	0.698
Cl(7)	4i	0.407	0	0.628
Cl(8)	8j	0.407	0.193	0.629
Cl(9)	8j	0.414	0.094	0.167

The projection will be rectangular with edges $a \sin\beta$ and b. Show that the Mo and Cl atoms form M_2Cl_{10} dimers, and that the dimers pack in the same manner as for the Nb_2Cl_{10} structure (Fig. 9.8). Because of this overall similarity, the two structures are described as isostructural even though the atomic coordinates (especially z) differ considerably.

9.10* Color polyhedra in Figure 9.9 to enhance the perspective. **Build a cardboard model of the pharmacosiderite structure using advice from the text.

9.11** Make cardboard or spoke models of the polyions in Figure 9.10. ***Look up crystal structures in the references listed in the text.

9.12 After studying Figure 9.11, close the book and reproduce the figure from memory. *Make cardboard models.

9.13* Color the $TcCl_4$ chain in Figure 9.12, and draw at least one of the three other chains. **Close the book and construct Figure 9.12 using the cell dimensions, space group, and atomic coordinates in the text. Build a model.

9.14** Look up the crystal structure of NbI_4 (Dahl and Wampler, 1962), whose edge-shared octahedral chain is strongly distorted because of coupling of unpaired electrons between the Nb atoms. ***Build a model.

9.15** The most recent refinement of the crystal structure of sillimanite (Winter and Ghose, 1979) gives: a = 7.488 Å, b = 7.681 Å, c = 5.777 Å, **Pbnm**: Al(1) 000; Al(2) 0.1417, 0.3449, 0.25; Si 0.1533, 0.3402, 0.75; O(1) 0.3605, 0.4094, 0.75; O(2) 0.3569, 0.4341, 0.25; O(3) 0.4763, 0.0015, 0.75; O(4) 0.1252, 0.2230, 0.5145. Draw diagrams of general positions and symmetry elements for the space group. Determine the coordinates for the general and special positions. Draw the c-axis projection of the sillimanite structure and compare with Burnham (1963) and Winter and Ghose (1979). Show that the sillimanite structure can be idealized into Figure 9.13 by removing the distinction between Al(2) and Si, and making small shifts in atomic coordinates. Show that **Pbam** (c = 2.89 Å) is a subgroup of **Pbnm** (c = 5.78 Å). Calculate interatomic distances and check with Winter and Ghose (1979, Table 6). ***Build a model.

9.16* Build either a ball or polyhedral model of olivine. Calculate the length of polyhedral edges, and show that unshared edges are longer than shared edges and that an edge shared between an octahedron and a tetrahedron is shorter than one shared between two octahedra. Check your answers with Birle et al. (1968).

9.17** Look up the following papers on the humite family: Ribbe et al. (1968), Gibbs and Ribbe (1969), Gibbs et al. (1970), Ribbe et al. (1971) and Robinson et al. (1973). Study the linkages of octahedra and tetrahedra in Figures 9.15c and 9.15d and locate the symmetry elements in the structures of norbergite (**Pbnm**) and chondrodite (**P2₁/b11**; note unusual orientation). ***Make similar abstract diagrams for humite (**Pbnm**) and clinohumite (**P2₁/b11**), and check against Papike and Cameron (1976, Fig. 4). Study the relation between cation repulsion and polyhedral distortion, and the chemical differences between the humite minerals.

9.18* The crystal structure of CdI_2 has a hexagonal cell with a = 4.24 Å, c = 6.84 Å, and the following coordinates: Cd 000; I ±($\frac{1}{3}\frac{2}{3}\frac{1}{4}$). Draw the c-axis projection of the unit cell. Calculate the Cd–I and I–I distances. Halve the average I–I distance to obtain an estimate of the ionic radius of I⁻, and subtract from the Cd–I distance to obtain an estimated radius for Cd^{2+}. Using these radii, draw a vertical section corresponding to that of Figure 9.16*b*.

9.19* Calculate the Cl–Cl distances in the $CdCl_2$ structure, using u = $\frac{1}{4}$, a = 3.85 Å, c = 17.46 Å (hexagonal cell). Estimate the ionic radii as in the preceding exercise.

9.20* Study Figure 9.17, and color the octahedral chains. Construct a clinographic projection of the rutile structure, using Figure 9.17 as a guide. Draw a cube as in Figure 3.5*b*. Shorten the c axis by the axial ratio 2.96/ 4.59. Place hexagons at the corners of the unit cell. Draw body diagonals pr', qs', and sq'. Check that they intersect at t, and place a hexagon there. Draw face diagonal pr, and measure its length. To locate oxygen positions d and e, measure re = pd = 0.305pr. To locate oxygen positions j and i, join the midpoints v and w of ss' and qq'. Check that vt = tw. Measure it = tf = 0.305vw. Draw octahedra around the Ti atoms, using parallel rulers to save time. Use heavy dots to mark shared edges. **Build a ball model.

9.21* The high-pressure polymorph of sanidine, $KAlSi_3O_8$, has the hollandite structure type with a = 0.938 nm, c = 0.274 nm, **I4/m**, atomic coordinates: K (2*b*); disordered $Al_{0.25}Si_{0.75}$ (8*h*) with u = 0.348, v = 0.167; O(1) u = 0.152, v = 0.208; O(2) u = 0.542, v = 0.152. Draw a c-axis projection like Figure 9.18. Calculate K–O and Al,Si–O distances. **Build a ball model.

9.22** Study Figure 9.19, and read about further complications in Bursill (1979).

9.23** After studying Figure 9.20, reconstruct the c-axis projection just from the space group, cell dimensions, and atomic coordinates given in the text. Calculate interatomic distances and check against values in the text.

9.24** After studying Figure 9.21, reconstruct the a-axis projection from the information in the text. Check that Pb–O distances are essentially equal.

9.25*** As a final severe challenge, look up the crystal structures of garnet (Geller, 1967; Novak and Gibbs, 1971), β-Mg_2SiO_4 (Moore and Smith, 1970), and mitridatite (Moore and Araki, 1977).

ANSWERS

9.1 2.36, 1.74 Å.

9.2 2.34, 3.45 Å; cuboctahedron.

9.3 P–O(1)=1.542, P–O(2)=1.528, P–O(3)=1.528, P–O(4)=1.538 Å; Al–O(1)=1.883, Al–O(2)=1.859, Al–O(3)=1.873, Al–O(4)=1.888, Al–O(5)=1.892, Al–O(6)=1.953; O(5)–H(1)=0.84, O(5)–H(2)=0.81; O(6)–H(3)=0.87, O(6)–H(4)=0.85; H(1)–O(5)–H(2)=110°; H(3)–O(6)–H(4)=111°. See Kniep and Mootz (1973) for diagrams.

9.4 M1–O(1)=2.04, M1–O(2)=1.96, M1–O(3)=2.19 Å; M2–O(1)=2.06, M2–O(2)=1.99, M2–O(2)'=1.84, M2–O(3)=2.18, M2–O(3)'=1.94. The projection of the unit cell is similar to that in Figure 9.1c. The **bmm** planes lie respectively at $x = 0,\frac{1}{2}$; $y = \frac{1}{4},\frac{3}{4}$; $z = 0,\frac{1}{2}$. The space-group symbol could also be given as **Bnna**, whose planes lie, respectively, at $x = \frac{1}{4},\frac{3}{4}$; $y = \frac{1}{4},\frac{3}{4}$; $z = \frac{1}{4},\frac{3}{4}$.

9.5 Check your answer with Wang et al. (1965). The stereoplot for crandallite (Blount, 1974, Fig. 2) should also be useful. Space-group diagrams and positions for **R$\bar{3}$m** are given on page 273 of *International Tables for X-Ray Crystallography*, Volume 1.

9.6 Representative symmetry elements for space group **Pbca** are given in Figure 9.22. A center of symmetry occurs at the origin and related positions. A dashed line joins each pair of S_2 atoms whose interatomic distance of 2.13 Å is essentially the same as 2.14 Å for pyrite. At first sight, each Pd

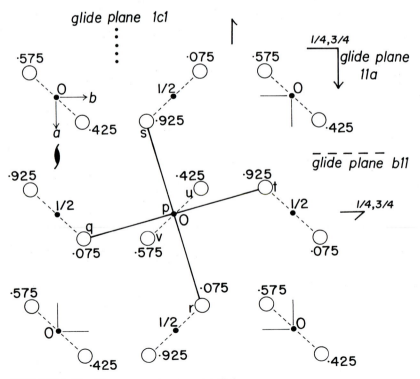

FIGURE 9.22 The answer to exercise 9.6.

atom lies in an octahedron of S atoms; thus atom p at height $z = 0$ has S neighbors at $z = 0.075$ (q and r), $z = -0.075$ (s and t), $z = 0.425$ (u) and $z = -0.425$ (v). However, the pv and pu distances (3.40 Å) are much longer than the pq and ps distances (2.30 Å) and the pr and pt distances (2.31 Å), and bonding must be almost entirely to the $q, r, s,$ and t atoms that lie almost at the corners of a square. This is interpreted chemically as the result of hybridization of dsp^2 orbitals of the Pd atom, and is the obvious reason for distortion of the PdS_2 structure from the more symmetric structure of pyrite. In turn, pyrite is derived from halite, as shown in Figures 8.7 and 8.8.

9.18   I–I (horizontal) 4.24, (across layer) 4.20_6, mean 4.22_3 Å. Cd–I 2.98_6 Å. Hence ionic radii are 2.11_2 and 0.87_4 Å. The radius ratio of 0.414 matches the ideal value for octahedral coordination. Circles should almost touch in the drawing.

9.19   Cl–Cl (horizontal) 3.85, (across layer) 3.66_2, mean 3.75_6. Cd–Cl 2.65_7 Å. Hence ionic radii are 1.87_8 and 0.77_9 Å.

9.21   K–O(1)=2.78, (Si,Al) = 1.80, 1.82, 1.88 Å.

Vale

"Dear Reader" sounds archaic, but this Victorian device is useful. Probably you will not have studied all the book and done all the exercises, and indeed this is not necessary to obtain a feeling for geometrical and structural crystallography. However, I hope that you have spent enough time to have developed a fascination for the elegant ways in which atoms and molecules link together to form solid materials.

To those readers who will not become professional crystallographers, may I respectfully suggest that your training in geometrical relationships and symmetry theory can enrich many aspects of life: listen to the harmony of the spheres in a fugue; look at the design of a Renaissance painting or the decoration of a Muslim building; examine the linkages of a nucleic acid and the enwrapping of a glycoprotein; question whether economy leads to overstrict regularity and boredom in modern buildings; and so on.

To those readers who will become professional crystallographers, or will use crystallographic methods in another scientific discipline, may I suggest that endless challenges await you. This book was deliberately designed to tolerate the needs and abilities of a wide range of readers, and all aspects can be treated by mathematical methods of greater rigor and compactness. Furthermore, only the simpler aspects of the packing and connectivity of structural units have been explored, and new types of crystal structures are being observed as yet more are being invented theoretically. Structures with stacking faults and other mistakes pose endless challenges to the theoretician, and are of great practical value. Biological phenomena depend on a subtle interplay between regularity and irregularity.

Finally, I hope that all of you will find that the neatness of the ways in which atoms and molecules fit together can help to provide assurance that there is some sense to this world, even in the face of so much human evidence to the contrary. Thank you for joining me in this exploration of the world of crystals. Please let me know about the inevitable deficiencies and omissions in this book so that I can try to improve it.

Bibliography

Books Listed in the Text

Bradley C.J. and Cracknell A.P. 1972. *The Mathematical Theory of Symmetry in Solids. Representation Theory for Point Groups and Space Groups.* Clarendon Press, Oxford. Comprehensive treatise.

Bragg Sir L., Claringbull G.F., and Taylor W.H. 1965. *Crystal Structures of Minerals.* Cornell University Press, Ithaca. Systematic monograph.

Buerger M.J. 1956. *Elementary Crystallography.* Wiley, New York. Thorough, not-so-elementary text.

Burns G. and Glazer A.M. 1978. *Space Groups for Solid State Scientists.* Academic Press, New York.

Cotton F.A. 1971. *Chemical Applications of Group Theory.* Wiley-Interscience, New York. Elegant treatment.

Coxeter H.M.S. 1963. *Regular Polytopes.* 2nd ed. Macmillan, New York. Rapidly moves to advanced topics.

Cundy A.M. and Rollett A.P. 1961. *Mathematical Models.* 2nd ed. Clarendon, Oxford. Especially good for polyhedra.

Deer W.A., Howie R.A., and Zussman J. 1978. *Single-chain Silicates,* vol. 2A, 2nd ed. of *Rock-forming Minerals.* Wiley, New York.

Fejes Tóth L. 1964. *Regular Figures.* Macmillan, New York. Thorough, elementary treatment.

Fischer W., Burzlaff H., Hellner E., and Donnay J.D.H. 1973. *Space Groups and Lattice Complexes.* National Bureau of Standards Monograph 134, Washington, D.C. Encyclopaedic compilation.

Goldschmidt V. 1913–1923. *Atlas der Kristallformen.* Winter, Heidelberg. Thousands of crystal drawings.

Henry N.F.M. and Lonsdale K. (Eds.) 1952. *International Tables for X-Ray Crystallography.* Kynoch Press, Birmingham. Standard compilation prepared for the International Union of Crystallography. Vol. I. "Symmetry Groups" is particularly useful here. A pilot issue of new "Symmetry Tables" has been circulated to selected crystallographers, but has not yet been published.

Hurlbut C.S., Jr. and Klein C. 1977. *Manual of Mineralogy*. 19th ed. Wiley, New York. Contains many drawings of crystal structures.

Kitaigorodskii A.I. 1961. *Organic Chemical Crystallography*. Consultants Bureau, New York. Trans. from Russian; Akademizdat, Moscow.

Kitaigorodsky A.I. 1973. *Molecular Crystals and Molecules*. Academic Press, New York.

Laves F. 1955. In *Theory of Alloy Phases*, pp. 124–198. American Society for Metals, Cleveland. Summary of structural relationships among alloys.

Macgillavry C.H. 1965. *Symmetry Aspects of M.C.Escher's Periodic Drawings*. A. Oosthoek's Uitgeversmaatschappij, Utrecht. Test your skill in locating symmetry elements in these cunning drawings. Buy a set of originals if you are a millionaire.

McKie D. and McKie C. 1974. *Crystalline Solids*. Nelson, London. An elementary survey covering broad aspects of crystallography and phase equilibria.

Megaw H.D. 1957. *Ferroelectricity in Crystals*. Methuen, London. Classical monograph.

Meier W.M. and Olson D.H. 1978. *Atlas of Zeolite Structure Types*. Polycrystal Book Service, Pittsburgh. Stereo diagrams of 4-connected nets.

Ore O. 1963. *Graphs and Their Uses*. New Mathematical Library, Random House, and L.W. Singer Co., New York.

Pearson W.B. 1972. *The Crystal Chemistry and Physics of Metals and Alloys*. Wiley–Interscience, New York. Fundamental treatise.

Phillips F.C. 1971. *An Introduction to Crystallography*. 4th ed. Oliver and Boyd, Edinburgh. Simple elegance.

Ribbe, P.H. (Ed.) 1974. *Sulfide Mineralogy*. Mineralogical Society of America, Short Course Notes, vol. 1.

Shubnikov A.V. and Belov N.V. 1956. *Colored Symmetry*. Macmillan, New York. Collected translations of fundamental papers.

Shubnikov A.V. and Koptsik V.A. 1974. *Symmetry in Science and Art*.

Plenum; New York. English translation of Russian monograph. Covers a wider range of topics than any other book on symmetry in a clear and rigorous manner. A "best buy."

Smith J.V. 1974. *Feldspar Minerals.* Springer, Heidelberg.

Thompson D'Arcy W. 1948. *On Growth and Form.* New abridged edition. Cambridge University Press, London. Victorian classic.

Wells A.F. 1970. *Models in Structural Inorganic Chemistry.* Oxford University Press, New York. Very useful set of exercises on model building.

Wells A.F. 1975. *Structural Inorganic Chemistry.* 4th ed. Clarendon Press, Oxford. The "bible." A reviewer called it the ideal book if stranded for life on a desert island. After 6 years, I have discovered only a few of its nuggets.

Wells A.F. 1977. *Three-Dimensional Nets and Polyhedra.* Wiley, New York. Specialized monograph that opens up an infinite field for mathematical research.

Weyl H. 1952. *Symmetry.* Princeton University Press, Princeton. Beautiful long essay by a "Renaissance man."

Wyckoff R.W.G. 1963–1971. *Crystal Structures.* 2nd ed., vols. 1–6. Interscience, New York. Encyclopaedic compilation.

Additional Books

Adams D.M. 1974. *Inorganic Solids.* Wiley, London. General survey with emphasis on chemical bonding.

Azároff L.V. 1960. *Introduction to Solids.* McGraw-Hill, New York. General introductory survey.

Bernal I., Hamilton W.C. and Ricci J.S. 1972. *Symmetry: A Stereoscopic Guide for Chemists.* Freeman, San Francisco. Useful diagrams.

Bhagavantan S. 1966. *Crystal Symmetry and Physical Properties.* Academic Press, New York. Detailed monograph.

Bloss F.D. 1971. *Crystallography and Crystal Chemistry.* Holt, Rinehart & Winston, New York. A standard undergraduate text.

Brindley G.W. and Brown G. (Eds.) 1980. *Crystal Structures of Clay Minerals and Their X-Ray Identification.* Mineralogical Society, Monograph no. 5, London.

Buerger M.J. 1960. *Crystal Structure Analysis.* Wiley, New York. Classical monograph.

Burdett J.K. 1980. *Molecular Shapes: Theoretical Models of Inorganic Stereochemistry.* Wiley, New York.

Dent Glasser L.S. 1977. *Crystallography and Its Applications.* Van Nostrand Reinhold, New York. Delightful introduction.

Dunitz, J. 1980. *X-ray Analysis and the Structure of Organic Molecules.* Cornell University Press, Ithaca. Authoritative review.

Evans R.C. 1964. *An Introduction to Crystal Chemistry.* 2nd ed. Cambridge University Press, Cambridge.

Gay P. 1972. *The Crystalline State—An Introduction.* Oliver & Boyd, Edinburgh. Undergraduate text.

Harary F. 1969. *Graph Theory.* Addison-Wesley, Reading, Mass.

Holden A. 1971. *Shapes, Space and Symmetry.* Columbia University Press, New York. Good pictures.

Kelly A. and Groves G.W. 1970. *Crystallography and Crystal Defects.* Addison-Wesley, Reading, Mass.

Krebs H. 1968. *Fundamentals of Inorganic Crystal Chemistry.* McGraw-Hill, London.

Lipson H. and Cochran W. 1966. *The Determination of Crystal Structures.* 3rd ed. Bell, London. Classical monograph.

Lockwood E.H. and Macmillan R.H. 1978. *Geometric Symmetry.* Cambridge University Press, London.

Loeb A.L. 1976. *Space Structures.* Addison-Wesley, Reading, Mass. Contains the important concepts about polyhedra.

Lyusternik L.A. 1963. *Convex Figures and Polyhedra.* Dover, New York. Useful theorems.

Megaw H.D. 1973. *Crystal Structures: A Working Approach.* Saunders, Philadelphia. A practical guide based on extensive research experience.

Nye J.F. 1957. *Physical Properties of Crystals.* Oxford University Press, London. Standard monograph.

Pauling L. 1940. *The Nature of the Chemical Bond.* Cornell University Press, Ithaca.

Povarennykh A.S. 1972. *Crystal Chemical Classification of Minerals.* Plenum, New York.

Schneer C.J. (Ed.) 1977. *Crystal Form and Structure.* Dowden, Hutchinson & Ross, Stroudsburg. Collected reprints of old classical papers.

Stout G.H. and Jensen L.H. 1968. *X-Ray Structure Determination: A Practical Guide.* MacMillan, New York. Reliable textbook.

Structure Reports (1940–   ). Oosthoek, Utrecht. Progressive summaries of new structures.

Terpstra P. and Codd L.W. 1961. *Crystallometry.* Longmans, London. Classical approach.

Tutton A.E.H. 1922. *Crystallography and Practical Crystal Measurement.* Vol. I. Macmillan, London. Old-fashioned monograph

Verma A.R. and Krishna P. 1966. *Polymorphism and Polytypism in Crystals.* Wiley, New York. A detailed monograph.

Walton A. 1978. *Molecular and Crystal Structure Models.* Wiley, New York; Ellis Horwood, Chichester. Practical guide with list of suppliers.

Wells A.F. 1956. *The Third Dimension in Chemistry.* Clarendon Press, Oxford. A charming essay.

Wenninger M. 1971. *Polyhedron Models.* Cambridge University Press, London. Good illustrations.

Wooster W.A. 1973. *Tensors and Group Theory for the Physical Properties of Crystals.* Clarendon Press, London. Detailed monograph.

Journal Articles Cited in the Text

Bayer G. 1978. "Zur Kristallchemie des Zirkons und des Scheelits." *Schweiz. Mineral. Petrog. Mitt.* **58** 111–126.

Belov N.V. 1956. In *Colored Symmetry* by A.V. Shubnikov, N.V. Belov, et al. Ed. W.T. Holser. Macmillan, New York.

Birle J.D., Gibbs G.V., Moore P.B. and Smith J.V. 1968. "Crystal structures of natural olivines." *Amer. Mineral.* **53** 807–824.

Blount A.M. 1974. "The crystal structure of crandallite." *Amer. Mineral.* **59** 41–47.

Boisen N.B., Jr. and Gibbs G.V. 1976. "A derivation of the 32 crystallographic point groups using elementary group theory." *Amer. Mineral.* **61** 145–165.

Boisen M.B., Jr. and Gibbs G.V. 1978. "A method for constructing and interpreting matrix representations of space-group operations." *Canad. Mineral.* **16** 293–300.

Britton D. and Dunitz J.D. 1973. "A complete catalogue of polyhedra with eight or fewer vertices." *Acta Cryst.* **A29** 362–371.

Brunner G.O. 1979. "The properties of coordination sequences and conclusions regarding the lowest possible density of zeolites." *J. Solid State Chem.* **29** 41–45.

Buerger M.J. 1954. "The stuffed derivatives of the silica structures." *Amer. Mineral.* **39** 600–614.

Buerger M.J., Dollase W.A., and Garaycochea-Wittke. 1967. "The structure and composition of the mineral pharmacosiderite." *Z. Krist.* **125** 92–108.

Burnham C.W. 1963. "Refinement of the crystal structure of sillimanite." *Z. Krist.* **118** 337–360.

Bursill L.A. 1979. "Structural relationships between β-gallia, rutile, hollandite, psilomelane, ramsdellite and gallium titanate type structures." *Acta Cryst.* **B35** 530–538.

Busing W.R. and Levy H.A. 1957. "Neutron diffraction study of calcium hydroxide." *J. Chem. Phys.* **26** 563–568.

Busing W.R. and Levy H.A. 1958. "A single-crystal neutron diffraction study of diaspore, AlO(OH)." *Acta Cryst.* **11** 798–803.

Byström A. and Byström A.M. 1950. "The crystal structure of hollandite." *Acta Cryst.* **3** 146–154.

Chabot B., Cenzual K., and Parthé E. 1981. "Nested polyhedra units: A geometrical concept for describing complicated cubic structures." *Acta Cryst.* **A37** 6–11.

Clark J.R., Appleman D.E., and Papike J.J. 1969. "Crystal–chemical characterization of clinopyroxenes based on eight new structure refinements." *Mineral. Soc. America Special Paper* **2** 31–50.

Colville A.A., Anderson C.P., and Black P.M. 1971. "Refinement of the crystal structure of apophyllite. I. X-ray diffraction and physical properties." *Amer. Mineral.* **56** 1222–1223.

Dahl L.F. and Wampler D.L. 1962. "The crystal structure of α-niobium tetraiodide." *Acta Cryst.* **15** 903–911.

Dollase W.A. and Baur W.H. 1976. "The superstructure of meteoritic low tridymite solved by computer simulation." *Amer. Mineral.* **61** 971–978.

Dowty E. 1980. "Computing and drawing crystal shapes." *Amer. Mineral.* **65** 465–471.

Elder M. and Penfold B.R. 1966. "The crystal structure of technetium (IV) chloride. A new AB_4 structure." *Inorg. Chem.* **5** 1197–1200.

Evans H.T., Jr. 1971. "Heteropoly and isopoly complexes of the transition elements of groups 5 and 6." In *Perspectives in Structural Chemistry* **4** 1–59 (J.D. Dunitz and J. Ibers, Eds.). Wiley, New York.

Evans H.T., Jr. 1974. "The molecular structure of the hexamolybdotellurate ion in the crystal complex with telluric acid, $(NH_4)_6[TeMo_6O_{24}]$. $Te(OH)_6 \cdot 7 H_2O$." *Acta Cryst.* **B30** 2095–2100.

Evans H.T., Jr., Gatehouse B.M., and Leverett P. 1975. "Crystal structure of the heptamolybdate (VI) (paramolybdate) ion, $[Mo_7O_{24}]^{6-}$, in the ammonium and potassium tetrahydrate salts." *J. Chem. Soc., Dalton Trans.* 505–514.

Figueiredo M.O. and Lima-de-Faria J. 1978. "Condensed models of structures based on loose packings." *Z. Krist.* **148** 7–19.

Fischer K. 1969. "Verfeinerung der Kristallstruktur von Benitoit $BaTi[Si_3O_9]$." *Z. Krist.* **129** 222–243.

Fischer W. 1968. "Kreispackungsbedingungen in der Ebene." *Acta Cryst.* **A24** 67–81.

Fischer W. 1971. "Existenzbedingungen homogener Kugelpackungen in Raumgruppen tetragonaler Symmetrie." *Z. Krist.* **133** 18–42.

Fischer W. 1973. "Existenzbedingungen homogener Kugelpackungen zu kubischen Gitterkomplexen mit weniger als drei Freiheitsgraden." *Z. Krist.* **138** 268–278.

Fischer W. 1974. "Existenzbedingungen homogener Kugelpackungen zu kubischen Gittenkomplexen mit drei Freiheitsgraden." *Z. Krist.* **140** 50–74.

Fischer W. 1976. "Eigenschaften der Heesch-Laves-Packung und ihres Kugelpackungstyps." *Z. Krist.* **143** 140–155.

Fischer W. and Koch E. 1978. "Limiting forms and comprehensive complexes for crystallographic point groups, rod groups and layer groups." *Z. Krist.* **147** 255–273.

Fischer W. and Koch E. 1979. "Geometrical packing analysis of molecular compounds." *Z. Krist.* **150** 245–260.

Frank F.C. and Kasper J.S. 1958. "Complex alloy structures regarded as sphere packings. I. Definitions and basic principles." *Acta Cryst.* **11** 184–190.

Frank F.C. and Kasper J.S. 1959. "Complex alloy structures regarded as sphere packings. II. Analysis and classification of representative structures." *Acta Cryst.* **12** 483–499.

Geller S. 1967. "Crystal chemistry of the garnets." *Z. Krist.* **125** 1–47.

Gibbs G.V. 1966. "The polymorphism of cordierite–I: The crystal structure of low cordierite." *Amer. Mineral.* **51** 1068–1087.

Gibbs G.V. and Ribbe P.H. 1969. "The crystal structures of the humite minerals: I. Norbergite." *Amer. Mineral.* **54** 376–390.

Gibbs G.V., Breck D.W., and Meagher E.P. 1968. "Structural refinement of hydrous and anhydrous synthetic beryl, $Al_2(Be_3Si_6)O_{18}$ and emerald, $Al_{1.9}Cr_{0.1}(Be_3Si_6)O_{18}$." *Lithos* **1** 275–285.

Gibbs G.V., Ribbe P.H., and Anderson C.P. 1970. "The crystal structures of the humite minerals. II. Chondrodite." *Amer. Mineral.* **55** 1182–1194.

Hazen R.M. and Finger L.W. 1979. "Crystal structure and compressibility of zircon at high pressure." *Amer. Mineral.* **64** 196–201.

Hellner E. 1965. "Descriptive symbols for crystal-structure types and homeotypes based on lattice complexes." *Acta Cryst.* **19** 703–712.

Hellner E. 1979. "The frameworks (Bauverbände) of the cubic structure types." *Structure and Bonding* **37** 61–140. (Ed. J.D. Dunitz et al., Springer, Berlin.)

Hellner E. and Koch E. 1981. "Cluster or framework considerations for the structures of Tl_7Sb_2, α-Mn, Cu_5Zn_8 and their variants $Li_{22}Si_{51}$, $Cu_{41}Sn_{11}$, $Sm_{11}Cd_{45}$, Mg_6Pd and Na_6Tl with octuple unit cells." *Acta Cryst.* **A37** 1–6.

Holser W.T. 1958a. "Relation of symmetry to structure in twinning." *Z. Krist.* **110** 249–265.

Holser W.T. 1958b. "Point groups and plane groups in a two-sided plane and their subgroups." *Z. Krist.* **110** 266–281.

Jeffrey G.A. and McMullan R.K. 1967. "The clathrate hydrates." *Progr. Inorg. Chem.* **8** 43–108.

Kay M.I.., Frazer B.C. and Almodovar I. 1964. "Neutron diffraction refinement of $CaWO_4$." *J. Chem. Phys.* **40** 504–506.

King R.B. 1969. "Chemical applications of topology and group theory. I. Coordination polyhedra. II. Metal complexes of planar unsaturated carbon systems." *J. Amer. Chem. Soc.* **91** 7211–7216, 7217–7223.

King R.B. 1970. "Chemical applications of topology and group theory. III. Relative interligand repulsions of coordination polyhedra. IV. Polyhedra for coordination numbers 10–16." *J. Amer. Chem. Soc.* **92** 6455–6460, 6460–6466.

King R.B. 1972. "Chemical applications of topology and group theory. V. Polyhedral clusters and boron hydrides." *J. Am. Chem. Soc.* **94** 95–103. "VI. Polyhedral water networks in clathrates and semiclathrates." *Theoret. chim. Acta (Berlin)* **25** 309–318.

King R.B. and Rouvray D.H. 1977. "Chemical applications of topology and group theory. 7. A graph-theoretical interpretation of the bonding topology in polyhedral boranes, carboranes, and metal clusters." *J. Amer. Chem. Soc.* **99** 7834–7840.

Kniep R. and Mootz D. 1973. "Metavariscite–a redetermination of its crystal structure." *Acta Cryst.* **B29** 2292–2294.

Koch E. and Fischer W. 1978a. "Types of sphere packings for crystallographic point groups, rod groups and layer groups." *Z. Krist.* **147** 21–38.

Koch E. and Fischer W. 1978b. "Types of sphere packings for crystallographic point groups, rod groups and layer groups." *Z. Krist.* **148** 107–152.

Laves F. 1930. "Die Bau-Zusammenhänge innerhalb der Kristallstrukturen." *Z. Krist.* **73** 202–265, 275–324.

Liebau F. 1972. "Silicon." In *Handbook of Geochemistry,* Ed. K.H. Wedepohl, pp. 14-A-1 to 14-A-32. Springer, Berlin.

Lima-de-Faria J. 1978. "Rules governing the layer organization of inorganic crystal structures." *Z. Krist.* **148** 1–5.

Loens J. and Schulz H. 1967. "Struktur Verfeinerung von Sodalit $Na_8 Si_6 Al_6 O_{12} Cl_2$." *Acta Cryst.* **23** 434–436.

Louisnathan S.J. 1969. "Refinement of the crystal structure of hardystonite, $Ca_2 ZnSi_2 O_7$." *Z. Krist.* **130** 427–437.

Louisnathan S.J. and Smith J.V. 1970. "Crystal structure of tilleyite: Refinement and coordination." *Z. Krist.* **132** 288–306.

Moore P.B. 1970a. "Structural hierarchies among minerals containing octahedrally coordinating oxygen. I. Stereoisomerism among corner-sharing octahedral and tetrahedral chains." *Neues Jahrb. Mineral. Mh.* 163–173.

Moore P.B. 1970b. "Crystal chemistry of the basic iron phosphates." *Am. Mineral.* **55** 135–169.

Moore P.B. and Araki T. 1977. "Mitridatite, $Ca_6 (H_2O)_6 [Fe^{III}_9 O_6 (PO_4)_9]$·

$3H_2O$. A noteworthy octahedral sheet structure." *Inorg. Chem.* **16** 1096–1106.

Moore P.B. and Smith J.V. 1970. "Crystal structure of β-Mg_2SiO_4: Crystal–chemical and geophysical implications." *Phys. Earth Planet. Interiors* **3** 166–177.

Niggli P. 1926, 1928. "Die topologische Strukturanalyse. I, II." *Z. Krist.* **65** 391–415; **68** 404–466. Derivation of 2D nets from circle packings.

Novak G.A. and Gibbs G.V. 1971. "The crystal chemistry of the silicate garnets." *Amer. Mineral.* **56** 791–825.

Nyman H. and Andersson S. 1979. "The elongated rhombic dodecahedron in alloy structures." *Acta Cryst.* **A35** 305–308.

Nyman H. and Hyde B.G. 1981. "The related structures of α-Mn, sodalite, Sb_2Tl_7, etc." *Acta Cryst.* **A37** 11–17.

Ohashi Y. and Finger L.W. 1978. "The role of octahedral cations in pyroxenoid crystal chemistry. I. Bustamite, wollastonite, and the pectolite-schizolite-serandite series." *Amer. Mineral.* **63** 274–288.

O'Keefe M. and Hyde B.G. 1980. "Plane nets in crystal chemistry." *Phil. Trans. Roy. Soc. London* **295** 553–622.

Pannhorst W. 1979. "Structural relationships between pyroxenes." *Neues Jahrb. Miner. Abh.* **135** 1–17.

Papike J.J. and Cameron M. 1976. "Crystal chemistry of silicate minerals of geophysical interest." *J. Geophys. Res.* **14** 37–80.

Paquette L.A., Balogh D.W., Usha R., Kountz D. and Christoph G.G. 1981. "Crystal and molecular structure of a pentagonal dodecahedrane." *Science* **211** 575–576.

Perrotta A.J. and Smith J.V. 1965. "The crystal structure of kalsilite, $KAlSiO_4$." *Mineral. Mag.* **35** 588–595.

Ribbe P.H. and Gibbs G.V. 1971. "Crystal structures of the humite minerals: III. Mg/Fe ordering in humite and its relation to other ferromagnesian silicates." *Amer. Mineral.* **56** 1155–1173.

Ribbe P.H., Gibbs G.V., and Jones N.W. 1968. "Cation and anion substitutions in the humite minerals." *Mineral. Mag.* **37** 966–975.

Robinson K., Gibbs G.V., and Ribbe P.H. 1973. "The crystal structures of the humite minerals, IV. Clino- and titanoclinohumite." *Amer. Mineral.* **58** 43–49.

Samson S. 1958. "The crystal structure of the intermetallic compound $Mg_3Cr_2Al_{18}$." *Acta Cryst.* **11** 851–857.

Samson S. 1965. "The crystal structure of the phase β-Mg_2Al_3." *Acta Cryst.* **19** 401–413.

Samson S. 1968. "The structure of complex intermetallic compounds." In *Structural Chemistry and Molecular Biology,* pp. 687–717, Ed. A. Rich and N. Davidson. Freeman, San Francisco.

Sands D.E. and Zalkin A. 1959. "The crystal structure of $MoCl_5$." *Acta Cryst.* **12** 723–726.

Shannon R.D. 1976. "Revised effective ionic radii and systematic studies of interatomic distances in halides and chalcogenides." *Acta Cryst.* **A32** 751–767.

Shannon, R.D. and Prewitt C.T. 1969. "Effective ionic radii in oxides and fluorides." *Acta Cryst.* **B25** 925–945.

Smith J.V. 1963. "Structural classification of zeolites." *Mineralogical Society of America Special Paper no. 1,* pp. 281–290.

Smith J.V. 1977. "Enumeration of 4-connected 3-dimensional nets and classification of framework silicates. I. Perpendicular linkage from simple hexagonal net." *Amer. Mineral.* **62** 703–709.

Smith J.V. 1978. "Enumeration of 4-connected 3-dimensional nets and classification of framework silicates. II. Perpendicular and near-perpendicular linkages from 4.8^2, 3.12^2 and $4.6.12$ nets." *Amer. Mineral.* **63** 960–969.

Smith J.V. 1979. "Enumeration of 4-connected 3-dimensional nets and classification of framework silicates. III. Combination of helix, and zigzag, crankshaft and saw chains with simple 2D nets." *Amer. Mineral.* **64** 551–562.

Smith J.V. and Bennett J.M. 1981. "Enumeration of 4-connected 3-dimensional nets and classification of framework silicates. IV. The infinite sets of ABC-6 nets; the Archimedean and σ-related nets." *Amer. Mineral.* **66** 777–788.

Smith J.V. and Yoder H.S. 1956. "Experimental and theoretical studies of the mica polymorphs." *Mineral. Mag.* **31** 209–234.

Smyth J.R. 1974. "The crystal chemistry of armalcolites from Apollo 17." *Earth Planet. Sci. Lett.* **24** 262–270.

Takéuchi Y. 1958. "A detailed investigation of the structure of hexagonal $BaAl_2Si_2O_8$ with reference to its α–β inversion." *Mineral. J. Japan* **2** 311–322.

Takéuchi Y. and Donnay G. 1959. "The crystal structure of hexagonal $CaAl_2Si_2O_8$." *Acta Cryst.* **12** 465–470.

Takéuchi Y., Kawada I., Irimaziri S. and Sadanaga R. 1969. "The crystal structure and polytypism of manganpyrosmalite." *Mineral J. Japan* **5** 450–467.

Taylor D. and Henderson C.M.B. 1978. "A computer model for the cubic sodalite structure." *Phys. Chem. Minerals.* **2** 325–336.

Wade K. 1976. "Structural and bonding patterns in cluster chemistry." *Adv. Inorg. Chem. and Radiochem.* **18** 1–66.

Wang R., Bradley W.F., and Steinfink H. 1965. "The crystal structure of alunite." *Acta Cryst.* **18** 249–252.

Weber L. 1929. "Die Symmetrie homogener ebener Punktsysteme." *Z. Krist.* **70** 309–327.

Wechsler B.A., Prewitt C.T., and Papike J.J. 1976. "Chemistry and structure of lunar and synthetic armalcolite." *Earth Planet. Sci. Lett.* **29** 91–103.

Winter J.K. and Ghose S. 1979. "Thermal expansion and high-temperature crystal chemistry of the Al_2SiO_5 polymorphs." *Amer. Mineral.* **64** 573–586.

Yagi T., Mao H.-K, and Bell P.M. 1978. "Structure and crystal chemistry of perovskite-type $MgSiO_3$." *Phys. Chem. Minerals* **3** 97–110.

Zalkin A. and Sands D.E. 1958. "The crystal structure of $NbCl_5$." *Acta Cryst.* **11** 615–619.

Zoltai T. 1960. "Classification of silicates and other minerals with tetrahedral structures." *Amer. Mineral.* **45** 960–973.

Index of Concepts

Important concepts are emphasized by **boldface** in both this index and the text. Other important references are shown in plain type, but these are deliberately selective and are not comprehensive.

Index of Minerals and Materials

The principal discussion is shown in **boldface**.

Index of Chemical Formulas

The elements in compounds with some ionic bonding are arranged in order of structural significance from weak cation via strong cation and strong anion to weak anion, whereas those in alloys and covalent compounds are arranged alphabetically. Each element symbol is taken as an alphabetical entity. The principal discussion is shown in **boldface.**

Index of Authors